# Lecture Notes on Multidisciplinary Industrial Engineering

**Series editor**

J. Paulo Davim, Department of Mechanical Engineering, University of Aveiro, Aveiro, Portugal

More information about this series at http://www.springer.com/series/15734

Kurian Antony · J. Paulo Davim
Editors

# Advanced Manufacturing and Materials Science

## Selected Extended Papers of ICAMMS 2018

Springer

*Editors*
Kurian Antony
Department of Mechanical Engineering
Muthoot Institute of Technology and Science
Kochi, Kerala
India

J. Paulo Davim
Department of Mechanical Engineering
University of Aveiro
Aveiro
Portugal

ISSN 2522-5022 ISSN 2522-5030 (electronic)
Lecture Notes on Multidisciplinary Industrial Engineering
ISBN 978-3-030-09449-2 ISBN 978-3-319-76276-0 (eBook)
https://doi.org/10.1007/978-3-319-76276-0



Printed on acid-free paper

This Springer imprint is published by the registered company Springer International Publishing AG part of Springer Nature
The registered company address is: Gewerbestrasse 11, 6330 Cham, Switzerland

# Contents

# Effect of Magnetic Field to Reduce Emissions and Improve Combustion Performance in a Spark-Ignition Engine

**Vishnu Sankar, Sanath M. Chandran, Tharun Tomy, Unni Raj, Vivek Samson and K. Ramachandran**

## 1 Introduction

Recently, researchers have made many devices for reducing exhaust emissions [1] and fuel consumption [2, 3] of an engine in a very productive manner. But it has few disadvantages like high cost [4] and some side effects [5] that affect engine performance. In general, burning of 1 kg fuel releases nearly 1.1 kg water vapor and 3.2 kg carbon monoxide [BP Australia Limited, 2000]. In addition, problems such as incomplete combustion, presence of unburned hydrocarbons, oxides of nitrogen, and sulphur dioxide etc. also affect the performance of the engine. The most common type of fuel saving is done by adding chemical compounds to the fuel [6, 7]. Unwanted exhaust emission is due to incomplete combustion of carbon dioxide, nitrogen dioxide and sulphur dioxide which results in higher exhaust gas emissions. Usage of fuel saving gadgets is also not recommended due to its high cost [8]. In order to overcome these difficulties, we propose a motor vehicle incorporated with a magnetic device, whose fuel consumption and emission characteristics are then analyzed. The present research works describes about the fabrication of a low cost bike associated with a magnetic device to reduce emission characteristics as well as fuel consumption. The strength of magnetic field lines needed was around 5000 Gauss which is provided by a powerful neodymium magnet. These magnets are cost effective, weightless, safe and environment friendly [9]. The fuel is exposed to a magnetic field by a magnetic source which is placed in

V. Sankar · S. M. Chandran (✉) · T. Tomy · U. Raj · V. Samson
Rajagiri School of Engineering and Technology, Kakkanad, Kerala, India
e-mail: Sanathmchandran@gmail.com

V. Sankar
e-mail: vishnusankar9369@gmail.com

K. Ramachandran
Holy Kings College of Engineering and Technology, Muvattupuzha, Kerala, India

 
K. Antony and J. P. Davim (eds.), *Advanced Manufacturing and Materials Science*, Lecture Notes on Multidisciplinary Industrial Engineering,
https://doi.org/10.1007/978-3-319-76276-0_1

fuel supply line to magnetize the fuel before it enters the engine cylinder. It is easy to attach the magnetic device on the fuel line because of its robustness. The magnetic field lines acting on the fuel disperses the hydrocarbon cluster into smaller particles thereby improving the efficiency of combustion [10]. Moreover, it also reduces unburned hydrocarbon present in the emission. Results are validated by conducting experiments in a petrol engine for comparing the effect that magnetic field has on fuel consumption and exhaust emissions.

## 2 Methodology

The main combustible elements present in a fuel are carbon, hydrogen, sulphur etc. Fuel must contain one or several of these elements so that combustion takes place. During combustion, the chemical energy of fuel will be converted into heat energy. For utilizing the energy of the fuel in its most usable form, transformation of fuel from one state to another is required. One way is to convert solid form to liquid/gases form [11]. Through this method energy of fuels is utilized more effectively and efficiently for various purposes. The straight run gasoline can be obtained either by distilling crude petroleum or by synthesis technique [12]. It contains undesirable compounds like unsaturated straight chain hydrocarbons and sulphur compounds. But, it has a boiling range of 40–120 °C. This leads to the unsaturated hydrocarbons getting oxidized and polymerized. This also leads to gum and sludge formation on the storage system. Sulphur compounds adversely affect tetraethyl lead, which is added to gasoline to improve its ignition properties. Moreover, it also leads to the internal combustion engine getting corroded. The sulphur compounds can be removed from gasoline with sodium plumbite, an alkaline solution [7]. When a gasoline fuel is burnt, it reacts with oxygen to produce heat. Two neodymium magnets having field strength of 1500–2000 Gauss and 2500–2800 Gauss used for making of magnetic field and strength of magnets is validated by Gaussmeter or magnetometer. Magnetic fuel saver can reduce the fuel consumption in a petrol engine. Fuels are in liquid form in the oil tank. It undergoes combustion when it is vaporized and mixed with air. In order to improve the combustion, fuel particles should be turned to finer tiny particles. Magnets will help to ionize the fuel [13, 14]. Fuels are classified from group of hydrocarbons. When hydrocarbon molecules flow along the direction of magnetic field, they change their orientation to a direction opposite to the applied magnetic field. As a result, the molecule configuration changes and intermolecular force between the molecules weakens. Thus, presence of magnetic field disperses the molecules into fine tiny particles thereby making the fuel less viscous.

## 3 Experimental Findings

Magnetic fuel treatment illustrates about the magnetic field interaction of hydrocarbon molecules of fuel with oxygen molecules. Liquid fuel consists of many organic compounds, predominantly hydrogen and carbon atoms. Densely packed structures called pseudo compounds are formed because of different types of attractive forces. These compounds can be further organized into clusters or association which is relatively stable. When the fuel mixes with the air, oxygen is not able to penetrate into the pseudo compounds. This hinders the availability of appropriate quantities of oxygen to the interior of the molecular groups (associations). This is the reason for incomplete combustion which leads to the formation of carbon deposits, carbon monoxide and large quantities of hydrocarbons being emitted into the environment. Researchers have proved [15] that external forces such as magnetic field can polarize the hydrocarbon fuel. The magnetic field produces a moment due to the movement of outer electrons in the hydrocarbon chain, which moves the electrons into a higher energy level. This causes the breakage of fixed valence electron which has a vital role in the bonding process of fuel. Nevertheless, hydrocarbon fuel becomes directionalized. This magnetic alignment permits rapid bonding within respective oxidizing media. It leads to complete combustion of hydrocarbon fuel. Hydrocarbon molecules interact with high magnetic field which leads to their breaking up into small associates with high specific surface [16]. This also leads to improved combustion. In accordance with Van Der Waals discovery of weak clustering force, hydrocarbons in magnetically energized fuel have a higher affinity towards oxygen which is the reason for optimal combustion in the engine chamber. The outcomes of fuel with high magnetic field are improved fuel combustion, increased engine power and reduction in fuel consumption. Improved fuel combustion can cause reduction of carbon, carbon monoxide and other hydrocarbon emissions. In the present work, objective was to understand the effect of magnetic field on combustion and emissions in a spark ignition engine. In order to fix magnet on fuel line, we need circular disk type magnets having dimension of diameter 1 inch, such that it should be around the fuel line. Geometry of the magnet helps for easy installation on fuel line. The magnetic installation is as shown in Fig. 1.

The performance tests were carried out for a single cylinder, four stroke petrol engine. The setup consists of an engine and an exhaust gas analyzer. The engine specification is detailed in Table 1. Initially, engine was made to run with petrol as fuel for all tests. The accurate measurement of the fuel consumption rate was taken by the Burette method. The fuel consumption rates were measured for different engines on different loading conditions and exhaust gases and emission from the engine were also analyzed and measured by exhaust gas analyzer during experimental tests as shown in Fig. 2. It measures exhaust gases such as hydrocarbons, carbon monoxide, oxides of nitrogen and carbon dioxide concentrations for each load. This process was repeatedly done with magnetic installation and without magnet installation. The results were then compared.

**Fig. 1** Magnet installed around the fuel line

**Table 1** Engine specification

| | |
|---|---|
| Manufacturer | Bajaj |
| Model | Bajaj CT100 |
| Displacement [cc] | 99.27 |
| Engine type | Single cylinder, Four-stroke |
| Maximum power [ps @ RPM] | 8.2[ps] @ 7500 [RPM] |
| Maximum torque [Nm @ RPM] | 8.05[Nm] @ 4500 [RPM] |
| Fuel delivery system | Carburetor |
| Top speed [km/h] | 110 [km/h] |
| Transmission | 4 speed constant mesh |
| Stroke × bore [mm] | 42 × 46 |

In the experiment, exhaust gases are analyzed at different engine speeds by varying the fuel supply using valve. Fuel from the fuel tank is conveyed to the magnetic device after it passes through the valve. Exhaust gasses after combustion from the engine is passed to the gas sensor. The gas sensor detects the exhaust gasses and it is given to the measuring device to detect the concentrations of various components. When hydrocarbon fuel ($CH_4$) is combusted, oxidation of hydrogen atoms will take place. Further carbon atoms are subsequently burned ($CH_4 + 2O_2 = CO_2 + 2H_2O$) which leads to carbon dioxide and water [17]. Since the time for oxidizing hydrogen atoms during high speed combustion process is very less, as in normal condition, only few carbon atoms will be partially oxidized which results in incomplete combustion. Oxygen molecules combine with hydrogen molecules readily. However, the reaction between carbon and oxygen is of too low energy. The optimal fuel efficiency (combustion efficiency) and performance is obtained by the application of magnet around the fuel line and it is first indicated as an increase in amount of carbon dioxide (CO) produced and it has been validated by the values in emission control device.

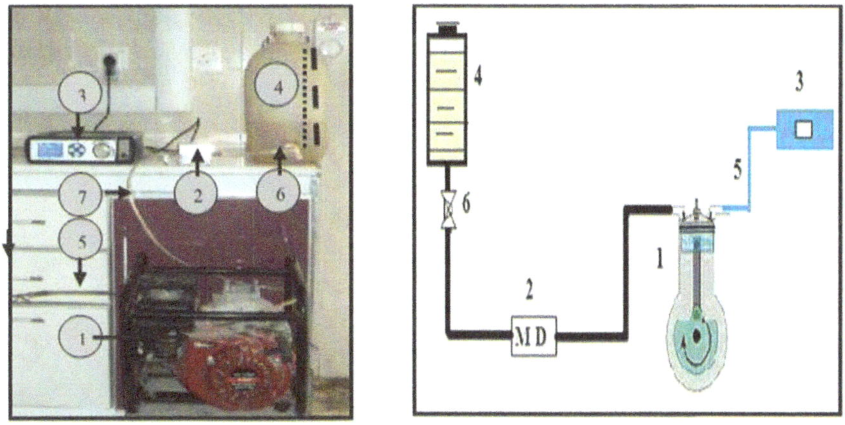


**Fig. 2** Experiment set-up (*1* Engine, *2* Magnetic device, *3* Measuring device for exhaust gases, *4* Fuel tank, *5* Gas sensor, *6* Valve and *7* Flexible tube)

## 4 Results and Discussions

The percentage of the exhaust gases which was measured during engine operation, for three different speeds, with and without magnetic field is shown in the graphs. Fuels which are basically Hydrocarbons possess a "cage like" structure, so that during the combustion process oxidizing inner carbon atoms is difficult. Furthermore they combine into much larger groups of pseudo compounds and they form clusters (associations). The entry of oxygen in the correct amount into the interior of these clusters is obstructed and it is this lack of oxygen into the cluster which hinders the proper or complete combustion [18]. The exhaust would theoretically consist of carbon dioxide, water vapour and nitrogen from air, which does not involve in the combustion. Practically, the exhaust emission consists of CO, $H_2$, HC, $NO_x$ and $O_2$ gases. Nevertheless, complete combustion of fuel will never be achieved and incomplete oxidized carbon is evident in the form of HC, CO or it will deposit on internal combustion engine chamber walls as black carbon residue. Hydrocarbon fuel molecules exposed to magnetic field tend to decluster, leading to smaller particles which are readily penetrated by oxygen. Thus it helps in better combustion. They become normalized & independent, distanced from each other, having bigger surface available for binding (attraction) with more oxygen for oxidation. According to van der Waals' theory of weak-clustering force, hydrocarbons are bound strongly with oxygen in such magnetized fuel, which ensures proper burning of mixture in the engine chamber. This results in the reduction of CO by 10% in Fig. 3.

It was found that the percentage of $CO_2$ increased by about 3% as shown in Fig. 4 but the percentage of HC was reduced by 5% as shown in Fig. 5. This is because Magnetic field helps to improve the combustion level, which leads to the reduction of exhaust emission of dangerous gaseous. The amount of unburned

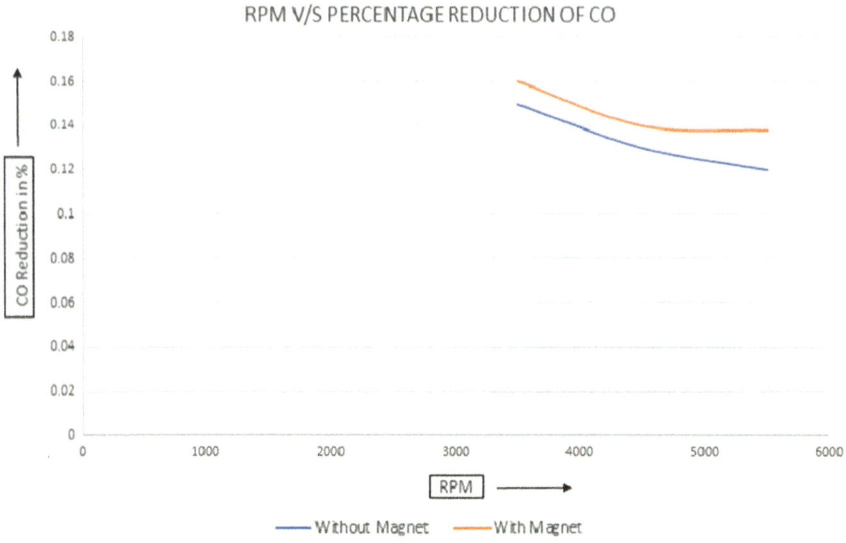


**Fig. 3** Graph plotted with RPM versus CO

hydrocarbon can also be reduced when combustion rate is improved. This system use oxygen for conversion of carbon monoxide to carbon dioxide and hydrocarbon to water and carbon dioxide. Magnetizer was used to develop strong flux fields to change the hydrocarbon molecule from its para state to higher energized ortho state. In the para $H_2$ molecule, the spin state of one atom relative to another is in opposite direction ("counter clockwise", "anti-parallel", "one up & one down"), rendering it diamagnetic. In the ortho molecule, spins are parallel ("clockwise", "coincident", "both up"), with same orientation for the two atoms. The ortho-hydrogen is more reactive than its para-hydrogen counterpart. This results in the reduction of emission and improved combustion performance when the magnetic field is applied.

It has been found that, today's hydro-carbon fuels deposit carbon residue which clogs the carburetor and the fuel injector, leading to reduction in efficiency and fuel wastage. It also leads to pinging, loss of effective horsepower, stalling and large reduction in mileage of the vehicle. In an internal combustion engine, power is obtained as a result of combustion of liquid fuel. For complete combustion to occur, the liquid fuel should get completely vaporized and then get mixed with air. If incomplete combustion takes place, it leads to harmful tail pipe emissions like carbon monoxide and oxides of nitrogen and sulphur, which reacts in the atmosphere to form smog. The fuel generally used in an internal combustion engine is compound of molecules. Each molecule consists of a number of atoms made up of number of electrons, which orbit their nucleus. These molecules which already possess magnetic movements also have positive and negative electrical charges. But these molecules are not properly aligned and hence the fuel cannot actively combine with oxygen during the combustion process. Hence it is necessary to ionize

**Fig. 4** Graph plotted with RPM versus Carbon dioxide

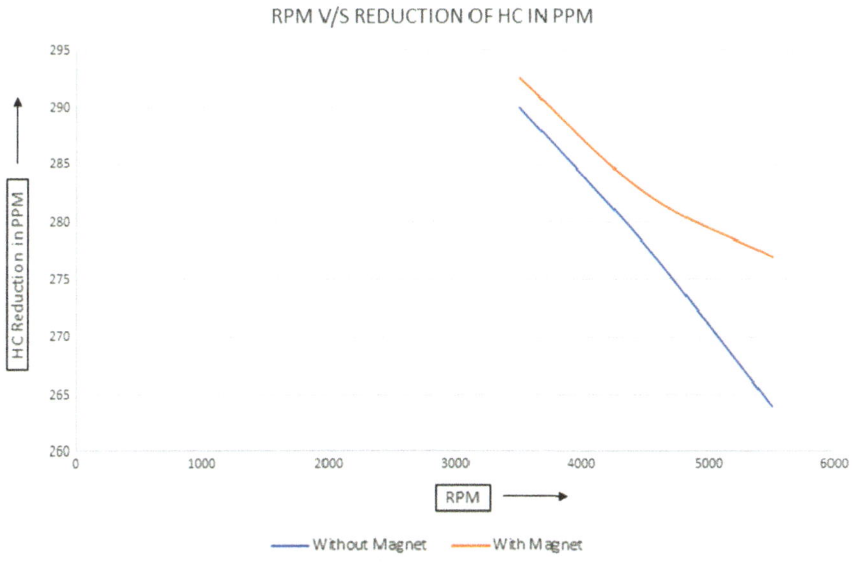


**Fig. 5** Graph plotted with RPM versus Hydro Carbon

and realign the fuel molecule or the hydrocarbon chain. This is achieved by the application of magnetic field. During magnetization of the fuel, the bonds between hydrocarbon chain breaks down which results in lowering its density, surface tension and, hence smaller particulars and droplets during atomization or injection within an internal combustion engine. Finer particles and droplets help in improved evaporation rates, better mixing of fuel and oxidizer, also improved promotion of oxidation. The net effect of this is better rate of combustion, an increase in power, and reduced pollutants. Increased oxidation of the hydrocarbon fuel has positive effects on combustion. Quicker and more complete oxidation results in more rapid and more complete combustion of the fuel.

## 5 Conclusion

Magnetizer is able to lower the NOx level. Since all oxygen is bound up with hydrocarbon fuel and formation of unwanted nitrogen compound is hindered. After burning stoichiometric fuel through proper magnetic lines of force, the engine will get maximum energy per litre along with lowest possible level of toxic emissions. The number of HC decreases with increase in mileage of $\sim$15–25%. Thus it leads to the reduction in emission and increased combustion efficiency. Since, fuel energizer increases combustion efficiency and thereby saves fuel; there is a reduction in CO emissions. Fuel associated with magnetic field is cost effective and hence it reduces the exhaust emission rate. Magnetic treatment does not need external energy source. The other key features of using magnet on automobiles are, decrease in fuel consumption per liter, higher initial torque, reduced knocking and detonation, smooth running, decrease smoke emission, faster A/C cooling, no fuel wastage and long term maintenance free engine.

**Fig. 6** Final view of spark-ignition engine with magnetic field

**Table 2** Cost Estimation

| Item | Price |
|---|---|
| Bike chases | 2500 |
| Engine | 3000 |
| Magnet | 300 |
| Mono shock | 800 |
| FZ chain lock | 45 |
| Silencer | 550 |
| CD unit | 180 |
| Petrol tap | 200 |
| Painting | 600 |
| Total | Rs. 8175 |

## References


1. Hricak RZ (1994) Air fuel magnetizer. U. S. Patent No. 5333133, 7
2. Cummins JCL (1976) Early IC and automotive engines. SAE paper 760604, Society of Automotive Engineers, Warrendale, Pennsylvania, USA pp 18–26
3. Mingdong S et al (1984) Study on the combustion efficiency of magnetized petroleum fuels. Chin Sci Bull
4. Keith H (1991) Magnetic fields raise fuel efficiency. New Scientist 131
5. Kiran RK, Muley KS (2006) Two stroke cycle engine performance and fuel treatment devices. Overdrive-an automobile magazine, New Delhi, 23–28
6. Nedunchezhian N, Dhandapani S (1999) Experimental investigation of cyclic variation of combustion parameters in a catalytically activated two stroke SI Engine combustion chamber. SAE-India, Paper 990014:1–16
7. Govindasamy P, Dhandapani S (2007) Experimental investigation of cyclic variation of combustion parameters in catalytically activated and magnetically energized two-stroke SI engine. J Energy Environ 6:45–59
8. Govindasamy P, Dhandapani S (2007) Experimental investigation of the effect of magnetic flux to reduce emissions and improve combustion performance in a two stroke catalytic-coated spark-ignition engine. Int J Automot Technol 8(5):533–542
9. Bowker RR (2000) Permanent magnet design guide. Magnet Sales and Manufacturing & Co, USA, pp 11–67
10. Chavan S, Jhavar P (2016) Effect of application of magnetic field on the efficiency of petrol engine. IRJET (Int Res J Eng Technol) 3(9):152–161
11. Yujuan C et al (1989) Reduction of viscosity of crude oil by a strong magnetic field and its application. Acta Petrolei Sinica
12. Yazhong Y (1990) The applied test of wax protection by intensive magnetism. Oil and gas storage and transportation (China)
13. Pera I, Pines P (1987) Magnetizing hydrocarbon fuels and other fluids. U. S. Patent No. 4716024
14. Okoronkwo CA, Nwachukwu CC Dr, Ngozi LC Dr, Igbokwe JO (2010) The effect of electromagnetic flux density on the ionization and the combustion of fuel (an economy design project). Am J Sci Ind Res 1(3):527–531, ISSN: 2153-649X. https://doi.org/10.5251/ajsir
15. Chavan S, Jhavar P (2016) Effect of application of magnetic field on emission of petrol engine. IJIR (Imperial J Interdisc Res) 2(10):2121–2128

16. Ugara V, Latude S, Muduafale K, Dihole A (2005) Performance of I.C engines under the influence of strong magnetic field. IOSR J Mech Civ Eng 11–17
17. Farisa AS, Hazim Mohammed SJ (2012) Effect of magnetic field on fuel consumption & exhaust gas emission in two stroke petrol engine. Sci Direct 18:327–338
18. Patel PM, Rathod GP, Patel TM (2014) Effect of magnetic field on performance and emission of single cylinder four stroke diesel engine. IOSR J Eng 4:28–34

# Experimental Investigation on Reverse Engineering Techniques Employing Response Surface Methodology for Freeform Surfaces

**G. Sreeram Reddy, V. V. Satyanarayana, Manzoor Hussain and J. Jagadesh Kumar**

## 1 Introduction

The reverse engineering concepts are particularly useful whenever the products original designs are lost and need to produce it using the surface characteristics thus reducing the product development lead time. In reverse engineering the CAD model created from physical component is rebuilt for freeform surface construction which is a NURBS model [1, 2]. The quality of the freeform surface so generated depends on the curvature and susceptible to level of noise contamination [3]. The curve characteristics are investigated by employing height decision method and achieved the data simplification [4]. By using non-contact and contact measurement techniques the freeform surface and the part machining of a scaled down sports car is studied for the effective utilization of reverse engineering concepts [5]. Design of experimental techniques is undertaken and deviation from CAD models, reproducibility and measurement uncertainty are evaluated for various configurations [6]. A methodology for reconstruction of a surface in the reverse engineering has been attained with cost reduction and improved computational efficiency [7]. Hierarchical space decomposition model and octree data structures are used for the scanned data in the construction of surface was found to be more robust and stable [8]. Parameter optimization study is undertaken for reconstruction of the surface by

G. Sreeram Reddy (✉) · V. V. Satyanarayana · J. Jagadesh Kumar
Department of Mechanical Engineering, Vidya Jyothi Institute of Technology, Hyderabad, Telangana, India
e-mail: sreeramgundeti@gmail.com

V. V. Satyanarayana
e-mail: vvs@vjit.ac.in

M. Hussain
Department of Mechanical Engineering, JNTUH College of Engineering, Sultanpur, Medak, Telangana, India

 
K. Antony and J. P. Davim (eds.), *Advanced Manufacturing and Materials Science*, Lecture Notes on Multidisciplinary Industrial Engineering,
https://doi.org/10.1007/978-3-319-76276-0_2

reverse engineering techniques [9] and found that the accuracy level depends on the number of triangles, point cloud data and noise reduction [10].

In the present investigation various responses pertaining to reverse engineering free form surface have been evaluated and the optimal factor conditions are identified.

# 2 Experimental Methodology

## 2.1 Component Selection

The component part employed in the investigation was machined using CNC mill. The part has the free form surface possessing sharp edges smooth enough to acquire a point cloud data by employing a Faro make laser scanner which has the capability to scan up to 19,200 points/s with 35 μm accuracy. The data acquired is a point cloud along the chosen scanned line and analyzed for various sample sizes.

## 2.2 Responses Measurement

The investigation involved varies from a sample size of 20% of population to whole population in steps of 20%. The responses evaluated are deviation, curvature deviation, memory, defects and process time. The offset distance between actual surface and reverse engineered model is taken in root mean square values for calculating deviation (Fig. 1), while curvature deviation is done with Gaussian curvature and original CAD model (Fig. 2). The defects include intersections, non manifold edges, small holes, tunnels, spike edges etc. (Fig. 3). Processing time and memory from the CAD model to reverse engineered surface construction have been identified from the computer system employed.

## 2.3 Input Factors and Their Levels

Three factors are chosen at 5 levels each to evaluate their effect on the responses under consideration. Trial runs are conducted while fixing the limits in the domain for the parameters (Table 1).

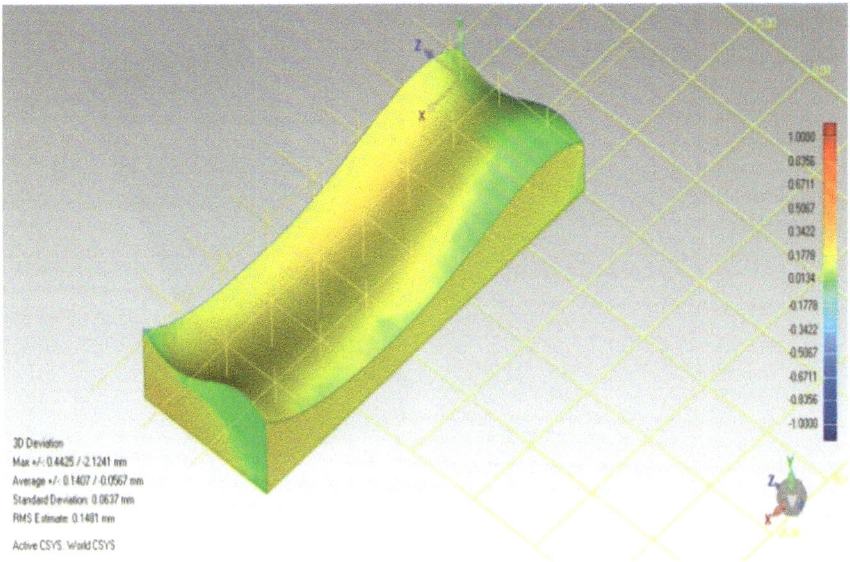


**Fig. 1** Deviation analysis

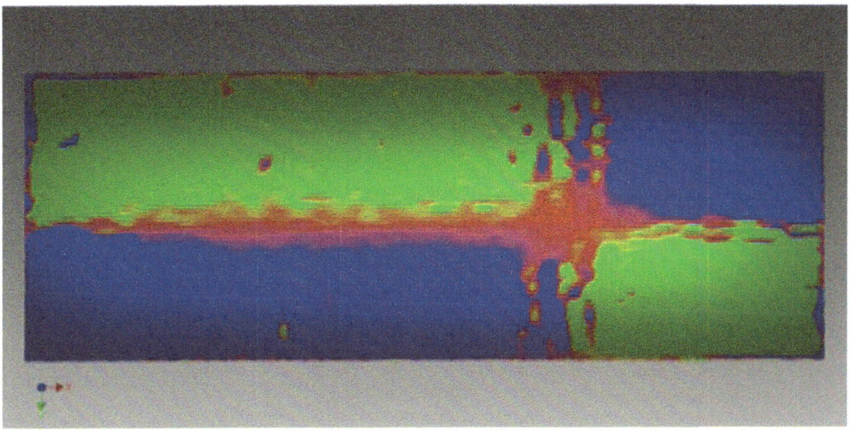

**Fig. 2** Gaussian curvature JPEG image

## 2.4 Design of Experiments

The experiments are conducted as per the central composite design of the response surface methodology [11] and a response polynomial is fitted and is in the form;

**Fig. 3** Model with defects statistics

**Table 1** Process parameters

| S. No | Parameters | Notation | −2 | −1 | 0 | 1 | 2 |
|---|---|---|---|---|---|---|---|
| 1 | Noise reduction level | A | 0 | 1 | 2 | 3 | 4 |
| 2 | Triangle count (%) | B | 20 | 40 | 60 | 80 | 100 |
| 3 | Smoothing levels | C | 0 | 2 | 4 | 6 | 8 |

$$Y = b_0 + b_1x_1 + b_2x_2 + b_3x_3 + b_4x_1^2 + b_5x_2^2 + b_6x_3^2 + b_1x_1x_2 + b_8x_2x_3 + b_9x_1x_3$$

The coefficients $b_0$, $b_1$ … are computed by the least square method and analysis of variance is performed by Yate's algorithm [12]. The analysis is conducted for 95% confidence level and 5% level of significance. The experimental matrix comprises of 20 runs for the factors at their levels is depicted in Table 2.

## *2.5 Results*

Experiments are conducted randomly to avoid the bias for various sample sizes. The results for all the responses at a sample size of 40% of population are given in Table 3.

**Table 2** Experimental matrix

| Parameters | Runs | | | | | | | | | | | | | | | | | | | |
|---|---|---|---|---|---|---|---|---|---|---|---|---|---|---|---|---|---|---|---|---|
| | 1 | 2 | 3 | 4 | 5 | 6 | 7 | 8 | 9 | 10 | 11 | 12 | 13 | 14 | 15 | 16 | 17 | 18 | 19 | 20 |
| A | −1 | 1 | 0 | −1 | 0 | 1 | 0 | −1 | 1 | 0 | 1 | 0 | 0 | 0 | 2 | 0 | 0 | −1 | 0 | −2 |
| B | −1 | 1 | −2 | −1 | 0 | −1 | 0 | 1 | 1 | 0 | −1 | 0 | 0 | 0 | 0 | 0 | 0 | 1 | 2 | 0 |
| C | 1 | 1 | 0 | −1 | 0 | −1 | 0 | −1 | −1 | 0 | 1 | 0 | 0 | 2 | 0 | 0 | −2 | 1 | 0 | 0 |

Table 3 Experimental results

| Runs | Deviation (mm) | Curvature deviation (%) | Defects | Memory (MB) | Processing time (s) |
|---|---|---|---|---|---|
| 1 | 0.036 | 62 | 9 | 4.63 | 8.79 |
| 2 | 0.0305 | 62.6 | 16 | 9.26 | 11.16 |
| 3 | 0.0404 | 62.3 | 0 | 2.31 | 9.48 |
| 4 | 0.0298 | 62.6 | 21 | 4.63 | 9.2 |
| 5 | 0.0294 | 63.3 | 0 | 6.94 | 10.12 |
| 6 | 0.0291 | 63.2 | 15 | 4.63 | 11.27 |
| 7 | 0.0294 | 63.3 | 0 | 6.94 | 9.28 |
| 8 | 0.0286 | 62.9 | 26 | 9.26 | 9.13 |
| 9 | 0.0278 | 63.2 | 15 | 9.26 | 10.43 |
| 10 | 0.0294 | 63.3 | 0 | 6.94 | 8.85 |
| 11 | 0.0357 | 62.2 | 16 | 4.63 | 10.23 |
| 12 | 0.0294 | 63.3 | 0 | 6.94 | 9.37 |
| 13 | 0.0294 | 63.3 | 0 | 6.94 | 9.46 |
| 14 | 0.0341 | 62.7 | 0 | 6.95 | 10.36 |
| 15 | 0.0304 | 62.7 | 0 | 6.94 | 12.49 |
| 16 | 0.0294 | 63.3 | 0 | 6.94 | 10.75 |
| 17 | 0.03 | 64.5 | 122 | 6.94 | 9.15 |
| 18 | 0.0302 | 62.2 | 13 | 9.26 | 9.55 |
| 19 | 0.028 | 63.3 | 0 | 11.5 | 9.25 |
| 20 | 0.0291 | 61.8 | 0 | 6.94 | 7.23 |

# 3 Analysis and Discussion

The deviation has varied from 0.028 to 0.0357 mm while the curvature deviation is in the range 61.8–64.5%. Largest number of defects to a tune of 122 is recorded while the processing time is in the range 7.23–11.27 s with a maximum memory of 11.5 MB. The deviation is invariant with the parameter combination triangle count

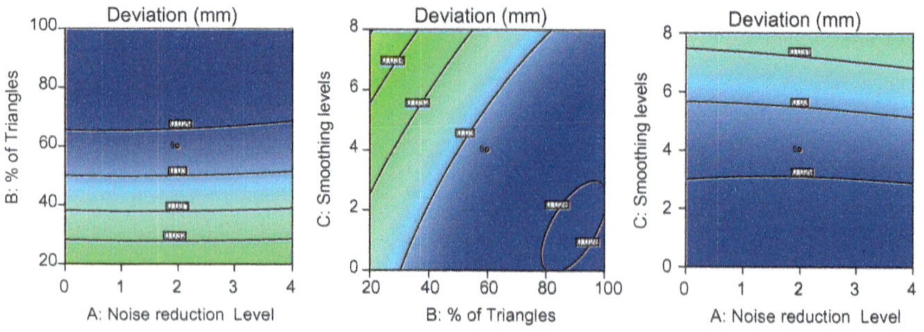


Fig. 4 Contour plots for deviation

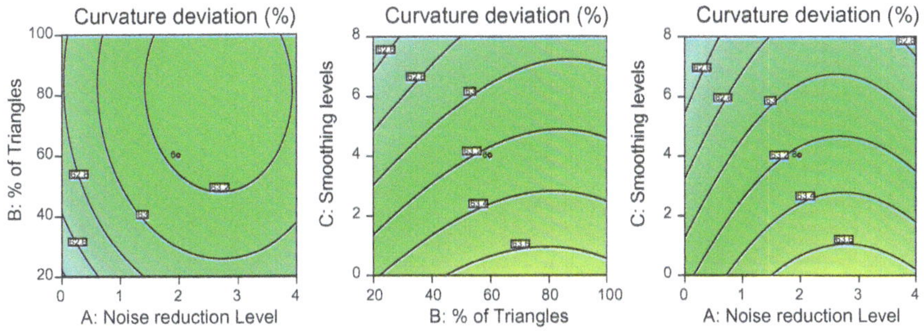


**Fig. 5** Contour plots for curvature deviation

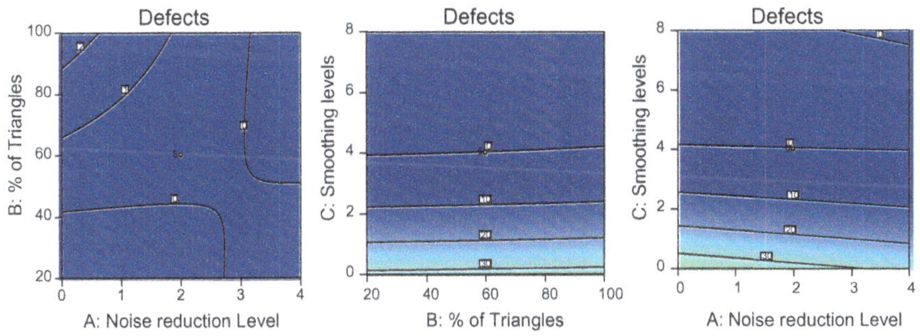


**Fig. 6** Contour plots for defects

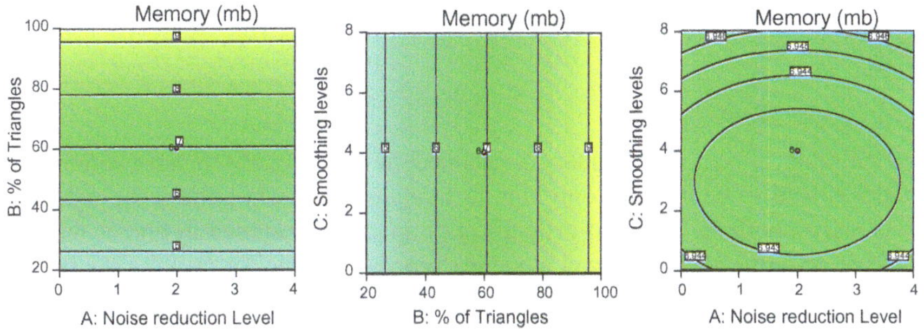


**Fig. 7** Contour plots for memory

and noise reduction level. A similar trend is found with the parameter combination smoothing level and noise reduction level (Fig. 4). The interaction of process parameters on curvature deviation is found to be following parabolic trends (Fig. 5) and harping at a maximum value of 63%. The contour graphs shown in Fig. 6 illustrates that the factor smoothing level alone is governing the occurrence of defects in the reverse engineered model and the other factors did not have any

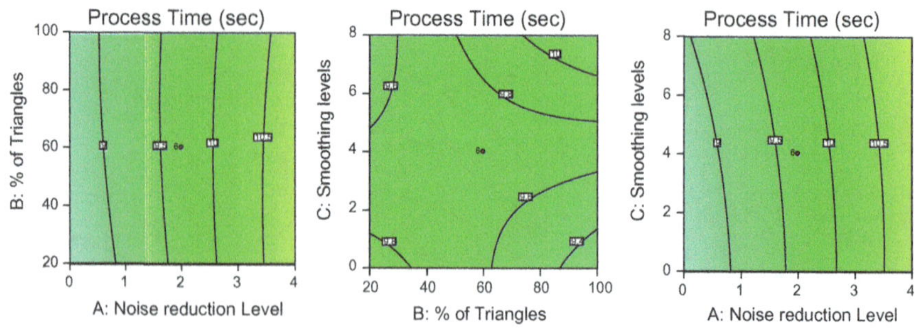


**Fig. 8** Contour plots for process time

**Table 4** ANOVA for deviation

| Analysis of variance table (partial sum of squares—type III) | | | | |
|---|---|---|---|---|
| Source | Sum of squares | DF | Mean square | F value |
| Model | 1.809E−004 | 9 | 2.010E−005 | 13.67 |
| (A)—Noise reduction level | 7.563E−008 | 1 | 7.563E−008 | 0.051 |
| (B)—% of triangles | 9.168E−005 | 1 | 9.168E−005 | 62.37 |
| (C)—Smoothing levels | 4.001E−005 | 1 | 4.001E−005 | 27.22 |
| AB | 3.125E−008 | 1 | 3.125E−008 | 0.021 |
| AC | 2.813E−007 | 1 | 2.813E−007 | 0.19 |
| BC | 9.031E−006 | 1 | 9.031E−006 | 6.14 |
| $A^2$ | 6.870E−008 | 1 | 6.870E−008 | 0.047 |
| $B^2$ | 3.411E−005 | 1 | 3.411E−005 | 23.21 |
| $C^2$ | 9.893E−006 | 1 | 9.893E−006 | 6.73 |
| Residual | 1.470E−005 | 10 | 1.470E−006 | |
| Total | 1.956E−004 | 19 | | |

recognizable effect. The memory consumed in building the reverse engineered model is said to be affected only due to triangle count while the noise reduction level and smoothing level had insignificant impact as illustrated in Fig. 7, with the increase of noise reduction level alone the processing time for creating the reverse engineered model has increased and invariant with other factors (Fig. 8). The data has been subjected to ANOVA and the 'F' ratios are computed for the significance of coefficients. A typical ANOVA for deviation has been given in Table 4 and the regression equations for all the responses have been enlisted in Table 5.

The analysis has also been carried out for all the sample sizes and observed that the curvature deviation has increased with small and large sample sizes while the value is lowest with medium sized samples. The curvature deviation is primarily governed by the parameter smoothing level and to a lesser extent by another parameter triangle count as shown in Fig. 9.

**Table 5** Regression equations

| S. No | Response | Regression equations |
|---|---|---|
| 1 | Deviation | $0.033258 - 1022202E{-}004 * C{-}6.64063E{-}006 * B * C + 7.27983E{-}007B^2 + 3.92045E{-}005 * C^2$ |
| 2 | Curvature deviation | $0.62529 + 4.51723E{-}003 * A + 1.63472E{-}004 * B{-}1.25646E{-}003 * C{-}7.66315E{-}004 * A^2{-}1.00625E{-}006 * B^2$ |
| 3 | Defects | $+33.67045 - 12.64489 * C + 0.95952 * C^2$ |
| 4 | Memory | $+3.46804 + 0.058227 * B{-}4.75852E{-}006 * B^2$ |
| 5 | Process time | $+8.86213 + 0.52153 * A$ |

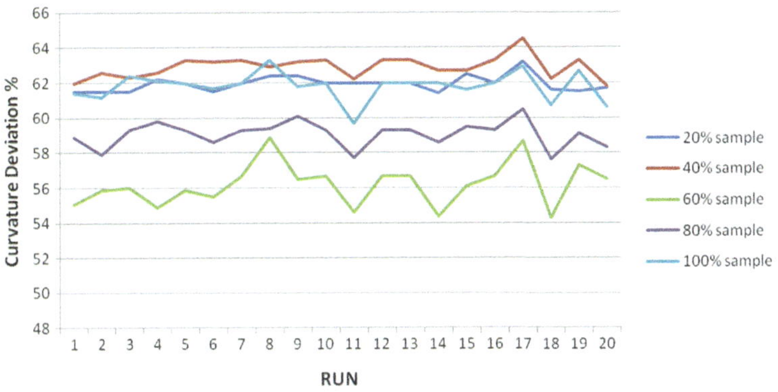


**Fig. 9** Variations of the curvature deviation with respect to runs for different % of sample sizes

The memory size and processing time have been increased in proportion with the sample size but governed by the factors triangle count and noise reduction respectively. The response in the study of defects is depending only on the parameter smoothing levels and is independent of sample size.

## 4 Conclusions

The curvature deviation is governed by all the parameters under study while the deviation is influenced by triangle count and smoothing levels only. The remaining responses namely memory, defects and processing time are affected by one factor each. The optimal parameter combination can be computed depending upon the responses undertaken.

## References


1. Chang KH, Magdum S, Khera S, Goel VK (2003) An advanced computer modeling and prototyping method for human tooth mechanics study. Ann Biomed Eng 31(5):621–631
2. Kruth JP, Kerstens A (1998) Reverse engineering modelling of free-form surfaces from point clouds subject to boundary conditions. J Mater Process Technol 76:120–127
3. Mousa MH (2011) Matching 3D objects using principle curvatures descriptors. In: IEEE Pacific Rim conference on communications, computers and signal processing (PacRim), Canada, pp 447–452. ISSN: 1555-5798
4. Chang DY, Chang YM (2002) A freeform surface modelling system based on laser scan data for reverse engineering. Int J Adv Manuf Technol 20:9–19
5. Piratelli-Filho A, Souza PHJ, Arencibia RV, Anwer N (2014) Study of contact and non-contact measurement techniques applied to reverse engineering of complex freeform parts. Int J Mech Eng Autom 1(3):166–175
6. Barini EM, Tosello G, De Chiffre L (2010) Uncertainty analysis of point-by-point sampling complex surfaces using touch probe CMMs DOE for complex surfaces verification with CMM. Precis Eng 34:16–21
7. Min Z (2011) A new approach of composite surface reconstruction based on reverse engineering. Procedia Eng 23:594–599
8. Azernikov S, Miropolsky A, Fischer A (2003) Surface reconstruction of freeform objects based on multiresolution volumetric method. J Comput Inf Sci Eng 3(4):334–338
9. Li Y, Huang X, Gong C, Wang K (2004) An engineering rules based parameterization approach for turbine blade reverse engineering. In: Proceedings of the Geometric Modeling and Processing (GMP'04)
10. Al-Ahmari AMA, Aalam J (2015) Optimizing parameters of freeform surface reconstruction using CMM measurement. J Meas 64(10):17–28
11. Montgomery C (1997) Design and analysis of experiments, 4th edn. Wiley, New York
12. Box GEP, Hunter JS, Hunter WG (2005) Design of experiments. Wiley, New York

# Effect of Process Parameters and Cryotreated Post Tempered Electrodes on the Radial Overcut of AISI 304 During EDM: A Comparative Study

**Munmun Bhaumik and Kalipada Maity**

## 1 Introduction

Electro discharge machining (EDM) is a kind of nonconventional machining process where the series of repetitive electrical discharge that occurs between the tool and workpiece are responsible for the material removal [1]. Muthukumar et al. [2] developed a model for radial overcut (ROC) using response surface methodology (RSM) during EDM operation of Incoloy 800. They concluded that lower value of voltage and current considerably minimize the ROC. Bhaumik et al. [3] established semi empirical model for material removal rate (MRR) and ROC using dimensional analysis and found a good correlation between experimental and numerical results. Later on, the effect of cryotreated double tempered tungsten carbide electrode on the dimensional accuracy of AISI 304 stainless steel during powder mixed EDM (PMEDM) was investigated. They reported that cryotreated double tempered electrode provides least ROC compared to untreated electrode [4].

Cryotreatment is the recent advancement in the domain of machining to enhance the tool life as well as to reduce the machining cost. Cryotreatment is a kind of heat treatment applied to a material at low temperatures [5]. Kumar et al. [6] reported that tool wear reduced by using cryotreated double tempered copper electrode compared to untreated copper electrode during EDM of Inconel 718. Jefferson and Hariharan [7] performed a comparison study using untreated and cryotreated electrodes during micro EDM (μ-EDM) of AISI 304 stainless steel. They reported 58% reduction of tool wear using cryotreated tungsten carbide tool followed by cryotreated brass and copper electrodes with a reduction of 51% and 35% respectively. Kalsi et al. [8] performed multi tempering after cryotreatment of tungsten carbide inserts. They reported that the wear behaviour of inserts significantly enhances for third tempering after cryotreatent.

M. Bhaumik (✉) · K. Maity
Department of Mechanical Engineering, NIT, Rourkela 769008, Orissa, India
e-mail: munmunbhaumik@gmail.com



K. Antony and J. P. Davim (eds.), *Advanced Manufacturing and Materials Science*, Lecture Notes on Multidisciplinary Industrial Engineering,
https://doi.org/10.1007/978-3-319-76276-0_3

From the above literature survey, it is apparent that machining performance is enhanced by using cryotreated post tempered electrodes. Dimensional accuracy plays an important role in the EDM products. For achieving the dimensional accuracy, the ROC should be as least as possible. The aim of this study is to evaluate the effect of process parameters (pulse on time, peak current, duty cycle, gap voltage and flushing pressure) and cryotreated post tempered tungsten carbide electrodes on the radial overcut (ROC) of machined AISI 304 stainless steel and compare them with the findings while using untreated electrode. The microstructural analysis has also been carried out on the machined surfaces.

## 2 Experimental Details

In this investigation, the EDM operation of AISI 304 stainless steel was executed in ELECTRONICA-ELECTRAPLUS PS 50 ZNC EDM machine. Experiments have been carried out using three different types of electrode i.e. untreated (UT), cryotreated single tempered (CT1) and cryotreated double tempered electrode (CT2). The level of machining parameters and the experimental outline are tabulated in Tables 1 and 2 respectively.

ROC of the machining holes were evaluated from the observed data by using this formula

$$ROC = \frac{D_h - D_t}{2} \quad (1)$$

$D_h$ and $D_t$ are diameters of machined hole and electrode.

Tungsten carbide electrodes were considered for the deep cryogenic treatment at a temperature of −193.15 °C for 24 h. Further, one electrode was subjected to single tempering, while the another one was subjected to double tempering at 200 °C for relieving stress induced by cryogenic treatment. The variation of temperature with time curve is depicted in Fig. 1.

**Table 1** Experimental conditions

| Working parameters | Description |
|---|---|
| Workpiece material | AISI 304 stainless steel |
| Electrode material | Tungsten carbide |
| Dielectric fluid | EDM oil |
| Peak current ($I_P$) (A) | 2, 4, 6, 8, 10 |
| Gap voltage ($V_g$) (V) | 40, 50, 60, 70, 80 |
| Pulse on time ($T_{on}$) (μs) | 50, 100, 150, 200, 300 |
| Duty cycle (r) (%) | 55, 60, 65, 70, 75 |
| Flushing pressure ($F_P$) (kgf/cm$^2$) | 0.2, 0.3, 0.4, 0.5, 0.6 |
| Machining time (min) | 10 |

**Table 2** Experimental outline for machining of AISI 304 stainless steel

| Sl. No | $I_P$ (A) | $V_g$ (V) | $T_{on}$ (μs) | r (%) | $F_P$ (kgf/cm$^2$) |
|---|---|---|---|---|---|
| 1 | 2 | 45 | 50 | 60 | 0.2 |
| 2 | 4 | 45 | 50 | 60 | 0.2 |
| 3 | 6 | 45 | 50 | 60 | 0.2 |
| 4 | 8 | 45 | 50 | 60 | 0.2 |
| 5 | 10 | 45 | 50 | 60 | 0.2 |
| 6 | 5 | 40 | 50 | 60 | 0.2 |
| 7 | 5 | 50 | 50 | 60 | 0.2 |
| 8 | 5 | 60 | 50 | 60 | 0.2 |
| 9 | 5 | 70 | 50 | 60 | 0.2 |
| 10 | 5 | 80 | 50 | 60 | 0.2 |
| 11 | 5 | 45 | 50 | 60 | 0.2 |
| 12 | 5 | 45 | 100 | 60 | 0.2 |
| 13 | 5 | 45 | 150 | 60 | 0.2 |
| 14 | 5 | 45 | 200 | 60 | 0.2 |
| 15 | 5 | 45 | 300 | 60 | 0.2 |
| 16 | 5 | 45 | 50 | 55 | 0.2 |
| 17 | 5 | 45 | 50 | 60 | 0.2 |
| 18 | 5 | 45 | 50 | 65 | 0.2 |
| 19 | 5 | 45 | 50 | 70 | 0.2 |
| 20 | 5 | 45 | 50 | 75 | 0.2 |
| 21 | 5 | 45 | 50 | 60 | 0.2 |
| 22 | 5 | 45 | 50 | 60 | 0.3 |
| 23 | 5 | 45 | 50 | 60 | 0.4 |
| 24 | 5 | 45 | 50 | 60 | 0.5 |
| 25 | 5 | 45 | 50 | 60 | 0.6 |

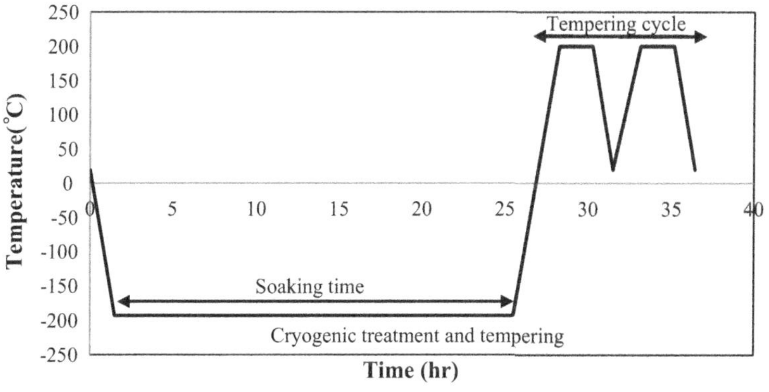


**Fig. 1** Temperature versus time for the cryogenic treatment and tempering process

## 3 Results and Discussion

Radial overcut (ROC) is the deviation between the machined hole produced during EDM operation and the electrode. ROC is an inherent feature in EDM process which is unavoidable. To achieve the accuracy of EDMed products, minimization of overcut is required. Therefore, it is essential to distinguish the influence of machining parameters on the overcut. Accordingly, by controlling the EDM parameters a perfect groove can be created during machining. In this investigation, the influence of various EDM parameters and cryotreated post tempered electrodes on the accuracy of ROC has been discussed.

Figures 2, 3, 4, 5, and 6 depicts the effects of process parameters, untreated and cryotreated post tempered electrodes on radial overcut. ROC enhances with the increase of pulse on time, peak current and gap voltage for all the electrodes. This occurs because increase of all the parameters the discharge energy increases resulting removal of high amount of molten material from the workpiece. Consequently, ROC enhances. The double tempered electrode exhibits less overcut than the other two electrodes. When peak current increases from 2 to 10 A, ROC reduced up to 11.1 and 28.8% using cryotreated single and double tempered electrodes related to untreated electrode respectively.

When pulse on time increases from 50 to 300 μs, ROC reduced up to 17.4% for single tempered and 22.85% for double tempered electrode respectively. ROC decreased up to 10.46% for single tempered and 18.18% for double tempered electrode, when gap voltage increases from 40 to 80 V. ROC is maximum at 75% duty cycle and minimum at 55% duty cycle. ROC decreased up to 20 and 30% for single and double tempered electrodes respectively, when duty cycle increases from 55 to 75%. ROC decreases with the increase of flushing pressure because of the

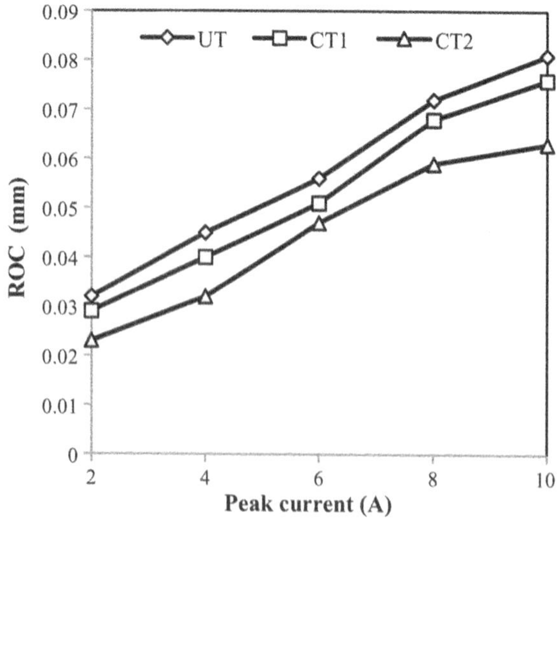


**Fig. 2** Effect of peak current on ROC

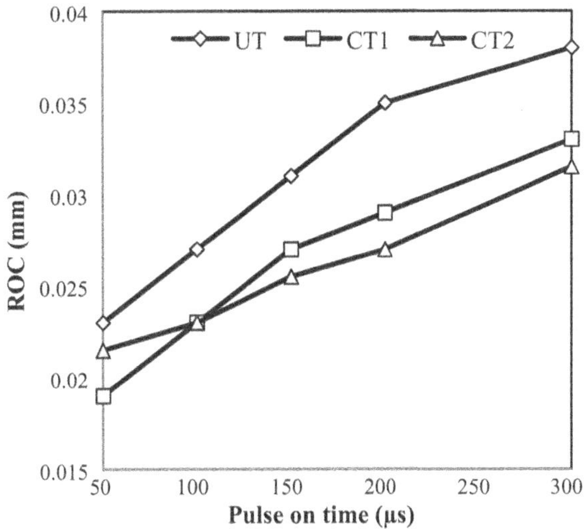


**Fig. 3** Effect of pulse on time on ROC

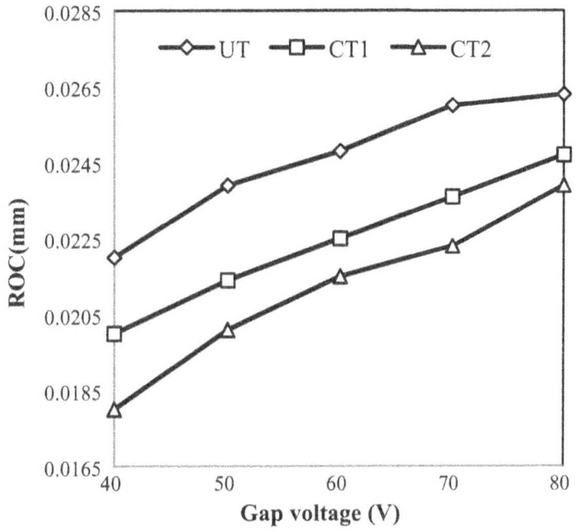


**Fig. 4** Effect of gap voltage on ROC

efficient removal of heat generated during sparking facilitating a reduction in the temperature around the surface of the electrode at a higher flushing pressure [9]. ROC for the untreated electrode is higher than the cryotreated post tempered electrodes. When flushing pressure increases from 0.2 to 0.6 kgf/cm$^2$, ROC reduced up to 22.2 and 16.66% for single and double tempered electrodes respectively.

The mechanical (hardness) and thermal property (thermal conductivity) of the tungsten carbide electrode enhances after cryotreatment because of the refinement of carbide grain size and grain boundary. During machining, the roundness of the

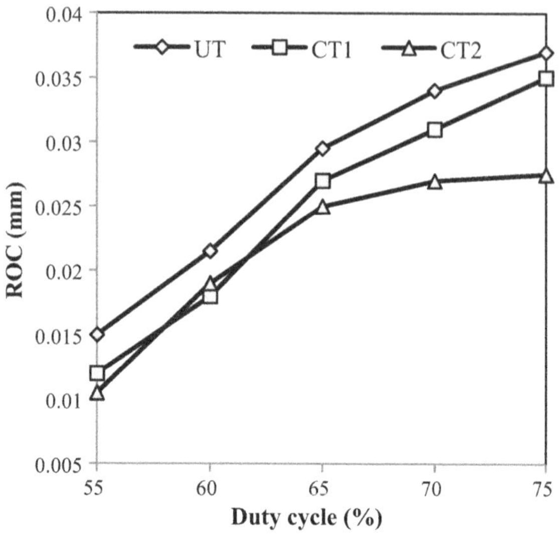


**Fig. 5** Effect of duty cycle on ROC

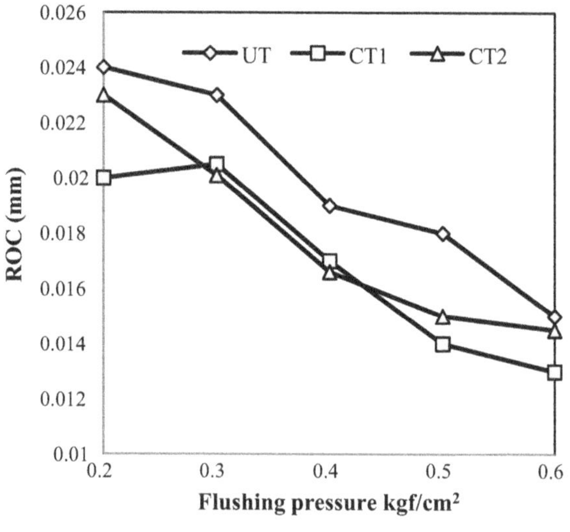


**Fig. 6** Effect of flushing pressure on ROC

untreated electrode gets damaged owing to poor thermal properties. The cryotreated electrodes possess high thermal conductivity compared to untreated electrode. Therefore, during sparking electrode surface going through less melting and vaporization compared to untreated electrode and the roundness of the electrode is maintained. Consequently, the cryotreated post tempered tungsten carbide electrodes provide better dimensional accuracy due to uniform sparking and proper maintenance of electrode shape than that of untreated electrode.

## 4 Microstructural Analysis

Figure 7a–c depict the SEM images of EDMed surface machined by using the untreated, cryotreated single and double tempered electrodes at $I_P = 6$ A, $V_g = 45$ V, $T_{on} = 50$ μs, $\tau = 60\%$, $F_P = 0.2$ kgf/cm$^2$ respectively. From the SEM images, it is observed that the machined surface has complex exterior such as cracks, craters, spherical particles, pores, pockmarks, and voids because of the rapid heating and quenching.

From these images, it is apparent that a lesser number of surface cracks and small craters are observed for the surface machined by cryotreated post tempered electrodes compared to untreated electrode. The recast layer thickness also found to be lesser for the cryotreated post tempered electrodes. The EDMed surface machined by cryotreated post tempered electrodes show comparatively smoother surface than the surface machined by the untreated electrode.

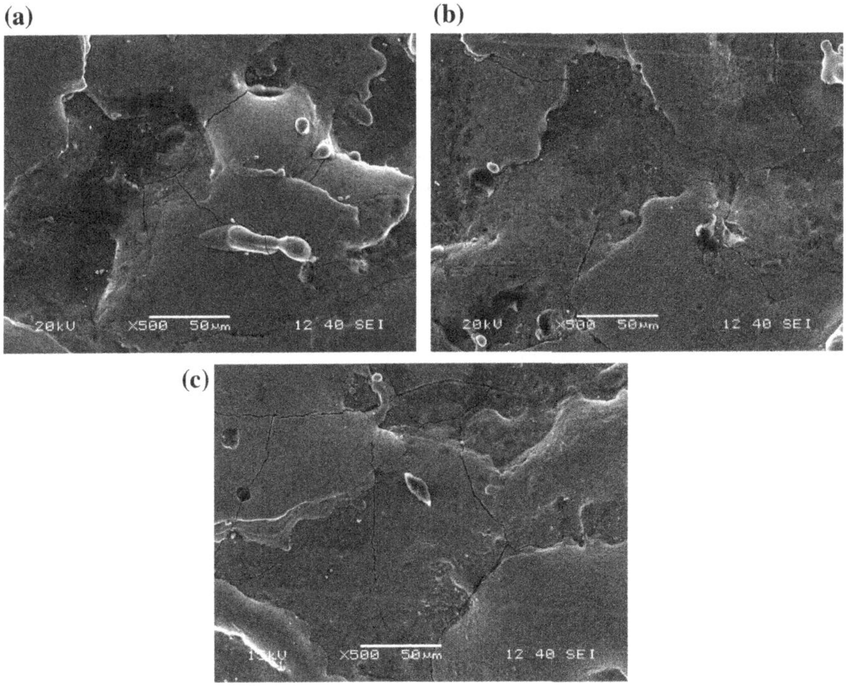


**Fig. 7** Surface texture machined by **a** untreated, **b** cryotreated single and **c** cryotreated double tempered tungsten carbide electrodes

## 5 Conclusions

The effect of process parameters (viz. pulse on time, peak current, duty cycle, gap voltage and flushing pressure) on radial overcut (ROC) during electro discharge machining (EDM) of AISI 304 stainless steel were studied using tungsten carbide electrodes (untreated, cryotreated single and double tempered). The conclusions based on the experimental results are as follows:

- Minimum ROC were observed at lower value of peak current, pulse on time, gap voltage and duty cycle and higher value of flushing pressure.
- Cryotreated post tempered electrodes significantly reduces the radial overcut.
- Comparatively smoother surface have been observed for cryotreated post tempered electrodes compared to the surface in the case of untreated electrode.

## References

1. Pandey PC, Shah HS (1980) Modern machining processes. Tata Mcgraw-Hill, New Delhi
2. Muthukumar V, Rajesh N, Venkatasamy R, Sureshbabu A, Senthilkumar N (2014) Mathematical modeling for radial overcut on electrical discharge machining of Incoloy 800 by response surface methodology. Procedia Mater Sci 6:1674–1682
3. Bhaumik M, Maity KP, Mohapatra KD (2016) Determination of material removal rate and radial overcut in electro discharge machining of AISI 304 using dimensional analysis. Appl Mech Mater 852:160–165
4. Bhaumik M, Maity KP (2017) Effects of process parameters and cryotreated electrode on the radial overcut of AISI 304 in SiC powder mixed EDM. Surf Rev Lett. https://doi.org/10.1142/S0218625X18500294
5. Gu K, Wang J, Zhou Y (2014) Effect of cryogenic treatment on wear resistance of Ti-6Al-4V alloy for biomedical applications. J Mech Behav Biomed 30:131–139
6. Kumar A, Maheshwari S, Sharma C, Beri N (2012) Machining efficiency evaluation of cryogenically treated copper electrode in additive mixed EDM. Mater Manuf Process 27:1051–1058
7. Jefferson JM, Hariharan P (2013) Machining performance of cryogenically treated electrodes in microelectric discharge machining: a comparative experimental study. Mater Manuf Process 28:397–402
8. Kalsi NS, Sehgal R, Sharma VS (2014) Effect of tempering during cryogenic treatment of tungsten carbide-cobalt bonded inserts. Bull Mater Sci 37(2):327–335
9. Chowdary YR, Yuvaraj C, Rao KP, Durgaprasad B (2012) Neural network for prediction of EDM of Al/SiC-graphite particulate reinforced hybrid composites. Int J Emerg Technol Adv Eng 2(12):730–739

# Shrinkage Porosity Reduction in Aluminium Alloy A356 Using Simulation Tool with Assisted Rapid Prototyping Casting

**Suresh Laxman Chittewar, Rajesh M. Metkar and Santosh Tanaji Ghutukade**

## 1 Introduction

The rate of solidification affects the microstructure of casting material which again influences the mechanical properties such as hardness, strength etc. The placement of riser plays an important role in casting and mainly depends upon the shape and size of casting used. Placement of riser also depends on thermal properties of casting material and mold design. The aluminium alloys (Al-alloys), are prone to defects, such as shrinkage, one of the chronic problems, which impact on the quality of the castings [1]. Due to the increasing use of aluminium castings, shrinkage characteristics are useful for improving mechanical properties of the castings. Prediction of the liquid metal during solidification has been important to consider the different modes of shrinkage and trace the evolution of the liquid metal free surface. With the rapid development of computer science and technology, many different kinds of casting simulation software have being sprung up all over the world. Similarly prototypes are very important for realization of concepts in design, manufacturing and analysis. The rapid prototyping processes can be broadly classified into process that uses laser and ones which does not [2]. The laser based processes requires high level of care and maintenance and the machinery is very costly as compared to non-laser based processes. Fused Deposition Modeling (FDM) is second most widely used rapid prototyping technology, after Stereo-lithography (uses laser). In FDM a plastic filament is unwound from a coil and supplies material to an extrusion nozzle which moves over the table in the

S. L. Chittewar (✉) · S. T. Ghutukade
Department of Mechanical Engineering, Annasaheb Dange
College of Engineering and Technology, Ashta, Sangli, India
e-mail: slc_mech@adcet.in

R. M. Metkar
Department of Mechanical Engineering, Government College of Engineering, Amravati, Amravati 444604, India



K. Antony and J. P. Davim (eds.), *Advanced Manufacturing and Materials Science*, Lecture Notes on Multidisciplinary Industrial Engineering,
https://doi.org/10.1007/978-3-319-76276-0_4

required geometry and deposits a thin bead of extruded plastic to form each layer of the required geometry [3]. Several materials are available for the process including Polylactic acid (PLA) and sand casting process, Acrylonitrile-Butadiene-Styrene (ABS) and investment casting wax.

From the existing and recent literature records it is found that the currently available casting solidification simulation software's have a lot of capabilities to analyse the defects. Various researches and their finding related to improvement in casting quality and reduction of shrinkage defects by means of simulation software has been mentioned in this paper. Rao [4] has studied on the simulation of the mould filling solidification of casting of green sand ductile iron casting sand and concluded that the use of casting simulation software like ProCAST can able to eliminate the defects like shrinkage, porosity etc. in the casting. It also improves yield of the casting and at the same time gating system can be optimize. Sutaria [5] worked on a new idea where optimization of casting feeding is done with the help of feed paths. The computation of feed-paths is done by method known as Vector Element Method (VEM). Ramu et al. [6] had developed a solid model of flywheel by using Pro-E and exported it to MAGAMA-5 simulation software to simulate the model and to develop the optimal design of pattern so as to avoid the foundry defects which will be helpful in reducing the production cost and also increase in the yield of casting. Hosseini et al. [7] studied the effect of cooling rate on the solidification parameters, microstructure, and mechanical property of LM13 alloy. To obtain different cooling rates, different mould configuration was used. The cooling rates and the solidification parameters were determined by using computer-aided thermal analysis method. Ravi [8] has discussed on the basics of casting simulation. He gave the advantages of using casting simulation software in foundry and what are the features of simulation tool why they should be used in foundry, simulation includes mold filling visualization, solidification analysis and to predict location of internal defects such as shrinkage porosity, misrun. Choudhari et al. [9] discussed the simulation process of casting solidification with the AutoCAST-X software of a intricate shape small size casting of LM6 (Al alloy) metal. With the help of AutoCAST software hotspots in the casting were located and also suggested optimum position for placing feeder in the casting. Application of exothermic sleeve was also studied by them as feedaids. The simulation study has shown the improvement in feeding yield and quality of the casting.

## 2 Methodology

The entire study has been carried out in three section viz. casting design calculation of Aluminium component (Al alloy A356), numerical simulation using AutoCAST-XI software and preparation of pattern using 3D printing process with the help of Fused Deposition Modelling (FDM) technique. The material here for pattern was taken Polylactic acid (PLA) and had done validation with experimental trial. Sand casting is used as a manufacturing process (Fig. 1).

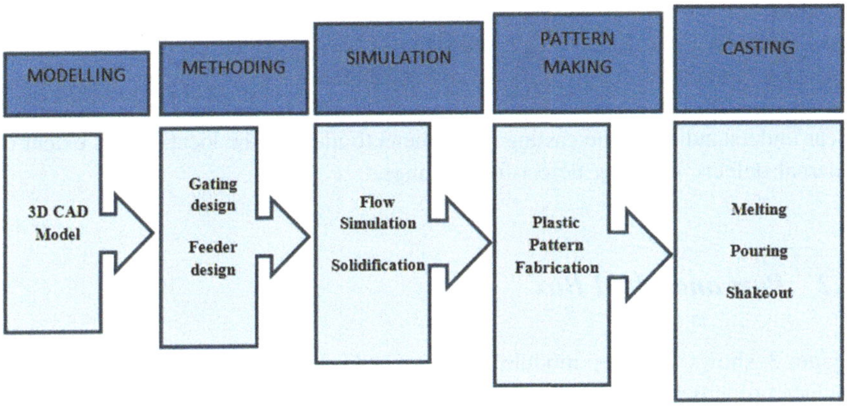


**Fig. 1** Rapid casting system process chain

## 2.1 3D Modelling, Methoding, Simulation

The first step is to create a 3D CAD model of the cast part in CAD software and export it to simulation software AutoCAST-XI (Fig. 2).

Simulation has been performed in AutoCAST-XI software according to the design dimensions obtained for pattern with allowances, gating system and feeder. Later on, the optimum location of the feeder was identified based on hot spot. The

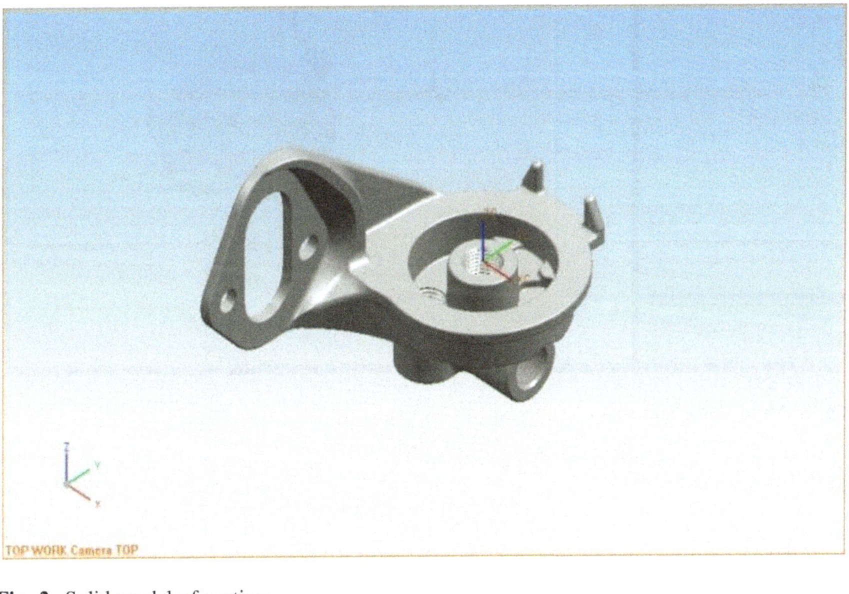

**Fig. 2** Solid model of casting

location of hot spot has been used as criterion to place the feeder. Based on this location as an input, feeder shape and size has been modeled using AutoCAST-XI software. Simulation based trials do not involve wastage of material, energy and labour, and do not hold up regular production. Computer simulation provides a clear understanding of the casting phenomena to identify the location and extent of internal defects, ensuring defect-free castings.

## 2.2 *Part and Mold Box*

Figure 3 shows the part module of AutoCAST-XI software. The sub modules include properties, thickness, holes, optimize. In this module properties are assigned to component manually or may be selected from database of software. Also thickness module shows global maximum and minimum thickness of component along with that holes are identified in the component. The mould box dimensions have been taken as 174.43 mm × 185.88 mm × 146.02 mm as shown in Fig. 3.

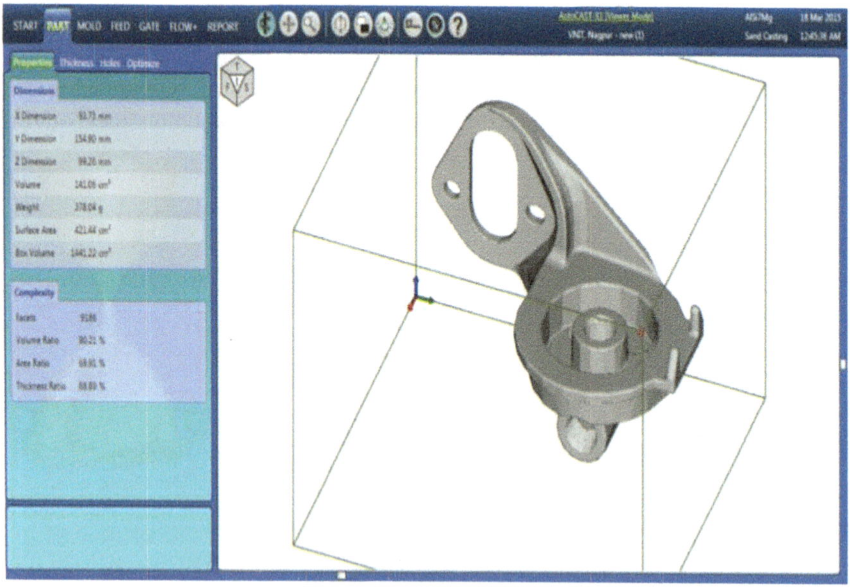


**Fig. 3** Mold shape and size

## 2.3 Feeder Design

The feed module enables designing and optimizing the feeders and feed aids to obtain maximum yield with desired quality. With the help of casting simulation software results are shown such as cooling animation, feed metal paths, and shrinkage porosity distribution. Here, the last solidifying region of the casting or hotspots in casting has been identified and feeder is placed exactly over the hotspot so that to provide feed metal during solidification (Fig. 4).

## 2.4 Solidification Simulation

Casting solidification is simulated to view the progress of cooling from casting surface to interior, and to predict the location of shrinkage defects such as porosity and cracks. This helps in verifying and optimizing the design of feeders, so that the desired quality and high yield are achieved.

Figure 5 shows that the hotspot inside casting is being shifted in the riser which means the last solidifying is now riser not the part.

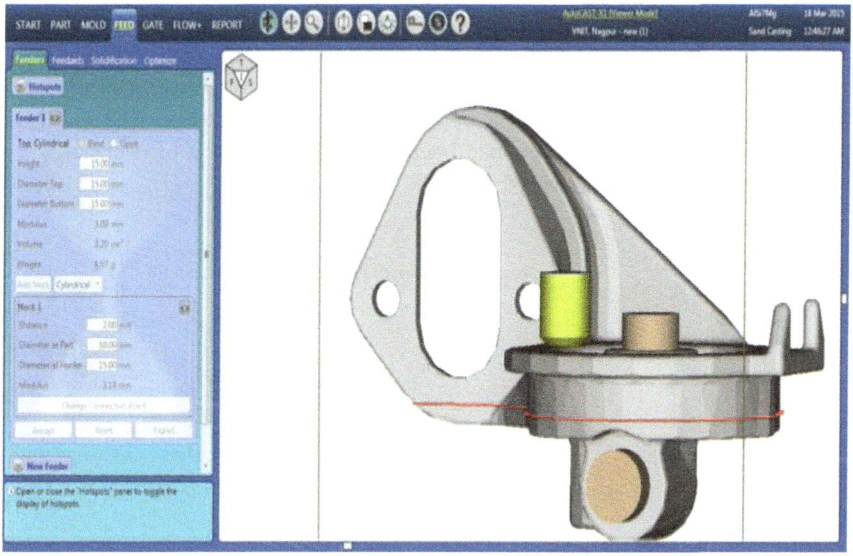


**Fig. 4** Feeder placement

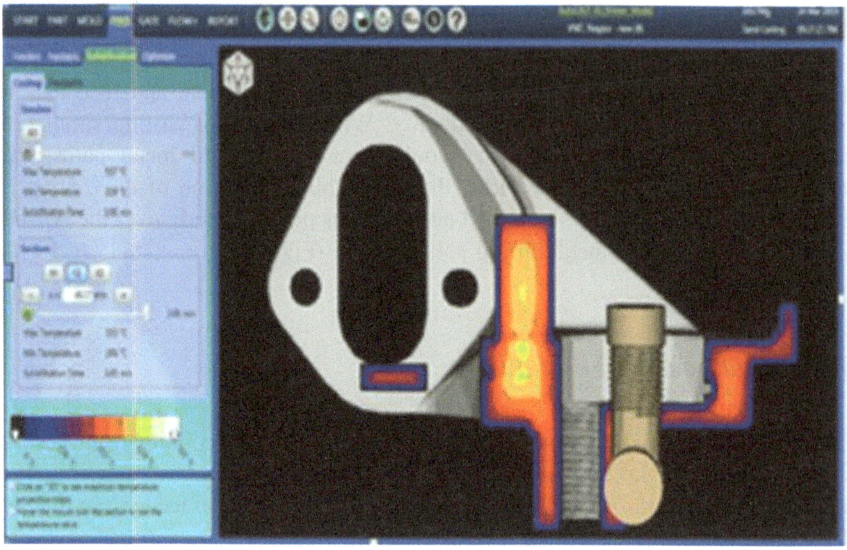

**Fig. 5** Hotspot shifting inside riser

## 2.5 *Shrinkage Porosities*

The shrinkage porosity was computed from the temperature and gradients using metal-specific process characteristics, which can be adjusted to calibrate the results with respect to the observed location of shrinkage porosity (Table 1). The shrinkage porosities at various riser heights had been calculated and effect of riser height on shrinkage cavity has been studied which is shown in figures.

**Table 1** Variation of shrinkage porosity w.r.t to riser height

| S. No. | Riser diameter | Riser height | Yield |
|---|---|---|---|
| 1 | 15 | 10 | 80.54 |
| 2 | 15 | 13 | 80.70 |
| 3 | 15 | 15 | 84.75 |
| 4 | 15 | 17 | 80.42 |
| 5 | 15 | 20 | 80.20 |
| 6 | 15 | 25 | 79.60 |

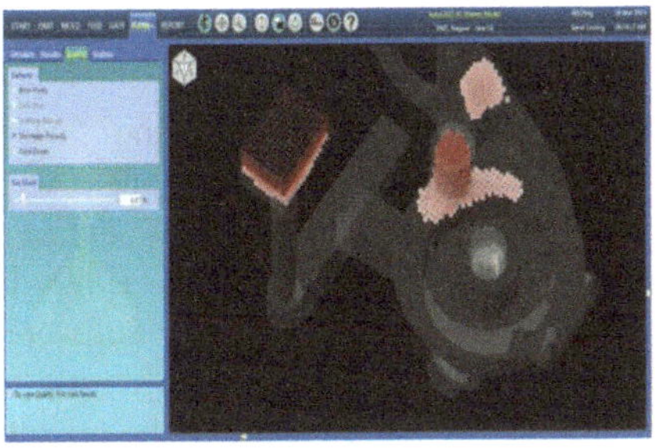

H = 10mm SP 3.75%

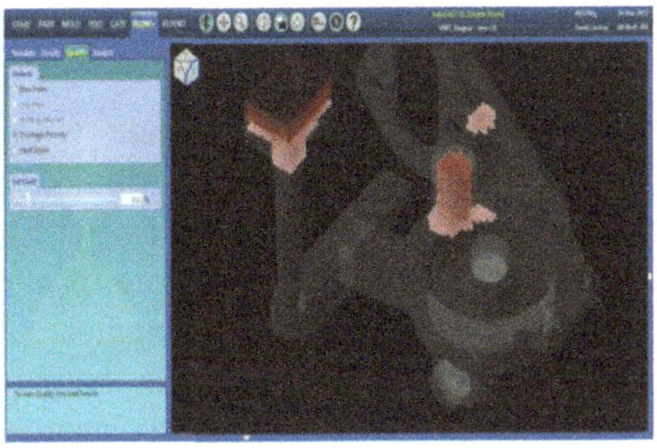

H = 15mm SP 3.023%

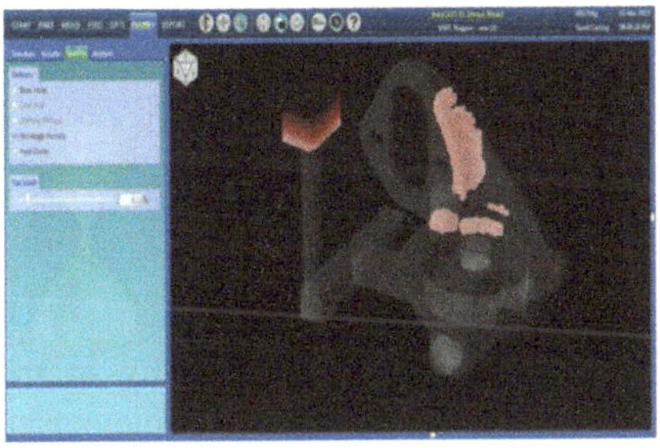

H = 20mm SP 4.11%

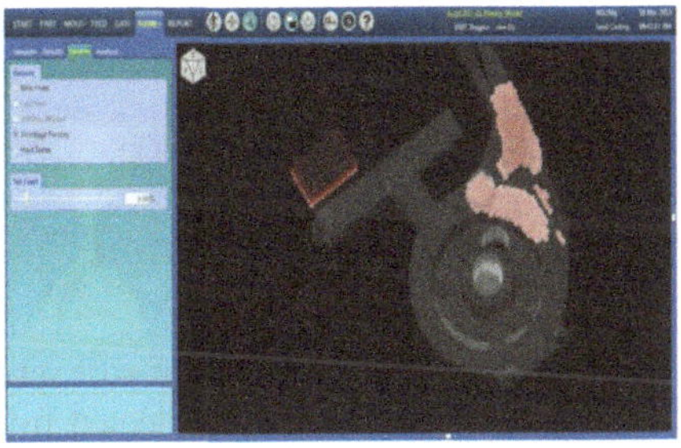

H = 25mm SP 4.88%

By changing riser height the shrinkage porosity values are also altered. The lowest value for shrinkage porosity is **3.023** for riser height **15 mm**. Similarly the effect of Riser height on yield was also studied. By taking number of iterations on simulation software the riser height were varied and accordingly the yield of casting component was noted. It was found that on riser height **15 mm** the casting has maximum yield of **84.75%** (Fig. 6).

**Fig. 6** Riser height versus yield plot

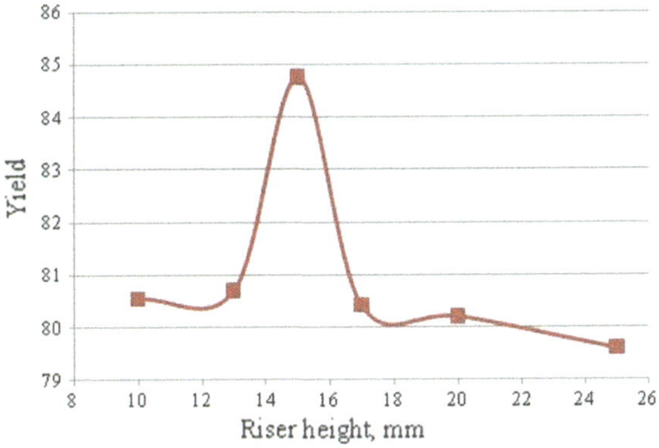

### 2.6 *Pattern Printing*

In this study pattern was made with rapid prototyping process using Fused Deposition Modelling (FDM) technique with PLA as pattern material. Pattern was fabricated by 3D Digital Systems Pvt Ltd. The CAD model of the part is sent to a tabletop 3D printer for fabricating of plastic pattern. The CAD model (STL file) is first processed offline in a computer containing an open source slicing software, which generates the G-code file hatching motion of the 3D printer. This file is copied into an SD card, which is inserted in the card reader attached to the printer. Then the printer prints according to G-codes of part that is saved in SD card, Semisolid plastic comes out of the nozzle and gets deposited on a platform layer by layer (Fig. 7).

## 3 Experimental Validation

The results obtained from software are then performed experimentally in the foundry. The optimal design selected from the software method was implemented in the foundry trail. The process consists of mould box preparation as shown in Fig. 8 thereafter followed by metal pouring and casting is allowed to solidify.

The casting removed from mould box it is then send to post processing work such as grinding, finishing, polishing etc. after all that process a finished casting product is obtained and can be used for specific applications as per requirement (Fig. 9).

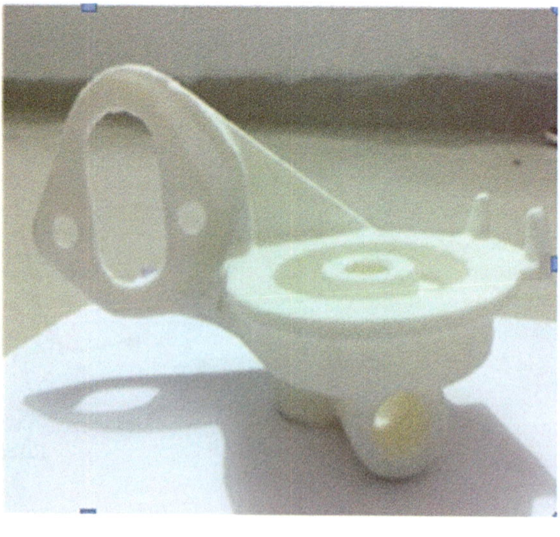

**Fig. 7** 3D printing of pattern

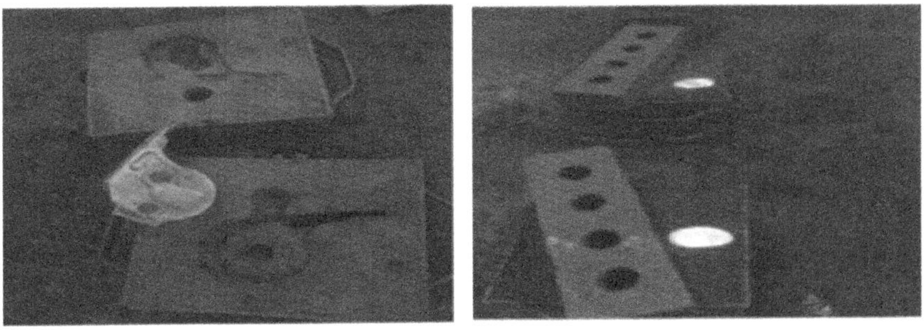

**Fig. 8** Casting process

**Fig. 9** Finished part

## 4 Conclusions

Based on the study of the results, the following conclusions are drawn:

1. The application of computer aided methoding, solid modeling, and casting simulation technologies in foundries allow to predict and minimize the casting defects.
2. Computer simulation tool allows rapid development of casting by reducing number of foundry trails.
3. It has been observed that by placing riser above hotspot there is reduction in shrinkage porosity in part.
4. Also optimum values of riser height and neck had been calculated with no. of iterations and their effects on yield have been studied.
5. By use of optimal design parameters the pattern is produced by rapid prototyping technique and casting is produced with maximum yield and minimum defects.

**Acknowledgements** The Authors would like to thank VNIT, Nagpur for providing their facilities to carry out the research work. I would like to express my gratitude to Mr. Prateek Lohia of 3D Labs India for providing 3D printed pattern.

## References

1. Sabau AS, Visvanathan S (2002) Microporosity prediction in aluminum alloy castings. Metall Mater Trans 33B:243–255
2. Pranjal J, Kuthe AM (2013) The manufacturing engineering society international conference. MESIC 2013, vol 63, pp 4–11
3. Khandelwal H, Ravi B (2014) Tabletop foundry for training, research and small enterprises. Indian Foundry J 7(2):23–29
4. Rao P, Chakaraverthi G (2011) Application of casting simulation. Int J Therm Technol 1
5. Sutaria M, Casting Simulation Case Study (2010) Shaft pin (cast iron-green sand casting). Indian Foundry J 56(12):53
6. Ramu T et al (2012) Modeling, simulation and analysis in manufacturing of a flywheel casting by S.G.Iron. Int J Materi Biomater Appl 2(4):25–28
7. Hosseini VA, Shabestari SG, Gholizadeh R (2013) Study on the effect of cooling rate on the solidification parameters, microstructure, and mechanical properties of LM13 alloy using cooling curve thermal analysis technique. Mater Des 50:7–14
8. Ravi B (2010) Casting simulation-best practices. Transactions of 58th Indian Foundry Congress, Ahmadabad, pp 1–6
9. Choudhari CM, Narkhede BE, Mahajan SK (2014) Modelling and simulation for optimum design and analysis of riser in sand casting with experimental validation. Appl Mech Mater. Trans Tech Publications, Switzerland 464:657–661

# Finite Element Analysis of Rail Vehicle Suspension Spring for Its Fatigue Life Improvement

**M. A. Kumbhalkar, D. V. Bhope and A. V. Vanalkar**

## 1 Introduction

The paper represents a case study over an investigation for fatigue failure response of primary inner suspension spring of a high-speed main line locomotive for goods hauling trains which has three motor on individual axle and is referred to as Co-Co frame assemblies, is main part of the locomotive. Total weight of the rail road vehicle is supported by the bogie frames and provides a means for transmission of the tractive effort to the rails. To absorb and isolate the superstructure from the shocks is an important function of frame caused by variations in the trackbed and hence suspension system minimizes the transmission of these shocks to the locomotive under frame [1].

The helical spring is the simplest element which is found in many mechanical systems. It makes it conceivable to maintain a tension or a force in a suspension system [2] of railway vehicle, to assimilate the shocks and to diminish the vibrations. Fatigue is the most well-known reason for failure in springs. Fatigue breakage by and large starts at the surface and the settled tensile stresses bring on additional development of the created cracks and prompt premature failure of the springs.

M. A. Kumbhalkar (✉)
Department of Mechanical Engineering, JSPM Narhe Technical Campus, Pune, Maharashtra, India
e-mail: manoj.kumbhalkar@rediffmail.com

D. V. Bhope
Department of Mechanical Engineering, Rajiv Gandhi College of Engineering, Research & Technology, Chandrapur, Maharashtra, India
e-mail: dvbhope@rediffmail.com

A. V. Vanalkar
Department of Mechanical Engineering, KDK College of Engineering, Nagpur, Maharashtra, India
e-mail: avanalkar@yahoo.co.in



K. Antony and J. P. Davim (eds.), *Advanced Manufacturing and Materials Science*, Lecture Notes on Multidisciplinary Industrial Engineering,
https://doi.org/10.1007/978-3-319-76276-0_5

**Fig. 1** Photographs of failed primary inner spring and arrangement of middle axle suspension spring near middle wheel

The helical suspension spring framework has a noteworthy significance on the operation of rail road vehicle, considering the effect of curving and tracking [3].

The suspension system of rail road vehicle has the main function to control isolation and shock absorption between the bogie and frame [4]. The function of primary springs of middle axle of each frame is also to permit free movement in lateral direction. A primary suspension mounted between wheelbase and bogie with inclined damper at end axles and linkage to restrict lateral movement at middle axle. The middle axle has an assembly of concentric suspension spring of inner and outer opposite handed spring to acquire load without damper. This paper focuses on premature fatigue failures of inner suspension spring due to dynamic effect. The failure region of middle axle primary inner suspension spring and arrangement for mounting on middle axle housing is shown in Fig. 1.

## 2 Fatigue Analysis

The middle axle primary inner suspension spring is subjected to variable loads and hence the fatigue analysis approach is used to investigate the failure of the spring using finite element method. A progression of forward and reverse loading portrayed a fatigue; where plasticity is initiated in each cycle. The fatigue life of a suspension spring can be communicated as the number of loading cycles for the initiation of crack and the number of cycles propagates that crack to failure [5]. A computational model for fatigue examination of suspension spring has been exhibited. A maximum and minimum load is often used for simulation of the cyclic loading in fatigue analyses on helical suspension spring [6].

The actual crack starts at the cause which develops gradually over the fatigue zone, with a typical growth rate. The progression marks shows up which is because of the varieties in the load that brought about relating varieties in the crack growth rate. Eventually, the crack reaches the point where the remaining material gets overstressed, and the overload zone results. The progression marks indicate how the crack has developed and are just present in fractures where there have been generous varieties in the component stress as the crack develops over the piece [7]. The

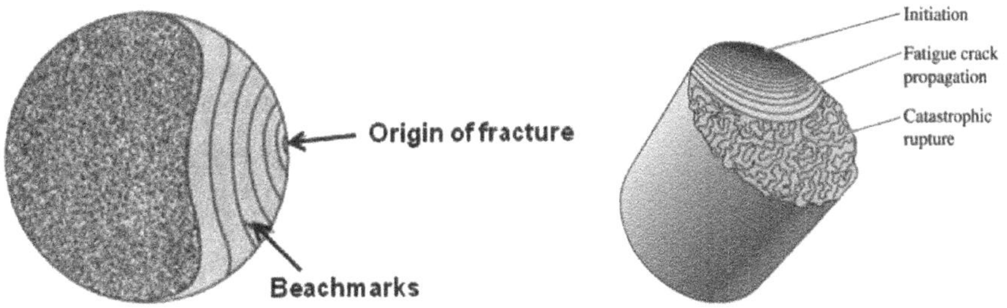


(a) Fatigue fracture with benchmark

(b) Fatigue crack initiation and propagation

**Fig. 2** Schematic representation of fatigue fracture surface

schematic representation of fatigue fracture surface showing crack origin and its progression is shown in Fig. 2.

This section discusses the finite element analysis of helical suspension spring of rail vehicle using a numerical tool ANSYS to find its fatigue life. Analytically, the forces acting on suspension springs and shear stresses induced are calculated for static condition and continued to fatigue analysis for displacement variation of 6–8 mm as per the observation of rubbing marks over dampers. Fatigue analysis has been carried in ANSYS considering the load ratio, ultimate and endurance shear limit for chrome vanadium material.

It is observed that the spring gets deflected at some instances as observed from rubbing marks over the end axle dampers as shown in Fig. 3. Because of this the additional force acts over the spring. But the band of polished surface on the damper indicates that, there is the displacement of the spring in the range of 6–8 mm which may cause the fatigue failure of spring. The maximum and minimum load corresponding to the additional deflection of 6–8 mm is given Table 1.

The basic S-N curve for fatigue analysis of suspension spring can be plotted for fully-reversed stress cycle for its alternating stress values. The typical loading condition with mean stress is shown in Fig. 4.

The accompanying conditions are utilized to characterize a stress cycle with both alternating and mean stress. The stress range is the mathematical contrast between the maximum and minimum shear stress in a cycle: The shear stress amplitude is one-half of the stress extend: The mean shear stress is the arithmetical mean of the maximum and minimum shear stress in the cycle [8]:

$$\Delta\tau = \tau_{\max} - \tau_{\min} \qquad \tau_a = \frac{\Delta\tau}{2} = \frac{\tau_{\max} - \tau_{\min}}{2} \qquad \tau_m = \frac{\tau_{\max} + \tau_{\min}}{2}$$

**Fig. 3** Rubbing and polishing marks over end axle damper

**Table 1** Parameters for fatigue analysis of inner suspension spring

| Particulars | Symbol and unit | Inner suspension spring | | |
|---|---|---|---|---|
| | | For 6 mm deflection | For 7 mm deflection | For 8 mm deflection |
| Minimum deflection | $\delta_{min}$ (mm) | 64.6 | 64.6 | 64.6 |
| Minimum load | $F_{min}$ (N) | 9351.46 | 9351.46 | 9351.46 |
| Maximum deflection | $\delta_{max}$ (mm) | 70.6 | 71.6 | 72.6 |
| Maximum load | $F_{max}$ (N) | 10,219.62 | 10,364.37 | 10,509.13 |
| Mean load | $F_m$ (N) | 9784.54 | 9857.92 | 9930.29 |
| Alternating load | $F_a$ (N) | 434.08 | 506.46 | 578.83 |
| Wahl's factor | K | 1.29 | 1.29 | 1.29 |
| Mean shear stress | $\tau_m$ (N/mm$^2$) | 626.46 | 630.77 | 634.72 |
| Variable shear stress | $\tau_a$ (N/mm$^2$) | 27.79 | 32.42 | 37.06 |
| Stress ratio | R | 0.92 | 0.90 | 0.89 |
| Equivalent shear stress | $\tau_{eq}$ (N/mm$^2$) | 688.07 | 702.66 | 717.89 |

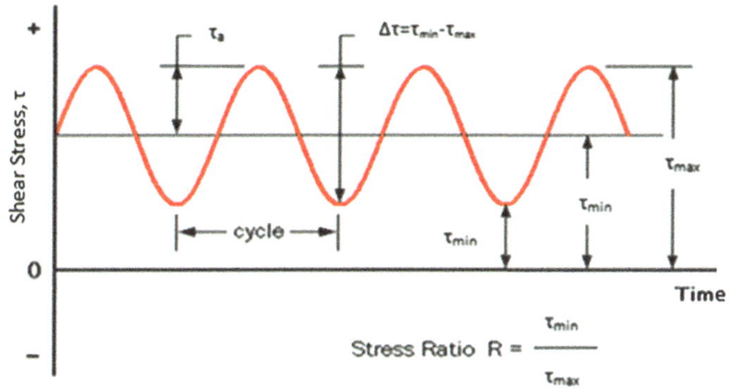


**Fig. 4** Typical cyclic loading parameters

Two proportions that are regularly characterized for the portrayal of mean shear stress are the shear stress ratio R and the amplitude ratio A [8]:

$$R = \frac{\tau_{\mathrm{min}}}{\tau_{\mathrm{max}}} \qquad A = \frac{\tau_a}{\tau_{\mathrm{m}}} = \frac{1-R}{1+R}$$

This method define various curves to connect the endurance limit on the alternating stress axis to the yield shear strength, $S_{ys}$, ultimate shear strength $S_{us}$, or true fracture shear stress $S_{fs}$ on the mean stress axis. A Soderberg criteria given below is mainly used for ductile material for S-N curve [3, 8].

$$\frac{FS.\tau_m}{S_{ys}} + \frac{FS.K_f.\tau_a}{S_{es}} = 1$$

where $K_{fs}$ is fatigue shear stress concentration factor and $\tau_{eq}$ is equivalent to allowable stress $S_{ys}/FS$ [8].

$$\frac{S_{ys}.K_f.\tau_a}{S_{es}} + \tau_m = \frac{S_{ys}}{FS} \qquad \frac{S_{ys}.K_f.\tau_a}{S_{es}} + \tau_m = \tau_{eq}$$

## 3 Finite Element Analysis for Fatigue Life of Suspension Spring

FE analysis of inner suspension spring has been carried out to find its fatigue life in ANSYS 12.0 with the help of 3-D, 10-node SOLID 187 element [9, 10]. A stress ratio is given as input for fatigue analysis in ANSYS and it is given in Table 1. Stress-Life (S-N) curve with low cycle and high cycle fatigue life has to be provided in material property of spring material. Hence S-N curve has been plotted for

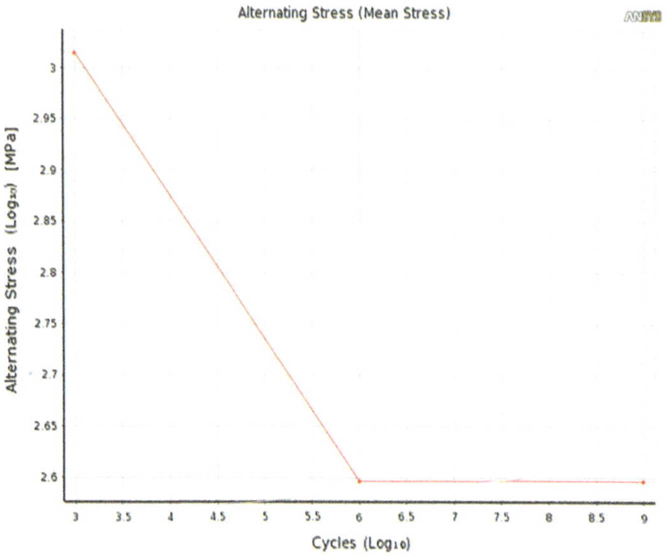


**Fig. 5** S-N curve for spring material and alternating stress values for fatigue life using ANSYS

ultimate shear strength of 1152.4 N/mm$^2$ and endurance shear strength of 395.6 N/mm$^2$ as shown in Fig. 5. Considering fatigue strength factor as unity, an equivalent alternating shear stress is determined and it is given in Table 1.

The finite element analysis revealed the fatigue life and factor of safety contours as shown in Fig. 6 and the results are tabulated in Table 2.

From Fig. 6 and from Table 2, it is observed that, the spring has finite life of $1.89 \times 10^4$ cycles. While examining the failed specimen as shown in Fig. 1, it has been observed that the cross section of failure resembles to that of fatigue failure is shown in Fig. 7 with the fatigue life. The fatigue life varies from 10 cycles to $10^5$ cycles for the zone nearer to the inner side of the coil which indicates finite life. The progressively increases for the cross section slightly away from the inside diameter but still this region is having the finite life. While the rest of the cross section has the life more than $10^6$ cycles and which corresponds to infinite life. Thus during the service of the spring the crack initiates at the inner side of the spring after very few cycles of operation and it propagates through the cross section. Upto $10^5$ cycles the length of crack growth is estimated to be 2–2.5 mm. After observation of cross section of failed spring it is seen that the polished surface at the failure zone has the band of approximately 3–3.5 mm. It has been found that the actual fatigue failure zone of spring is more than fatigue zone observed in FE results.

The reasoning for this difference is, when the crack initiates at the inner surface, the stress on rest of the cross section increases which further decreases the fatigue life. As the crack proceeds there is possibility that the actual crack growth will be

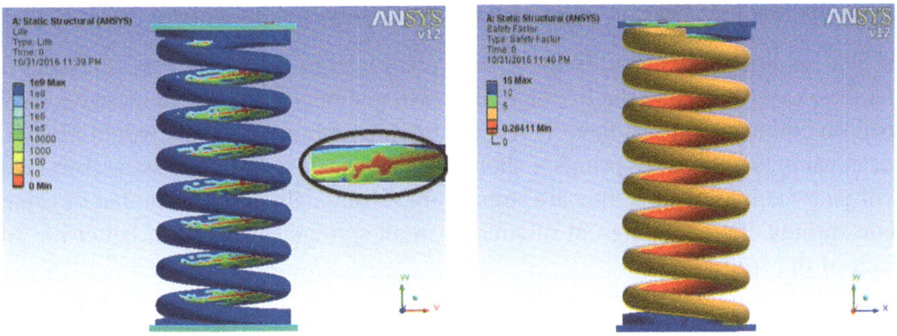


Fig. 6 FE analysis of inner spring for its fatigue life and safety factor

Table 2 Fatigue life and factor of safety for inner suspension spring

| Particulars | Symbol and unit | Inner suspension spring | | |
|---|---|---|---|---|
| | | For 6 mm deflection | For 7 mm deflection | For 8 mm deflection |
| Factor of safety | F.S. | 1.17 | 1.13 | 1.09 |
| Fatigue life | Cycles | $1.89 \times 10^4$ | $1.62 \times 10^4$ | $1.40 \times 10^4$ |

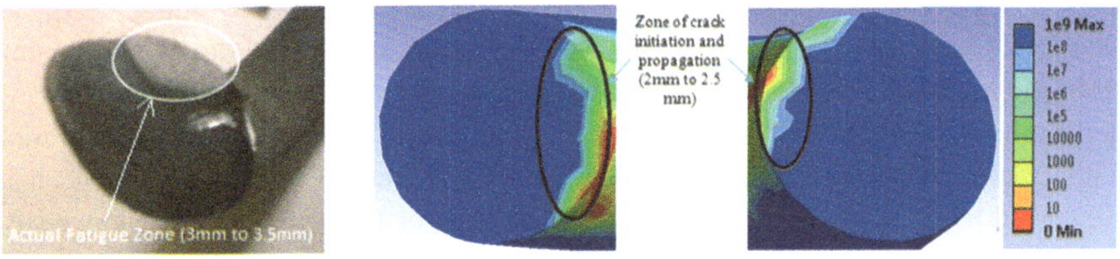


Fig. 7 FE result for fatigue zone of middle axle inner suspension spring showing crack initiation and propagation of 2–2.5 mm

more as compared to theoretical FE results due to progressive increase in stress magnitude. This analysis reveals that the spring fails due to fatigue failure as it is having finite life.

## 4 Modifications Suggested in Suspension System of Rail Vehicle

The present investigation reveals that the inner suspension spring fails due to fatigue failure. To avoid the failure of spring, following modifications are suggested.

i. **Reduction of Load on Middle Axle Inner Suspension Spring**

In first modification, the addition of shim shall lead to decrease in pre-stressing of middle axle springs and the life of springs will get enhanced. Due to increase in the load on end axle springs, there is rise in stress magnitude. But this would likewise mean using the end axle springs to their expanded limit as the shear stresses in them will increment as of now they are stresses on substantially lesser than that of center axle springs. This evening-out of stresses in all springs should yield better life for each of the spring.

By increasing the height of end axle spring and reducing forces on middle axle spring, the life of inner suspension spring is observed to be enhanced and shown in Fig. 8. Also the fatigue zone represents decrease in crack propagation length for shim addition of 5, 6 and 10 mm as shown in Fig. 9.

ii. **Change in the Damping Coefficient of Inclined Damper**

The second modification revealed that the primary suspension has maximum displacement of 6–8 mm contributed from the dynamic analysis observed as per the polished marks on damper. So there is need to reduce this dynamic amplitude of spring by modifying damping coefficient of an existing damper to increase its fatigue life. The existing damper has damping coefficient of 50,000 N s/m. Hence by considering frequency ratio of 1 and excitation amplitude of 1.3 mm, the displacement amplitude response is determined for varying damping coefficient as shown in Fig. 11. From Fig. 11, it is observed that when damping coefficient is increased from 100,000 N s/m to 250,000 N s/m, displacement amplitude reduces from 3.58 mm to 2.21 mm and this reduces dynamic amplitude of suspension spring. The fatigue analysis is carried out corresponding to the dynamic spring deflection of 7, 5.27, 4.42 and 3.97 mm and given in Table 3 and its fatigue zone with crack propagation obtained by FE analysis is shown in Fig. 10. Hence it is suggested to increase damping coefficient of damper to increase life of inner suspension spring i.e. from $1.62 \times 10^4$ cycles to $6.67 \times 10^6$ cycles.

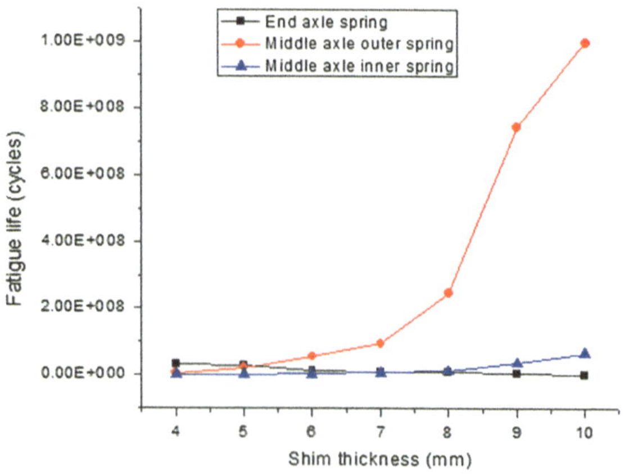


**Fig. 8** Fatigue life of inner suspension spring by adding shim at end axle spring

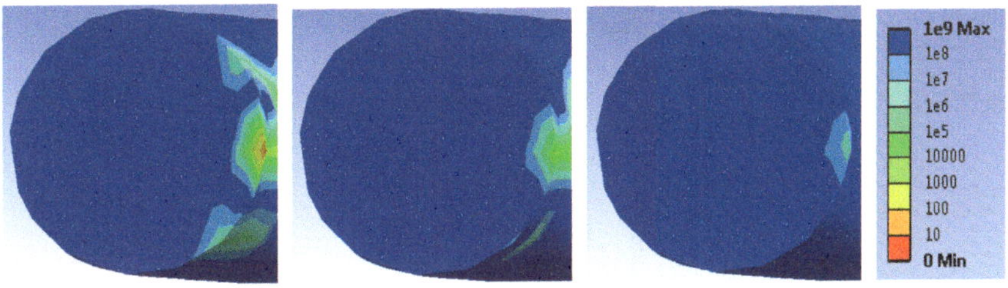


**Fig. 9** Fatigue zone of middle axle inner suspension spring for 5, 6 and 10 mm shim addition

**Table 3** Fatigue life of inner suspension spring for modification of damping coefficient for second and third suggested modification

| Modifications suggested | Particulars | Inner suspension spring | | | |
|---|---|---|---|---|---|
| | | For 7 mm deflection | For 5.27 mm deflection | For 4.42 mm deflection | For 3.97 mm deflection |
| | Damping coefficient (N s/m) | 100,000 | 150,000 | 200,000 | 250,000 |
| Second modification | Fatigue life (cycles) | $1.62 \times 10^4$ | $1.94 \times 10^5$ | $7.64 \times 10^5$ | $6.67 \times 10^6$ |
| Third modification | Fatigue life (cycles) | $2.34 \times 10^7$ | $7.68 \times 10^7$ | $8.1 \times 10^7$ | – |

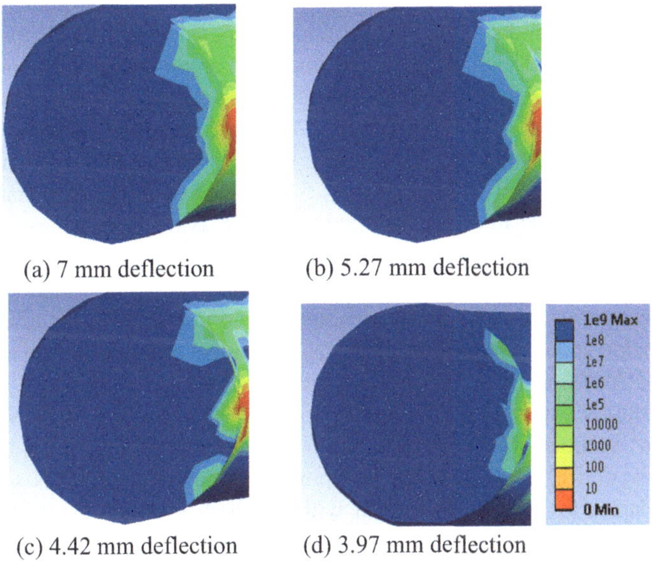


**Fig. 10** Fatigue zone for dynamic middle axle inner suspension spring deflection of 7, 5.27, 4.42 and 3.97 mm

Fig. 11 Displacement amplitude of suspension spring for increase in damping coefficient

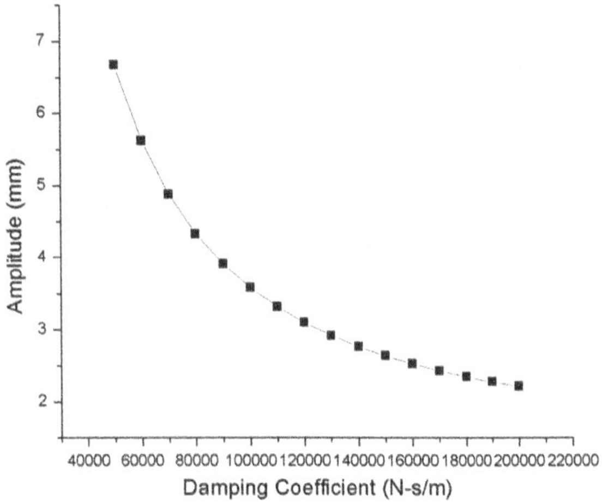


Thus the damping coefficient of 250,000 N s/m will reduce the dynamic amplitude to 1.99 mm and the expected life of inner suspension spring will be approximately $6.67 \times 10^6$ cycles which is considered to be infinite life. Hence damping coefficient of 250,000 N s/m is recommended for existing suspension (Fig. 11).

iii. **Provision of Shim Along with Change in Damping Coefficient of Inclined Damper**

Third modification discusses, the shim provision on end axle box housing reduces the shear stress on middle axle primary inner suspension spring, also increase in damping coefficient of inclined damper reduces the amplitude of suspension spring and both modification increases its fatigue life as discussed earlier. As discussed in first modification, the provision of shim of 5–7 mm thick at end axle box housing is highly recommended for increase in fatigue life of inner suspension spring but due to the stringent rules and regulations only 5 mm thick shim is incorporated at end axle box housing. Due to this failure rate of middle axle inner primary suspension spring has been reduced by 88% without affecting other suspension springs. So it is felt necessary to study the effect of combination of both the modifications to increase fatigue life of inner suspension spring and due to this, the advantages of shim provision and modification of damping coefficient can be achieved effectively. From the calculation, it is observed that the modification of damping coefficient from 100,000 N s/m to 250,000 N s/m leads to shear stress of 640 N/mm$^2$ and 590 N/mm$^2$ without and with shim provision respectively as shown in Fig. 12. Hence, provision of shim along with damping coefficient of 150,000 N s/m reduces the shear stress to 598 N/mm$^2$ which will definitely improve the fatigue life of spring and given in Table 3 and the contour plot for fatigue life is shown in Fig. 13. Thus the provision of shim along with modification of damping coefficient of damper is strongly advisable to avoid failures of middle axle primary inner suspension spring.

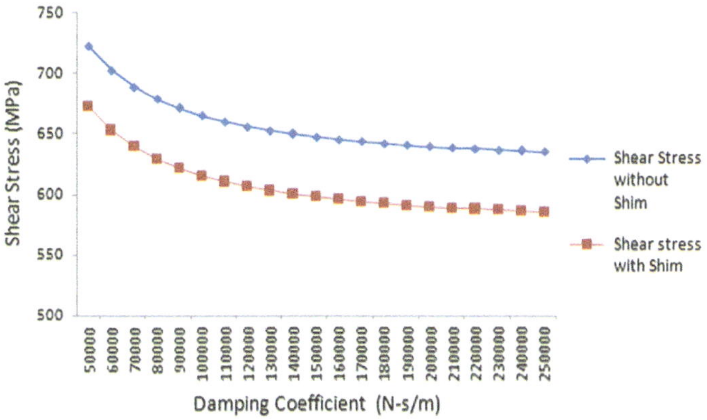


**Fig. 12** Shear stress in inner suspension spring provision of shim along with change in damping coefficient of inclined damper

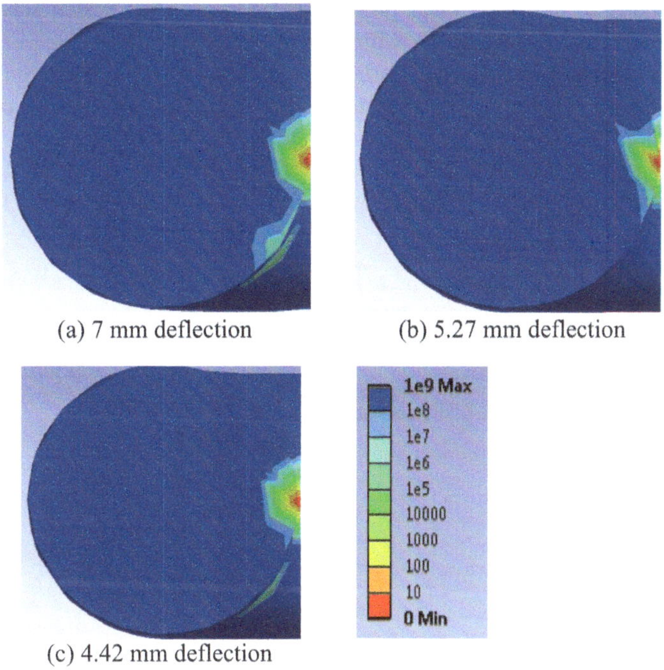


**Fig. 13** Fatigue zone for dynamic middle axle inner suspension spring deflection of 7, 5.27, 4.42 and 3.97 mm

iv. **Design Modification of Primary Suspension Spring**

The fourth modification suggested, for improving the fatigue life of primary springs it is necessary to reduce the load shared by middle axle suspension springs by decreasing the stiffness of spring. This can be achieved by increasing the mean diameter of middle axle suspension spring. Though increase in mean diameter has the effect on increase in stress magnitudes but the reduction of stiffness reduces the load shared by the spring and will lead to lower stress values which may improve its fatigue life. The suspension springs of WAG-9 rail vehicle are mounted on axle box provide for each wheel. An axle box has collars to mount inner and outer spring according to coil and outer diameter of spring. Accordingly, a WAG-9 rail vehicle axle box housing has a provision to mount outer spring with maximum 221 mm outer diameter and maximum 36 mm coil diameter and to mount inner spring with maximum 110 mm outer diameter and maximum 18 mm inner diameter. A schematic representation of spring mounting in axle box housing of WAG-9 rail vehicle is shown in Fig. 14.

The force distribution and static stress analysis is carried out for the modified mean diameter of inner and outer primary suspension spring which are compared with results given in Chapter "Experimental Investigation on Reverse Engineering Techniques Employing Response Surface Methodology for Freeform Surfaces" for the case of rail vehicle moving on straight track. From analytical analysis, it is observed that the shear stresses in the spring slightly reduces for inner and outer suspension spring and increases for end axle spring. The shear stress and fatigue life for the modified suspension spring is given in Table 4 and contour plot for fatigue

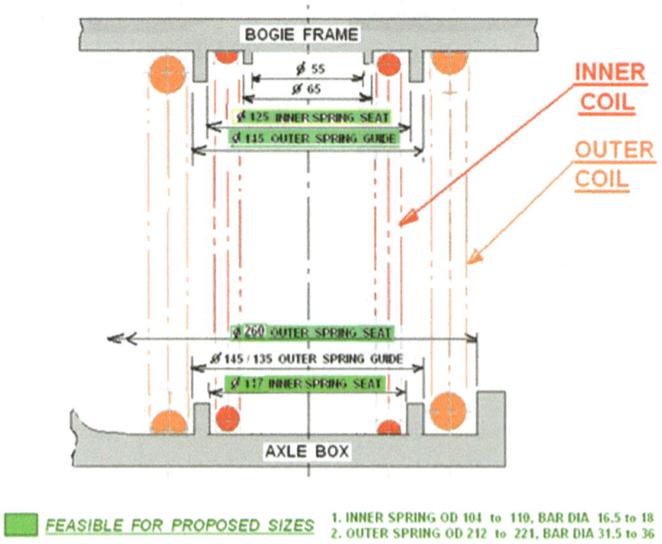


**Fig. 14** Schematic representation of spring mounting in axle box of WAG-9 rail vehicle

**Table 4** Comparison of shear stresses and fatigue for existing and proposed suspension spring for increase in mean diameter

| Particulars | Unit | Existing parameters | | | Proposed for WAG-9 | | |
|---|---|---|---|---|---|---|---|
| | | Middle axle | | End axle spring | Middle axle | | End axle spring |
| | | Outer spring | Inner spring | | Outer spring | Inner spring | |
| Force (W) | N | 31268.7 | 9351.46 | 40451.38 | 28050.3 | 7936.68 | 42757.68 |
| Deflection (δ) | mm | 65.6 | 64.6 | 45.6 | 69.26 | 67.26 | 49.26 |
| Shear stress | $N/mm^2$ | 583 | 598.36 | 530.98 | 541.33 | 539.96 | 561.25 |
| Fatigue life | Cycle | $3.2 \times 10^6$ | $1.62 \times 10^4$ | $4.64 \times 10^7$ | $5.48 \times 10^7$ | $5.8 \times 10^4$ | $5.3 \times 10^7$ |

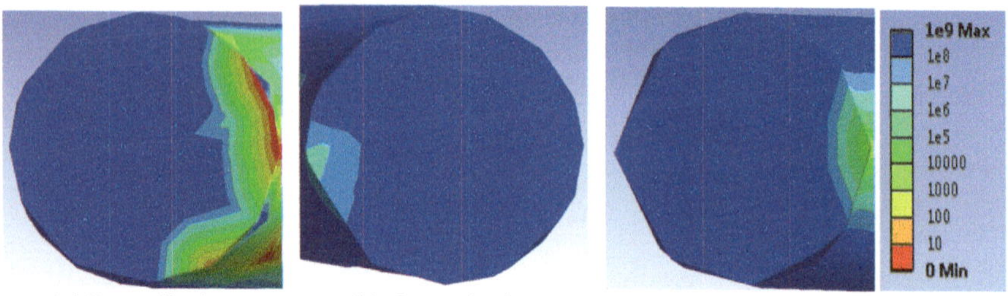


(a) Inner Spring (b) Outer Spring (c) End Axle Spring

**Fig. 15** Contour plot for fatigue life of primary suspension spring

life is shown in Fig. 15 which shows very little improvement in shear stress and fatigue life as compared to the original spring and hence, this modification is not advisable.

## 5 Conclusion

The fatigue analysis reveals that the middle axle inner suspension spring has a finite life of $1.89 \times 10^4$ cycles which clearly indicates that the spring fails because of fatigue failure observed from the cross section of the failed spring with crack initiation at inside diameter. It has been also roughly estimated that the spring failure occurs within 90 days and it has been also confirmed from loco shed authorities.

To reduce the failure rate of inner suspension spring and hence to increase its fatigue life, four types of design modifications are suggested. Out of these four suggested modifications, the first modification of provision of shim for end axle box housing is highly recommended as it leads to least changes in the existing suspension system, while the second and third modifications requires changes in the dampers with required damping coefficients and leads to new design and development of dampers. This modification is also advisable but it is not in the jurisdiction of Electric Loco Shed to replace dampers immediately without the necessary permissions from rail authorities. The fourth modification is not recommended as it marginally improves the fatigue life of middle axle inner suspension spring and in finite life zone.

## References

1. Iwnicki S (2006) Handbook of railway vehicle dynamics. CRC Press, Taylor & Francis, USA
2. Michalczyk K (2009) Analysis of helical compression spring support influence on its deformation. Arch Mech Eng LVI(4)

3. Puff R, de Bortoli MGD, Bosco R Jr (2010) Fatigue analysis of helical suspension springs for reciprocating compressors. In: International compressor engineering conference, paper 1989
4. Harak SS, Sharma SC, Harsha SP (2014) Structural dynamic analysis of freight railway wagon using finite element method. Procedia Mater Sci 6:1891–1898
5. Sangid MD (2013) The physics of fatigue crack initiation. Int J Fatigue 57:58–72
6. Fajdiga G, Sraml M (2009) Fatigue crack initiation and propagation under cyclic contact loading. Eng Fract Mech 76:1320–1335
7. Sachs NW (2005) Understanding the surface features of fatigue fractures: how they describe the failure cause and the failure history. J Fail Anal Prev 2(5):11–15
8. Shigley JE, Mischke CR (2006) Mechanical engineering design. TATA McGraw Hill, Eighth Edition, ISBN: 0−390−76487−6, pp 292–318
9. Prawoto Y, Ikeda M, Manville SK, Nishikawa A (2008) Failure analysis of automotive suspension coil springs. In: AIST steel properties & applications conference, Detroit, MI, USA, 35–48
10. Abidin MIZ, Mahmud J, Latif MJA, Jumahat A (2013) Experimental and numerical investigation of SUP12 steel coil spring. Procedia Eng 68:251–257

# Analysis of Optimized Roller Burnishing Parameters Using ANSYS

**Priyanka S. Yadav and Dayanand A. Ghatge**

## 1 Introduction

Burnishing process is a superficial plastic deformation process which is used to improve surface integrity properties by producing fine grain size in the surface region of various materials. Burnishing is a cold working process, in which the material is displaced from peaks to valleys of the surface profile by planetary motion of a tool over the bored or turned surface. Due to the compressive force applied by the tool over the work piece, compressive stresses are developed over the workpiece surface. Compressive stresses can be also over the work piece surface by processes like laser shot peening, shock peening, etc. but the stresses were found to be reduced when introduced to heat. This thermal relaxation of compressive stresses makes the component life shorter and the performance of workpiece reduces. Thus, burnishing was introduced as a process that could impart thermally unvarying surface compressive stresses [1].

## 2 Development of Compressive Stresses Over the Work Piece Surface

In roller burnishing, a hard roller is pressed against a rotating cylindrical work piece and parallel to the axis of work piece. The pressure of tool is thus applied over the surface of the work piece. This creates stress over the surface of the work piece. When

P. S. Yadav (✉)
Faculty of Engineering, Yashoda Technical Campus, Satara, India
e-mail: priyankaydv26@gmail.com

D. A. Ghatge
Department of Mechanical Engineering, Karmaveer Bhaurao Patil College of Engineering, Satara, Maharashtra, India



K. Antony and J. P. Davim (eds.), *Advanced Manufacturing and Materials Science*, Lecture Notes on Multidisciplinary Industrial Engineering,
https://doi.org/10.1007/978-3-319-76276-0_6

this stresses exceeds the yield strength of the material, it results in the plastic flow of the material from the peaks of the surface irregularities into the valleys. This induces thermally stable compressive residual stresses over the surface of the work piece [1].

## 3 Finite Element Analysis

**Introduction:** Finite element analysis is a computerized method for predicting how a product reacts when it is exposed to the practical forces, vibrations and other physical effects. It shows whether a work piece will break, crack, wear out or work safely for the purpose and at the working conditions for which it was designed [2].

**ANSYS:** In this research work the finite element analysis is carried out by ANSYS software. ANSYS is a complete CAE (Computer Aided Engineering) package. It is mostly useful in finite element analysis. The finite element method is a numerical method that involves complicated physics, geometry, boundary conditions. In finite element method, a given domain is viewed as a collection of sub domains and over each sub domain the governing equation is approximated by any of the traditional vibrational methods. ANSYS can be used for analysis of structures, stress analysis, fluid dynamics, thermal analysis, etc. [2].

**Loading conditions:** The loading conditions in the present experimentation are dynamic as the point of application of force changes continuously. The use of ANSYS is done for stress analysis and deformation obtained of the work piece. The work piece used for experimentation is drawn as model in ANSYS and then analysed (Fig. 1).

The experimentation parameters are shown in Table 1.

The experimentation is carried out at various combinations of process parameters and the results obtained with the help of Taguchi design of experiment approach and main effects plot is shown (Figs. 2 and 3; Table 2).

The analysis by ANSYS is now done for the optimized parameter conditions. The material properties for EN9 steel are as follows:

Density: 7.85e−006 kg $mm^{-3}$

| Young's modulus (MPa) | Poisson's ratio | Bulk modulus (MPa) | Shear modulus (MPa) |
|---|---|---|---|
| 2.06e+005 | 0.28 | 1.5606e+005 | 80,469 |

**ANSYS results:**

**ANSYS for optimum set found for surface roughness:** The force applied during this stage is 498.17 N. The optimum parameter set found for best surface roughness value is as follows:

| Speed (rpm) | Feed (mm/revolution) | Depth of penetration (mm) | Number of passes |
|---|---|---|---|
| 500 | 0.08 | 0.3 | 3 |

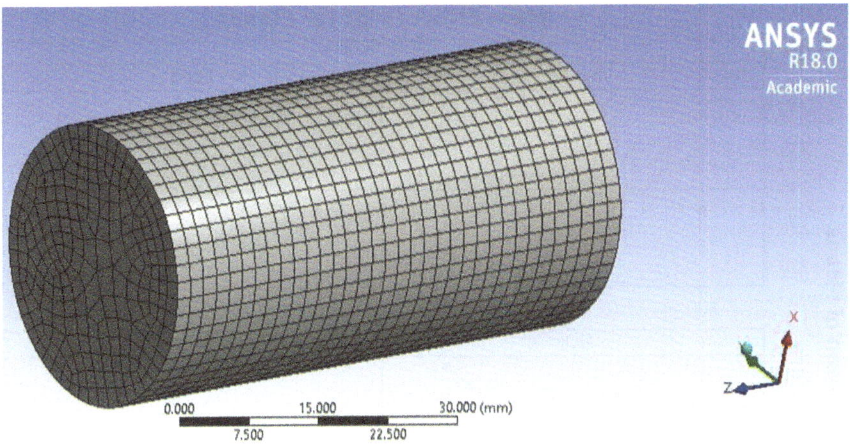


**Fig. 1** Model to be analysed

**Table 1** Process parameters and their levels [3, 4]

| Factors | Level 1 | Level 2 | Level 3 |
|---|---|---|---|
| Spindle speed (N) | 500 | 800 | 1100 |
| Feed (f) | 0.08 | 0.10 | 0.12 |
| Depth of penetration (DOP) | 0.1 | 0.2 | 0.3 |
| Number of passes (NOP) | 1 | 2 | 3 |

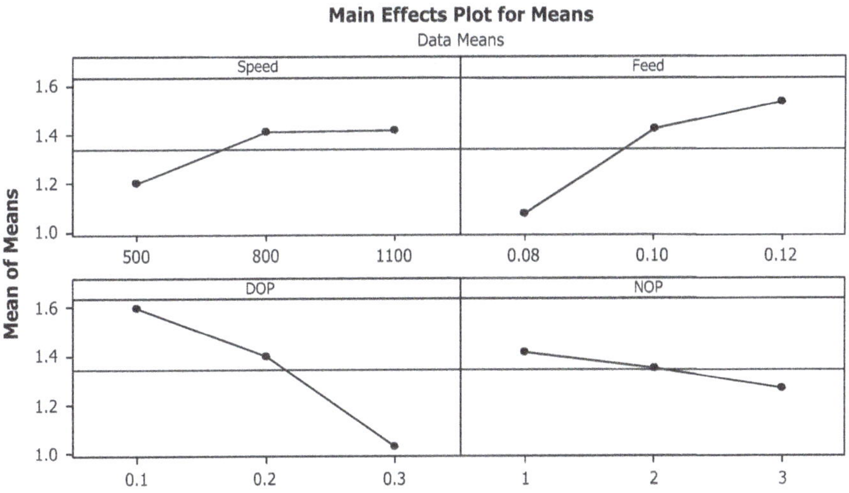


**Fig. 2** Main effect plots for means of surface finish

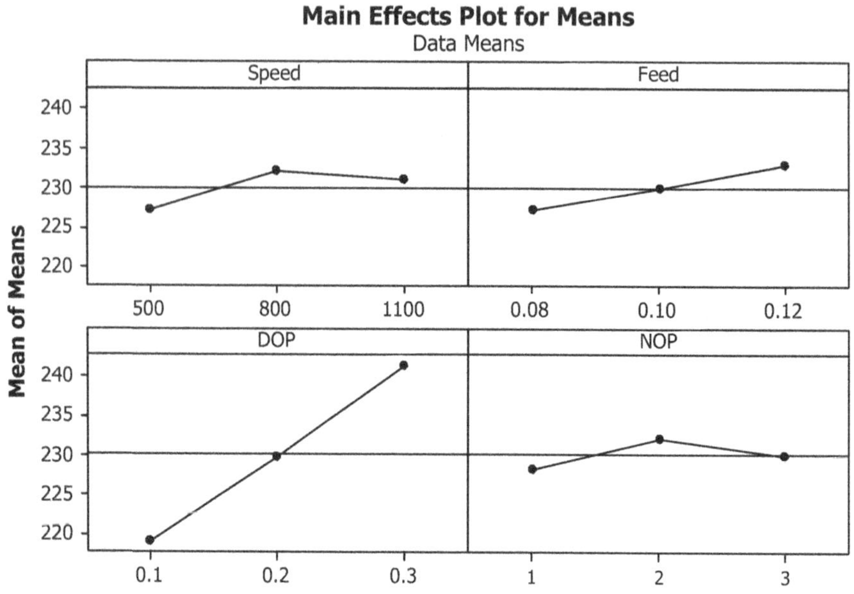


**Fig. 3** Main effects plot for means of surface hardness

**Table 2** Optimized parameter set

| Response parameter | Speed (rpm) | Feed (mm/ revolution) | Depth of penetration (mm) | Number of passes |
|---|---|---|---|---|
| Surface roughness | 500 | 0.08 | 0.3 | 3 |
| Surface hardness | 800 | 0.12 | 0.3 | 2 |

Results: 1. Maximum stress developed: 0.28139 MPa; 2. Maximum deformation occurred: 1.2936 mm (Figs. 4, 5 and 6).

**ANSYS for optimum set found for surface hardness:** The force applied during this stage is 616.83 N. The optimum parameter set found for best surface roughness value is as follows:

| Speed (rpm) | Feed (mm/revolution) | Depth of penetration (mm) | Number of passes |
|---|---|---|---|
| 800 | 0.12 | 0.3 | 2 |

Results: 1. Maximum stress developed: 0.3648 MPa; 2. Maximum deformation occurred: 1.7986 mm (Figs. 7, 8 and 9).

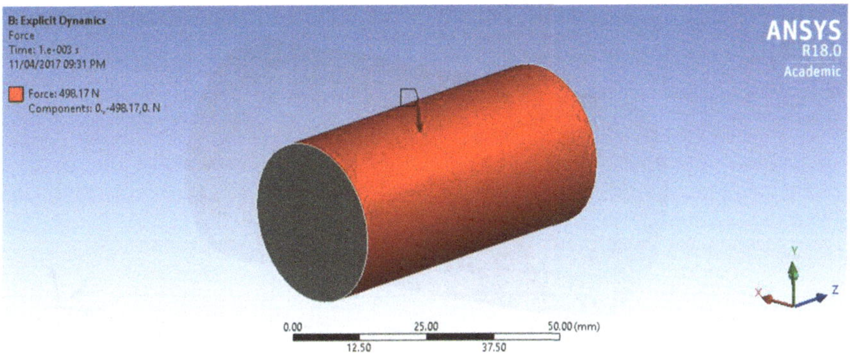


**Fig. 4** Force applied for surface roughness parameter set

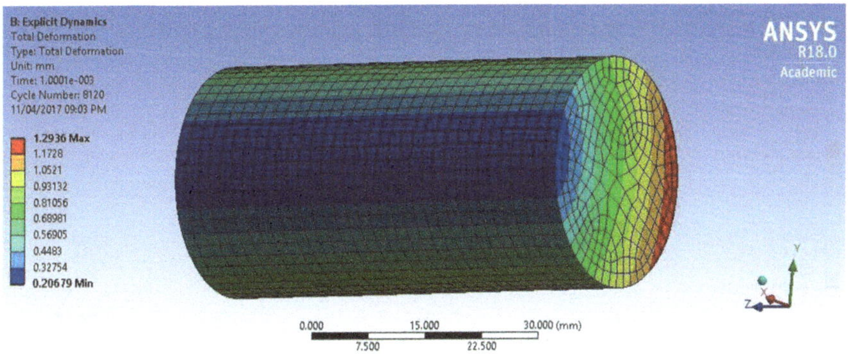


**Fig. 5** Result for deformation occurred for surface roughness parameter set

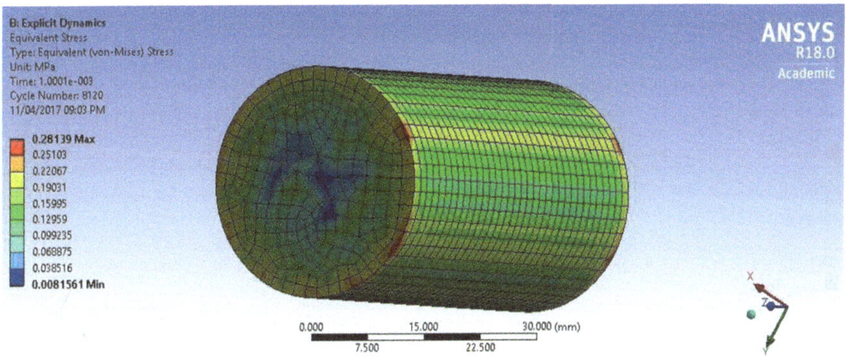


**Fig. 6** Result for stresses developed for surface roughness parameter set

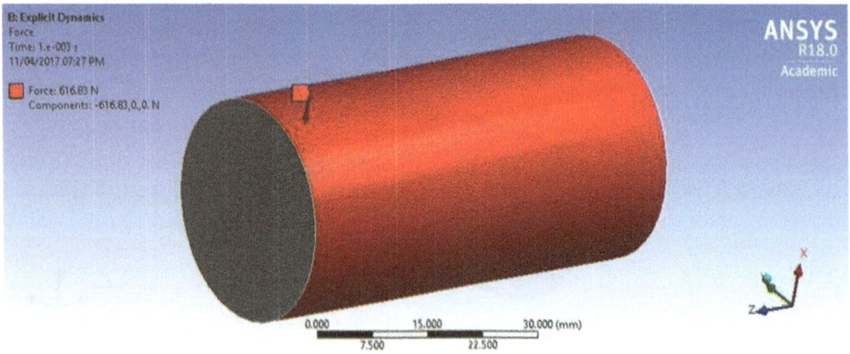


**Fig. 7** Force applied for surface hardness parameter set

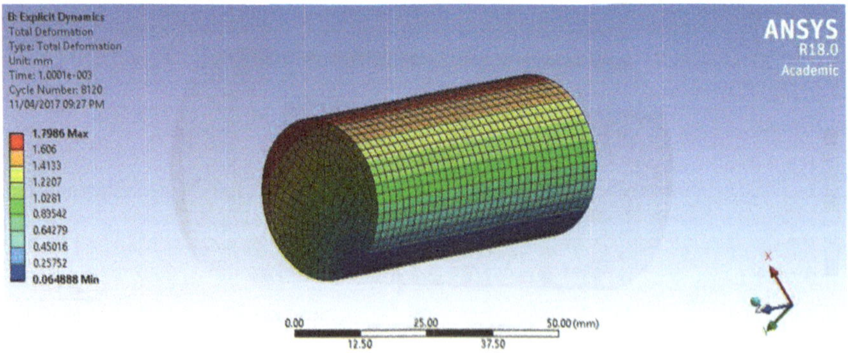


**Fig. 8** Result for deformation occurred for surface hardness parameter set

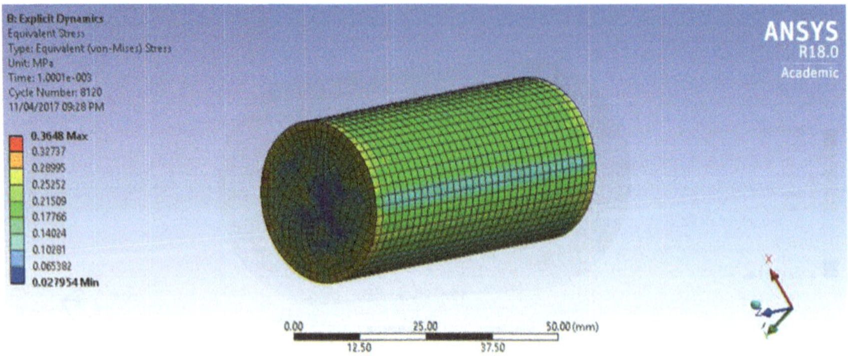


**Fig. 9** Result for stresses developed for surface hardness parameter set

## 4 Conclusion

1. As there is very less deformation obtained thus the work piece dimensions are not much affected. So the work piece remains within tolerance limit.
2. We know that, lower the stress developed on the work piece surface, higher is the work piece fatigue life. Thus, the component fatigue life can be improved by burnishing process.

## References

1. http://shodhganga.inflibnet.ac.in/bitstream/10603/8463/10/10_chapter%201.pdf
2. https://www.autodesk.in/solutions/finite-element-analysis
3. Deshmukh A, Patil RN (2015) Analysis and optimization of roller burnishing process on cylindrical surface micro hardness of aluminium alloy. Int J Innov Res Sci Eng Technol 4 (7):6044–6055
4. Qureshi NM, Patil Vaibhav B, Teli Basavaraj D, Mohite Radhika S, Patil Sonal S (2015) Analysis of effect of ball and roller burnishing processes on surface roughness on EN8 steel. Int J Eng Res Technol (IJERT) 4(6):311–315

# Resistance Spot Welding of Cold Rolled Mild Steel with Filler Metal

Sushil T. Ambadkar and Deepak V. Bhope

## 1 Introduction

Car designers today seek materials with the very best stiffness, mass reduction, and safety performance. Safety and crashworthiness, are being preferably considered for material selection, other factors being mass saving, formability, weldability, corrosion resistance, fatigue resistance, cost etc. Resistance spot welding has been the dominant welding process employed by the automotive factories for joining of steel sheet for body structures because of its inherently low cost and high speed. Spot welds are the primary method of joining automotive structural components. Mild steel is an excellent candidate for car body structural applications due to weight savings, enhanced crashworthiness, tough mechanical and fire-resistant properties with good formability. Efforts are going on to improve spot weld properties by improving microstructure, phases of the weldments by changing techniques of welding, welding different metals and newer welding designs. Resistance spot welding of dissimilar metal and subsequent changes in microstructure and response to mechanical testing are under investigation.

Advanced high strength steels (AHSS) like dual phase (DP) steel and transformation-induced plasticity (TRIP) steel offers excellent strength and formability, thus exploring the probability for improved crash performance with weight reduction. The mechanical properties of these steels are controlled by the martensite and the ferrite volume fractions [1, 2]. But these steels are uneconomical and hence are used in higher end vehicles only. Also, it has been stated that spot

S. T. Ambadkar (✉)
Department of Mechanical Engineering, Government College of Engineering, Chandrapur, MS, India
e-mail: sushipme@gmail.com

D. V. Bhope
Department of Mechanical Engineering, Rajiv Gandhi College of Engineering, Research and Technology, Chandrapur, MS, India



K. Antony and J. P. Davim (eds.), *Advanced Manufacturing and Materials Science*, Lecture Notes on Multidisciplinary Industrial Engineering,
https://doi.org/10.1007/978-3-319-76276-0_7

welding of AHSS steels has weldability issues due to the relatively higher alloying content in these steels [3]. Diminished mechanical performance with interfacial failures are problems encountered in TRIP steel weldments [4].

Looking at this, techniques to increase strength of traditional mild steel in spot welding can be explored to improve its crash performance. Whereas weldability of mild steel has been verified, same cannot be said about AHSS steels. Alternate technique/method to increase strength of mild steel spot welds can be explored so as to maintain its unique position in automotive industry due to its excellent mechanical behaviour. One of such effort was investigated by adding filler metal to cold rolled mild steel spot weldment. Its composition and carbon percentage was determined and test coupon was selected by referring literature. The material was cut and surface prepared by removing dirt, oxide layer, for spot welding. The spot welding was carried out on K J Thermoweld make spot welder. Experimentation was carried out to determine the range of filler metal to be added for investigations. The weld cycle was kept constant and spot welding was carried out with different quantities of filler metal. Tensile testing was carried out and response to these testing was investigated with load-displacement curves.

## 2 Experimental

The materials used in the present study are typical for automotive body applications and were 1.25-mm-thick CR Mild Steel sheets. The chemical composition of this material is shown in Table 1.

Tensile tests were carried out at a nominal cross-head movement speed of 1 mm/min to obtain the baseline tensile stress-strain curves of the base steel sheets. A Resistance spot welder with specifications as mentioned below was used to weld the specimen. The welder is K J Thermoweld make and its detailed specifications are as follows (Table 2).

The cycle in the spot welder can be controlled with the help of parameters as mentioned in Table 3. The machine was microcontroller based and energy levels were programmable.

Weld coupons were 70 mm long and 25 mm wide. Spot welding was done for a 1.25 mm thick sheet. The electrode was made from C15000 copper alloy and has a 5 mm face diameter. The water flow in the electrode was maintained at 4 lit/min in all experimentations (Fig. 1).

The spot welding schedules for experimentation is shown in Table 4. The parameters for experimentation is mentioned in setting parameter column in Table 4. The experimentation was carried out to determine the extent filler metal in mg to be added for further investigations. The filler metal was added from 30 to 80 mg. The schedule followed is mentioned in Table 4 and was maintained unchanged with addition of filler metal. Specimens were also welded without filler metal with same unchanged parameters. Specimens were then tested and load-displacement curve plotted to determine breaking strength. The failure mode

**Table 1** Composition of CR mild steel

| Sample identity | C% | Si% | Mn % | P% | S% | Cr% | Mo% | Ni% | Al% | Cu% | Nb% | Ti% | V% |
|---|---|---|---|---|---|---|---|---|---|---|---|---|---|
| Cold rolled mild steel | 0.047 | 0.011 | 0.25 | 0.0060 | 0.0181 | 0.036 | 0.0070 | 0.0286 | 0.052 | 0.047 | <0.005 | 0.0012 | <0.003 |

**Table 2** Spot welder specification

| Machine Model K J THERMOWELD | Unit | TSP 30 |
|---|---|---|
| Rating KVA @ 50% duty cycle | KVA | 30 |
| Max. available current (short circuit) @ throat depth: 460 mm | K.Amp | 15 |
| Optimum weldability{@ 460 throat depth} | | |
| Minimum | mm | 0.3 + 0.3 |
| Maximum | mm | 1.5 + 1.5 |
| Machine dimensions (Approx.L × W) | mm | 1020 × 520 |
| Height | | 1500 |
| Net Wt. (for 460 mm throat) (+30 kg. for 200 mm throat (approx) | Kgs. | 450 |
| Shipping Wt. (For 460 throat) (+ 30 kg. for 200 throat) (approx) | Kgs. | 525 |

**Table 3** Spot welder parameters

| Function of soft button as indicated | **Red Indication** | **Description** |
|---|---|---|
| Squeeze time | SQZ | 00–99 cycle |
| Preheat time | W1 | 00–99 cycle |
| Energy level | W1 | 00–99% programmable |
| Cool (I) time | C1 | 00–99 cycle |
| Slope for W2 | SLP | 0–99 cycle |
| Weld Time | W2 | 00–99 cycle |
| Weld energy (% heat setting) | W2 | 00–99% programmable |
| Cool (II) time | C2 | 00–99 cycle |
| Anneal time | W3 | 00–99 cycle |
| Energy level | W3 | 00–99% programmable |
| Hold time | HLD | 00–99 cycle |
| Off time | OFF | 00–99 cycle |

was also observed and was interpreted by visual inspection. Some of the spot weld specimens were cut across the cross section normal to the length and mounted for macrostructure and microstructure observations following standard metallographic procedures.

## 3 Result and Discussion

Investigations are carried out by adding small quantity of filler metal from 30 to 120 mg in spot weldments. Weld cycle and joint specifications are kept constant and only filler metal quantity was varied. Variation in is strength, plasticity is subsequently determined using stress-strain curve. The average breaking strength is observed to be maximum for 30 mg of filler metal. A linear reduction in breaking

**Table 4** Machine variables for experimentation

| Function of soft button as indicated | Red indication | Description | Setting parameter |
|---|---|---|---|
| Squeeze time | SQZ | 00–99 cycle | 04 |
| Preheat time | W1 | 00–99 cycle | 0 |
| Energy level | W1 | 00–99% programmable | 0 |
| Cool (I) time | C1 | 00–99 cycle | 34 |
| Slope for W2 | SLP | 0–9 cycle | 02 |
| Weld time | W2 | 00–99 cycle | 37 |
| Weld energy (% Heat Setting) | W2 | 00–99% programmable | 20 |
| Cool (II) time | C2 | 00–99 cycle | 28 |
| Anneal time | W3 | 00–99 cycle | 0 |
| Energy level | W3 | 00–99% programmable | 0 |
| Hold time | HLD | 00–99 cycle | 05 |
| Off time | OFF | 00–99 cycle | 05 |

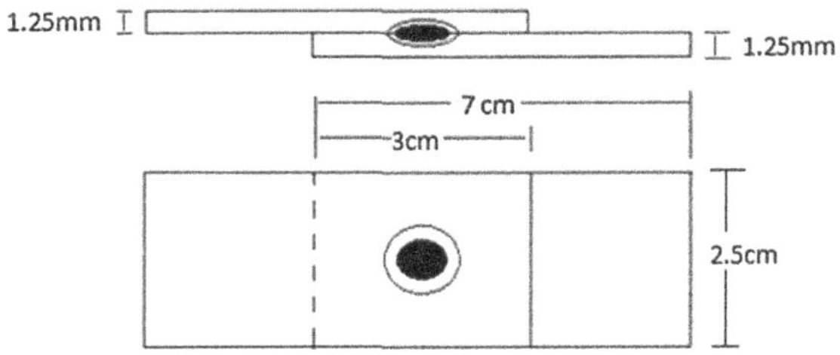


**Fig. 1** Test coupon

strength is clearly visible with subsequent addition of filler metal. The failure mode is predominantly button pullout for majority of specimen with 30 mg of filler metal as shown in Fig. 2. The plasticity was found to be maximum with the addition of 30 mg of filler metal with gradual reduction with further addition of filler metal.

Addition of 30 mg of filler metal lead to increase in breaking strength which is 7.029% greater than breaking strength of weldment with no filler metal. Thus 30 mg of filler metal produce optimum results and hence considered to be the optimal quantity in the given weld cycle. Two distinct failure modes were observed during mechanical testing: interfacial fracture and nugget pullout.

The average breaking strength is observed to be 10.81 kN corresponding to addition of 30 mg of filler metal. The nugget diameter for specimen 6 with maximum strength of 11.25 kN is found to be 5.26 mm and failure mode button pullout. So the addition of 30 mg of filler metal improved nugget diameter and breaking strength of the weldment. A gradual reduction in nugget diameter and depth of penetration with increase in filler metal was observed resulting in reduction in strength (Figs. 3, 4; Table 5).

The weld cycle was constant during experimentation. This ensured constant temperatures throughout experimentation, uniform solidification, uniform cooling

Fig. 2 Resistance spot welder

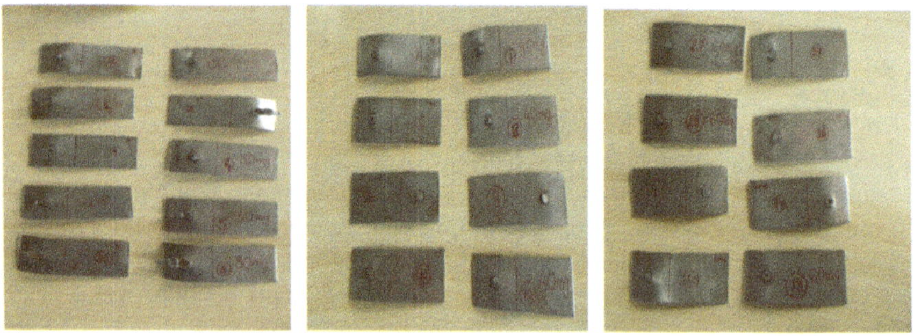

Fig. 3 Few tested specimens

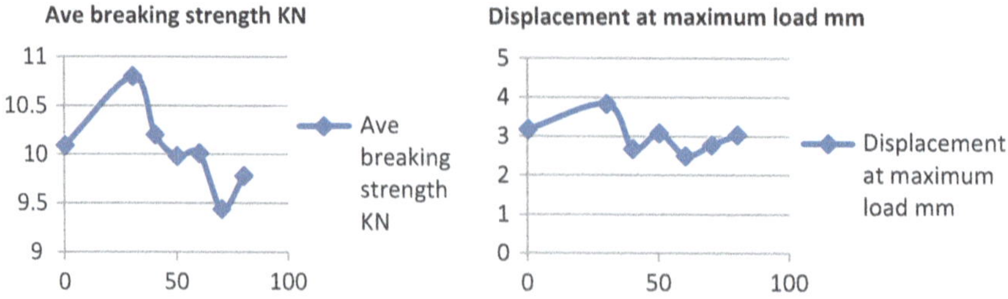


Fig. 4 Variation of average breaking strength and displacement with filler metal

**Table 5** Results obtained with filler metal

| Specimen no. | Weight of filler material by weight (mg) | Breaking load kN | Average breaking load kN | Type of failure | Displacement at max load (mm) | Average value (mm) | Nugget diameter (mm) |
|---|---|---|---|---|---|---|---|
| 3a | 0 | 10.06 | 10.01 | Interfacical | 3.3 | 3.2 | 4.13 |
| 6a | 0 | 9.96 | | | 3.1 | | 3.9 |
| 2 | 30 | 10.44 | 10.81 | Pullout and interfacial | 3.5 | 3.85 | 4.57 |
| 3 | 30 | 10.89 | | Button pull out | 3.9 | | 4.75 |
| 5 | 30 | 10.65 | | Button pull out | 3.6 | | 5.01 |
| 6 | 30 | 11.25 | | Button pull out | 4.4 | | 5.26 |
| 7 | 40 | 10.47 | 10.21 | Pullout and interfacial | 2 | 2.7 | 4.46 |
| 8 | 40 | 10.26 | | Pullout and interfacial | 3.8 | | 4.12 |
| 9 | 40 | 9.81 | | interfacial | 2.1 | | 3.76 |
| 10 | 40 | 10.32 | | interfacial | 2.9 | | 4.23 |
| 11 | 50 | 10.38 | 9.99 | interfacial | 2.3 | 3.1 | 4.8 |
| 12 | 50 | 10.23 | | Pullout and interfacial | 3.1 | | 4.25 |
| 22 | 50 | 10.20 | | interfacial | 2.5 | | 3.67 |
| 23 | 50 | 8.88 | | interfacial | 3 | | 3.24 |
| 24 | 50 | 10.17 | | Pullout and interfacial | 4.3 | | 4.2 |
| 25 | 50 | 10.08 | | interfacial | 3.4 | | 4.6 |
| 14 | 60 | 10.05 | 10.02 | Pullout and interfacial | 1.8 | 2.52 | 4.4 |
| 26 | 60 | 10.2 | | Pullout and interfacial | 2.8 | | 4.27 |
| 27 | 60 | 10.14 | | interfacial | 2.8 | | 3.21 |
| 28 | 60 | 9.69 | | interfacial | 2.7 | | 3.25 |
| 15 | 70 | 9.42 | 9.45 | interfacial | 2.3 | 2.8 | 3.15 |
| 16 | 70 | 9.36 | | interfacial | 4.3 | | 3.11 |
| 29 | 70 | 8.97 | | interfacial | 2.4 | | 3.23 |
| 30 | 70 | 9.72 | | interfacial | 2.2 | | 3.04 |
| 31 | 70 | 9.78 | | interfacial | 2.8 | | 3.19 |

(continued)

**Table 5** (continued)

| Specimen no. | Weight of filler material by weight (mg) | Breaking load kN | Average breaking load kN | Type of failure | Displacement at max load (mm) | Average value (mm) | Nugget diameter (mm) |
|---|---|---|---|---|---|---|---|
| 17 | 80 | 9.9 | 9.79 | Pullout and interfacial | 3 | 3.06 | 3.71 |
| 18 | 80 | 9.75 | | interfacial | 3.1 | | 3.65 |
| 32 | 80 | 9.57 | | interfacial | 3 | | 3.25 |
| 33 | 80 | 9.69 | | interfacial | 2.8 | | 3.55 |
| 34 | 80 | 10.05 | | interfacial | 3.4 | | 3.56 |

due to water cooled electrode and hence no changes in the phases were observed due to the addition of filler metal. No significant effect on grain size, shape, distribution was observed due to the addition of filler metal and these were observed to be same as that of weldment without filler metal. Three separate microstructural zones were visible in joint region: (i) Weld nugget (ii) Heat affected zone and (iii) Base metal. The heat affected zone must have experienced a maximum temperature and a cooling rate which are inversely proportional to its distance from the fusion line during welding. The fusion boundary in heat affected zone consisted of martensite and ferrite. This can be attributed to quenching effect due to water cooled copper electrodes with short welding cycle [5]. The microstructure of the region away from fusion boundary consisted of ferrite and pearlite (Figs. 5, 6).

## 4 Determination of Filler Metal Quantity for Investigation

Invetigations were carried out by adding small quantity of filler metal from 30 to 80 mg in spot weldments. Weld cycle and joint specifications were kept constant and only filler metal quantity was varied. Variation in is strength, plasticity was

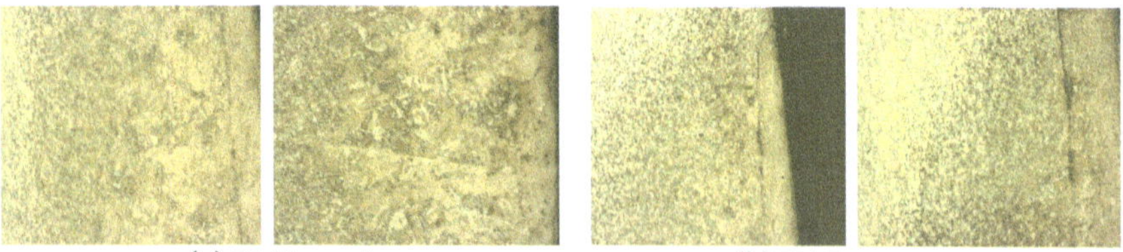

**(a) Weld metal** **(b) HAZ**

**Fig. 5** Weld metal and HAZ of specimen no 2, 8 (Filler metal 30, 40 mg)—Weld metal shows Widmanstatten ferrite, Normal ferrite (Whitish) and Pearlite (blackish), HAZ shows martensite (black) and ferrite

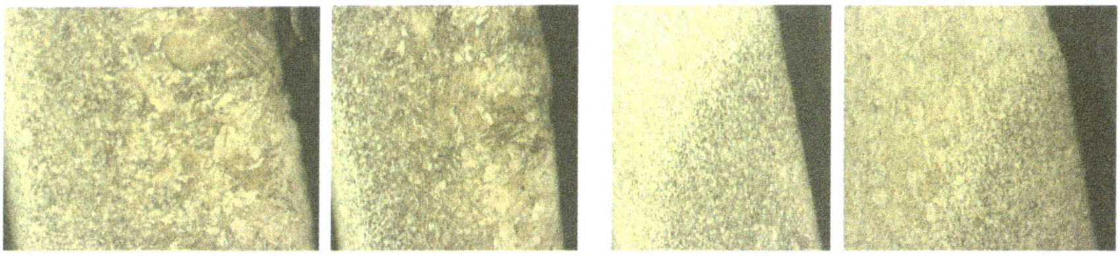

(a) Weld metal (b) HAZ

**Fig. 6** Weld metal and HAZ of specimen no 27, 60 (Filler metal 60 mg, No filler metal)—Weld metal shows Widmanstatten ferrite, Normal ferrite (Whitish) and Pearlite (blackish), HAZ shows martensite (black) and ferrite

subsequently determined using stress-strain curve. A direct corelation between addition of filler metal and average breaking strength was observed. The average breaking strength was observed to be maximum for 30 mg of filler metal. A gradual reduction in breaking strength was clearly visible with subsequent addition of filler metal. The failure mode was predominantly button pullout for majority of specimen with 30 mg of filler metal.

Thus 30 mg of filler metal produced optimum results and hence found to be the desired quantity in the given weld cycle.

## 5 Investigation into Effect of Addition of Filler Metal on Mechanical Properties of Resistance Spot Weldments

Mechanical properties like breaking strength/load carrying capacity and plasticity were investigated by plotting stress-strain curves corresponding to small quantity of filler metal from 30 to 80 mg in spot weldments. These properties were also invstigated for weld specimen without filler metal. Addition of 30 mg of filler metal lead to increase in breaking strength which was 7.029% greater than breaking strength of weldment without filler metal. It was seen that increasing filler metal (30 mg) leads to increase in the maximum load and failure energy of joint primarily due to increase in fusion zone and its depth penetration. Two separate failure modes were observed during mechanical testing: interfacial fracture and nugget pullout. Thus increase in energy absorption capability lead to enhanced vehicle crashworthiness due to filler metal. The interfacial failure results in fracture through nugget, whereas in pullout mode, nugget withdrawal from one sheet results failure. The reliability of the spot welds during vehicle lifetime is excellent if pullout failure mode is guaranteed and hence process parameters should ensure button pull out failure [6, 7].

Increase in breaking strength with addition of filler metal was associated with improvement in failure mode as button pull out as pointed out by Pouranvari et al. [7]. Specimen with increasing quantity of filler metal showed interfacial failure or combination of pull out and interfacial mode with inferior mechanical properties. The plasticity was found to be maximum with the addition of 30 mg of filler metal with gradual reduction with further addition of filler metal.

## 6 Investigation into the Changes in Nugget Diameter with Addition of Filler Metal in Resistance Spot Weldments

Invetigations were carried out by adding small quantity of filler metal from 30 to 80 mg in spot weldments. The average breaking strength was observed to be 10.81 KN corresponding to addition of 30 mg of filler metal. The nugget diameter for specimen 6 with maximum strength of 11.25 kN was found to be 5.26 mm and failure mode button pullout. So the addition of 30 mg of filler metal improved nugget diameter and breaking strength of the weldment. There was reduction in nugget diameter and penetration with increase in filler metal resulting in reduction in strength. At a given current in a cycle, filler metal can be accommodated optimally, the quantity of which depends on plate thickness. This addition of 30 mg of filler metal contributes in fusion zone and resulted in increase in fusion zone size and penetration. Diffusion controls the penetration of filler metal through the base metal during welding. This was in accordance with the results obtained by Hasanbasoglu and Kacar [8]. Increase in load-bearing area due to increase in nugget size in the joint resulted in improved failure load for the joint [9]. A direct correlation between joint tensile load and nugget diameter in the case of the spot welded specimen is observed [10]. Expulsion of metal may begin then due to excess availability of filler metal, resulting in weak joint and higher areas of stress concentration. But no expulsion of metal was visible in present experimentation.

## 7 Summary

The main results in this study are listed as follows;

1. It was observed that there is increase in strength of spot weldments with the addition of filler metal. The maximum breaking load was observed to be 11.25 kN with the addition of 30 mg of filler metal. This strength is maximum for any quantity of filler metal. The breaking strength is observed to be maximum at 10.44 kN without the addition of filler metal. The average increase in breaking strength is found to be 7.029% with the addition of 30 mg of filler metal. Gradual reduction in breaking strength is observed with increase in addition of filler metal.
2. Gradual transformation from button pullout mode to interfacial failure mode was visible with increase in filler metal. All failure modes for the weldments without

filler metal are observed to be interfacial. The specimen with 30 mg of filler metal fail with button pull out mode. All other specimen with filler metal in excess of 30 mg is found to fail by interfacial failure mode.
3. The plasticity of the specimens is found to be varying with the addition of filler metal. The addition of 30 mg of filler metal produce average displacement at maximum load, a measure of plasticity at 3.85 mm. This is found to be decreasing with further addition of filler metal in the weldment. Spot weldments without filler metal give average displacement at maximum load to be 3.2 mm.
4. Superior mechanical properties and desired button pullout mode are obtained with the addition of 30 mg of filler metal. Hence 30 mg of filler metal is found to be optimum quantity of filler metal in the given weld cycle.

## References


1. Rathbun RW, Matlock DK, Speer JG (2003) Fatigue behavior of spot welded high-strength sheet steels. Weld J 82(8):207–218
2. Matlock DK, Krauss G, Ziaebrahimi F (1984) Strain hardening of dual phase steels: an evaluation of the importance of processing history. In: Krauss G (ed) Deformation, processing, and structure, ASM International, Materials Park, Ohio, pp 47–87
3. Smith S, den Uijl N (2006) Resistance spot welding of advanced high strength steels for the automotive industry. In: The 4th international seminar on advances in resistance welding. Wels, Austria, pp 30–60
4. Khan MI, Kuntz ML, Biro E, Zhou Y (2008) Microstructure and mechanical properties of resistance spot welded AHSS. Mater Trans JIM 49(7):1629–1637
5. Li MV, Dong D, Kimchi M (1998) SAE technical paper 982278. SAE International, Warrendale, PA
6. Chao YJ (2003) Sci Technol Weld Join 8:133–137
7. Pouranvari M, Asgari HR, Mosavizadeh SM, Marashi PH, Goodarzi M (2007) Sci Technol Weld Join 12:217–225
8. Hasanbasoglu A, Kacar R (2007) Resistance spot weldability of dissimilar materials (AISI 316L-DIN EN 10130-99 steels). Mater Des 28:1794–1800
9. Ozyurek DA (2008) An effect of weld current and weld atmosphere on the resistance spot weldability of 304L austenitic stainless steel. Mater Des 29:597–603
10. Sun DQ, Lang B, Sun DX, Li JB (2007) Microstructures and mechanical properties of resistance spot welded magnesium alloy joints. Mater Sci Eng A 46:494–498

# Study on Squeeze Casting of Aluminum Matrix Composites—A Review

L. Natrayan and M. Senthil Kumar

## 1 Introduction

Composite materials have led to increase the rate of development in engineering field, Metal matrix composite (MMC) are generally reinforced with other metal, ceramic organic compounds. Reinforcements significantly improved the properties such as high tensile strength, toughness, hardness, low density and good wear resistance compared to base metal. It has increasing attentiveness to fabricate composites at low cost [25], commonly AMMCs are used now automobile, airplane, aerospace then many supplementary fields [17], silicon carbide (SiC), graphite (Gr) and aluminium oxide ($Al_2O_3$) are most generally used reinforcements. Al alloy-SiC reinforcement gets increases the toughness, thickness, ductile strength then wear resistance [23]. $Al_2O_3$ reinforcement has good compressive strength and wear resistance [29]. Gr used as the solid lubricant, between these materials, SiC and Gr particles reinforcement in hybrid composite gives low friction coefficient and high wear resistance [5, 8, 14]. Al/SiC/Gr hybrid composite revealed that the presence of SiC particles has improves the strength and hardness, compensate the wearying properties of Gr [12, 14]. AMMC properties particle distribution has shown a very spirited role and is better by intensive shearing.

L. Natrayan (✉) · M. Senthil Kumar
School of Mechanical and Building Sciences, VIT University,
Chennai 600127, Tamilnadu, India
e-mail: natrayanphd@gmail.com



K. Antony and J. P. Davim (eds.), *Advanced Manufacturing and Materials Science*,
Lecture Notes on Multidisciplinary Industrial Engineering,
https://doi.org/10.1007/978-3-319-76276-0_8

## 2 Squeeze Casting Technique

Though the idea of squeeze casting periods back towards 1800 [3, 33]. Till 1931 squeeze casting research was not accompanied [24]. Squeeze casting methods remains mixture of the closed die forging plus gravity die casting. Applied pressure in closed die halves metals has changed to solidifies. The functional pressure then the sudden interaction of melted metal through the die surface generates quick heat transfer that crops a porous free casting by mechanical properties imminent the twisted product. The Squeeze casting technique has various names such as extrusion casting, liquid metal, Squeeze forming.

Its offers low operating costs, low shrinking porosity, high metal yield, excellent surface finish. Premium quality castings and higher rates of heat removal through the metal mold boundary are found due to the close interaction among the mold and liquid metal. Squeeze casting process is illustrated schematically in Fig. 1 [33]. The method is essentially separated hooked on two types: indirect and direct method. Squeeze pressure is applied through the die-closing punch this action is direct process, such the Squeeze pressure is applied as after closing die by the secondary ram this method is indirect process, particular feature of the squeeze casting technique over predictable die casting method.

- Squeeze casting method has suitable prospective for critical application due to internal soundness obtained under pressure in solidification process.
- Absence of shrinkage porosity.
- Wrought alloys as well as casting alloys can stay squeeze casting to finish suitable for long freezing alloys too.
- Squeeze casting has earlier cycle times.
- Good dimensional reproducibility is possible with Combination of thin die coatings and high quality of reusable dies, applied corresponding pressure to die is recycled to form the components.
- In squeeze casting has formed forging quality of the components.
- Dimensional precision, extraordinary grade of surface finish, clear shape.

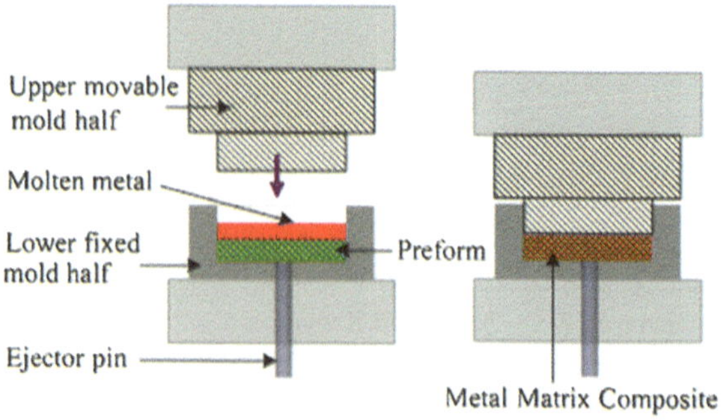


**Fig. 1** Squeeze casting process [33]

- Squeeze forming consequences an extraordinary amount of modification in structure of the composite. Particle dimension reduced towards the amount of 50% of that usually gravity cast composite produce better microstructure.
- In squeeze casting method material molded as exhibit higher toughness and good equiaxed grain structure then material formed through gravity casting.
- Because of the better microstructure, mechanical properties shown Major improvement.

Compared the mechanical and microstructure properties, Al-8% SiC particles has fabricated through chill casting, sand casting and squeeze casting methods [19]. Chill and sand casting products has increased in the grain size of microstructure, therefore chill and sand cast used in less quality parts for requiring engineering and non-engineering applications, though squeeze cast yields might remain used in as cast state in industrial requests demanding high excellence parts.

## 3 Single and Multiple-Reinforced AMC

SiC particle reinforced Al-MMC increased the tensile strength, hardness and density, but impact toughness decreased [18, 20]. Weak matrix bonding, clustering elements and atom cracking are the most affected behaviour of composites, the impact behaviour of all materials has not actual significant in the test temperature [27].

AA 7034/SiC/15p-PA and Al 7034/SiC/15p-UA MMCs microstructures, modulus, strength and ductility decreased with an increase in temperature. The cyclic fatigue life supplementary prominent aimed at under-aged microstructure, at ageing condition to growing the load fraction its gets higher fatigue strength [30]. AMMC based on pre ceramic-polymer-bonded SiC with polymer content 1.25 wt% discussed provided appropriate constancy to the preforms towards support composite dispensation. PMS subsequent discusses the preferred strength to the SiC preforms on Al-SiC composites mechanical properties gets without damaging [28]. Squeeze cast reinforcement of $Al_2O_3$ and SiC exhibit better mechanical and physical properties, high ultimate tensile and impact strength and hardness, such as low coefficient of thermal expansion. It can used potential lightweight material in automobile filed, the results found with addition of Al-SiC reinforcement gets low wear rate then reinforcement of $Al_2O_3$ [34]. Element clustering on flow behaviour of Al reinforced SiC particle, results found the plastic deformation is affected desperately and mechanical reaction of the matrix is high than the elastic response [2]. Particle clustered microstructure experiences grater ratio on particle crack then chance dispersal [21]. Effect of vary the volume fraction in Al-$Al_2O_3$ particles results found that increase in volume of fraction and decreased the fissure hardiness. It has decrease due to the inter-particle spacing among nucleated micro cavities [9]. Al reinforcement of SiC/Gr found that tensile strength has decreased with increased in Gr particle size, elastic module and tensile strength both has depend on size of the Gr particle [6, 11].

## 4 Mechanical Properties of AMC with Various Reinforcement

$SiC/Al_2O_3$ particles shown density decreased due to porous of composite, the porosity in matrix has an effect on the particle size. The tensile strength and porosity of particle reinforced metal matrix composites improved owing to decreasing particle size and unit volume has growing surfaces area of particles [1].

$Al_2O_3/SiC$ reinforcements improved the tensile strength of aluminum alloy composites. Figure 2 shows the composite reinforcement obtained for $Al_2O_3/SiC$ particle 10 vol. %. Tensile strength and porosity are dependent on reinforcement size. In composite expose addition to particulate in base alloy to increase the tensile strength in reinforcements [32].

$Al_2O_3/SiC$ particles in aluminum matrix during deformation cause the interface to crack and debonding since the matrix undergoes plastic flow while the particles do not deform. Tensile elongation has decreased with increased in the $Al_2O_3/SiC$ particle size in Al alloys based composites. Tensile strength increased with decrease in ductility, with decreasing particle size [4, 11, 31] (Fig. 3; Table 1).

Moderate ductility and higher tensile strength has accomplished in the decrease of reinforcement particle size, the reinforcement of constant volume fraction, SiC small particle size delivers extra boundary area, it helps by way of nucleation locations aimed at the grain development through squeeze casting process and through the ensuing heat treatment the grain has growth. Hence, it has commonly unexpected grain size of the reinforcement has decreased as well as the consequence of a smaller amount capture length between ceramic particles in addition to

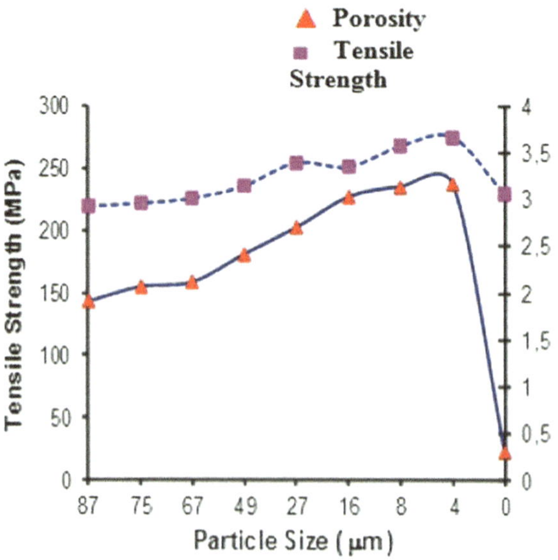


**Fig. 2** Tensile strength and porosity increase depending on $Al_2O_3/SiC$ particle size [32]

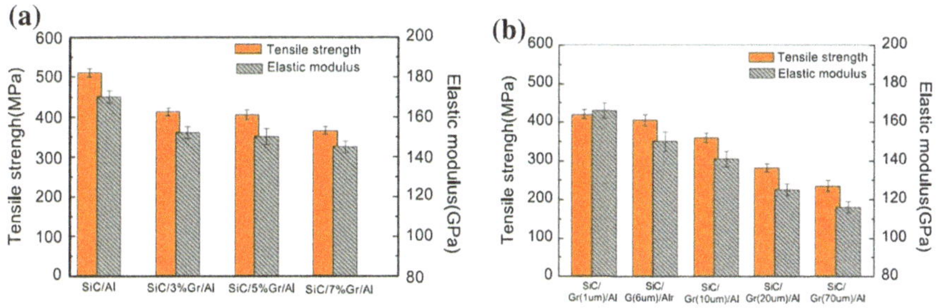


**Fig. 3** Effects of the volume fraction and size of Gr particles on mechanical properties of Al/SiC/Gr composites [11]

**Fig. 4** Tensile strength and elastic modulus in different Particle size of Gr with Al6061/40% SiC [26]

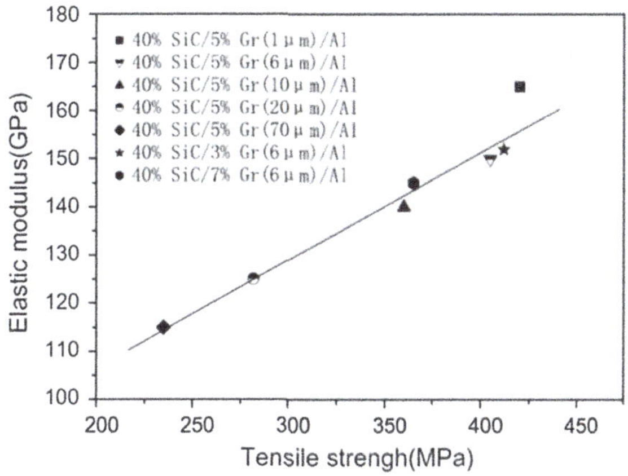


get higher mechanical strength in metal matrix result due to the higher dislocation [15, 31] (Fig. 5).

Since the discussion Al/SiC/Gr composites obviously establish the particles size of graphite particles and the volume fraction has affected extraordinary in elastic modulus and the tensile strength has depends on particle size of the graphite and volume fraction. Figure 4 shows the correlation among the elastic modulus and tensile strength of Al/SiC/Gr MMCs. In graph tensile strength has decrease as well as elastic modulus also decreases [26].

## 5 Tribological Behaviour of al MMCs

Wear is the tolerant loss of the material owing to frictional resistance among the interaction surfaces [22]. Changed reinforcements have dissimilar effect on the tribological behavior of AMMC. Different researches have been made to fabricate

**Table 1** Mechanical properties of Al 6061 based alloys with and without various reinforcements [7]

| Alloy | Young's modulus. MPa × $10^3$ | Yield stress. MPa | Tensile strength. MPa | Elongation. % |
|---|---|---|---|---|
| 6061 (T6) | 69 | 286 | 310 | 12 |
| 6061 + 15% SiC (T6) | 91 | 342 | 364 | 3.2 |
| 6061 + 20% SiC (T4) | 98 | 405 | 460 | 7.0 |
| 6061 + 25% SiC (T4) | 115 | 430 | 515 | 4.0 |
| 6061 + 10% $Al_2O_3$ (T6) | 81 | 296 | 338 | 7.5 |
| 6061 + 15% $Al_2O_3$ (T6) | 87 | 317 | 359 | 5.4 |
| 6061 + 20% $Al_2O_3$ (T6) | 98 | 359 | 379 | 2.1 |

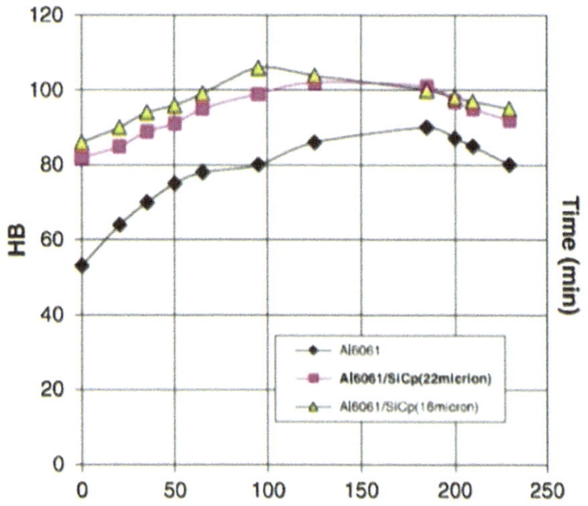


**Fig. 5** Hardness value of the composite and unreinforced material versus the again time [15]

such Al-MMC with high wear resistance. Wear and friction performance is dependent on various tribological factors.

The wear intensity increases by increase of sliding distance. Through raising the SiC wt%, wear decreases with increase in 20 wt% of SiC particles (Fig. 6), again raising the wt% of SiC particles such as 20–40%, wear also increases [16]. Increase the reinforcement volume fraction the normalized wear rate gets decreases, and then limits subsequently this develops seeming remains around 20% [4, 13, 31]. The normalized coefficient of friction reached the comparable need on the reinforcement volume fraction, the wear ratio and particle size reduces to increase the reinforcement [35] (Fig. 7).

The friction coefficient has decreases due to increase of SiC content, friction coefficient increases after 30% of SiC content [13]. The authors have reported Al/SiC/Gr MMCs are statistically the dry sliding friction behaviour. Friction coefficient has mostly affected in load factor shadowed through sliding speed. Increase the load and sliding distance is to coefficient of friction also gets increased [10].

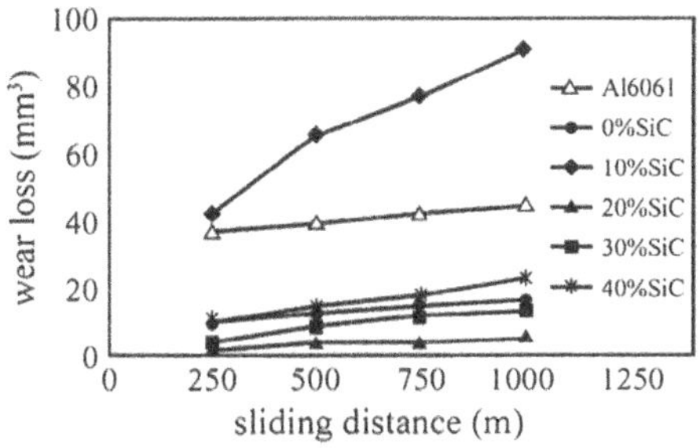


**Fig. 6** sliding distance and wear intensity with various wt % of SiC in Al/SiC/Gr hybrid composites [16]

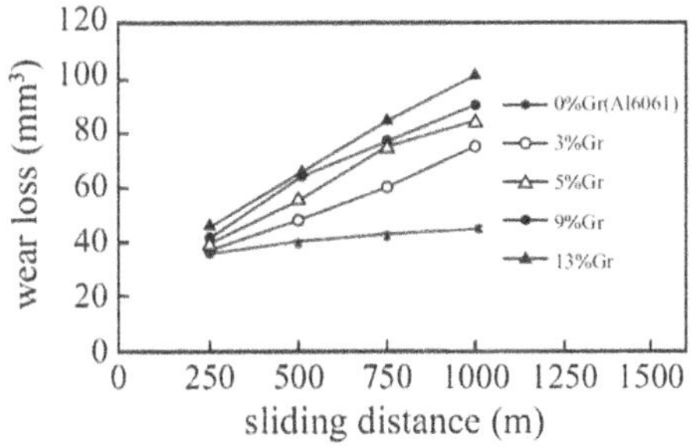


**Fig. 7** Sliding distance and wear loss and for different wt % of graphite content [4]

# 6 Conclusion

The following are conclusions on the review of Squeeze casting:

Aluminum based metal matrix reviewed for different reinforcements resulted in fine micro-structures with higher strength components, good surface texture and low levels of porosity. Aluminium alloys with reinforcement of $Al_2O_3$ increased the tensile strength and hardness along with ductility. SiC reinforced matrix improved the tensile and hardness behavior up to 10 wt% and decreased the tensile strength and hardness with the increasing wt% due to agglomeration of the hard ceramic particles that leads to porosity. SiC reinforced AMMCs reported better wear resistance than $Al_2O_3$ reinforced MMCs. Addition of graphite particles increased the wear resistance. The coefficient of friction has reduced and wear resistance gets increase due to increased the reinforcement on the wear behaviour of MMCs.

# References

1. Altinkok N, Koker R (2004) Neural network approach to prediction of bending strength and hardening behaviour of particulate reinforced (Al-Si-Mg)-aluminium matrix composites. Mater Des 25:595–602
2. Balasivanandha prabhu S, Karunamoorthy L, Kathiresan S, Mohan B (2006) Influence of stirring speed and stirring time on distribution of particles in cast metal matrix composite. Mater Process Technol 171:268–273

3. Chernov DK (1878) Reports of the Imperial Russian Metallurgical Society
4. Doel TJA, Bowen P (1996) Tensile properties of particulate-reinforced metal matrix composites. Compos Part A Appl Sci Manuf 27(8):655–665
5. Ge D, Gu M (2010) Mechanical properties of hybrid reinforced aluminium based composites. Materials Lett 49:334–339
6. Gui MC (2000) Microstructure and mechanical properties of cast (Al-Si)/SiCp composites produced by liquid and semisolid double stirring process. Mater Sci Technol 16(5):556–563
7. Gupta M, Srivatsan TS, Mohamed FA (1993) Microstructural evolution and mechanical properties of SiC/$Al_2O_3$ particulate-reinforced spray-deposited metal-matrix composites. J Mater Sci 28:2245–2259
8. Hollinggrak J (1819) Casting metals, UK Patent 4371
9. Kannan S, Kishawy HA (2006) Surface characteristics of machined aluminium metal matrix composites. Int J Mach Tools Manuf 46:2017–2025
10. Kumar S, Panwar RS, Pandey OP (2013) Effect of dual reinforced ceramic particles on high temperature tribological properties of aluminum composites. Ceram Int 39:6333–6342
11. Leng Jinfeng (2008) Mechanical properties of SiC/Gr/Al composites fabricated by squeeze casting technology. Scripta Mater 59:619–622
12. Leng J, Jiang L, Zhang Q, Wu G, Sun D, Zhou Q (2008) Study of machinable SiC/Gr/Al composites. J Mater Sci 43:6495–6499. https://doi.org/10.1007/s10853-008-2974-6
13. Lim SC (1998) Recent developments in wear mechanism maps. Tribol Int 31:87–97
14. Mahdavi S, Akhlaghi F (2011) Effect of SiC content on the processing, compaction behavior, and properties of. Al6061/SiC/Gr hybrid composites, J Mater Sci 46:1502–1511. https://doi.org/10.1007/s10853-010-4954-x
15. Mahdavi S, Akhlaghi F (2011) Effect of the SiC particle size on the dry sliding wear behavior of SiC and SiC-Gr-reinforced Al 6061 composites. J Mater Sci 46:7883–7894. https://doi.org/10.1007/s10853-011-5776-1
16. Mahdavi S, Akhlaghi F (2011) Effect of SiC content on the processing, compaction behavior and properties of Al6061/SiC/Gr hybrid composites. J Mater Sci 46(5):1502
17. Narayana Murty SVS, Nageswara Rao B, Kashyap BP (2003) On the hot working characteristics of 6061Al–SiC and 6061–$Al_2O_3$ particulate reinforced metal matrix composites. Compos Sci Technol 63(1):119–135
18. Natrayan L et al (2017) An experimental investigation on mechanical behaviour of SiCp reinforced Al 6061 MMC using squeeze casting process. Inter J Mech Prod Engi Res Develop 7(6):663–668
19. Ozben T, Kilickap E, Cakir O (2008) Investigation of mechanical and machinability properties of SiC particle reinforced Al-MMC. Mater Process Technol 198:220–225
20. Ozden S, Ekici R, Nair F (2007) Investigation of impact behaviour of aluminium based SiC particle reinforced metal–matrix composites. Compos Part A 38:484
21. Park BG, Crosky AG, Hellier AK (2008) Fracture toughness of microsphere $Al_2O_3$–Al particulate metal matrix composites. Compos B 39:1270–1279
22. Prasad SV, Asthana R (2004) Aluminum metal–matrix composites for automotive applications: tribological considerations. Tribol Lett 17:445–453
23. Previtali B, Pocci D, Taccardo C (2008) Application of traditional investment casting process to aluminium matrix composites. Compos Part A 39:1606–1617
24. Raji A (2010) A comparative analysis of grain size and mechanical properties of Al-Si alloy components produced by different casting methods. AU J T 13(3):158–164
25. Ralph B, Yuen HC, Lee WB (1997) The processing of metal matrix composites—an overview. J Mater Proc Technol 63:339–353
26. Seyed Reihani SM (2006) Processing of squeeze cast Al6061–30vol% SiC composites and their characterization. Mater Des 27:216–222
27. Srivatsan TS, Hajri Meslet Al, Vasudevan VK (2005) Cyclic plastic strain response and fracture behavior of 2009 aluminum alloy metal-matrix composite. Int J Fatigue 27:357–371

28. Sujan D, Oo Z, Rahman ME, Maleque MA, Tan CK (2012) Physio-mechanical properties of Aluminium metal matrix composites reinforced with $Al_2O_3$ and SiC. Int J Eng Appl Sci 6 (8):678–681
29. Suresha S, Sridhara BK (2010) Effect of silicon carbide particulates on wear resistance of graphitic aluminum matrix composites. Mater Des 31:4470–4477
30. Thunemann M, Beffort O, Kleiner S, Vogt U (2007) Aluminum matrix composites based on preceramic-polymer-bonded SiC preforms. Compos Sci Technol 67:2377
31. Tian J, Shobu K (2004) Fracture strength of melt-infiltrated SiC-mullite composite. J Mater Sci 39:3751–3755
32. Vencl A, Rac A, Bobic I (2004) Tribological behaviour of Al-based MMCs and their application in automotive industry. Tribol Ind 26:31–38
33. Welter VG (1931) Z Metallkd 23:255
34. Peng Z, Fuguo L (2010) Effects of particle clustering on the flow behavior of SiC particle reinforced Al metal matrix composites. Rare Met Mater Eng 39:1525
35. Zou XG, Miyahara H, Yamamoto K, Ogi K (2003) Sliding wear behaviour of Al-Si-Cu composites reinforced with SiC particles. Mater Sci Technol 19(11):1519–1526

# Development and Investigation of Microstructure and Mechanical Properties of Cast and Hot Forged AL2014-$TIB_2$ In Situ Composite by Stir Casting

**M. Senthil Kumar, R. D. Hemanth, B. Ashok Kumar, M. Venkateswarlu and P. Kuppan**

## 1 Introduction

The most economical method available for fabrication of aluminum matrix composites (AMC) is stir casting, in recent times enormous research is involved in developing metal matrix composites. AMCs are prepared by combining advanced materials of two or more different types by analyzing the tailored properties [1, 2]. Exclusive properties of AMCs are good wear resistance, high structural efficiency, high specific strength, low density, very good electrical and thermal properties [3, 4]. Hence, it finds its applications in infrastructure, automotive, military and aerospace industries [5]. Stir casting method is preferred compared to the other processes due to its simplicity, flexibility and its application in large volume production [6–8]. Literature review carried out exhibited very few studies have been done on aluminum matrix using $TiB_2$ as reinforcement in evaluating the microstructure and mechanical properties. This paper mainly focuses on investigating the mechanical properties (hardness, density, impact strength and tensile strength) and microstructure of Al-$TiB_2$ samples prepared by stir casting technique.

M. S. Kumar (✉) · R. D. Hemanth · B. A. Kumar
School of Mechanical and Building Sciences, VIT University,
Chennai 600127, Tamilnadu, India
e-mail: msv305@yahoo.co.in

M. Venkateswarlu
Research and Engineering Centre, Amara Raja Batteries Ltd,
Karkambadi, Tirupati 517520, Andhra Pradesh, India

P. Kuppan
School of Mechanical Engineering, VIT University,
Vellore 632014, Tamilnadu, India

 
K. Antony and J. P. Davim (eds.), *Advanced Manufacturing and Materials Science*,
Lecture Notes on Multidisciplinary Industrial Engineering,
https://doi.org/10.1007/978-3-319-76276-0_9

## 2 Experimental Details

A12014 alloy and raw material selected was $TiO_2$ and $B_2O_3$ for this investigation to obtain 10 and 15 wt% In situ reactant of $TiB_2$ particles through the following reaction

$$10Al + 3TiO_2 + 3B_2O_3 = 5Al_2O_3 + 3TiB_2$$

1 kg of raw materials (Al2014) was melted using electric furnace (Fig. 1) of 6 KW at 850 °C in a graphite crucible. The scum powder is added into the mold to settle the impurities. Commercially available Hexa-chloroethane tablets were added to dissolve the gases and in turn remove the porosity of the final cast product. To achieve a fine vortex, molten metal was agitated by using mechanical stirrer. Ceramic coated stainless blades were rotated at a speed of 350 rpm for 15 min. A fine vortex was created to disperse uniformly the $TiO_2$ and $B_2O_3$ reinforcements. The composite melt was maintained at a constant temperature of 850 °C for 25 min which allowed for the In situ reaction to occur. Molten metal is poured into the mould (Fig. 2b) and upon solidification; the composite billets were removed from the mould.

In order to perform studies on hot forging, billets were hot forged Billets were pre-heated at 400 °C in muffle furnace for 3 h and finally subjected to hand forging process on the anvil. The stir casted and hot forged billets were machined to obtain suitable shape and size as per the ASTM test standards.

**Fig. 1** Electric furnace

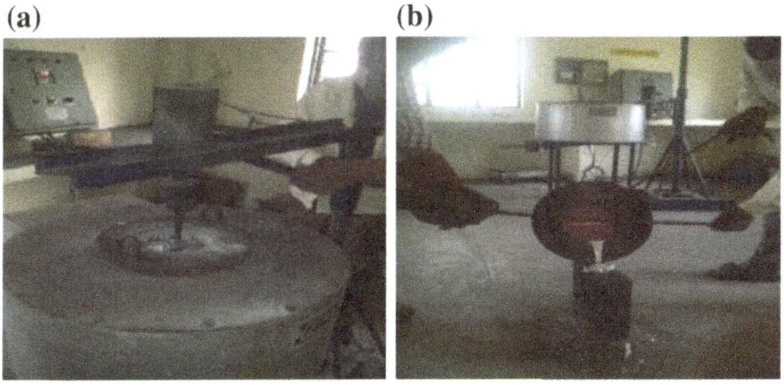

Fig. 2 **a** Stirring **b** Pouring the molten metal

## 3 Results and Discussions

## 4 Microstructure Studies

Optical micrographs (Fig. 3) revealed that during solidification dendrites are formed and the alloying elements of Al2014 such as Cu and Mg are higher than that of their solubility limit. As a result, the intermetallic phases like $CuAl_2$, $CuMgAl_2$ and $Mn_3SiAl_2$ etc., that forms around the dendrites during casting enables grain refinement and strengthening of alloy indicated as black regions. Cast Al2014-10% $TiB_2$ and forged Al2014-10%$TiB_2$ samples indicate homogenous distribution of reinforcement with minimal porosity in the matrix alloy. Homogenous distribution of reinforcement is observed within the matrix. Good bonding and wettability is confirmed by the existence of $TiB_2$ particles inside the dimples. Micrograph observation showed much shallower dimples. The $TiB_2$ particles are well dispersed in Al matrix in an equal manner under both forged and cast conditions. However, on comparison, $TiB_2$ particles are more uniformly distributed in the forged composited, this can be ascribed to thermo mechanical deformation during forging.

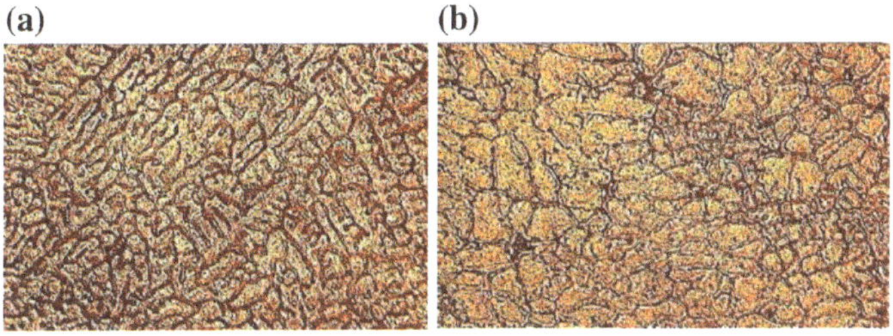

**Fig. 3** Microstructure (**a**) Casted Al2014-15%$TiB_2$ (**b**) Forged Al2014-15%$TiB_2$

## 5 Brinell Hardness Test

Hardness measurements were done by applying 150 kgf load on a universal mechanical tester (UMT) for a duration of 10 s. Hardness was measured at five different locations and average was taken in order to reduce the probability of error in measurement due to the indenter resting on the harder particles. The diameter of the indentation was measured by using simple microscope. Brinell hardness number for Al2014 and developed composites are shown in Fig. 4. Hardness of the cast and In situ reinforced $TiB_2$ with 15 wt% composites showed the highest hardness. This results confirm that the bonding strength increased with the addition of reinforcement. Increase in reinforcement content in the matrix alloy subsequently increased the hardness of composite. In situ forged composite possess higher hardness compared with the cast composites.

## 6 Density

Density of Al2014 composites were measured by using Archimedes principle. Figure 5 shows the variation in density of cast and cast composites. It is observed that increasing the reinforcement content in the matrix alloy, increased the density of composite for cast as well as forged composites. Al2014-15%$TiB_2$ forged composite exhibited the highest density. The porosity (Fig. 6) of the composites were evaluated using the relation:

$$\%\,\text{porosity} = \{(\rho_T - \rho_{EX}) \div \rho_T\} \times 100\%$$

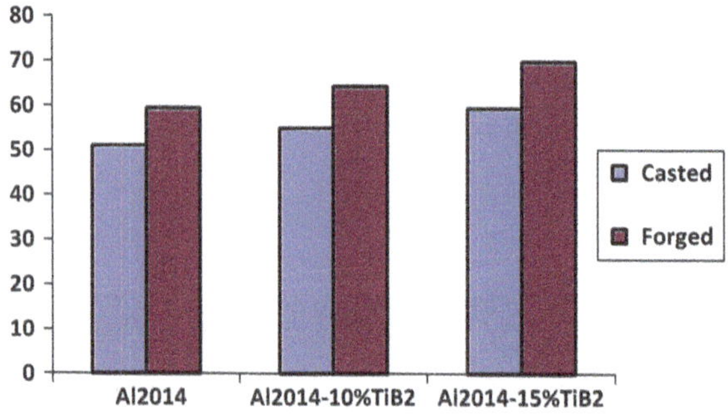


**Fig. 4** Variation in Brinell hardness of Al2014 alloys and its composites

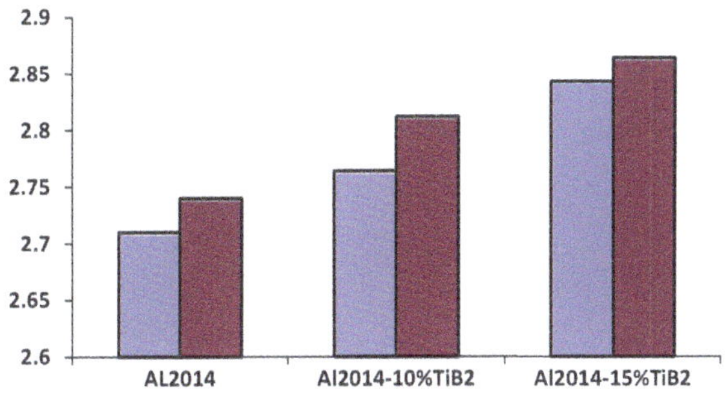


**Fig. 5** Variation in densities of Al2014 alloys and its composites

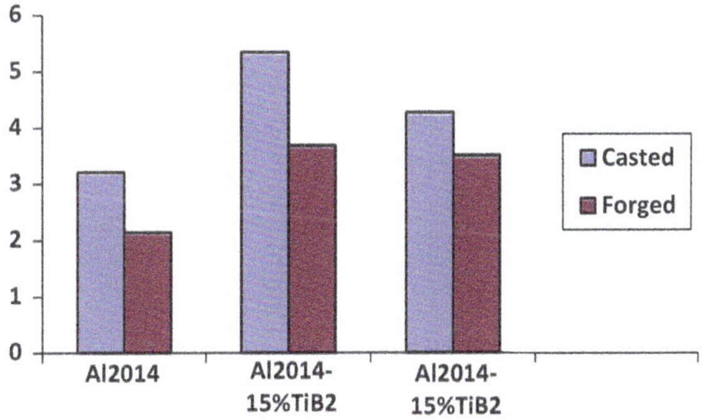


**Fig. 6** Variation in porosities of Al2014 and Al2014-$TiB_2$ composite

## 7 Impact Strength

Impact strength evaluation was performed on Al2014 base alloy, Al2014-10%$TiB_2$ and Al2014-15%$TiB_2$ composites in an impact testing machine. The specimens used for this test were prepared as per ASTM standards. This test was conducted by applying 100 kgf load on Al2014 and developed composites to calculate the impact strength. Figure 7 shows the variation of the impact strength levels of Al2014 alloy and composite with $TiB_2$ at 10% and 15 wt%. The impact strength increased with

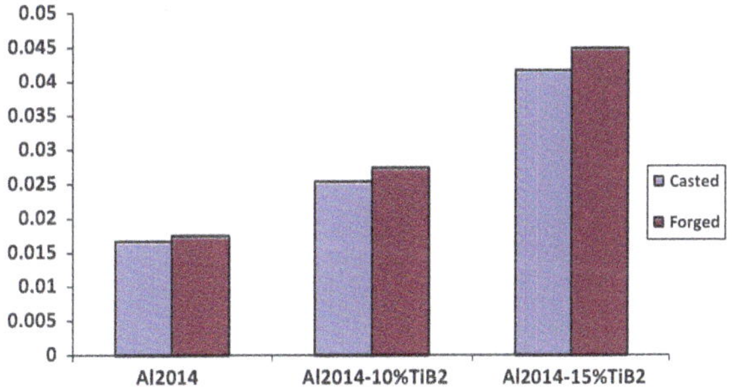


**Fig. 7** Variation in Impact Strengths of Al2014 and Al2014-$TiB_2$ composite

Fig. 8 Specimen after impact test

increase in composition and was found to be highest for forged Al2014-15%$TiB_2$ composite (0.047 Unit). Figure 8 shows the specimens after impact test was performed.

## 8 Tensile Strength

The tensile tests were conducted on composite samples using universal testing machine as per ASTM standard. Figure 9 shows the tensile strength of cast and forged metal matrix composites. Tensile strength was found to increase with increasing wt% of reinforcements, subsequently it showed better tensile strength with forged composites. The results clearly define that the interfacial bonding between the matrix and reinforcement plays a crucial role in defining the tensile properties. The higher hardness and superior mechanical properties of $TiB_2$

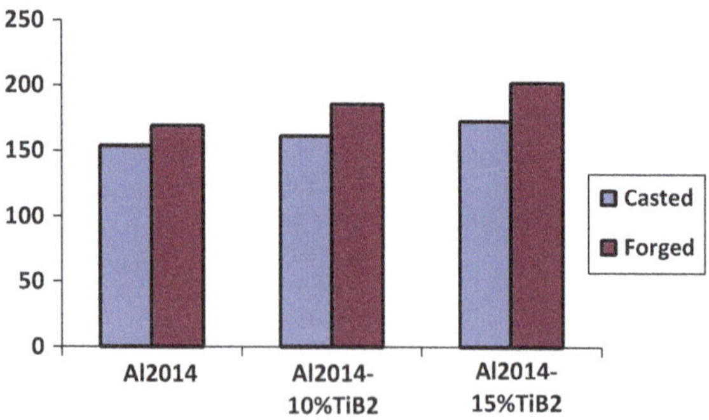


Fig. 9 Variation of micro hardness of tensile strength Al2014 composite with $TiB_2$

particles contributed to the improved ultimate tensile strength. Tensile strength of Al2014-10%TiB2 and Al2014-15% $TiB_2$ forged composites exhibited the highest compared to the cast composites. Al2014-15% $TiB_2$ forged composite reported the highest tensile strength of 201.92 unit.

## 9 Conclusion

Al2014-$TiB_2$ composites were successfully fabricated by using liquid metallurgical Process as cast and forged Al2014 and Al2014-$TiB_2$ composites. The following were the conclusions.

- Composite density increased with increasing wt% of reinforcements
- Homogenous distribution of $TiB_2$ particles was observed in both as cast and hot forged composites
- Hardness, tensile and impact strength of composites increased with the increasing wt% of $TiB_2$
- Mechanical properties of forged composites were high compared with cast composites.

The results of this research on Al2014-$TiB_2$ composite showed improved mechanical properties compared to Al2014 alloy.

## References

1. Rajan TP, Pillai RM, Pai BC (1998) Review reinforcement coatings and interfaces in aluminium metal matrix composites. J Mater Sci 33(14):3491–3503
2. Jayamathi M, Seshan S, Kailas SV, Kumar K, Srivatsan TS (2004) Influence of reinforcement on microstructure and mechanical response of a magnesium alloy. Curr Sci 87(9):1218–1231
3. Mandal D, Dutta BK, Panigrahi SC (2006) Microstructure and mechanical properties of Al–2 Mg alloy base short steel fiber reinforced composites prepared by vortex method. J Mater Sci 41(15):4764–4770
4. Zhang DL, Brindley C, Cantor B (1993) The microstructures of aluminium alloy metal matrix composites manufactured by squeeze casting. J Mater Sci 28(8):2267–2272
5. Surappa MK (2003) Aluminium matrix composites: challenges and opportunities. Sadhana 28:319–334
6. CanKurnaz S (2003) Production of saffil fibre reinforced Zn-Al (ZA 12) based metal matrix composites using infiltration technique and study of their properties. Mater Sci Eng A 346:108–115
7. Xing Chen, Chengxiao Yang, Leding Guan, Biao Yan (2008) $TiB_2/Al_2O_3$ ceramic particle reinforced aluminium fabricated by spray deposition. Mater Sci Eng A 496:52–58
8. Ramesh CS, Pramod S, Keshavamurthy RA (2011) Study on microstructure and mechanical properties of Al 6061–$TiB_2$ in-situ composites. Mater Sci Eng A 528(12):4125–4132

# Mechanical Characterization and Comparative Evaluation of the Different Combination of Natural and Glass Fiber Reinforced Hybrid Epoxy Composites

**C. M. Meenakshi and A. Krishnamoorthy**

**Nomenclature**

GFG Glass and Flax Fiber Reinforced Epoxy Composite Laminate.
GSG Glass and Sisal Fiber Reinforced Epoxy Composite Laminate.
GKG Glass and Kenaf Fiber Reinforced Epoxy Composite Laminate.
GAG Glass and Aloe Vera Fiber Reinforced Epoxy Composite Laminate.

## 1 Introduction

Asian researchers are showing interest in cellulose fiber reinforced plastics, advantage of these fibers as reinforcement is their low density, which is an important factor of consideration in the structural industry. Widely there are many research works are ongoing in this regard, [1] has proved that banana/bamboo hybrid fiber reinforced epoxy composite has a good impact strength, [2] based on many reviews saying that plant fibers are good replacement for synthetic fiber in terms of cost, density, and renewability and $Co_2$ emission. [3] has found that jute–epoxy composites have good tensile and flexural strength compare to jute-polyester composites [4] found out that sisal/Glass fiber composite is performing well with the tensile load and [5] saying research and development of biodegradable composites will lead to the use of suitable local materials with possible local manufacturing of products through local technological capabilities meeting not only local but global demand also [6] have quoted *that* hybridization with the glass fiber showed a positive response on the tensile, flexural, impact and water absorption properties of the hybrid composites.

C. M. Meenakshi (✉) · A. Krishnamoorthy
Department of Mechanical Engineering, Sathyabama Institute of Science and Technology, Chennai, Tamil Nadu, India
e-mail: cmmeenakshi84@gmail.com



K. Antony and J. P. Davim (eds.), *Advanced Manufacturing and Materials Science*, Lecture Notes on Multidisciplinary Industrial Engineering,
https://doi.org/10.1007/978-3-319-76276-0_10

In this work Flax, Sisal, Kenaf, and Aloe Vera fibers are used along with glass fiber in epoxy resin; four types of composite laminates are prepared and compared for their mechanical strengths.

## 2 Materials and Methodology

**Materials and Laminate Fabrication**. In this work the resin considers is Epoxy resin (LY 556) and the hardener used is Araldite HY-951, the matrix is prepared by mixing the resin and hardener in the ratio of 10:1 and the natural fibers used are flax, sisal Kenaf and Aloe Vera fibers in the form of biaxial mats brought from weavers association Chennai. The glass fiber used is 600 Gsm biaxial glass fiber mat from Sakthi fibers, Chennai.

Even though there are many advanced composite laminate preparation methods like vacuum bag molding methods are available till hand lay-up method is consider as one of the simpler and widely used methods for thermoset laminate preparation, the same is followed here in this work. Initially, the Glass, natural fiber mats and are cut into 300 × 300 mm sizes, Epoxy and Hardener are mixed in the ratio of 10:1. In the approximate weight fraction of 30 W% (10% Natural fiber and 20% Glass fiber) fiber and 70 W% resin, different types of laminates are prepared with the combination of Glass-Flax-Glass, Glass-Sisal-Glass, Glass-Kenaf-Glass and Glass-Aloe vera-Glass in each combination 2 laminates are prepared. Highest care has been given to produce a uniform and homogeneous composite laminate of 300 × 300 × 3 mm. The 3 mm thickness is achieved by laying up alternate layers of resin and fiber mat in the above-mentioned order. The fabricated specimens are shown in Fig. 1 and the detail combination scheme is given in Table 1.

## 3 Mechanical Testing

Composite laminates prepared are cut into appropriate smaller samples of ASTM standards, for various test procedure and the test results are obtained.

**Tensile Test**. The tensile test is carried out in accordance with ASTM D-638 standards on a Universal testing machine (UTM) UTN-60 with minimum graduation 1 kN and maximum capacity of 600 kN at room temperature. The test is carried out in such a way till the sample gets fractured by gradually applying the load and the ultimate tensile strength, elongation at the peak load and ultimate stress are found out. The load Vs displacement graph is auto generated by the machine. The samples before and after test are presented in Fig. 2.

**Flexural Test**. The flexural test is carried out in a UTM on Samples cut in accordance with ASTM D-790 standard the testing procedure is as per the three-point bending test method by placing the specimen on the universal testing machine and applying load till the specimen fracture and break. Results are

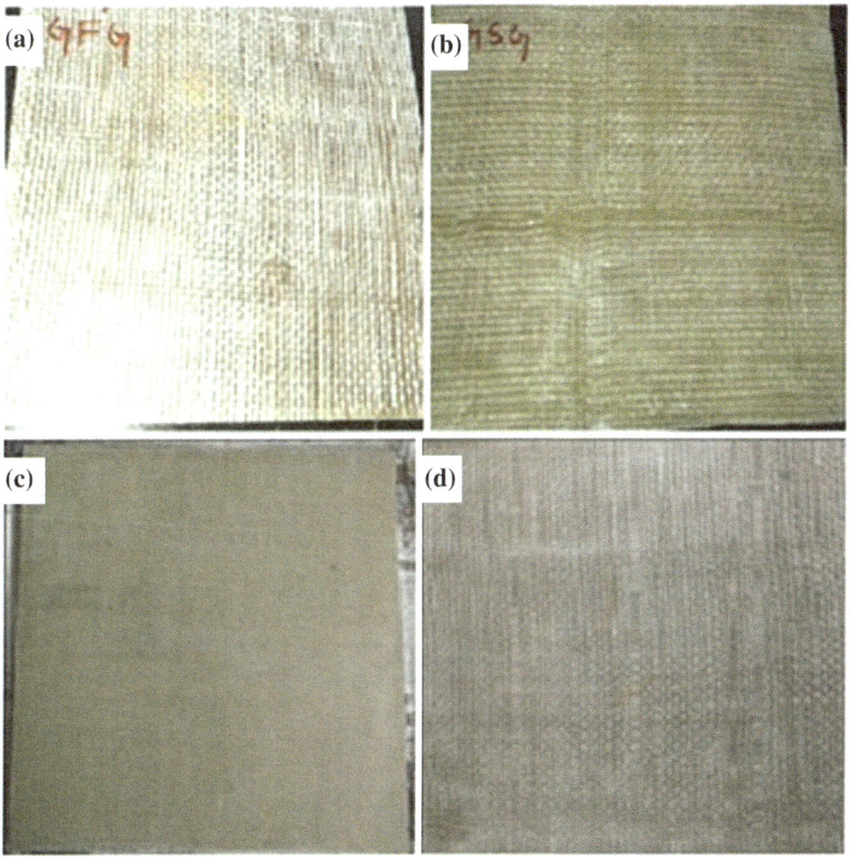


**Fig. 1** Sample Laminates of different type of composite **a** for GFG **b** GSG **c** GAG **d** GKG

**Table 1** Type of composites and their configuration

| S. No | Nomenclature | Composite type | Layer 1 | Layer 2 | Layer 3 |
|---|---|---|---|---|---|
| 1 | GFG | Glass and Flax Fiber Reinforced Epoxy | Glass Fiber Mat | Flax Fiber Mat | Glass Fiber Mat |
| 2. | GSG | Glass and Sisal Fiber Reinforced Epoxy | Glass Fiber Mat | Sisal Fiber Mat | Glass Fiber Mat |
| 3. | GKG | Glass and Kenaf Fiber Reinforced Epoxy | Glass Fiber Mat | Kenaf Fiber Mat | Glass Fiber Mat |
| 4. | GAG | Glass and Aloe Vera Fiber Reinforced Epoxy | Glass Fiber Mat | Aloe Vera Fiber Mat | Glass Fiber Mat |

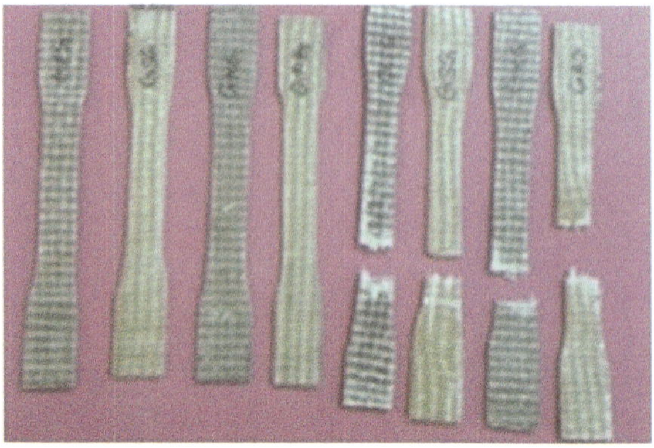

**Fig. 2** Tensile test specimens

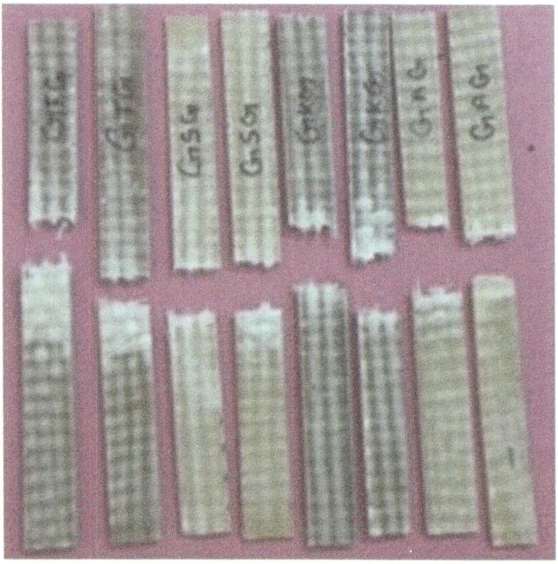

**Fig. 3** Flexural test specimens

compared and flexural strength of the materials is identified. The samples before and after test are presented in Fig. 3.

**Impact Test**. Impact energy is the energy that the specimen absorbs once a sudden load is applied. Izod impact testing is an ASTM standard method of determining the impact resistance. Material specimens are prepared to confirm to ASTM D4812. A pendulum from a particular height is released which strike the specimen and breaks it. From the amount of energy spent to break the specimen, its impact energy can be found out. Samples after test are shown in Fig. 4.

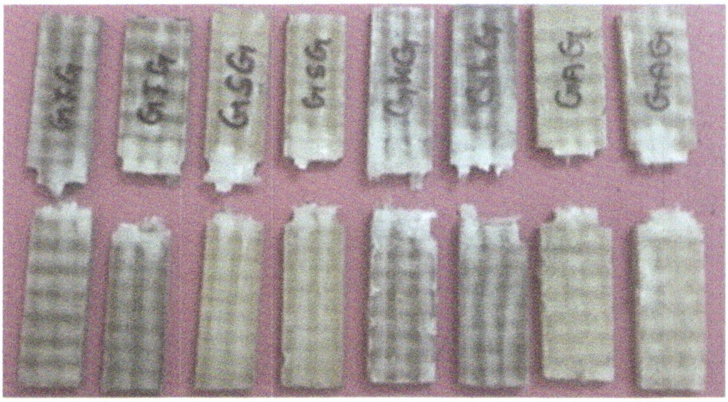

Fig. 4 Impact test specimens

# 4 Result and Discussion

**Tensile Test**. Figure 5 shows the variation in tensile strength among the four types of composite laminates and Table 2 give the tensile strength value. The tensile test results show that Glass–Kenaf fiber reinforced composite is having a high tensile strength of 91 N/mm$^2$ in an average which is 42% higher than the tensile strength (52 N/mm$^2$) of Glass–Flax composite. The average tensile strength value of glass-sisal and glass-aloe Vera is 62 and 67 N/mm$^2$ respectively. The auto-generated load versus Displacement graph of tensile test is shown in Fig. 6. And the breaking stress values are 52.978, 62.938, 92.208 and 69.99 N/mm$^2$ respectively in the order of GFG, GSG, GKG and GAG. Also, the ultimate load applied, after which the sample failed, for each type is 2.23, 2.66, 3.60, and 2.70 kN in the same order above. From the tensile test result of this work, the overall understanding is that the Glass–Kenaf fiber reinforced with Epoxy resin is giving a good tensile strength compared to other natural fibers considered in this work. The theoretical value of tensile strength of all the composites is obtained using the formula [3]:

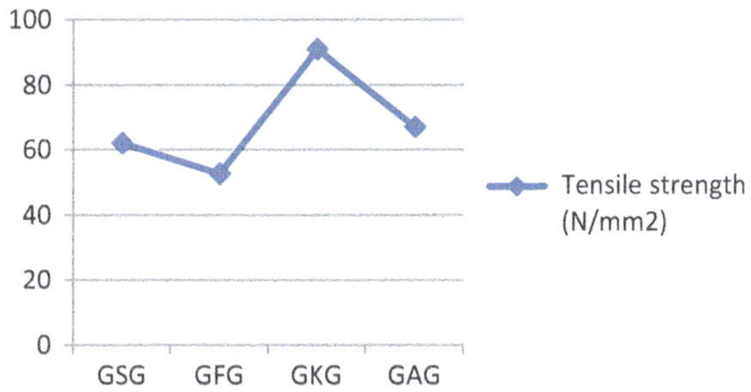


Fig. 5 Comparison of tensile strength

**Table 2** Overall mechanical performance of different composite laminate

| S. No | Composite type | Tensile strength (N/mm$^2$) | Flexural strength (N/mm$^2$) | Impact energy (J/m) |
|---|---|---|---|---|
| 1 | GSG | 62.4 | 204.95 | 17 |
| 2 | GFG | 52.978 | 155.334 | 13.4 |
| 3 | GKG | 91.41 | 255.442 | 16.4 |
| 4 | GAG | 67.488 | 199.268 | 15.6 |

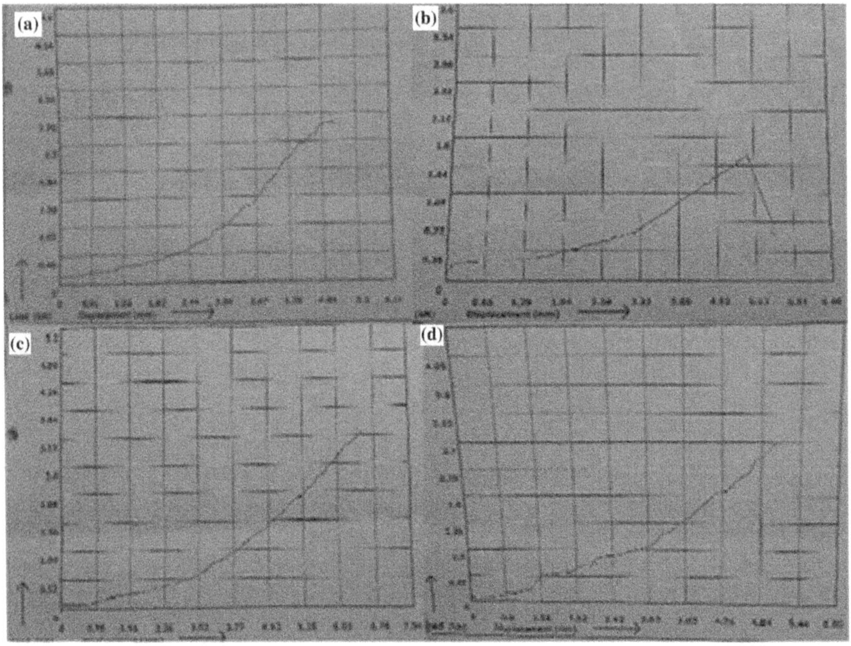

**Fig. 6** Sample load versus displacement graph from tensile test **a** for GSG **b** GFG **c** GKG **d** GAG

$$\sigma_t = P/bh$$

where,

P-Ultimate load of the specimen, b-Initial width of the specimen, h-Initial thickness of specimen.

The value obtained for different composites are $\sigma_t$ = 64 N/mm$^2$ for GSG, 53 N/mm$^2$ for GFG, 90 N/mm$^2$ for GKG and 66 N/mm$^2$ for GSG respectively.

**Flexural Test**. The resulted value of flexural test performed in UTM is given in Table 2 and the variation among values for different composites is shown in Fig. 7. Overall the flexural load withstanding capacity of natural fiber reinforced composites are very good due to their elastic nature. Out of the samples tested the

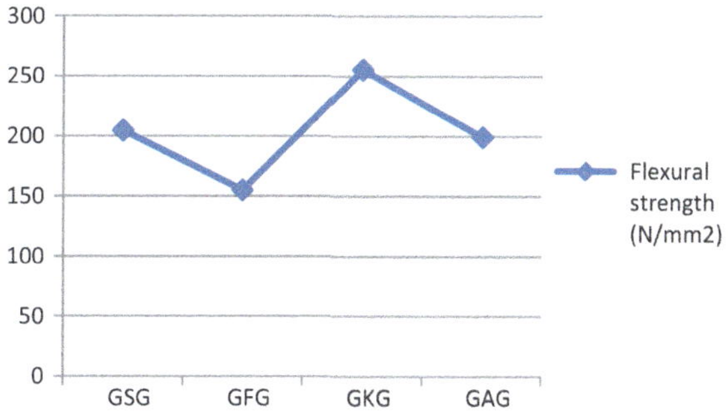


**Fig. 7** Comparison of flexural strength

flexural strength of Glass–Kenaf Composite is very high (255 $N/mm^2$) and the values of the other composites are as in Table 2. Flexural strength of the material depends not only on the brittle or ductile nature of the fiber also on the resin, epoxy exhibits a ductile behavior which contributes to the good flexural strength of the epoxy –Fiber reinforced composites, the variation in flexural strength of the hybrid composites is due to the variation in stiffness of the natural fibers used as the other two member resin and glass fiber are common in all the laminates prepared. Kenaf–Glass fiber reinforced composite is showing good flexural strength due to the high stiffness value of kenaf with the resin and glass fiber and flax glass is showing a lower value due to the lower stiffness value of flax fiber and so on. The theoretical values of flexural strength of all the composites are obtained using the formula [3, 7] :

$$\sigma_f = 3PL/2bh^2$$

where,

P-Ultimate load of the specimen, L-Span length of the specimen, b-Width of the specimen, h-Thickness of specimen.

The value obtained for different composites are $\sigma_f$ = 200 $N/mm^2$ for GSG, 153 $N/mm^2$ for GFG, 254 $N/mm^2$ for GKG and 202 $N/mm^2$ for GAG respectively.

**Impact Test**. The impact test is carried out to find the toughness of the composite laminates prepared; the test is carried out on ASTM D4812 sized samples. The pattern of failure observed in impact testing is fiber breakage and delamination. The value of impact resistance for each composite laminate is given in Table 2 and the comparison are shown in Fig. 8. Generally, higher cellulose content in natural fiber allows for more fracture to happen. Kenaf–Glass and Sisal–Glass fiber reinforces epoxy laminates shows better impact energy value (17 and 16.3 $J/mm^2$) than the other two laminates tested, Flax-glass laminate is the inferior player among the four laminates.

The pictorial comparison of overall mechanical properties of the composites laminates prepared is shown in Fig. 9 which is a proof that Kenaf-Glass composite laminate exhibits a superior performance to other types.

Fig. 8 Comparison of impact strength

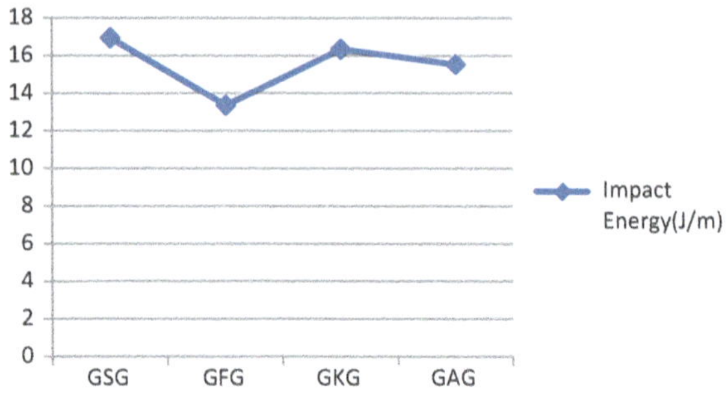


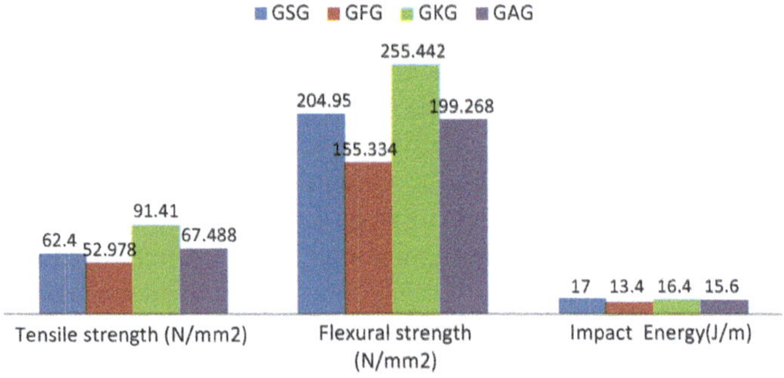


**Fig. 9** Comparisons of overall mechanical strength

## 5 Conclusion

This particular work is carried out to find out the mechanical strength of glass and natural hybrid fiber reinforced epoxy composite and the following conclusions are made.

- This works shows that successful fabrication of homogeneous composite laminates is possible by hand lay-up methods.
- Experimentation and theoretical values of tensile and flexural strength are coinciding with each other which show the accuracy of the experimentation followed.
- The various test results of the all hybrid composites infer that Kenaf/Glass fiber reinforced Epoxy composite is showing good mechanical strength having 40–50% higher value to the least performing member flax-glass composite.
- The flexural strength values of all the hybrid fiber reinforced composites are reasonably good.

From the above absorption, it is understood that hybrid fiber composites can be used as an alternate for glass fiber composites depend on the strength required by which the synthetic content in polymeric matrix composites can be reduced.

## References


1. Ramachandran M et al (2016) Experimental study of bamboo using banana and linen fiber reinforced polymeric composites. Perspect Sci. https://doi.org/10.1016/j.pisc.2016.04.063
2. Saxena M et al (2011) Composite materials from natural resource: recent trends and future potentials. Adv Compos Mater www.intechopen.com—Analysis and man-made materials
3. Gopinath A, Kumar MS, Elayaperumal A (2014) An experimental investigations on mechanical properties of jute fiber reinforced composites with polyester and epoxy resin matrices. Procedia Eng 97:2052–2063
4. Ramesh M, Palanikumar K, Hemachandra Reddy K (2013) Comparative evaluation of properties of hybrid glass fiber-sisal/jute reinforced epoxy composites. Procedia Eng 51:745–750
5. Satyanarayana KG et al (2009) Biodegradable composites based on lingo cellulosic fibers- an overview. Prog Polym Sci 34:982–1021
6. Nunna S et al (2014) A review on mechanical behavior of natural fiber based hybrid composites. J Reinf Plast Compos. https://doi.org/10.1177/0731684412444325
7. Gupta MK, Srivastava RK (2014) Tensile and Flexural properties of sisal fiber reinforced epoxy composites: an comparison between unidirectional and mat form of fibers. Procedia Mater Sci 5:2434–2439

# Use of Polylactic Acid (PLA) Material for Asymmetric Spur Gear Manufactured by 3-D Printing

**Laxman B. Abhang and Akshay Govardhan Dekhane**

## 1 Introduction

The concept of an asymmetric spur gear drive is that large and small pressure angles are considered for the drive and coast sides. The two profiles of gear tooth are performing different functions for most gear drives. The one side carrying higher workload than the other. The coast profile is unloaded or very lightly loaded during operating the drive for given time period. Asymmetric gear is made by using the two involutes of different base circle diameter i.e. base circle diameter of coast side and base circle diameter of drive side. Mostly the pressure angle on drive side is modified so that we can get more contacting area while the tooth meshing. One of the important advantage of asymmetric gears is reduction of contact stress on the drive flanks, resulting in higher load capacity. Such types of benefits of asymmetric gears are well known. One of the moto of asymmetric gear teeth design is to improve performance of the primary drive profiles than the performance off the opposite coast profiles. The optimum selection of the pressure angles for the driving and coast side of the teeth is important for the design of gear drives of asymmetric geometry. Benefits of asymmetric tooth side surfaces enables to increase the load carrying capacity and long life to the drive tooth side. So, the geometry and design of asymmetric spur gears gives an important problem. There are few articles of involute gears with asymmetric (buttress) teeth. They consider the small pressure angle profile (as a rule 20°) for the drive side and high pressure angle profile for the coast side teeth. Such an approach enables to decrease the bending stresses and keeps contact stresses on the same level as for symmetric teeth with equal pressure

L. B. Abhang
Pravara Rural Engineering College, Loni, Ahmednagar, Maharashtra, India
e-mail: lbabhang@gmail.com

A. G. Dekhane (✉)
Vishwabharati Academy's College of Engineering, Ahmednagar, Maharashtra, India
e-mail: agdekhane@gmail.com

 
K. Antony and J. P. Davim (eds.), *Advanced Manufacturing and Materials Science*,
Lecture Notes on Multidisciplinary Industrial Engineering,
https://doi.org/10.1007/978-3-319-76276-0_11

angle. Asymmetry of the tooth increases the contact ratio beyond the conventional gears' limits. Recently, a more interest has been observed in the involute gear drives of asymmetric tooth profiles. These gear drives can be used in automotive transmissions, wind turbines, gear pumps, lifting and other units in which the loading is unidirectional of movement. Its use tends to many advantages, the most significant of which is the increase of the carrying capacity of the gear drive in the driving direction of movement. Asymmetric gear tooth profile optimization leads to significant contact and bending stress reduction. This stress reduction gives the higher load carrying capacity, extended lifetime, reduced vibration, higher efficiency, higher reliability, reduced cost [1].

**Methodology:**
For the above work proposed, the methodology is as follows:

- Study of difference between symmetric and asymmetric spur gears
- Profile of the asymmetric spur gear
- Calculation of design parameters of the involute spur gear with asymmetric teeth
- Modeling of the asymmetric gear in the 3D software (CATIA)
- Finite Element Analysis of the gear for bending stress analysis
- Manufacturing of the asymmetric spur gear by 3D printing method
- Experimental analysis of the gear model
- Comparison between the FEA and experimental analysis
- Result & Discussion

## 2 Design & Modeling

By calculating different parameters,
Number of teeth = Zp = Zg = 32
$\alpha c = 20°$
$\alpha d = 25°, 30°, 35°$
Material Assumed: Polylactic Acid (PLA)
Thickness of gear = 9 mm
Ft = 189.342 N

### *2.1 Modelling of Gear in CATIA V5 R20*

The design of the gear is done by using the mathematical formulae referred from the gear design book. The materials used for the gear is poly lactic acid (PLA). Therefore, for the analysis purpose, the model has to do on the modeling software and analysis will be carried out by FEA software.

For the modelling as well as analysis CAD/CAE software's used:

1. CATIA V5 R20
2. ANSYS Workbench 16.0

## 2.2 Process of Drawing

(1) Open CATIA V5.
(2) Go to Part Design.
(3) Select plane and sketch tool.
(4) Draw dedendum circle of diameter 83.35 mm, base circle of diameter 89.35 mm and addendum circle of diameter 95.36 mm.
(5) Draw axis line at angle 4.50 and second axis line at an angle 250 as shown in figure.
(6) Draw a circle from dedendum circle coinciding the axis line.
(7) By using quick trim command, trim the data.
(8) Similarly, follow the step (7) with pressure angle 20° and again trim the unwanted circle.
(9) Draw the lines & Exit workbench.
(10) Using "pad" command, add thickness 9 mm as per design (Fig. 1).

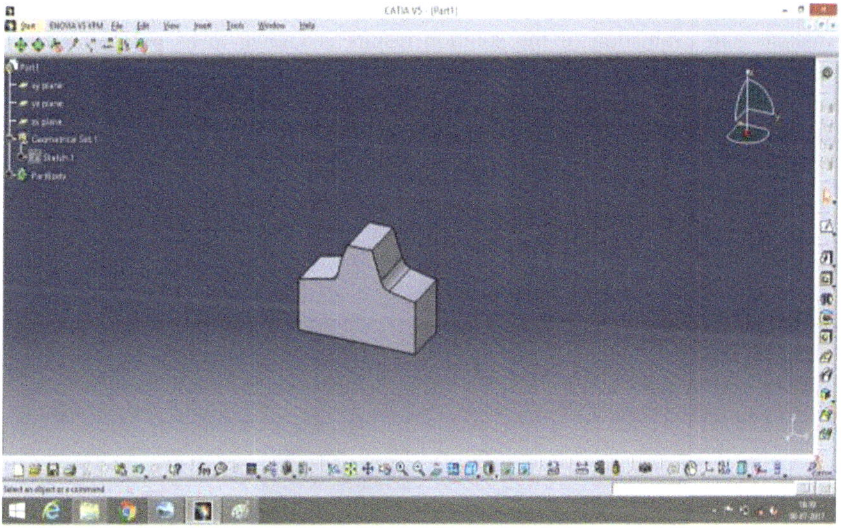

**Fig. 1** CATIA drawing

## 3 Fea of Asymmetric Spur Gear

ANSYS has developed product lines that allow you to make the most of your investment and choose which product works best in your environment. ANSYS is a Finite Element Analysis (FEA) code mostly used in the Computer-Aided Engineering (CAE) area.

### 3.1 Pre-processing

The pre-processing step is described as defining the model and includes

- Define the geometric domain of the problem.
- Define the element type(s) to be used.
- Define the material properties of the model.
- Define the geometric properties of the elements (length, area).
- Define the element connectivity's (mesh the model).
- Define the physical constraints (boundary conditions).
- Define the loadings (Figs. 2 and 3).

### 3.2 Post Processing

Analysis and evaluation of the solution results is known to as post processing. Postprocessor software is having sophisticated routines used for sorting, printing, and plotting selected results from a finite element solution [11]. Examples of operations that can be accomplished include:

- Sort element stresses in order of magnitude.
- Animate dynamic model behaviour.
- Produce colour-coded temperature plots (Figs. 4, 5 and 6).

**Fig. 2** Imported model

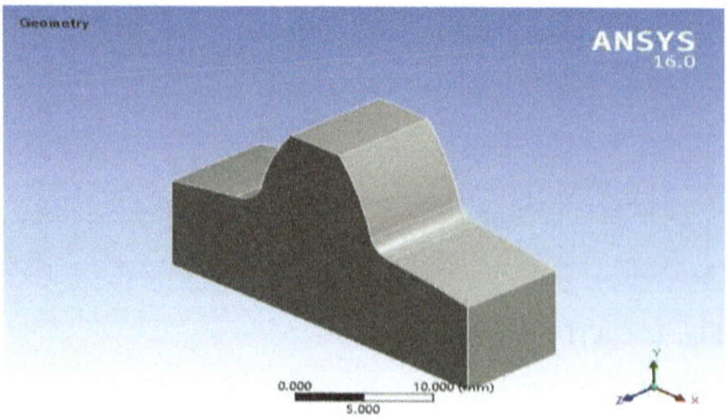

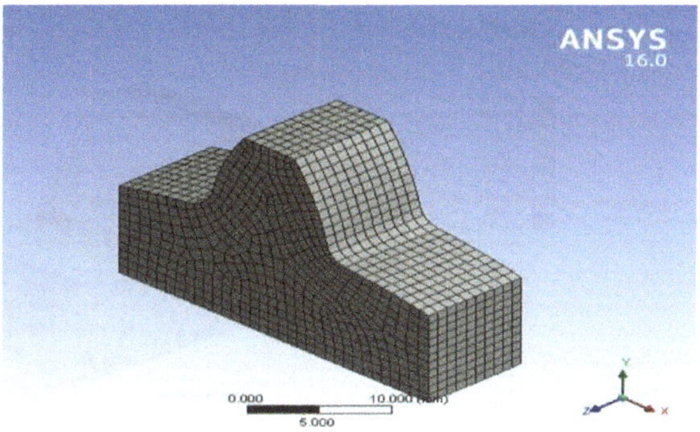


**Fig. 3** Tooth meshing

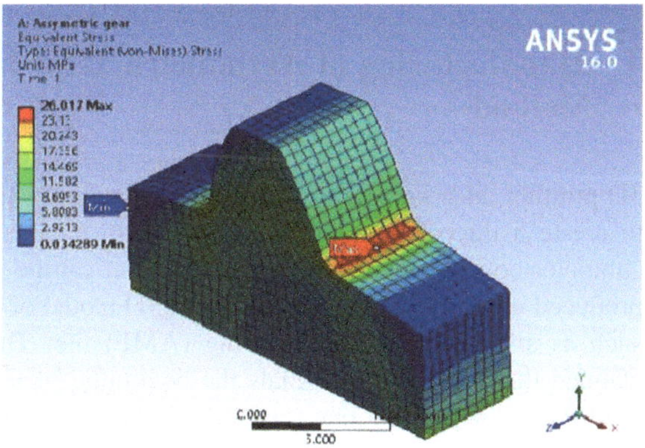


**Fig. 4** Equivalence stress for angle 20°–25°

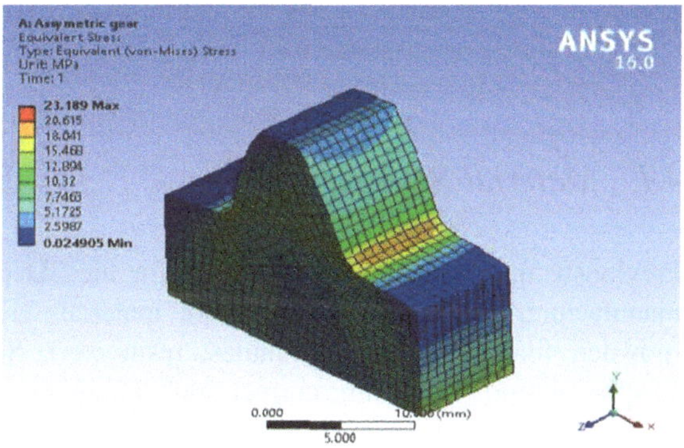


**Fig. 5** Equivalence stress for angle 20°–30°

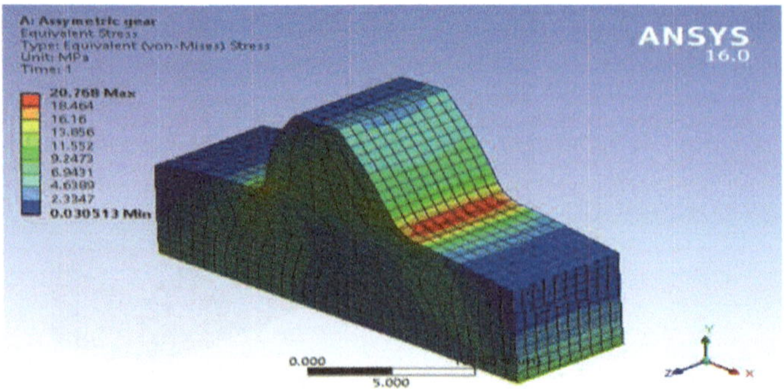


**Fig. 6** Equivalence stress for angle 20°–35°

# 4 Manufacturing of Asymmetric Gear by 3D Printing Method

3D printing, also known as additive manufacturing (AM), refers to processes used to create a three dimensional object in which layers of material are built under computer control to create an object. Object can be of any shape or geometry are produced using digital model data from 3D model or another electronic data source such as an Additive manufacturing (AMF) file. The futurologist Jeremy Rifkin claimed that 3D printing signals the beginning of third industrial revolution, succeeding the production line assembly that dominated manufacturing starting in the late 19th century. The term 3D printing originally referred to a process that deposits a binder material on to a powder bed with inkjet printer heads layer by layer. Now a days, this term is used in popular and wider variety of additive manufacturing processes.

## *4.1 Material Selection*

Previously the selection material used for the 3D printing is major task for the manufacturer. Now a wide variety of materials are supplied in different states (powder, filament, pellets, granules, resin etc.). Specific materials is used for specific platforms resulting correct applications (an example would be the dental field) with material properties that more precisely suitable for the application. However, there are now so many proprietary materials from the many different 3D printer vendors to cover them all here. Instead, this paper gives the most popular type of material in a more generic way. And also a couple of materials which stand out (Tables 1 and 2).

Nylon, or Polyamide, is mostly used materials in powder form with the sintering process or in filament form with the FDM process. It is having strong enough,

**Table 1** Properties of PLA material

| | |
|---|---|
| Density | 1.34 g/cm$^3$ |
| Yield load | 0.57 kN |
| Ultimate load | 0.82 kN |
| Yield strength | 46.22 N/mm$^2$ |
| U.T.S. | 66.49 N/mm$^2$ |
| % Elongation | 14.57 |
| Shore hardness | 40D, 41D, 45D |

**Table 2** Data of material selection

| Material | Density (g/cm$^{3)}$ | Yield strength (N/mm$^{2)}$ | U.T.S. | % Elongation | Hardness |
|---|---|---|---|---|---|
| PLA | 1.34 | 46.22 | 66.4 | 14.57 | 40D |
| ABS | 1.04 | 61 | 42.5 | 23 | 103 |
| Nylon | 1.14 | 85 | 76 | 2 | 88 |
| Epoxy resin | 1.56 | – | 85 | 0.8 | 78.3D |
| Acrylic | 1.19 | – | 74 | 3 | 100 |

flexibility and durable plastic material which has proved applicable for 3D printing. Naturally, it is in white colour but it can be coloured in pre-printing or post-printing. This material can also be mixed (in powder format) with powdered aluminium to produce other common 3D printing material for sintering process—Alumide.

ABS is one of the most commonly used plastic for 3D printing and other processes, and is widely used on the entry-level FDM 3D printers in filament form. It is also having strong enough and available in a wide range of colours. We can have ABS in filament form from a number of non-proprietary sources, that is the another reason why it is too much popular.

PLA is one more option, a bio-degradable plastic material that has gained traction with 3D printing. It may be used in resin format for DLP/SL processes also we can use it in filament form for the FDM. It comes in a variety of colours, including transparent, which has proven to be a very useful option for some applications of 3D printing.

A specially developed LayWood used for entry-level extrusion 3D printers. It may come in filament form and is a wood/polymer composite.

We can use metals, ceramics, paper, biomaterials and other materials for 3D printing process. But the very important part for the 3D printing is selection of right material for your print job. For the same consideration, there are numerous factors to consider, including strength, flexibility, accuracy, and special conditions the material may require in order to print correctly. Mostly 3D printer manufacturers are using ABS and PLA, but the differences between these two aren't immediately apparent. And the question is, which one is best optimum for your 3D printing needs? First, let's know about what exactly the two materials are. These two are comes in category of thermoplastics. Thermoplastics become malleable when superheated, thus it is allowed to you to mould and sculpt them into different size and shapes before cooling. However, you can repeat the process without disturbing the integrity of the material. While both are used for making objects in 3D printing via similar processes, ABS and PLA differ in some means, and therefore some printers will only catch ABS or PLA—or both, depending on the machine at hand. By observing the different properties about the details of ABS and PLA, one can say that there are some similarities between the two. Both requires a dry location for storing purpose, and furthermore, each is susceptible to moisture. Both ABS and PLA gives smell while printing, as heating the thermoplastic resulting off fumes. ABS will smell like hot plastic and may require ventilation while PLA may be smell mildly sweet, but the strength of these smells is going to be dependent on your printer. However, there are also major differences between these two thermoplastics. ABS is going to give your components better structural integrity and will be more suited to mechanical use given the material can better withstand the elements, but it will also require specific types of printers and printing surfaces. On the other side, PLA will give you more precise prints and better aesthetic quality, also more flexibility with printing conditions if you can do without the strength and resilience of ABS.

So, I have selected Polylactic Acid (PLA) for manufacturing of asymmetric gear with the help of 3D printing manufacturing process (Figs. 7 and 8).

**Fig. 7** 3D manufactured asymmetric gear

**Fig. 8** 3D manufactured asymmetric gear meshing

## 5 Experimental Stress Analysis

In this study, for the testing of bending stress in the gear tooth "Computerized Universal Testing Machine (UTM)" is used. I have carried out experimentation of this project in 'Ahmednagar Auto and engineering association' in Testing Lab. The capacity of this UTM is 100 KN (Figs. 9 and 10).

The main component of this machine are hydraulic grip, hand control, digital display, ball screw and dustcover, coloumn, nutund adjustable gap nut, compression plate, sensor, oil cylinder, base etc.

For loading and clamping purpose, the gear component should be mounted in such a way that the tooth of a gear must come along the downward direction of tool so that line loading on gear tooth occurred accurately. For this purpose it is necessary to cut the gear so that cut section should clamp properly in the UTM. The following figure shows the cutting of gear according to requirement.

It becomes easy to mount the cut section of gear on the UTM. A clamping device is also required to clamp properly and the component should not move from its position even if there is high loading. In Fig. 11, the gear component is simply

**Fig. 9** Computerized universal testing machine

**Fig. 10** Gear cutting

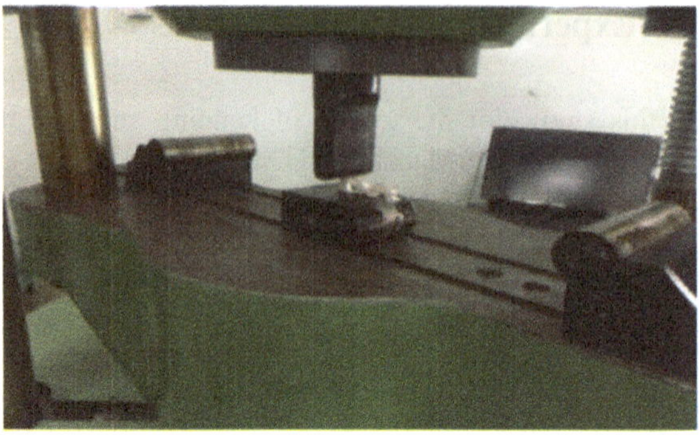

**Fig. 11** Loading on gear tooth

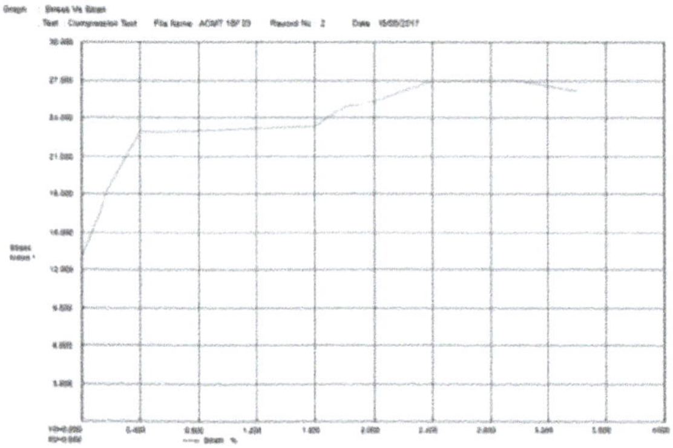

**Graph 1** Stress versus strain

**Table 3** Comparison of results obtained by ANSYS software and experimentation

| S. No. | Pressure angles | Stress obtained by ANSYS N/mm$^2$ | Stress obtained by experimentation in N/mm$^2$ |
|---|---|---|---|
| 1 | 20°–25° | 26.01 | 27.00 |
| 2 | 20°–30° | 23.18 | 24.70 |
| 3 | 20°–35° | 20.76 | 22.30 |

mounted on the plate and in Fig. 11, the component is clamped by using clamping device and the loading is started by using loading tool (Graph 1).

The 3300 N load is applied on the component as there is line loading preferred for the spur gear testing. The readings are noted for different values of loads. Different values of stress are obtained for different values of loads. Digital display shows the values of stress at particular loading.

The computerized universal testing machine gives the graphs of load versus displacement, stress versus displacement and stress versus strain.

The combine results obtained is tabulated and a report made by the testing engineer in the autocluster, ahmednagar is attached below. Mr. Prasad helped me for clamping and adjusting the machine component and getting the results. Firstly I have calculated the values of stress obtained by using ANSYS software and then values of stress shown by computerized universal testing machine. These values are tabulated in the following table for comparison between software results and experimentation (Table 3).

## 6 Conclusion

From this research and by observing results obtained by using ANSYS software and experimental results by using computerized universal testing machine, I have concluded that as the pressure angle on drive side increases, the bending stress decreases and bending load capacity increases.

## References


1. Kapelevich A (1998) Geometry and design of involute spur gears with asymmetric teeth. Received Apr 1998; received in revised form 15 Oct 1998; accepted 23 Oct 1998
2. Alipiev O (2011) Geometric design of involute spur gear drives with symmetric and asymmetric teeth using the realized potential method. Department of Theory of Mechanisms and Machines vol 46. University of Ruse, Bulgaria, p 32
3. Kapelevich AL, Shekhtman YV (2003) Direct gear design: bending stress minimization. September/October 2003 Gear Technology
4. Ingole GB. Analysis of stress relieving features of asymmetric spur gear. Novateur Publication' Int J Innov Eng Res Technol [IJIERT]. In: ICITDCEME'15 Conference Proceedings. ISSN: 2394-3696
5. Kapelevich AL, Shekhtman YV. Rating of asymmetric tooth gears. American Gear Manufacturers Association, Alexandria
6. Dhaduti SC, Sarganachari SG. Review of composite asymmetric spur gear. Int J Eng Res 4 (2):73–75. ISSN: 2319-6890 (online), 2347-5013 (print)
7. Prajapati JM, Vaghela PA (2013) Comparison of result of analytical and modeling software for critical section thickness, tip thickness of asymmetric spur involute gear tooth. Int J Eng Res Technol (IJERT) 2(11). ISSN: 2278-0181
8. Ekwaro-Osire S, Durukan I, Alemayehu FM (2009) An inverse problem technique for spur gears with asymmetric teeth. In: Proceedings of the SEM annual conference 1–4 June 2009. Society for Experimental Mechanics Inc, USA
9. Karagi SR, Patil PS (2014) Effect of pressure angle and back up ratio on asymmetric spur gear tooth bending stress, vol 3, no 4. ISSN: 2278-0149
10. Marimuthu P, Muthuveerappan G (2014) Effect of addendum height and teeth number on asymmetric normal contact ratio spur gear based on load sharing. J Mech Eng 2(4):132–136. https://doi.org/10.13189/ujme.2014.020402
11. Agrawal S (2012) Evaluation of bending stress at fillet region of an asymmetric gear with a hole as stress relieving feature using a FEA software ANSYS. Int J Comput Appl 51(8) (0975–8887)

# Tribological Study of AA2024-$B_4C$ Composites Using ANOVA

**Bhargavi Rebba, N. Ramanaiah and S. Rajesh**

## 1 Introduction

Because of their excellent combination of low density and high thermal conductivity, Aluminium alloys are widely used in automobile, aerospace and mineral processing components [1, 2]. Nevertheless, they suffer from poor tribological properties. To prevail over this, hard reinforcement phases, such as particulates, fiber or whiskers are well known for their high-specific strength, have been uniformly distributed [3]. Composites have bestowed revolutionary materials around the world for structural, mechanical, automobile as well as other critical applications where the property value needs to be tailored [4]. Particle reinforced aluminium metal matrix composites are a class of composite materials which are having desirable properties including low density, high specific stiffness, high specific strength, controlled co-efficient of thermal expansion, increased fatigue resistance and superior dimensional stability at high temperatures etc. [5, 6]. These materials have emerged as the important class of advanced materials giving engineers the opportunity to use the material properties according to their requirement. Radically these materials differ from the conventional engineering materials from the perspective of homogeneity. Particulate reinforced metal matrix composites have drive into a new path to produce high strength and high wear-resistant

B. Rebba (✉)
Department of Mechanical Engineering, Usha Rama
College of Engineering, Telaprolu, Andhra Pradesh, India
e-mail: rebba.bhar2@gmail.com

N. Ramanaiah
Department of Mechanical Engineering, Andhra University
College of Engineering, Visakhapatnam, Andhra Pradesh, India

S. Rajesh
Department of Mechanical Engineering, SRKR College
of Engineering, Bhimavaram, Andhra Pradesh, India

 
K. Antony and J. P. Davim (eds.), *Advanced Manufacturing and Materials Science*,
Lecture Notes on Multidisciplinary Industrial Engineering,
https://doi.org/10.1007/978-3-319-76276-0_12

materials by introducing hard ceramic particles and solid lubricant in the metal matrix [7]. Introduction of ceramic particles such as SiC, $Al_2O_3$, TiC, $B_4C$, and $ZrO_2$ to matrix metal enhances hardness and thermal shock resistance [8].

## 2 Experimental Details

### *2.1 Materials*

AA2024 used for the matrix is procured from perfect metal works, Bangalore, Karnataka, India, particles of $B_4C$ (particle size approximately 40 μm) were purchased from Sigma Aldrich, Bangalore, Karnataka, India and The chemical composition of the matrix alloy is given in the Table 1.

### *2.2 Sample Preparation*

AA2024-$B_4C$ composite specimens were fabricated through liquid metallurgy method to ensure uniform distribution of the reinforcements. Five various compositions were cast by varying the wt% of $B_4C$ particles by increasing the wt% by 1. Calculated amount of base metal was preheated at 450 °C for 3 h before melting. The preheated matrix alloy was first heated above the liquidus temperature to melt completely. It was then slightly cooled below the liquidus temperature to maintain the slurry in the semi solid state. Magnesium ribbons were added to the molten metal to improve the wettability of the particles reinforced in the metal matrix. $B_4C$ particles which were preheated at 500 °C were then added to the molten metal slowly and stirring was performed at a constant speed of 300 rpm for 10 min. The molten metal was then poured into the preheated cast iron moulds to minimize the casting defects. SEM and XRD analyses were performed on the composite samples fabricated. The microstructures, SEM images and EDS are presented in [9].

### *2.3 Design of Experiment*

Analyzing wear procedure in composites is a complicated task which involves a number of operating variables and consequently, understanding how the wear characteristics of the composites are affected by different operating variables is very

**Table 1** Composition of matrix metal

| Con | Cu | Mg | Si | Fe | Mn | Zn | Ti | Cr |
|---|---|---|---|---|---|---|---|---|
| % | 4.29 | 1.29 | 0.07 | 0.20 | 0.54 | 0.03 | 0.06 | 0.01 |

**Table 2** Process parameters and their levels

| Parameter | Levels | | | | |
|---|---|---|---|---|---|
| | 1 | 2 | 3 | 4 | 5 |
| Applied load | 10 | 20 | 30 | 40 | 50 |
| % Reinforcement | 0 | 1 | 2 | 3 | 4 |
| Sliding distance | 500 | 1000 | 1500 | 2000 | 2500 |
| Sliding velocity | 1.5 | 3 | 4.5 | 6 | 7.5 |

much essential. Design of experiment (DOE) is a very effective and useful statistical tool to model and analyze the influence of process parameters on output performance. Prudent selection of the control factors is the most important stage while using the DOE. By using Taguchi's L25 orthogonal array, in the present work, four parameters (process variables), i.e., applied load, % reinforcement, sliding distance and sliding velocity at five levels are considered as listed in the Table 2. The study of influence of these parameters on the wear behaviour of the AMC is done using this L25 orthogonal array experimental procedure. The objective of experiment is to minimize wear rate and coefficient of friction. The Signal to Noise (S/N) ratio, which condenses the multiple data points within a trial, depends on the type of characteristic being evaluated. The S/N ratio characteristics can be classified into three varieties, viz. "nominal is the best", "larger the better" and "smaller the better" characteristics. In this study, "smaller the better" characteristic was opted to analyze the dry sliding wear resistance. The S/N ratio for wear rate and coefficient of friction using "smaller the better" characteristic is as follows, given by Taguchi:

$$\mathrm{S/N} = -10\log_{\mathrm{n}}^{1}\sum_{i=1}^{n}\mathrm{Y}_{\mathrm{i}}^{2}$$

where $y_1$, $y_2$ … $y_n$ are the response of coefficient of friction and wear and n is the number of observations. The response table for signal to noise ratios reflects the average of selected characteristics for each level of the factor. This table presents the ranks based on the delta statistics, which compares the relative value of the effects. S/N ratio is a response which combines repetitions and the effect of noise levels into one data point. Analysis of variance of the S/N ratio is carried out to identify the statistically significant parameter.

## 2.4 *Wear Test Procedure*

The wear test specimens are shown in Fig. 1. All the tests were proposed to be done at room temperature. The initial weight of the specimen was measured in a single pan electronic weighing machine with a least count of 0.0001 g. The experiments

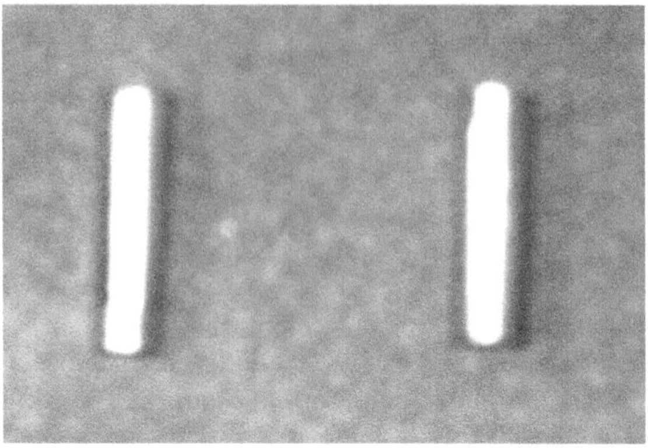

**Fig. 1** Wear test sample

were conducted as per the standard orthogonal array L25. The selection of the orthogonal array is dependent on the condition that the DOF for the orthogonal array should be greater than or at least equal to sum of those of wear parameters. In the present investigation an L25 orthogonal array was chosen and the experimental plan is shown in Table 3. The pin was held against the counterface of a rotating EN 31 hardened to 60 HRc grounded to 1.6 Ra surface roughness steel disc with a wear track diameter 60 mm. The first column was assigned to the applied load (AL), second column was assigned to the % reinforcement (% R), Third column was assigned to sliding distance (SD), and fourth column was assigned to sliding velocity (SV). The output to be studied is wear rate and coefficients of friction of the test samples are repeated three times corresponding to 25 tests. The weight of the specimen after performing the test was also measured to calculate the wear of the AMCs.

## 2.5 *Analysis of Results*

The focus of the experimental plan is to find the important factors and the combination of factors influencing the wear procedure to achieve the minimum wear and coefficient of friction. The experiments were developed based on an orthogonal array, with a target of relating the influence of sliding speed, applied load and sliding distance. These design parameters are distinct and intrinsic feature of the process that influence and determine the composite performance. The results of the experiments performed are given in Table 4. The experimental results are analyzed with the help of MINI-TAB15, statistical analysis software which is being extensively used in many fields of engineering research. Taguchi recommends analyzing the S/N ratio using conceptual approach that involves plotting the effects and identifying visually the significant factors.

**Table 3** Experimental plan

| S. No. | AL | % R | SD | SV |
|---|---|---|---|---|
| 1 | 1 | 1 | 1 | 1 |
| 2 | 2 | 1 | 2 | 2 |
| 3 | 3 | 1 | 3 | 3 |
| 4 | 4 | 1 | 4 | 4 |
| 5 | 5 | 1 | 5 | 5 |
| 6 | 2 | 2 | 3 | 1 |
| 7 | 3 | 2 | 4 | 2 |
| 8 | 4 | 2 | 5 | 3 |
| 9 | 5 | 2 | 1 | 4 |
| 10 | 1 | 2 | 2 | 5 |
| 11 | 3 | 3 | 5 | 1 |
| 12 | 4 | 3 | 1 | 2 |
| 13 | 5 | 3 | 2 | 3 |
| 14 | 1 | 3 | 3 | 4 |
| 15 | 2 | 3 | 4 | 5 |
| 16 | 4 | 4 | 2 | 1 |
| 17 | 5 | 4 | 3 | 2 |
| 18 | 1 | 4 | 4 | 3 |
| 19 | 2 | 4 | 5 | 4 |
| 20 | 3 | 4 | 1 | 5 |
| 21 | 5 | 5 | 4 | 1 |
| 22 | 1 | 5 | 5 | 2 |
| 23 | 2 | 5 | 1 | 3 |
| 24 | 3 | 5 | 2 | 4 |
| 25 | 4 | 5 | 3 | 5 |

## 2.6 *Results of Statistical Analysis of Experiments*

The obtained results were analyzed using the software MINITAB 15 specifically used for DOE applications. Table 4 shows the experimental results for wear rate and coefficient of friction. To analyze the quality characteristics, the experimental values are transformed into signal to noise ratio shown in Table 5. The influence of control parameters such as applied load, reinforcement percentage, sliding distance, and sliding speed on wear rate and coefficient of friction has been analyzed using signal to noise response table. The ranking of process parameters using signal to noise ratios obtained for different parameter levels for wear rate and coefficient of friction are given in Tables 6 and 7 respectively. The control factors are statistically significant in the signal to noise ratio and it could be observed that the sliding distance is a dominant parameter on the wear rate and coefficient of friction

**Table 4** Experimental results

| S. No. | AL | % R | SD | SV | WL | COF |
|---|---|---|---|---|---|---|
| 1 | 10 | 0 | 500 | 1.5 | 0.0056 | 0.579 |
| 2 | 20 | 0 | 1000 | 3 | 0.0071 | 0.598 |
| 3 | 30 | 0 | 1500 | 4.5 | 0.0119 | 0.657 |
| 4 | 40 | 0 | 2000 | 6 | 0.0137 | 0.672 |
| 5 | 50 | 0 | 2500 | 7.5 | 0.0164 | 0.753 |
| 6 | 20 | 1 | 1500 | 1.5 | 0.0112 | 0.651 |
| 7 | 30 | 1 | 2000 | 3 | 0.0153 | 0.74 |
| 8 | 40 | 1 | 2500 | 4.5 | 0.0186 | 0.782 |
| 9 | 50 | 1 | 500 | 6 | 0.0058 | 0.579 |
| 10 | 10 | 1 | 1000 | 7.5 | 0.0043 | 0.565 |
| 11 | 30 | 2 | 2500 | 1.5 | 0.0192 | 0.797 |
| 12 | 40 | 2 | 500 | 3 | 0.0067 | 0.592 |
| 13 | 50 | 2 | 1000 | 4.5 | 0.0101 | 0.643 |
| 14 | 10 | 2 | 1500 | 6 | 0.0061 | 0.585 |
| 15 | 20 | 2 | 2000 | 7.5 | 0.0087 | 0.625 |
| 16 | 40 | 3 | 1000 | 1.5 | 0.0096 | 0.639 |
| 17 | 50 | 3 | 1500 | 3 | 0.0129 | 0.663 |
| 18 | 10 | 3 | 2000 | 4.5 | 0.0108 | 0.648 |
| 19 | 20 | 3 | 2500 | 6 | 0.0125 | 0.661 |
| 20 | 30 | 3 | 500 | 7.5 | 0.0051 | 0.575 |
| 21 | 50 | 4 | 2000 | 1.5 | 0.0179 | 0.765 |
| 22 | 10 | 4 | 2500 | 3 | 0.0148 | 0.687 |
| 23 | 20 | 4 | 500 | 4.5 | 0.0049 | 0.57 |
| 24 | 30 | 4 | 1000 | 6 | 0.0053 | 0.579 |
| 25 | 40 | 4 | 1500 | 7.5 | 0.0075 | 0.604 |

followed by applied load, sliding speed and reinforcement percentage. Figures 2, 3, 4 and 5 shows the influence of process parameters on wear rate and coefficient of friction graphically.

## 2.7 *Analysis of Variance (ANOVA)*

Results are analyzed by analysis of variance (ANOVA) with a confidence limit of 95%. By performing ANOVA, it can be decided which independent factor dominates over the other and the percentage contribution of that particular independent variable. The factors AL (applied load), SD (sliding Distance) and SV (sliding velocity) are significant as their p-value is less than 0.005. The percentage contribution of the process parameters on the wear loss and coefficient of friction of the

**Table 5** Signal to noise ratios of wear loss & COF

| S. No. | WL | COF | S/N for WL | S/N for COF |
|---|---|---|---|---|
| 1 | 0.0056 | 0.579 | 45.0362 | 4.74,643 |
| 2 | 0.0071 | 0.598 | 42.9748 | 4.46598 |
| 3 | 0.0119 | 0.657 | 38.4891 | 3.64869 |
| 4 | 0.0137 | 0.672 | 37.2656 | 3.45261 |
| 5 | 0.0164 | 0.753 | 35.7031 | 2.4641 |
| 6 | 0.0112 | 0.651 | 39.0156 | 3.72838 |
| 7 | 0.0153 | 0.74 | 36.3062 | 2.61537 |
| 8 | 0.0186 | 0.782 | 34.6097 | 2.13586 |
| 9 | 0.0058 | 0.579 | 44.7314 | 4.74643 |
| 10 | 0.0043 | 0.565 | 47.3306 | 4.95903 |
| 11 | 0.0192 | 0.797 | 34.334 | 1.97083 |
| 12 | 0.0067 | 0.592 | 43.4785 | 4.55357 |
| 13 | 0.0101 | 0.643 | 39.9136 | 3.83578 |
| 14 | 0.0061 | 0.585 | 44.2934 | 4.65688 |
| 15 | 0.0087 | 0.625 | 41.2096 | 4.0824 |
| 16 | 0.0096 | 0.639 | 40.3546 | 3.88998 |
| 17 | 0.0129 | 0.663 | 37.7882 | 3.56973 |
| 18 | 0.0108 | 0.648 | 39.3315 | 3.7685 |
| 19 | 0.0125 | 0.661 | 38.0618 | 3.59597 |
| 20 | 0.0051 | 0.575 | 45.8486 | 4.80664 |
| 21 | 0.0179 | 0.765 | 34.9429 | 2.32677 |
| 22 | 0.0148 | 0.687 | 36.5948 | 3.26087 |
| 23 | 0.0049 | 0.57 | 46.1961 | 4.8825 |
| 24 | 0.0053 | 0.579 | 45.5145 | 4.74643 |
| 25 | 0.0075 | 0.604 | 42.4988 | 4.37926 |

**Table 6** Response table for S/N ratios of wear loss

| Level | AL | RP | SD | SV |
|---|---|---|---|---|
| 1 | 42.52 | 39.89 | 45.06 | 38.74 |
| 2 | 41.49 | 40.40 | 43.22 | 39.43 |
| 3 | 40.10 | 40.65 | 40.42 | 39.71 |
| 4 | 39.64 | 40.28 | 37.81 | 41.97 |
| 5 | 38.62 | 41.15 | 35.86 | 42.52 |
| Delta | 3.90 | 1.26 | 9.20 | 3.78 |
| Rank | 2 | 4 | 1 | 3 |

AMCs is also calculated and has been presented in the Tables 8 and 9. The contribution of sliding distance is highest followed by applied load, sliding velocity, and % reinforcement. The results clearly indicate that increase in load & sliding distance levels increased the wear loss and increase in % of reinforcement & sliding velocity levels decreased the wear loss.

**Table 7** Response table for S/N ratios of COF

| Level | AL | RP | SD | SV |
|---|---|---|---|---|
| 1 | 42.52 | 39.89 | 45.06 | 38.74 |
| 2 | 41.49 | 40.40 | 43.22 | 39.43 |
| 3 | 40.10 | 40.65 | 40.42 | 39.71 |
| 4 | 39.64 | 40.28 | 37.81 | 41.97 |
| 5 | 38.62 | 41.15 | 35.86 | 42.52 |
| Delta | 3.90 | 1.26 | 9.20 | 3.78 |
| Rank | 2 | 4 | 1 | 3 |

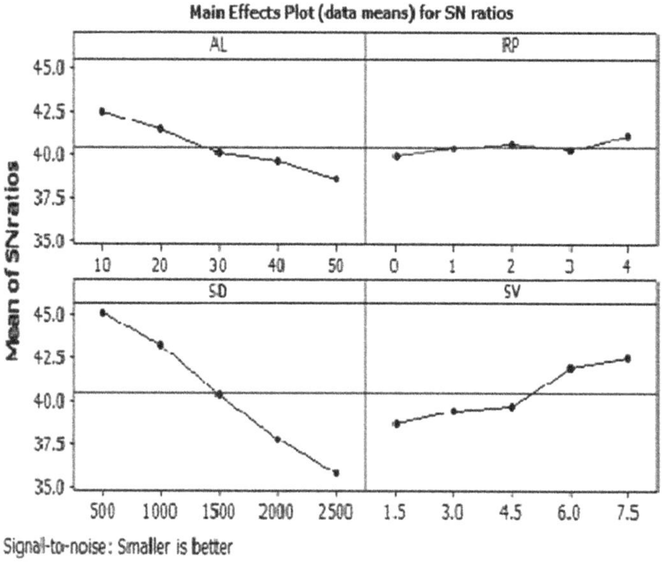


**Fig. 2** Main effects for plot for S/N ratios—wear rate

## 2.8 *Optical and SEM Images of the Wear Tracks*

Figures 6 and 7 shows the worn out surfaces of the pin at 20 N AL, 0 RP, 1000 m SD, and 3 m/s SV. Large, distinct parallel and continuous grooves are formed in the zero-reinforced aluminum pin. These grooves indicate the occurrence of extensive plastic deformation. The depth of the grooves is also higher, suggesting severe abrasion. This characteristic is more predominant in brittle fractures with continuous plow marks. Figures 8 and 9 depicts the worn out surface of the pin at 10 N AL, 4% RP, 1500 m SD, and 3 m/s SV. As the reinforcement increases to a higher value, the large distinct grooves are reduced to fine scratches along the sliding direction and are visible on the surface shown. Many researchers [10, 11] described

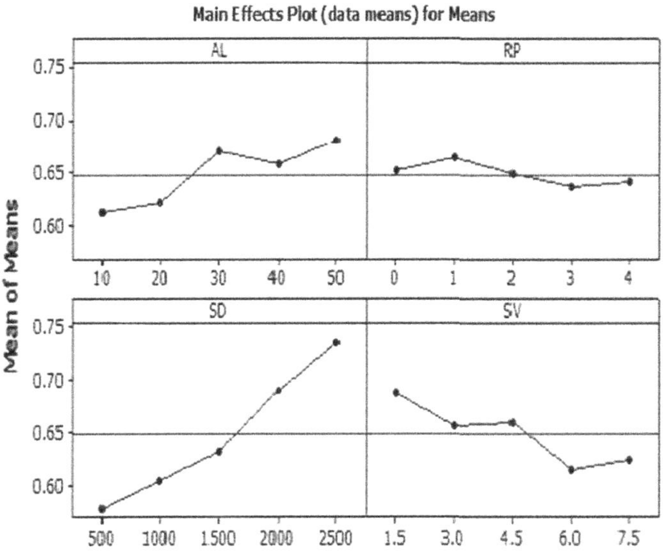


**Fig. 3** Main effects for plot for means—wear rate

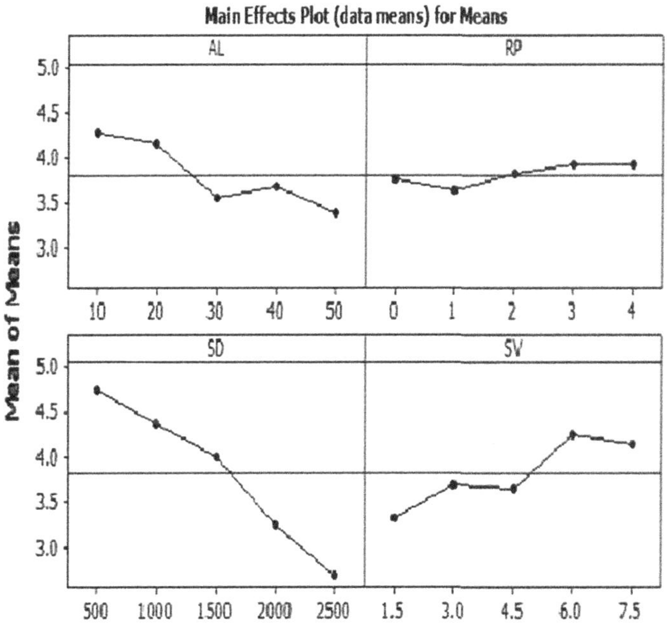


**Fig. 4** Main effects for plot for S/N ratios—COF

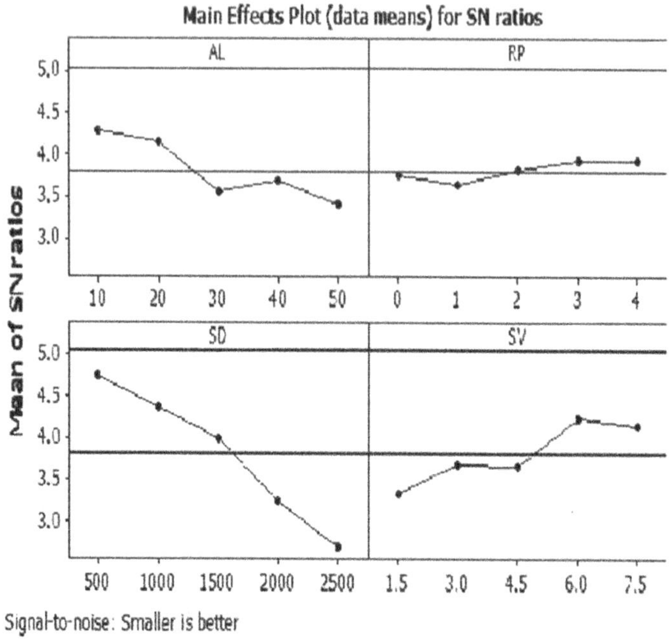


**Fig. 5** Main effects for plot for means—COF

**Table 8** ANOVA for wear loss

| Source | Seq SS | DF | Variance | Calculated F-value | % Contribution |
|---|---|---|---|---|---|
| AL | $64e^{-4}$ | 4 | $173e^{-4}$ | 21.625 | 13.2 |
| % R | $e^{-4}$ | 4 | $129e^{-4}$ | 16.125 | 9.82 |
| SD | $32e^{-4}$ | 4 | $817e^{-4}$ | 102.125 | 62.18 |
| SV | $65e^{-4}$ | 4 | $164e^{-4}$ | 20.5 | 12.48 |
| Error | $68e^{-4}$ | 8 | $8e^{-5}$ | | 2.32 |
| Total | $52e^{-3}$ | 24 | | | 100.00 |

**Table 9** ANOVA for COF

| Source | Seq SS | DF | Variance | Calculated F-value | % Contribution |
|---|---|---|---|---|---|
| AL | 0.013514 | 4 | 0.00337 | 7.67 | 11.20 |
| % R | 0.010848 | 4 | 0.00271 | 6.775 | 8.99 |
| SD | 0.079680 | 4 | 0.01992 | 45.27 | 66.03 |
| SV | 0.014861 | 4 | 0.00371 | 8.43 | 12.31 |
| Error | 0.001776 | 8 | 0.00044 | | 1.47 |
| Total | 0.120674 | 24 | | | 100.00 |

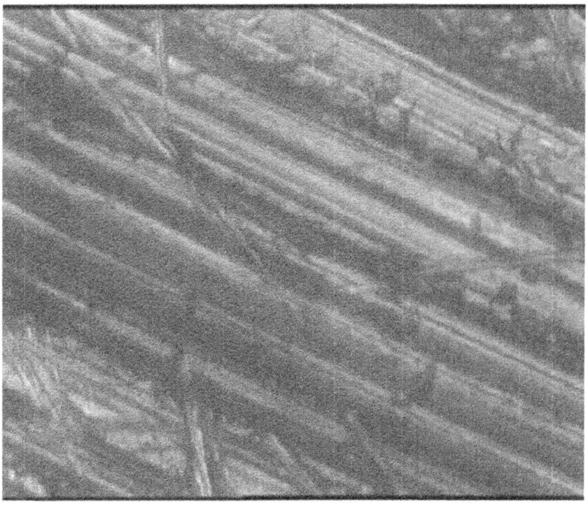

**Fig. 6** Wear track of unreinforced metal

**Fig. 7** Wear track of 4% AA2024-$B_4C$

the dominance of oxidative wear during the dry sliding wear of aluminum MMCs. The projected $B_4C$ particles in the pin dig into the surface of the counter face. The Fe content of the digged surface of the counter face may react with the aluminum of the pin and form $Fe_2O_3$. Moreover, the $B_4C$ particles crush against the very minute particles, resulting in a layer called the mechanical mixed layer (MML). The oxide layer of the lubricant phenomena may be enhanced by the increase in the RP of $B_4C$, suggesting that the worn surface is oxidized. This phenomenon results in low WVL and causes the delaminating wear mechanism.

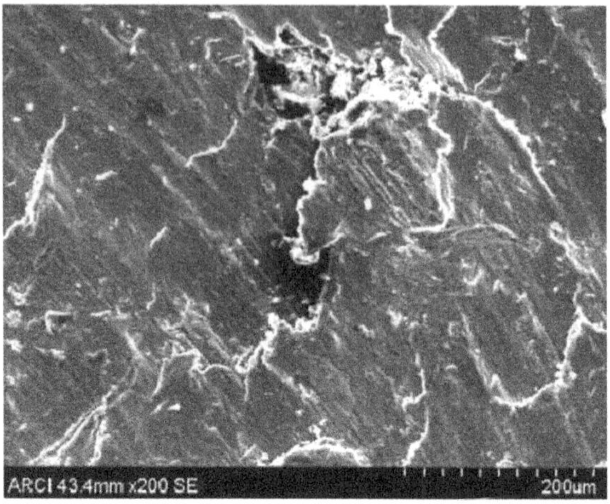


**Fig. 8** SEM image—Wear track of 0% RP

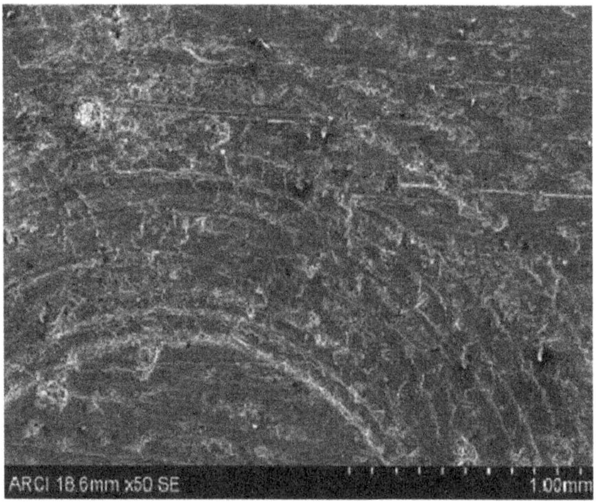


**Fig. 9** SEM image—Wear track of 4% RP

# 3 Conclusions

Dry sliding experiments were performed on aluminum MMCs with reinforcement of $B_4C$ particulates of up to 4% reinforcement with an increment of 1% using a pin-on-disc experimental setup. Following are the conclusions drawn from the investigation.

1. From both, statistical analysis using S/N ratios and ANOVA, it is concluded that the sliding distance is the highest contributing factor for wear in the AMC followed by applied load, sliding velocity and reinforcement percentage.

2. COF is also influenced predominantly by sliding distance followed by applied load, sliding velocity and reinforcement percentage.
3. The wear loss and COF increased with the increase in sliding distance and applied load.
4. The wear loss and COF decreased with the increase in sliding velocity and % reinforcement.

## References


1. Kumar S, Balasubramanian V (2008) Developing a mathematical model to evaluate wear rate of AA 7075/SiCp powder metallurgy composites. Wear 264:1026–1034
2. Brown KR, Venie MS, Woods RA (1995) The increasing use of aluminium in automotive applications. JOM 47:20–23
3. Nair SV, Tien JK, Bates RC (1985) SiC reinforced aluminium metal matrix composites. Int Mater Rev 30(6):275–290
4. Yamasaki T, Zheng YJ, Ogino Y, Terasawa M, Mitamura T, Fukami T (2003) Formation of metal–TiN/TiC nanocomposite powders by mechanical alloying and their consolidation. Mater Sci Eng A 350A:168
5. Ibrahim IA, Mohamed FA, Lavernia EJ (1991) Metal matrix composites. Rev J Mater Sci 26:1137–1157
6. Sinclair I, Gregson PJ (1997) Structural performance of discontinuous metal matrix composites. Mater Sci Technol 3:709–725
7. Zhan Y, Zhang G (2003) Graphite and SiC hybrid particles reinforced copper composite and its tribological characteristic. J Mater Sci Lett 22:1087–1089
8. Liang YH, Wang HY, Yang YF, Wang YY, Jiang QC (2008) Evolution process of the synthesis of TiC in the Cu−TiC−C system. J Alloy Compd 452:298–303
9. Rebba B, Ramanaiah N (2014) Studies on mechanical properties of Al 2024-$B_4C$ composites. Int J Adv Mater Manuf charact 4(1)
10. Basvarajappa S, Chandramohan G (2005) Wear studies on metal matrix composites: a Taguchi approach. J Mater Sci Technol 21(6):845–850
11. Ravikiran A, Surappa MK (1997) Effect of sliding speed on wear behaviour of A356 Al-30 wt.% SiCp MMC. Wear 206(1, 2):33–38

# A Review on Effects of Various Parameters in Modulated Tool Path (MTP) Machining

P L. Jayananthan and B. Aravind Kumar

## 1 Introduction

Research on continuous and entangling of chips is plentiful over the decade. Many methods were developed to avoid Nesting of chips, which are usually common with ductile materials. But every method has its own limitations. The researchers at UNC, Charlotte came with Modulated Tool Path (MTP) which utilizes the machine axes through G-code commands to avoid nesting of chips. MTP can be used with any depth of cut, any disparate material (even for pyrophoric materials like depleted uranium), with all types of cutting tools and with all part shapes without any additional hardware. User selectable chip lengths and lower cutting temperatures are the ancillary features of MTP machining.

MTP correlates sinusoidal oscillations with conventional tool path with the use of machine axes and G codes to build segmented chips. From the works of Smith et al. [1], Fig. 1 gives an exaggerated representation of MTP. Other waveforms apart from sinusoid can also be used to produce tool path oscillations. The repeated interpretation of programmed tool path oscillations tangential to the direction of work piece movement leads to the formation of a segmented chip with each oscillation. This paper aims to give a succinct review on effects of various parameters used in MTP over the erstwhile years of research in order to gain a better understanding of various selection parameters to gear up towards the implementation of MTP in machining operations.

P.L. Jayananthan (✉) · B. Aravind Kumar
Department of Mechanical Engineering, SRM University,
Kattankulathur, Chennai, Tamil Nadu, India
e-mail: palajayanan@gmail.com

 
K. Antony and J. P. Davim (eds.), *Advanced Manufacturing and Materials Science*,
Lecture Notes on Multidisciplinary Industrial Engineering,
https://doi.org/10.1007/978-3-319-76276-0_13

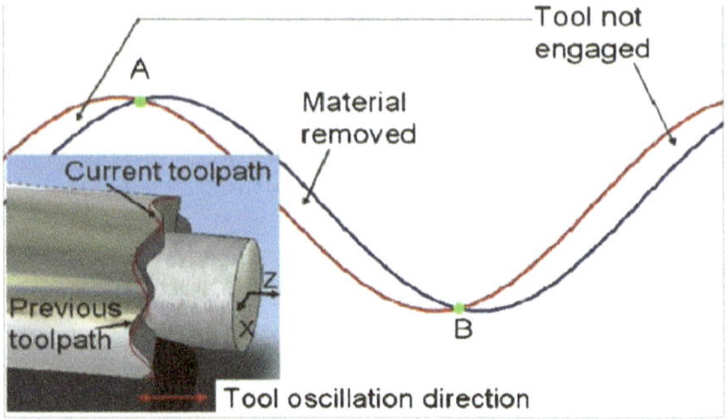


**Fig. 1** Exaggerated representation of MTP

## 2 Parameter Selection Considerations

MTP performance relies on various parameter selections—among them, *the Phase shift between Successive cuts, Oscillation amplitude relative to the global feed direction (Raf), Oscillation per spindle revolution (OPR) and user-defined chip length* are highly significant. Figure 2 gives a time domain simulation of MTP over successive passes to give a clear idea of chip breaking [1].

**Phase shift**. The oscillation of the tool should be correctly phased with the spindle revolution so that at a point of time the phase between them becomes zero and the chip breaks. Otherwise, the two subsequent passes will coincide and continuous chips will be produced. Figure 3 shows examples of chip breaking parameter in which the blue line of 4.5 Hz is correctly phased and produces broken chips, whereas the red line of 5 Hz is not correctly phased which leads to continuous chips [2].

**Amplitude ratio (Raf)**. Amplitude ratio (Raf) is defined as the oscillation amplitude relative to the global feed per revolution. It is defined as the ratio of the tool oscillation amplitude (A) and the global axis feed per revolution (f). Figure 4 [1] suggests that the minimum Raf required to produce broken chips is 0.5. The range of phase shift which creates broken chips for Raf = 1, is in the range of

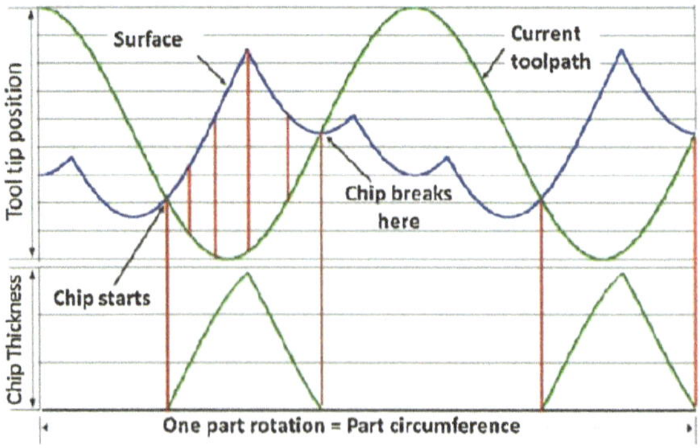


**Fig. 2** Time domain simulation over successive passes

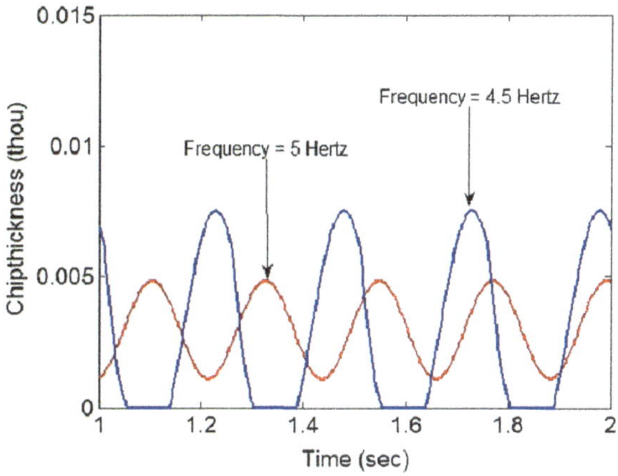


**Fig. 3** Chip thickness versus time [2]

1.047–5.236 [3]. The combination of low feed and high Raf can lead to a large range of phase shifts which lead to broken chips, but it is not useful as increasing range of phase shift demands larger machine axes which is not possible. In Fig. 5 it is observed that for higher values of Raf, there is minimum contact time between the tool and the workpiece. Because larger oscillation amplitude will leave the cut for a longer period of time [4]. As shown in Fig. 6 [2] Raf will only influence the chip thickness and it will not influence the chip length [3, 5]. If the value of Raf

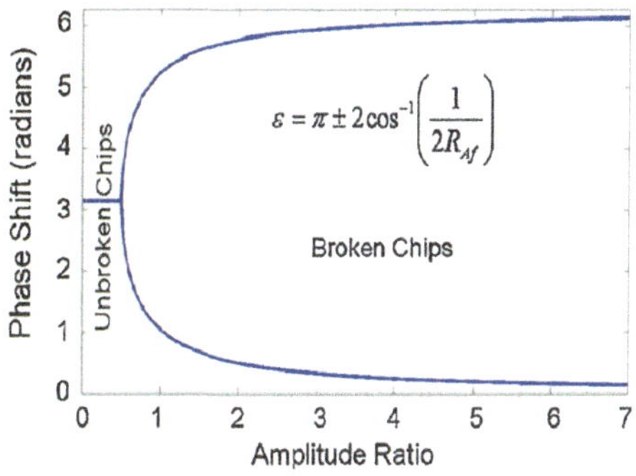


**Fig. 4** Phase shift versus amplitude ratio [2]

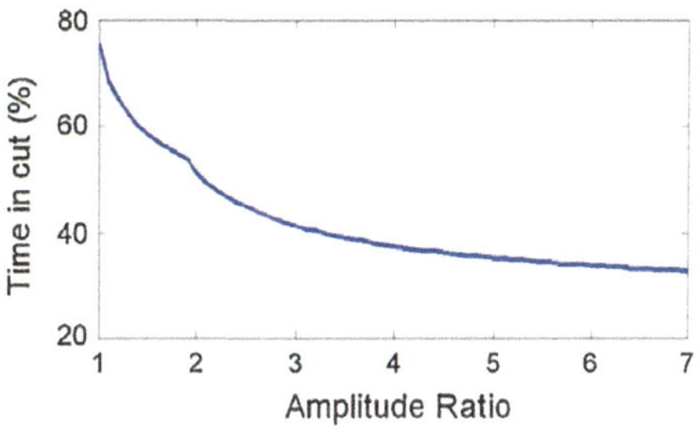


**Fig. 5** Time in cut with respect to Raf [3]

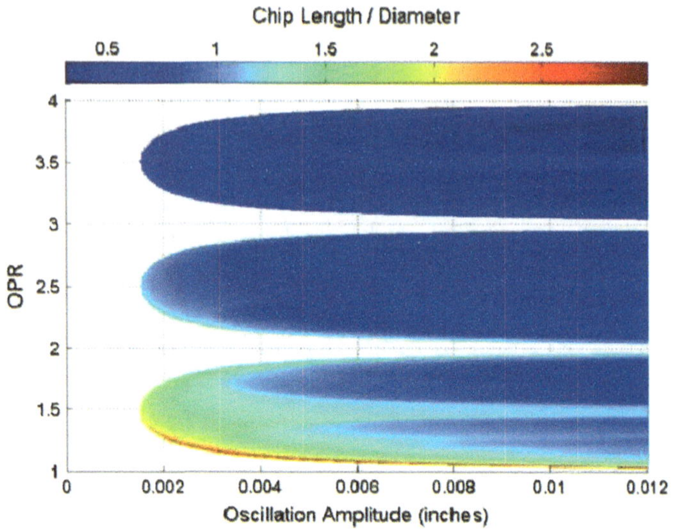


**Fig. 6** Parameter map relating OPR, Raf and chip length to diameter [2]

increases, the chip thickness will also increase, and Raf will not influence the length of the chip, but it will vary the size of the chip by a factor of two over a range of zero to four. As the value of Raf increases, the number of chips required will decrease. So the number of chips produced while MTP machining will depend on the value of Raf.

**Oscillations of the Tool Per Work Piece Revolution (OPR)**. OPR is a combination of spindle speed and oscillation frequency. It is observed that the value of OPR influences the chip length, increase in tool frequency will generally decrease the chip length [2]. From Fig. 6, it can be observed that the chip length shows a mirror trend at every integer values of OPR and at decreasing chip length the value of OPR increases. The values of OPR and Raf will change for each part diameter. For increasing part diameters we need increasing values of OPR. The user has to select the values of OPR and Raf depending on the part diameter and broken chip length. Selection of values of OPR and Raf depends on surface finish and kurtosis values [3].

**Chip length**. The chip length is defined as the circumferential distance of the material flows past the tool while the tool is engaged in cutting. The combination of Raf and phase shift will determine the chip length. The local diameter of the workpiece influence the chip length, the increase in local diameter will lead to larger chip lengths as the angular displacement of the workpiece from the time the tool enters and exit the work piece is same regardless of the local diameter [3]. For lower values of waves per revolution (wpr) it is difficult to get uncut chip length shorter than 6 in. As the frequency of oscillation increases the chip gets shorter. We can conclude that a small number of waves per revolution (wpr) will give maximum benefit. The chip length and uncut chip length are given by [3, 5]:

$$Chip\,length = \frac{\pi D * (2Raf + 1)}{OPR * Raf} \tag{1}$$

$$Uncut\,chip\,length = r\theta \tag{2}$$

where R is the radius of path, θ is the angle of rotation of the part that creates one chip, D is the diameter of the part.

## 3 Miscellaneous Effects of MTP in Machining

**Temperature**. The Temperature rise in machining depends on the length of metal cut in continuous segments rather than on the length of non cutting intervals [6, 7]. MTP machining reduces the length of metal cut in continuous segments which restricts the maximum temperature produced during machining. Smith et al. [4, 8] used Phoenix infrared high speed thermal imaging camera to measure the temperature at interrupted conditions. The infrared intensity measured by the camera was calibrated with a cutting insert and a curve fit of temperature-intensity data. This temperature-intensity data was made to find the relationship between temperature and infrared intensity using the Eq. (3).

$$T = 76.5\ln(I) - 535.12 \tag{3}$$

where temperature (T) is in Celsius and I is the infrared intensity of the camera. The tests were performed with 1018 carbon steel at a frequency of 6.8 Hz and by varying instantaneous rate. The results of the tests carried out by Smith et al. is tabulated in Table 1 [4]. From the table, it is inferred that interrupted cuts have the lower temperature when compared to continuous cuts. Conventional cuts range from 37.6 to 48.7 °C and interrupted cuts range from 33.5 to 41.0 °C.

**Tool Wear**. Temperature of the tool and the feed rate factors which influence the tool wear mostly. Smith et al. [2] conducted the tests with high speed steel inserts to find the effect of MTP machining in Tool wear. A conventional turning pass is compared with two MTP passes with Raf 2.5 and 1.23. Two digital microscopes were setup to view the wear along the tool profile and along the crater. Pictures were taken and compared at every fourth pass at a depth of cut of 0.025 per pass. Finally, it was observed that conventional cut has higher wear along its tool profile and deeper crater wear when compared with MTP machining. The insert machined with 2.5 Raf shows lesser crater wear when compared with other two. Based on the results it is observed that for increasing value of Raf there will be lesser tool wear and crater wear along the tool. Also, the amount of time the insert is engaged in the cut has a greater influence over the tool wear than the instantaneous feed rate.

**Surface Texture and Residual stress**. The surface finish created by a conventional turning operation leads to barber pole surface texture scallops along the length of the workpiece. Due to the nature of MTP, the machined surface has a

**Table 1** Thermal imaging results

| Raf | A (mm) | Time in cut (%) | Maximum instantaneous feed (mm/rev.) | Max. temp (°C) |
|---|---|---|---|---|
| 6.50 | 0.3302 | 37.2 | 3.006 | 35.8 |
| 5.50 | 0.2794 | 38.0 | 2.544 | 33.5 |
| 4.50 | 0.2286 | 39.1 | 2.082 | 41.0 |
| 3.50 | 0.1778 | 41.0 | 1.619 | 34.0 |
| 2.75 | 0.1397 | 43.3 | 1.273 | 35.6 |
| 2.00 | 0.1016 | 47.5 | 0.925 | 37.5 |
| 1.50 | 0.0762 | 53.6 | 0.694 | 35.8 |
| 1.15 | 0.0584 | 66.2 | 0.528 | 35.1 |
| Conventional | | 100.0 | 0.051 | 37.6 |
| Conventional | | 100.0 | 0.051 | 39.6 |
| Conventional | | 100.0 | 0.051 | 41.6 |
| Conventional | | 100.0 | 0.051 | 47.0 |
| Conventional | | 100.0 | 0.051 | 48.6 |

raised threadlike structure, because the surface finish is determined by the composite impact of multiple machining passes that occur in opposite direction with changing feed rate. Based on the tests conducted by Smith et al. [2] it is concluded that selection of different oscillation parameters influence the surface texture and both the conventional machining and the MTP machining results in a similar type of surface texture. The results also imply that MTP process did not influence the surface texture of the workpiece.

Schmidt et al. [9] performed tests to compare residual stress of materials machined with conventional and MTP machining using different types of materials like Aluminum 6061-T6, Tantalum 97Ta3W, Stainless steel Nitronic33 and Inconel718. The plots for normal and shear stresses for each $\phi$ rotation did not provide a clear pattern as lattice spacing is not consistent for both types of machining. Among the materials used for testing, tantalum 97Ta3W showed minimum compressive residual stress in MTP machining while compared to conventional machining. Therefore MTP process can be considered for obtaining predictable compressive residual stress for Tantalum 97Ta3W.

## 4 Limitations to MTP Machining

**Dynamic Limitation of Machining Axis**. The deviation of final amplitude and frequency from the initially assigned amplitude and frequency values over the machining time is known as dynamic limitation of machining axis. Each Computer Numerical Control (CNC) machine has its own frequency range and oscillation amplitude over which it can reliably execute the MTP programmed commands.

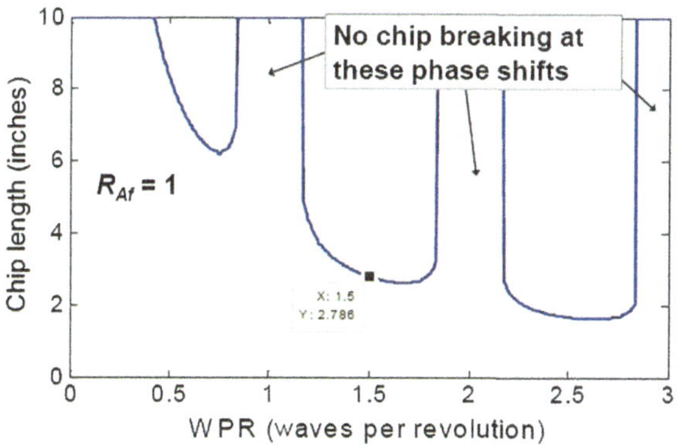


**Fig. 7** Chip length versus WPR for Raf = 1 [1]

Dynamic limitation will influence the surface quality and profile accuracy of the final material. Cross axis coupling and axes synchronization are two major tests performed to determine the dynamic limitation of the machine. In cross axis coupling one axis is oscillated and the deviations are measured from the desired null response on another axis. In axes synchronization oscillation motion commands were given on both axes and their response is plotted against each other which leads to an X-Z plot of 45°. The axes synchronization error is calculated as the deviation of the X-Z plot from 45° line. Figure 7 shows the axis synchronization errors detected with sensor nest test. The work of Berglind and Ziegert [10] clearly explains that operations outside the desired range of amplitude and frequency of a specific CNC machine can be corrected by introducing required motion delays in the machine response (Fig. 8).

**Chip Compression Ratio**. Chip compression ratio is the ratio of uncut chip thickness to cut chip thickness. It is known that cut chip thickness ratio is always greater than uncut chip thickness. As the volume of the material remains constant, the thickness of the chip increases and the length of the chip decreases. The chip

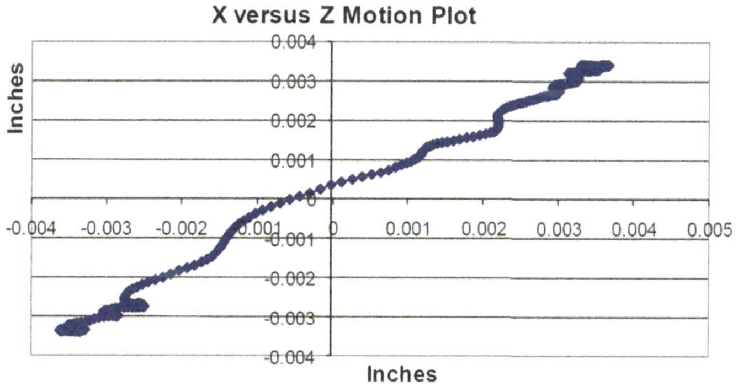


**Fig. 8** Axes synchronization errors detected with sensor-nest test

compression ratio is calculated by measuring the length of the chip and comparing it with the predicted values. The average compression ratio from the results was given as 3.71 [1].

## 5 Conclusion

Entangling chips and high machining temperature are the major problems during machining, MTP will be considered as an effective alternative to conventional machining as it works with all conditions, do not need any external device and the codes for MTP can easily be programmed with normal G codes of machine axis. This paper indicates how the MTP parameters like oscillation frequency, amplitude ratio, and phase shift and chip length should be selected. It is suggested that for effective MTP machining the phase shift between oscillation of tool path and spindle revolution should be correctly phased. The amplitude ratio selected should be above 0.5 and it should not be of large range due to its demand on machine axes. Chip length depends on OPR, as higher values of OPR lead to shorter chip length. And the chip length depends on the local part diameter. Lower temperatures in the range of 33.5–41.0 °C were obtained for MTP machining and for conventional machining the higher temperature should be in the range of 37.6–48.6 °C. So, MTP machining is consistently cooler than conventional machining. It is observed that for higher values of Raf the crater wear and flank wear decreases along the tool and MTP machining exhibit lower tool wear when compared with conventional machining. The Surface texture of both the machined surfaces resembled the same. Dynamic limitation of machine axes can be controlled by introducing motion delays in the oscillating tool path programming. On the whole, the MTP machining process is better than the conventional machining process in all the aspects.

## References

1. Woody BA et al (2008) Assessment of the process parameters and their effect on the chip length when using CNC toolpaths to provide chip breaking in turning operations. In: Proceedings of the 3rd ASME MSEC conference, Evanston, IL, USA
2. Smith KS et al (2010) Simulating the effect of modulated tool-path chip breaking on surface texture and chip length. No. Y/DX-2887. Oak Ridge Y-12 Plant (Y-12), Oak Ridge
3. Berglind L, Ziegert J (2013) Chip breaking parameter selection for constant surface speed machining. In: ASME 2013 international mechanical engineering congress and exposition, vol 2. American Society of Mechanical Engineers
4. Smith S et al (2009) Measurement of tool temperature using thermal imaging in interrupted cut turning. In: Proceedings of the ASME 2009 international manufacturing science and engineering conference, West Lafayette, IN, USA
5. Berglind L, Ziegert J (2015) Modulated tool path (MTP) machining for threading applications. Procedia Manuf 1:546–555
6. An A (1992) Tool temperatures in interrupted metal cutting. J Eng Ind 114:127

7. Palmai Z (1987) Cutting temperature in intermittent cutting. Int J Mach Tools Manuf 27 (2):261–274
8. Smith S et al (2009) Temperature control and machine dynamics in chip breaking using CNC toolpaths. CIRP Ann Manuf Technol 58(1):97–100
9. Schmidt P et al (2016) Residual surface stress: comparing traditional and modulated tool path machining processes. Mater Sci Technol 32(14):1471–1483
10. Barkman WE et al (2010) Modulated tool-path chip breaking for depleted uranium machining operations. No. Y/DX-2883. Oak Ridge Y-12 Plant (Y-12), Oak Ridge

# Machining of Difficult to Cut Materials by Electrochemical Discharge Machining (ECDM) Process: A State of Art Approach

**Manpreet Singh and Sarbjit Singh**

## 1 Introduction

The miniaturized products has gained acceptability in the field of micro electro mechanical system (MEMS), micro fluidic and 'Lab-on-a-chip' devices [1, 2]. There are numerous manufacturing techniques used to generate miniaturized products such as electrical discharge machining (EDM), wire electrical discharge machining (WEDM), laser beam machining (LBM) etc. [3, 4]. However, the process capabilities of these processes were dependent upon material properties. The research fraternity all over the world are making sincere efforts to develop processes, which can handle variety of materials irrespective of their mechanical or chemical properties. ECDM process is one of the promising machining process that can handle various materials regardless of their material properties [5].

Initially, this process was attempted to machine glass by Karafuji and Suda in 1968 [6], later this process shows potential for machining of composites [7], steel plates [8], ceramics [9] and super-alloys [10]. As shown in Fig. 1a, the ECDM setup consists of two electrodes i.e. tool electrode and auxiliary electrode known as cathode and anode respectively. The work piece to be machined is placed underneath the tool electrode and dipped in electrolyte. The tool electrode is partially dipped up to few mm in electrolyte, while the anode electrode is positioned at few centimetres away from cathode electrode. The electrolysis process started, when a voltage is applied through a DC power source. Because of electrolysis process, the hydrogen gas bubbles generates on cathode electrode as shown in Fig. 1b. The size

M. Singh (✉) · S. Singh
Department of Mechanical Engineering, PEC University of Technology, Chandigarh 160012, India
e-mail: gurayamanpreet5800@gmail.com



K. Antony and J. P. Davim (eds.), *Advanced Manufacturing and Materials Science*, Lecture Notes on Multidisciplinary Industrial Engineering,
https://doi.org/10.1007/978-3-319-76276-0_14

of hydrogen bubbles increases with increases in current density of the tool electrode. Therefore, a casing of hydrogen gas layer i.e. a thin hydrogen gas film is generated along the vicinity of the tool electrode as represents in Fig. 1c. This hydrogen gas film behaves as an insulating layer between cathode electrode and electrolyte [11]. As the applied potential over electrodes crosses the breakdown strength of gas film, the discharge induced at cathode electrode is shown in Fig. 1d. Therefore, the work piece in close region of tool electrode get melting, vaporizing and thermally eroding due to spark energy [12].

With the development of ECDM process, the researchers come out with different hypothesis/philosophy on mechanism of ECDM. Ghosh et al. [13] observed that electrical discharge is formed due to switching phenomenon, not due to disruption of insulating gas film. Jalali et al. [14] observed machining mechanism as mixture of local heating and chemical etching. Therefore, researchers with different ideas on ECDM process gives different names to the machining process i.e. electrochemical spark machining (ECSM) by Tandon et al. [7], electro erosion dissolution machining Khairy et al. [8], electrochemical arc machining (ECAM) by Crichton et al. [15], micro electrochemical discharge machining by Langen et al. [16] and spark assisted chemical engraving (SACE) by Fascio et al. [17]. The primary aim of this study is provide comprehensive knowledge and literature regarding various materials machined on ECDM process. The authors believe that information provided in this paper is quiet useful in future for machining of advance materials.

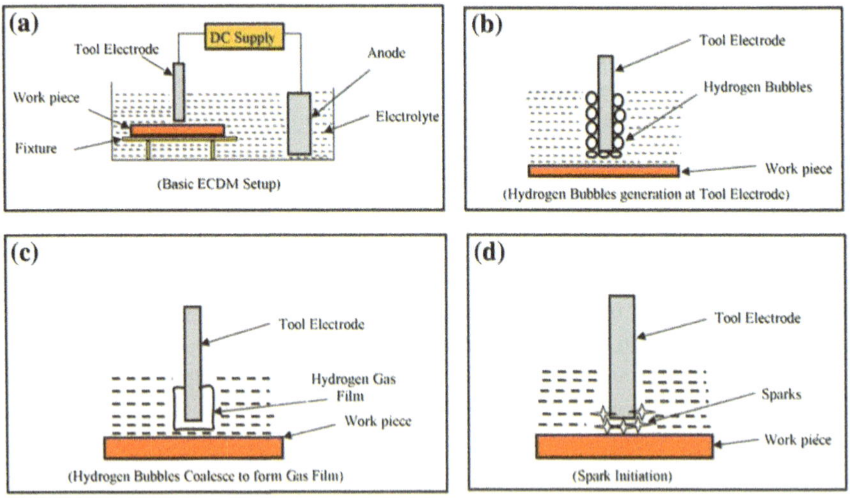


**Fig. 1** Mechanism of ECDM process **a** basic ECDM setup, **b** hydrogen bubbles generations at tool electrode, **c** hydrogen bubbles coalesce to form gas film, **d** spark initiation

## 2 Various Developments and Statistical Analysis of ECDM Process

This review article highlights machining behaviour of various difficult to cut materials machined by ECDM process. The various historical developments of ECDM process reported in past years are shown in Fig. 2a. Historical developments represent various phases of ECDM process for evolution from machining of glass to other engineering difficult-to-cut materials. The historical developments of ECDM process is classified into three phases.

In First phase (1968–1999) micro-holes were drilled on glass, steel plate and carbon fibre epoxy composites. The quality of the drilled hole in terms of cracks, heat affected zone, taper, overcut etc. were not as desired by the researchers. In the second phase (1999–2010) the researchers used hybridization of the ECDM process to improvement in output quality characteristics. They attempted in the domain of micro-machining with powder mixed electrolyte [18], ultrasonic vibrated electrolyte [19] and analysis of gas film formation during the process [20].

Third phase (2010–2017) includes further enhancement in process capability of ECDM process by incorporating hybridization of the hybrid ECDM process. In this phase the researchers attempted the effect of magnetic assistance during ECDM process [21], grinding aided ECDM [22] and rotary tool electrode ECDM [23] etc. for machining of difficult-to-cut materials.

Developments in ECDM process includes micro machining of glass, steel, ceramics, composites and super-alloys with better output characteristics. The statistical analysis of various materials machined by ECDM process for last ten years are represents in Fig. 2b. The figure shows that ECDM process is widely used for machining of non-conductive materials like glass, composites and ceramics. On the other hand, machining of conductive materials like steel, super-alloys and MMC

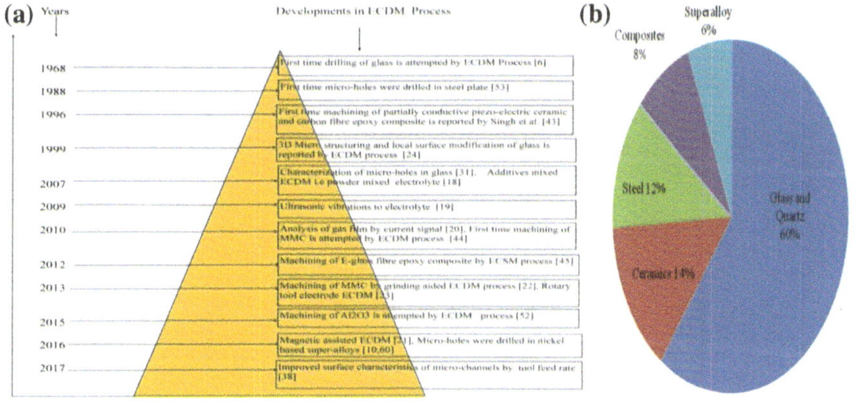


**Fig. 2** **a** Historical developments in ECDM process, **b** statistical analysis of last 10 years for various materials machined by ECDM process

**Table 1** Overview of difficult-to-cut materials machined by ECDM

| Authors name & year | Material | Process used | Output response | Findings | Generic inferences |
|---|---|---|---|---|---|
| Wuthrich R, Hof LA, Lal A, Fujisaki K, Bleuler H, MandinPh, Picard G (2005) | Glass | SACE | • Critical voltage<br>• Gas film thickness | • Gas film thickness is decreased with addition of surfactant to electrolyte | • Gas film thickness can be controlled with change in physical properties of electrolyte |
| Wuthrich R, Despont B, Maillard P, Bleuler H (2006) | Glass | SACE | • MRR<br>• Overcut | • Improved MRR and machining at higher depths | • Machining at higher depth is possible with vibration assisted tool electrode in ECDM |
| Han MS, Min BK, Lee SJ (2007) | Borosilicate glass | Additives mixed ECDM | • Surface roughness | • Surface integrity of machined samples can be improved with powder mixed electrolyte | • TWR can be controlled with change in density of electrolyte |
| Coteata M, Slatineanu L, Dodun O, Ciofu C (2008) | Steel sheet | ECDM | • Tool wear rate (TWR) | • Electrolyte density has significant influence on TWR as compared to other factors such as voltage and tool electrode diameter | • Surface characteristics of E-glass-fibre-epoxy composite are very poor due to incomplete cutting |
| Manna A, Narang V (2012) | E-glass- fibre-epoxy composite | ECSM | • MRR<br>• Overcut | • Machining of E-glass-fibre-epoxy composite is poor due to incomplete | • Difficult-to-cut materials can be easily machined with triplex hybridization of ECDM |

(continued)

**Table 1** (continued)

| Authors name & year | Material | Process used | Output response | Findings | Generic inferences |
|---|---|---|---|---|---|
| | | | | cutting and burning of fibres | |
| Liu JW, Yue TM, Guo ZN (2013) | Metal matrix composite (MMC) | Grinding-aided ECDM | • MRR<br>• Surface roughness | • Enhanced MRR and surface quality | • TWR is directly related to material and their composition |
| Behroozfar A, Razfar MR (2016) | Alumina and silica | ECDM | • TWR | •TWR is lower in tungsten carbide tool electrode as compared to steel and brass tool electrode | |
| Malik A, Manna A (2016) | E-glass-fibre-epoxy composite | Wire ECSM | • MRR<br>• Spark gap width | • Machining rate is significantly enhanced with applied voltage and wire speed | • Micro-slicing of E-glass-fibre-epoxy composite is successfully carried out by travel wire ECSM<br>• Machining characteristics is improved with change in inner structure of tubular electrode |
| Yan Z, Zhengyang X, Jun X, Di Z (2016) | Super-alloy | Tube electrode ECDM | • MRR<br>• Recast layer | •MRR is increased with increase in inner diameter of tubular electrode | |

has also been attempted by several researchers. Table 1 shows the machining overview of various difficult-to-cut materials with different findings and generic inferences.

## 3 Materials Machined on ECDM Process

This section elaborates machining behaviour of various difficult-to-cut materials machined on ECDM. This section includes machining behaviour of glass, composites, ceramics, steel and super-alloys by ECDM process.

### *3.1 Glass*

Glass is widely used in MEMS technology due to its transparency and chemical resistance properties [24]. At starting stage of ECDM, machining of glass was attempted [6]. Later on this method was used for machining of other materials such as composites, ceramics etc. The idea of 3D micro structuring and local surface modification on glass surface with ECDM has been firstly investigated by Fascio et al. [24]. The machining behaviour of glass with ECDM has been explored Wuthrich et al. [25–28]. The Wuthrich observed two regimes in drilling of glass i.e. firstly known as discharge regime (200–300 μm depth) and secondly hydrodynamic regime (depth higher than 300 μm). Machining characteristics of glass in discharge regime were superior to hydrodynamic regime. It was because of limited electrolyte availability in hydrodynamic regime. The lacks of electrolyte availability at higher machining depth were controlled with vibration assisted tool electrode in ECDM [29]. Bhondwe et al. [30] investigated machining of soda lime glass by ECDM process and showed that MRR is momentously influenced by electrolyte concentration and duty factor. In other studies machining of glass have been reported with different input characteristics such as applied voltage and depth [31], powder mixed electrolyte [18] and current signal [20].

Yang et al. [32] observed effect of surface roughness of tool electrode on ECDM performance in machining of quartz. Similarly, studies on machining of glass and quartz have been reported by researchers [33–36]. Machining of glass with various constraints such as vibrated tool electrode [37], magnetic assisted ECDM [21], tool feed rate [38] has been investigated by the research fraternity.

### *3.2 Composites and Ceramics*

Composites have gained reputation in the field of aerospace, automobile and electronic industries owing to their high temperature, wear resistant and high strength-to-weight ratio properties [39–42]. However, superior properties of these materials make it more difficult to frame into desired shape by traditional machining techniques. Therefore, various research groups investigated machining of composites by advanced manufacturing process i.e. ECDM process. Initially, machining of partially conductive material carbon fibre epoxy composite by ECSM process have been reported by Singh et al. [43]. They concluded that tool feed rate is key factor that will affect quality characteristics of machined surface. Liu et al. [44] successfully carried out machining of particulate reinforced metal matrix composite (MMC) with ECDM process. Liu et al. [22] reported machining of particulate reinforced MMC with superior surface characteristics by grinding aided ECDM process. Similarly, various research groups investigated machining of E-glass, fibre-epoxy composite by ECSM process. Machining surface of E-glass, fibre-epoxy composite in micro drilling is poor due to burning and heating [45]. On the other hand, micro slicing is successfully carried out on E-glass, fibre-epoxy composite by travel wire ECSM process [46].

In modern industry ceramics attained lots of interest owing to their numerous applications in industrial field like computer parts, heat exchanger and internal combustion engine [47]. Machining of partially conductive piezoelectric ceramic and silicon nitride ceramics has been reported by Singh et al. [43] and Sarkar et al. [9] respectively. They revealed that applied voltage is momentous factor that will affect the material removal rate (MRR) and HAZ. The modelling of the ECDM process has been attempted by Shanmukhi et al. [48] during micro-drilling of silicon nitride ceramics. Machining of aluminium oxide ceramic ($Al_2O_3$) have been investigated by various research groups by ECDM process with different developments [49–52]. In machining of aluminium oxide ceramic spark is directly controlled with change in shape of tool electrode tip [49]. Micro-holes were successfully drilled in aluminium oxide ceramic with use of abrasive tool electrode [50], electrolyte flow method in machining [51]. Micro-slicing of aluminium oxide ceramic surface is effectively done by travel wire ECSM process [52].

## *3.3 Steel and Super-Alloys*

With reference to process capability of ECDM, various researchers investigated machining of steel by ECDM process. Initially, machining of steel with wire ECAM process has been reported by 1988 [53]. In this study, machining accuracy is enhanced with coaxial flushing of electrolyte. Coteata et al. [54–56] observed tool wear mechanism in drilling of small holes on steel plate by ECDM process. Tool wear rate (TWR) is significantly influenced by solution density, applied voltage and tool diameter. Krotz et al. [57] drilled micro-holes in steel plate of 1 mm thickness by micro-ECAM process. Huang et al. [58] successfully drilled micro-holes in stainless steel plate with rotary tool electrode in ECDM. At high rotating speed, TWR is decreased owing to increase in heat dissipation capacity by centrifugal action.

Nickel based super-alloys are widely used in air craft engine owing to their better mechanical and heat resistant properties [59]. A novel technique tube electrode high-speed electrochemical discharge drilling (TEHECDD) is effectively used for machining of film cooling holes in super-alloys [10, 23, 60]. In these studies, MRR is appreciably increased with increase in inner diameter of tubular tool electrode.

## 4 Conclusions and Future Scope

ECDM is the proficient hybrid machining process which can broadly acceptable for machining of conductive and non-conductive difficult-to-cut materials. The exploration had done on ECDM is significantly focussed on machining of glass followed by ceramics, steel, composites and super-alloys. Moreover, it revealed that large class of new advent materials can also machined by ECDM irrespective to their properties. Indeed, even known since presence of ECDM, this process stays in a lab for research work. Therefore, this process require lots of exploration for transforms it from lab work to industrial application. The essential requirement for conversion of this process to industrial application is reproducible machining. In ECDM process, the gas film phenomenon is a key element that responsible for conversion of ECDM process to reproducible machining. Therefore, lots of efforts are required for control of stability and dynamics of gas film.

In the recent years, various research groups enhanced process capability of ECDM by triplex hybridization i.e. MMC is successfully machined by grinding aided ECDM process. The other triplex hybridization of ECDM such as vibration assisted ECDM, magnetic assisted ECDM and rotary tool electrode ECDM have been successfully attempted by several researchers. It concluded that triplex hybridization of ECDM process is proficient way to machine difficult-to-cut materials.

## References

1. Judy JW (2001) Micro electro mechanical systems (MEMS): fabrication, design and applications. Smart Mater Struct 10:1115–1134
2. Haebrle S, Zengerle R (2007) Microfluidic platforms for lab-on-chip applications. Lab Chip 7:1094–1110
3. Liu Q, Zhang Q, Zhu G, Wang K, Zhang J, Dong C (2016) Effect of electrodes size on the performance of micro EDM. Mater Manuf Process 31:391–396
4. Geng X, Chi G, Wang Y, Wang Z (2014) Study on micro rotating structure using micro wire electrical discharge machining. Mater Manuf Process 29:274–280
5. Chak SK, Rao PV (2007) Trepanning of $Al_2O_3$ by electrochemical discharge machining (ECDM) process using abrasive electrode with pulsed DC supply. Int J Mach Tools Manuf 47:2061–2070
6. Karafuji H, Suda K (1968) Electrical discharge drilling of glass. CIRP Ann 16:416–419

7. Tandon S, Jain VK, Kumar P, Rajurkar KP (1990) Investigation into machining of composites. Precis Eng 12:227–238
8. Khairy ABE, Mcgeough JA (1990) Die-sinking by electro erosion-dissolution machining. CIRP Ann Manuf Technol 39:191–195
9. Sarkar BR, Doloi B, Bhattacharya B (2006) Parametric analysis on electrochemical discharge machining of silicon nitride ceramics. Int J Adv Manuf Technol 28:873–881
10. Yan Z, Zhengyang X, Jun X, Di Z (2016) Effect of tube electrode inner diameter on electrochemical discharge machining of nickel-based superalloy. Chin J Aeronaut 29 (4):1103–1110
11. Goud M, Sharma AP, Jawalkar CS (2016) A review on material removal mechanism in electrochemical discharge machining (ECDM) and possibilities to enhance the material removal rate. Precis Eng 45:1–17
12. Antil P, Singh S, Manna A (2017) Electrochemical discharge drilling of SiC reinforced polymer matrix composite using Taguch's grey relational analysis. Arab J Sci Eng https://doi.org/10.1007/s13369-017-2822-6
13. Ghosh A (1997) Electrochemical discharge machining: principle and possibilities. Sadhana 22 (3):435–447
14. Jalali M, Maillard P, Wuthrich R (2009) Toward a better understanding of glass gravity-feed micro-hole drilling with electrochemical discharges. J Micromech Microeng 19:045001 (7 pp)
15. Crichton IM, Mcgeough JA (1985) Studies of discharge mechanism in electrochemical arc machining. J Appl Electrochem 15:113–119
16. Langen H, Breguet JM, Bleuler H, Renaud PH, Masujawa T (1998) Micro electrochemical discharge machining of glass. Int J Electr Mach 3:65–69
17. Fascio V, Wuthrich R, Bleuler H (2004) Spark assisted chemical engraving in the light of electrochemistry. Electrochem Acta 49:3997–4003
18. Han MS, Min BK, Lee SJ (2007) Improvement of surface integrity of electro-chemical discharge machining process using powder-mixed electrolyte. J Mater Process Technol 191L:224–227
19. Han MS, Min BK, Lee SJ (2009) Geometric improvement of electrochemical discharge micro-drilling using an ultrasonic-vibrated electrolyte. J Micromech Microeng 19:065004 (8 pp)
20. Cheng CP, Wu KL, Mai CC, Yang CK, Hsu YS, Yan BH (2010) Study of gas film quality in electrochemical discharge machining. Int J Mach Tools Manuf 50:689–697
21. Hajian M, Razfar MR, Etefagh AH (2016) Experimental study of tool bending force and feed rate in ECDM milling. Int J Adv Manuf Technol. https://doi.org/10.1007/s00170-016-9860-1
22. Liu JW, Yue TM, Guo ZN (2013) Grinding-aided electrochemical discharge machining of particulate reinforced metal matrix composite. Int J Adv Manuf Technol 68:2349–2357
23. Zhang Y, Xu Z, Zhu Y, Zhu D (2016) Effect of tube-electrode inner structure on machining performance in tube-electrode high-speed electrochemical discharge drilling. J Mater Process Technol 231:38–49
24. Fascio V, Wuthrich R, Viquerat D, Langen H (1999) 3D micro structuring of glass using electrochemical discharge machining (ECDM). In: International symposium on micro mechatronics and human science, pp 179–183
25. Wuthrich R, Fascio V, Bleuler H (2004) A stochastic model for electrode effects. Electrochem Acta 49:4005–4010
26. Wuthrich R, Comninellis C, Bleuler H (2005) Bubble evolution on vertical electrodes under extreme current densities. Electrochem Acta 50:5242–5246
27. Wuthrich R, Hof LA, Lal A, Fujisaki K, Bleuler H, Mandin P, Picard G (2005) Physical principles and miniaturization of spark assisted chemical engraving (SACE). J Micromech Microeng 15:S268–S275
28. Wuthrich R, Spaelter U, Wu Y, Bleuler H (2006) A systematic characterization method for gravity-feed micro-hole drilling in glass with spark assisted chemical engraving (SACE). J Micromech Microeng 16:1891–1896

29. Wuthrich R, Despont B, Maillard P, Bleuler H (2006) Improving the material removal rate in spark-assisted chemical engraving (SACE) gravity-feed micro hole drilling by tool vibration. J Micromech Microeng 16:N28–N31
30. Bhondwe KL, Yadava V, Kathiresan G (2006) Finite element prediction of material removal rate due to electro-chemical spark machining. Int J Mach Tools Manuf 46:1699–1706
31. Maillard P, Despont B, Bleuler H, Wuthrich R (2007) Geometrical characterization of micro-holes drilled in glass by gravity-feed with spark assisted chemical engraving (SACE). J Micromech Microeng 17:1343–1349
32. Yang CK, Cheng CP, Mai CC, Wang AC, Hung JC, Yan BH (2010) Effect of surface roughness of tool electrode materials in ECDM performance. Int J Mach Tools Manuf 50:1088–1096
33. Yang CK, Wu KL, Hung JC, Lee SM, Lin JC, Yan BH (2011) Enhancement of ECDM efficiency and accuracy by spherical tool electrode. Int J Mach Tools Manuf 51:528–535
34. Paul L, Hiremath SS (2013) Response surface modelling of micro holes in electrochemical discharge machining process. Procedia Eng 64:1395–1404
35. Nguyen KH, Lee PA, Kim BH (2015) Experimental investigation of ECDM for fabricating micro structures of quartz. Int J Precis Eng Manuf 16(1):5–12
36. Jui SK, Kamraj AB, Sundaram MM (2013) High aspect ratio micromachining of glass by electrochemical discharge machining (ECDM). J Manuf Process 15:460–466
37. Jiang B, Lan S, Ni J (2014) Experimental investigation of drilling incorporated electrochemical discharge machining. In: Proceeding of the ASME international manufacturing science and engineering conference, Detroit, MI, USA
38. Goud M, Sharma AK (2017) On performance studies during micromachining of quartz glass using electrochemical discharge machining. J Mech Sci Technol 31(3):1365–1372
39. Speer W, Es-Said OS (2004) Applications of an aluminium-beryllium composite for structural composite for structural aerospace components. Eng Fail Anal 11:895–902
40. Kunze JM, Bampton CC (2001) Challenges to developing and producing MMCs for space applications. JOM 53:22–25
41. Zweben C (2005) Advanced electronics packaging material. Adv Mater Process 163:33–37
42. Singh S, Singh I, Dvivedi A (2013) Multiobjective optimization in drilling of Al6063/10% SiC metal matrix composite based on grey relational analysis. Proceed Inst Mech Eng Part B J Eng Manuf 227(12):1767–1776
43. Singh YP, Jain VK, Kumar P, Agrawal DC (1996) Machining piezoelectric (PZT) ceramics using an electrochemical spark machining (ECSM) process. J Mater Process Technol 58:24–31
44. Liu JW, Yue TM, Guo ZN (2010) An analysis of the discharge mechanism in electrochemical discharge machining of particulate reinforced metal matrix composites. Int J Mach Tools Manuf 50:86–96
45. Manna A, Narang V (2012) A study on micro machining of e-glass-fibre-epoxy composite by ECSM process. Int J Adv Manuf Technol 61:1191–1197
46. Malik A, Manna A (2016) An experimental investigation on developed WECSM during micro slicing of e-glass fibre epoxy composite. Int J Adv Manuf Technol 85:2097–2106
47. Schwartz MM (1995) Engineering applications of ceramic materials. American society for metals Source book. Metals Park, Ohio
48. Shanmukhi K, Vundavilli PR, Surekha B (2015) Modeling of ECDM micro-drilling process using GA- and PSO-trained radial basis function neural network. Soft Comput 19:2193–2202
49. Bhattacharya B, Doloi BN, Sorkhel SK (1999) Experimental investigation into electrochemical discharge machining (ECDM) of non-conducting materials. J Mater Process Technol 95:145–154
50. Chak SK, Rao PV (2008) The drilling of $Al_2O_3$ using pulsed DC supply with a rotary abrasive electrode by the electrochemical discharge process. Int J Adv Manuf Technol 39:633–641
51. Dhanvijay MR, Ahuja BB (2014) Micromachining of ceramics by electrochemical discharge process considering stagnant and electrolyte flow method. Procedia Eng 14:165–172

52. Manna A, Kundal A (2015) An experimental investigation on fabricated TW-ECSM setup during micro slicing of nonconductive ceramic. Int J Adv Manuf Technol 76:29–37
53. Hofy HE, Mcgeough JA (1988) Evaluation of an apparatus for electrochemical arc wire-machining. J Eng Ind 110:119–123
54. Coteata M, Slatineanu L, Dodun O, Ciofu C (2008) Electrochemical discharge machining of small diameter holes. Int J Mater Form 1:1327–1330
55. Coteata M, Ciofu C, Slatineanu L, Munteanu A, Dodun O (2009) Establishing the electrical discharge weight in electrochemical discharge drilling. Int J Mater Form 2(1):673–676
56. Coteata M, Schulze HP, Slatineanu (2011) Drilling of difficult-to-cut steel by electrochemical discharge machining. Mater Manuf Process 26:1466–1472
57. Krotz H, Roth R, Wegener K (2013) Experimental investigation and simulation of heat flux into metallic surfaces due to single discharges in micro-electrochemical arc machining (micro-ECAM). Int J Adv Manuf Technol 68:1267–1275
58. Huang SF, Liu Y, Li J, Hu HX, Sun LY (2014) Electrochemical discharge machining micro-hole in stainless steel with tool electrode high-speed rotating. Mater Manuf Process 29:634–637
59. Akhtar W, Sun J, Sun P, Chen W, Saleem Z (2014) Tool wear mechanism in machining of Nickel based super-alloys: a review. Front Mech Eng 9(2):106–119
60. Zhang Y, Xu Z, Zhu D, Xing J (2015) Tube electrode high-speed electrochemical discharge drilling using low-conductivity salt solution. Int J Mach Tools Manuf 92:10–18

# Development of Super Hydrophobic Surfaces for Oil Spill Separation

**Alex K. Babu, Alwinson Kuriakose Geevarghese, Althaf Easa, Anthea Judette Maxi Fernandez, Rakesh Reghunath, Soney C. George and R. Asaletha**

## 1 Introduction

The rise of twentieth century has witnessed considerable increase in the usage of petroleum and oil products in the global sector. In the present scenario, the oil and petroleum industries present worldwide are innumerable. The major reasons include advancement in the field of transportation and automobile industries, rising needs of ever increasing population, lack of advancement in technologies promoting usage of ecofriendly sources of energy such as electric power, solar energy etc. This leads to the continuous consumption of conventional sources of energy leading to

A. K. Babu (✉) · A. K. Geevarghese · A. Easa · A. J. M. Fernandez · R. Reghunath
Department of Mechanical Engineering, Amal Jyothi College of Engineering Kanjirapally, Kanjirapally, Kerala, India
e-mail: lifewithalex95@gmail.com

A. K. Geevarghese
e-mail: alwinsonkk@gmail.com

A. Easa
e-mail: althafeasa@gmail.com

A. J. M. Fernandez
e-mail: antheafernandez56@gmail.com

R. Reghunath
e-mail: rakeshreghunath@amaljyothi.ac.in

S. C. George
Department of Basic Science, Amal Jyothi College of Engineering, Kanjirapally, Kerala, India
e-mail: soneycgeorge@amaljyothi.ac.in

R. Asaletha
Department of Basic Science, Cochin University College of Engineering Kuttanad, Kuttanad, India
e-mail: asalethar@cusat.ac.in



K. Antony and J. P. Davim (eds.), *Advanced Manufacturing and Materials Science*, Lecture Notes on Multidisciplinary Industrial Engineering,
https://doi.org/10.1007/978-3-319-76276-0_15

exhaustion. The consumption of natural resources at such an alarming rate has not only depleted the same but has other implications too. The major impact is nothing but the pollution of natural resources, especially water and soil. The most dangerous effect is that it makes even the air we breathe intoxicated. Recently, frequent water pollution, particularly oil spillage, chemical leakage, organic water effluents from industries have caused severe ecological problems. The absence of an efficient method of treatment may cause long term harm to the environment and ecological system [1, 2].

The treatment of polluted water resource by the oil spills has become a serious matter of concern. The conventional methods used for the treatment are:

- Bioremediation [3–5]
- Chemical catalytic decomposition [6, 7]
- Absorption and separation [8–10].

But it is tedious to handle the vast industrial pollution or oil spill by bioremediation and catalytic decomposition owing to its less efficiency. The method of physical sorption by utilizing porous materials could lead to a simple, quick and efficient separation of harmful organic content. This resulted in the concentration of research in making of cheap, practical sorbents for separation of oil spill from water bodies. The polar property of water can be efficiently utilized for the separation of organic liquid matter. The harmful organic contents are nonpolar and this property is exploited for the absorption of same by developing a non-polar surface. As water is polar, it will get repelled by non-polar porous material i.e., the surface will be super hydrophobic [11, 12] (contact angle >150°). The separation of oil and organic solvents directly from water can be performed by the naturally available porous absorbent materials such as natural fibers, saw dust etc. But they have got some limitations such as low absorption capacities, deficient selectivity and poor recyclability, ease of availability, ease of fabrication, chemical as well as physical stability and inertness to various organic solvents.

All these aspects have inspired scientists across the world to develop new porous absorbent materials possessing high absorption capacity, high porosity, large specific surface area, high selectivity, chemical inertness and excellent recyclability so that they can perform oil-water separation effectively. In order to serve this purpose, super hydrophobic and superoleophilic (oil contact angle <5°) materials have been developed which can absorb oil and organic contamination while repelling water completely. Many sorbent materials such as cross-linked polymers [13, 14] and resins, fibers [15], polymer gels [16], organic-inorganic hybrid [17], silica [18] and Nano composites [19] have been developed extensively. However, these materials have shown low sorption capabilities along with poor selectivity and their high cost and complicated fabrication process have limited their use.

Recent studies of novel oil sorbents have focused on super hydrophobic sponges and sponge-like materials that exhibited superior oil sorption capacity. In the study conducted by Gupta [20], He and Tai, melamine sponge is used as the core absorbing material and it is treated with reduced graphene oxide. This coating has helped in achieving super hydrophobicity. But the major drawbacks were the difficulty in synthesizing graphene, huge expense of the material and its poor recyclability. Gui et al. [21] reported the synthesis of Carbon Nano Tube sponges by chemical vapour deposition technique. CNTs were self-assembled into a 3 dimensionally interconnected framework. The obtained super hydrophobic sponge has the following features of having Contact angle >156°, Sorption capacity of the sponge is up to 180 times its own weight and Porosity >99%. The Drawbacks of the sponge are High cost of the CNTs and the Complex preparation processes. Zhao et al. [22] used ultralight nitrogen doped graphene and CNT as the coating material. The superhydrophobic sponge thus obtained were Ultralight, super elastic aerogel [23], with Density <1 mg/cc, Porosity 99.9% and Absorption capacity up to 200–900 times its own weight. But the limitations of the sponge were the high cost and Complex preparation procedure. Nguyen et al. [24] integrated solution-exfoliated graphene onto melamine sponges for oil absorbency, this was using a dip coating method. This process made the sponge superhydrophobic andit showed excellent absorption capacities up to 165 times of the initial weight. In particular melamine sponges have been shown to be an excellent scaffold for producing high-performance oil absorbents with 3D porous structure, as it is composed of a micro scale interconnected framework with good compressibility and is easy to chemically surface modify. Chen et al. [25] directly carbonized melamine sponges at 1800 °C and successfully prepared superhydrophobic sponges with extraordinary absorbency. Gao et al. [26] in his Experiment immersed melamine sponge in the $MoS_2$ nanosheet solution with ethanol by a squeezing and vacuum degassing procedure, and $MoS_2$ nanosheets were assembled on the MF sponge into a 3D $MoS_2$ sponge with a colour change from white to black. Finally, the coated sponge was directly dried to yield the SMS.

With the demand of finding an ideal sorbent material having the above mentioned desirable properties, we have come across melamine sponge which is used as the core absorbent material. The sponge is made super hydrophobic and superoleophilic by special surface treatment technique using octadecyltrichlorosilane and acetone. The surface treatment can alter the wettability and absorption efficiency of the material and thus the sponge can be benefitted to separate polar and non-polar solvents from polar or non-polar mixture.

## 2 Materials Required

The commercially available white melamine sponge, a density of 0.075–0.085 g/$cm^3$ and porosity of 99.44%, was used as the sponge sample, since melamine sponge exhibited superior oil absorption capacity [27]. The oils and organic solvents used in different tests and analysis include acetone, toluene, butanol octadecyltrichlorosilane, lubricating oil, petrol, and diesel.

# 3 Experimental Section

Raw samples of melamine sponges were treated with octadecyltrichlorosilane and these treated sponges were heated to a temperature of 120 °C. The procured melamine sponge was cut into 2.5 × 2.5 × 3 cm $^3$ pieces and immersed in the solution of octadecyltrichlorosilane in toluene (0.5 wt%) for 5, 10, 20, 30 and 45 min. Subsequently the sponge was removed from the solution and squeezed to extract the absorbed solution. Thereafter they were repeatedly washed in fresh toluene by the sorption-squeezing process and finally dried at 120 °C for 1 h. Experimental procedure for developing super hydrophobic sponge is shown in Fig. 1.

In this study, melamine sponges were treated with octadecyltrichlorosilane and the treated sponges were heated to a temperature of 120 °C. commercially available melamine sponges were cut into 2.5 × 2.5 × 3.0 cm$^3$ cube pieces and where immersed in a solution of octadecyltrichlorosilane in toluene (0.5 wt%) for 5, 10, 20, 30 and 45 min. The absorption capacities of the treated sponge are measured by immersing the sponge in different solutions of petrol, diesel [28], lubrication oil and butanol. The weights of the sponge were taken before the experiment and were immersed in the said solution for a certain fixed time interval. The immersed sponge was weighed, and was squeezed to remove the solution and is immersed again for the next time interval. The same cycle is repeated for rest of the solutions. In order for the treated sponge to be commercially useful, it is necessary to obtain the number of times the treated sponge can be used. The weight of the sponge is taken and is then immersed into different solutions like petrol, diesel, lubricating oil and butanol for 6–10 cycles for equal intervals of time.

The saturated sponge is weighed for the amount of solution absorbed, and is then squeezed manually until it is void of the absorbed solution, and it is therefore used for the next cycle. Hydrophobic sponge developed and a chemical for testing is shown in Fig. 2. The treated melamine sponges were characterized by Fourier Transform Infrared Spectroscopy. Fourier transform Infrared spectra were measured in the spectral range of 600–4000 cm$^{-1}$ on a Nicollet 380 FT-IR, KBr tablet, Thermo Electron Corporation spectrometer. Sample is mixed with a mulling agent such mineral oil and pressed between plates made of sodium chloride Sodium chloride is used because it has no IR absorptions. Glass or plastic plates would have IR absorptions of their own, Sodium chloride plates are good from 4000 to

**Fig. 1** Experimental procedure for sponge preparation

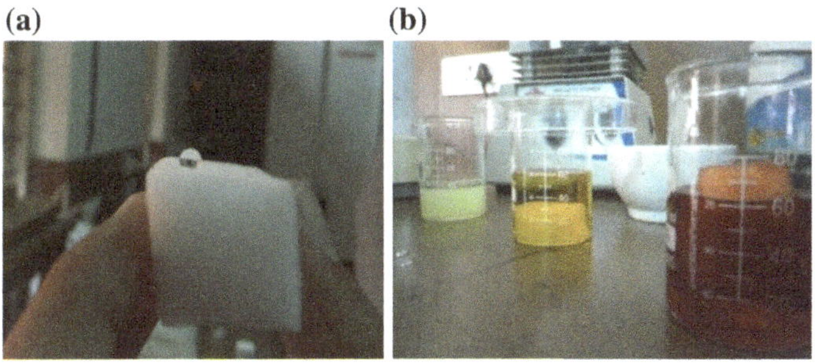


**Fig. 2** **a** Chemical treated sponge, **b** solutions for testing

650 $cm^{-1}$ and below 650 $cm^{-1}$ they begin to absorb, so Potassium bromide plates can be used in place of sodium chloride and are transparent to 400 $cm^{-1}$, but they are more expensive. Downside of this method is absorptions due to the mineral oil and Sample is mixed with solid potassium bromide and pressed into a pellet under high pressure. There is no absorptions from mulling agent and it only works for solids. Finally the Sample is dissolved in carbon tetrachloride and pressed between salt plates. The morphologies of melamine sponges were observed by scanning electron microscopy. The Ultra high resolution scanning electron microscopy (S-5200) was used for the analysis. Contact Angle measurement was conducted using optical tensiometers. The hydrophobicity is measured using the analysing the contact angle.

## 4 Results and Discussion

Melamine sponges used in this study has an open cell form like structure. It mainly consist formaldehyde-melamine-sodium bisulfite copolymer, which is widely used as an abrasive cleaner and as soundproofing material. Melamine sponge used here has a density of 7–9 $mg/cm^3$, high porosity of 99.5%, high elasticity. All these are ideal characteristics for sorption materials. The silanization was performed by dipping the melamine sponge into an octadecyltrichlorosilane solution at room temperature for 5–30 min, followed by a drying step at 120 °C for 1 h.

### *4.1 Sorption Capacity Studies*

The weight absorption capabilities of solvents were measured for all four types of sponges and shown in Fig. 3 and, it was found that diesel exhibited highest value of weight absorption capability for all types of sponges, as shown in. By contrast, Petrol exhibited lowest value of weight absorption capability since it is highly volatile compared to other solvents. It is inferred that weight of the sponge increases with increase with immersion time of the sponge and it reaches a constant value where further increase in weight does not take place, as the sponge has reached its maximum equilibrium state where further absorption of the solvent is not possible.

## 4.2 Recyclability Studies

In order to obtain the number of times the treated sponge can be used the recyclability tests were conducted and the result was plotted in the following graph shown in Fig. 4. Weight of the sponge absorbed after each 2-min interval was plotted against the number of cycles which implies that even after a number of uses, the rate of absorption of the sponge remains almost constant even after an average of eight cycles.

## 4.3 Contact Angle Studies

The hydrophobicity of the melamine sponge materials was characterized by water contact angle measurements. The hydrophobicity of the silanized melamine sponge was characterized by water contact angle measurements. As shown in Fig. 5, the

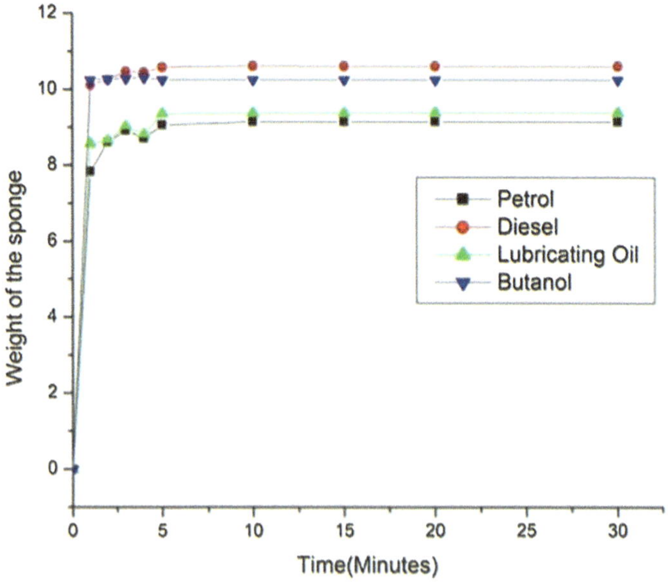


**Fig. 3** Absorption studies in different solvents

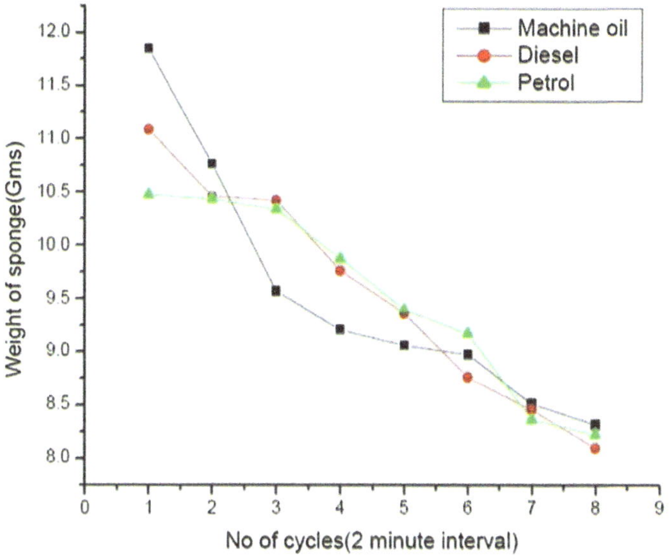


**Fig. 4** Recyclability studies in different solvents

water contact angle of the treated melamine sponge notably increased as a function of silanization time, from 0° for the melamine sponge to 30° ± 1.5° after 5 min of silanization. After 10 min, the value reached 70° ± 1.2° and a maximum of 109° ± 1.1° after 30 and 45 min, indicating the super hydrophobic characteristics of the functionalized sponges.

For analyzing the stability of the drop formed on the surface of the sponge, variation in the contact angle formed was measured for a contact angle of 45 min 30 min. It was found that for 45 min silanization the contact angle measured shows more stability when comparing with 30 min silanization as shown in Fig. 6. Experimental setup for measuring contact angle and measured angle is shown in Figs. 7 and 8 respectively.

## *4.4 FTIR Analysis*

The formations of self-assembled monolayer on the sponge skeleton were characterized by FTIR analysis. Molecules with symmetric bonds such as $N_2$, $O_2$, or $F_2$ do not absorb in the infrared since bond stretching [29] does not change the dipole moment of the molecule. The FTIR spectrum of the treated sponge is given in Fig. 9 The two primary modes of vibration are stretching and bending [30]. Stretching modes are typically of higher energy than bending modes and often divided into two a symmetric and asymmetric stretch; the asymmetric stretch is usually of higher energy. The different inferences from the spectrum of the

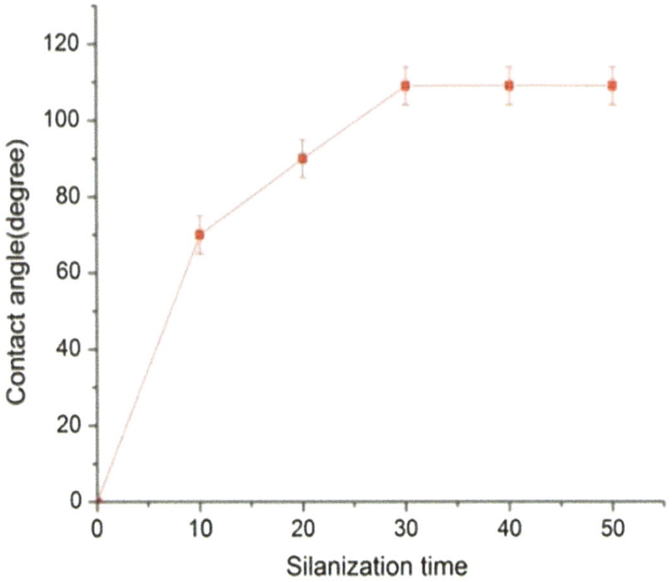


Fig. 5 Effect of silanization time on contact angle

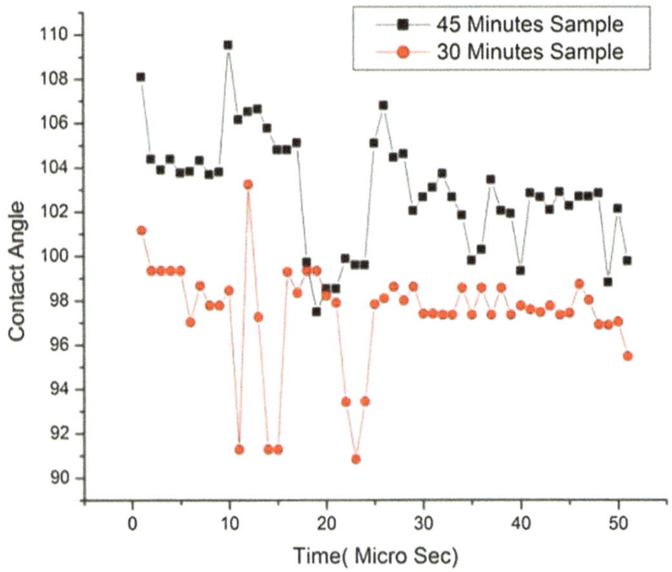


Fig. 6 Stability analyses for contact angle

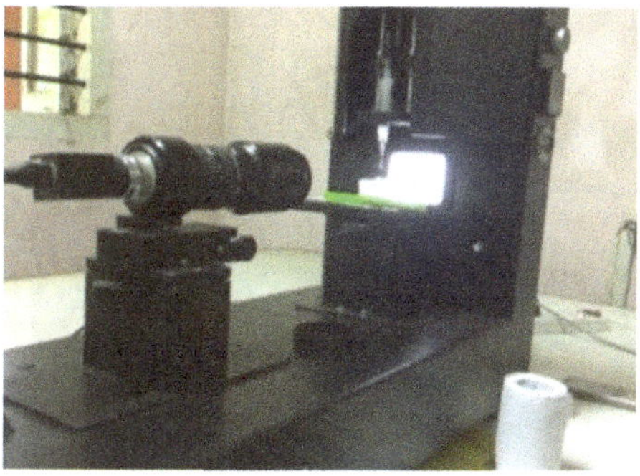

**Fig. 7** Experimental setup for contact angle measurement

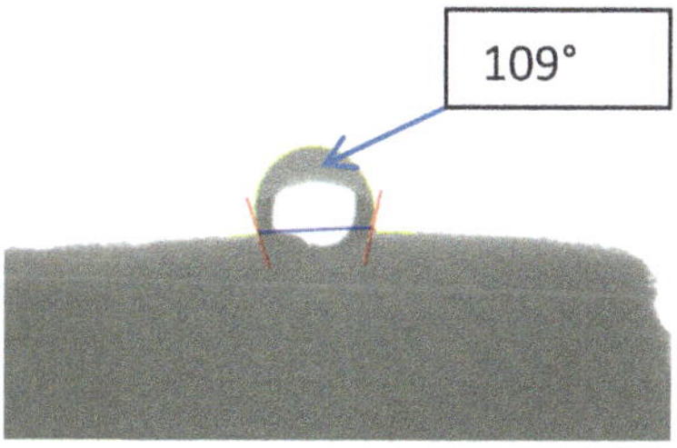


**Fig. 8** Measuring contact angle

melamine sponge are: The N–H bond (3400 $cm^{-1}$) disappears in the treated sponge. Formation of Si–N (873 $cm^{-1}$), Si–O (1100 $cm^{-1}$) and Si–C (1275 $cm^{-1}$). Formation of C–H stretching (3000–2840 $cm^{-1}$), $CH_2$ bending (1465 $cm^{-1}$), $CH_3$ bending (1375 $cm^{-1}$) and $CH_2$ (four or more $CH_2$ groups) rocking at (720 $cm^{-1}$).

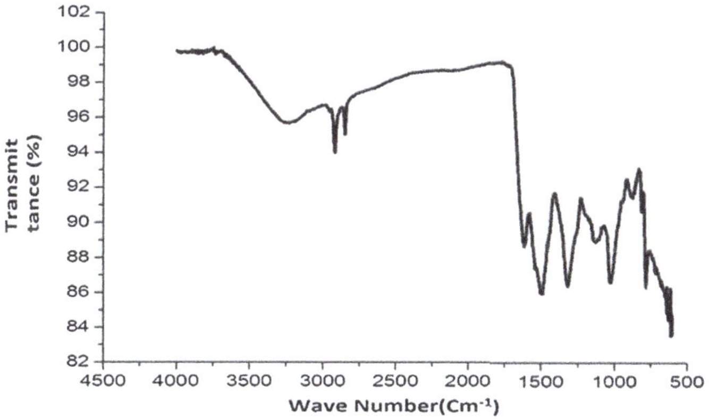


**Fig. 9** FTIR spectrum of treated sponge

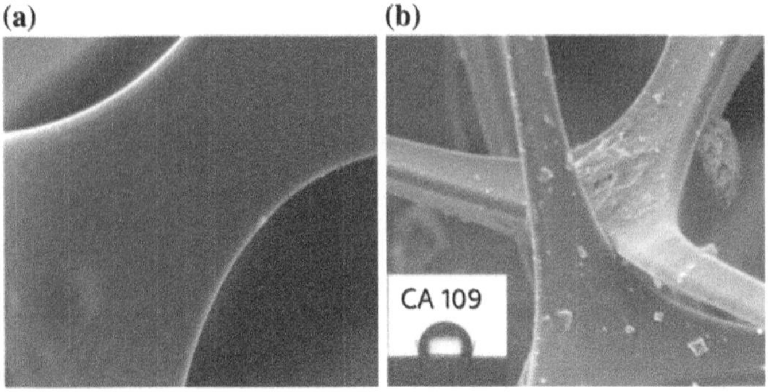


**Fig. 10** SEM image of **a** untreated sponge, **b** treated sponge

## *4.5 Morphological Studies*

Morphology of sponge surfaces were examined using scanning electron microscopy before and after silanization of the sponge as shown in Fig. 10. Generally, the wet ability of the surface depends on the surface chemical composition and the surface topographical microstructure [31]. Melamine sponges consist of a three dimensional hierarchical and porous structure with pore size in the range of 100–150 μm with smooth skeletons of average diameter approximately 10 μm. After silanization a thin layer of the alkylsilane self-assembled monolayer, which finally led to the superhydrophobicity of the treated melamine sponge.

# 5 Conclusion

In this study, we explored the experimental process to fabricate the hydrophobic sponges and analyse its surface, microstructure, hydrophobicity, absorption capacity and reusability. The melamine sponges were treated with octadecyltrichlorosilane, after the treatment, the characteristic of melamine sponge was changed from hydrophilic to hydrophobic which is attributed to the removal of branched NH bonds in melamine which decreased the $NH_2$ functional groups and an increase in oxygen containing functional groups, resulted in a decrease of the polarity. The silanization occurred through secondary amine groups attached to the surface of sponge skeletons with alkylsilane compounds, forming a layeron their surface. The maximum contact angle which was obtained from the experiment was about 109°. The sponge is easier to fabricate by this method compared to other methods and less time consuming. It could be bought into more practical applications specifically it finds applications can be in oil spills and also can be designed for the industrial ducts so that the oil content in the water could be reduced thereby reducing water pollution. In summary, a one-step, facile, and cost-effective method was reported for the fabrication of robust, superhydrophobic sponges by silanization of commercial melamine sponges via a solution-immersion process.

## References

1. Irons DB, Peterson CH, Rice SD, Short JW, Esler D, Bodkin JL, Ballachey BE (2003) Long-term ecosystem response to the Exxon Valdez. Oil Spill Sci 302:2082–2086
2. Phillips L, Schaum J, Cohen M, Perry S, Artz R, Draxler R, Frithsen JB, Heist D, Lorber M (2010) Screening level assessment of risks due to dioxin emissions from burning oil from the BP deepwater horizon gulf of mexico spill. Environ Sci Technol 44:9383–9389
3. Mohajeri S, Zahed MA, Aziz HA, Isa MH, Mohajeri L (2010) Optimal conditions for bioremediation of oily seawater. Bioresour Technol 101(24)
4. Schinner F, Margesin R, (2001) Biodegradation and bioremediation of hydrocarbons in extreme environments. Appl Microbiol Biotechnol 56(5–6)
5. McDonagh M, Swannell RP, Lee K (1996) Field evaluations of marine oil spill bioremediation. Microbiol Rev 60(2):
6. Mullin JH, Buist I, Potter S, Nedwed T (2011) Surfactants to contract and thicken oil spills in pack ice for in situ burning. Cold Reg Sci Technol 67(1–2):3–23
7. Redmond MC, Kujawinski EB, Kido Soule MC, Valentine DL, Boysen AK, Longnecker K (2011) Fate of dispersants associated with the deepwater horizon oil spill. Environ Sci Technol 45(4):1298–1306
8. Mercier L, Walcarius A (2010) Mesoporous organosilica adsorbents: nanoengineered materials for removal of organic and inorganic pollutants. J Mater Chem 20(22):4478–4511
9. Zhu X, Zhang Z, Ren G, Yang J, Wang K, Xu X et al (2012) A novel superhydrophobic bulk material. J Mater Chem 22(38):20146–20148
10. El-Saeed SM, Farag RK (2008) Synthesis and characterization of oil sorbers based on docosanyl acrylate and methacrylates copolymers. J Appl Polym Sci 109(6):3704–3713
11. Gao D, Cao L, Hu HH (2007) Design and fabrication of micro-textures for inducing a superhydrophobic behavior on hydrophilic materials. Langmuir 23(8):4310–4314
12. Crego-Calama M, Li XM, Reinhoudt D (2007) What do we need for a superhydrophobic surface? A review on the recent progress in the preparation of superhydrophobic surfaces. Chem Soc Rev 36(8):1350–1368
13. Huang ZM, Shan GR, Xu PY, Weng ZX (2003) Oil-absorption function of physical crosslinking in the high-oil-absorption resins. J Appl Polym Sci 90(14):3945–3950
14. Zhou XM, Cheng CZ (2010) Synthesis and characterization of a novel high-oilabsorbing resin. J Appl Polym Sci 115(6):3321–3325
15. Inagaki M, Kawahara A, Konno H (2002) Sorption and recovery of heavy oils using carbonized fir fibers and recycling. Carbon 40(1):105–111
16. Ono T, Sugimoto T, Shinkai S, Sada K (2007) Lipophilic polyelectrolyte gels as superabsorbent polymers for nonpolar organic solvents. Nat Mater 6(6):429–433
17. Yuan J, Liu X, Akbulut O, Hu J, Suib SL, Kong J, Stellacci F (2008) Superwetting nanowire membranes for selective absorption. Nat Nanotechnol 3(6):332–336
18. Yang Y, Sayari A, Hamoudi S (2005) Applications of pore-expanded mesoporous silica. 1. Removal of heavy metal cations and organic pollutants from wastewater. Chem Mater 17 (1):212
19. An Y, Tao S, Wang Y (2011) Superwetting monolithic $SiO_2$ with hierarchical structure for oil removal. J Mater Chem 21(32):119

20. Tai NH, Gupta S, He WD (2016) A comparative study on superhydrophobic sponges and their application as fluid channel for continuous separation of oils and organic solvents from water. Compos Part B 101:99–106
21. Gui X, Li H, Wang K, Wei J, Jia Y, Li Z, Fan L, Cao A, Zhu H, Wu D (2011) Recyclable carbon nanotube sponges for oil absorption. Acta Mater 59:4798–4804
22. Zhao Y, Hu C, Hu Y, Cheng H, Shi G, Qu L (2012) A versatile, ultralight, nitrogen-doped graphene framework. Angew Chem Int Ed 51:11371–11375
23. Sun H, Xu Z, Gao C (2013) Multifunctional, ultra-flyweight, synergistically assembled carbon aerogels. Adv Mater 25:2554–2560
24. Nguyen DD, Tai NH, Lee SB, Kuo WS (2012) Superhydrophobic and superoleophilic properties of graphene-based sponges fabricated using a facile dip coating method. Environ Sci Technol 5:7908
25. Chen S et al (2013) Elastic carbon foam via direct carbonization of polymer foam for flexible electrodes and organic chemical absorption. Environ Sci Technol 6:2435
26. Gao X, Wang X, Ouyang X, Wen C (2016) Flexible superhydrophobic and superoleophilic MoS sponge for highly efficient oil-water separation. Scientific reports 1–8
27. Pham VH, Dickerson JH (2014) Superhydrophobic silanized melamine sponges as high efficiency oil absorbent materials. ACS Appl Mater Interfaces
28. Liu T et al (2015) The preparation of superhydrophobic graphene/melamine composite sponge applied in treatment of oil pollution. J Porous Mater 22:1573–1581
29. Arkles B (1977) Tailoring surfaces with silanes. Chemtech 7:766–778
30. Merline A, Vukusic DJ, Abdala S (2013) Melamine formaldehyde: curing studies and reaction mechanism. Polym J 45:413–441
31. Liu K, Tian Y, Jiang L (2013) Bio-inspired superoleophobic and smart materials: design, fabrication, and application. Prog Mater Sci 58:503–556

# FEA Study of the Multiple Structural Orientations on Selective Laser Melted Cobalt Chrome Open-Porous Scaffolds

**Aby K. Abraham and V. G. Sridhar**

## 1 Introduction

This paper presents the manufacturing of the customized bone porous scaffold and its compressive property evaluation employing FEM. Additive producing processes manufacture end-use elements from CAD models by supplemental material layer by layer, and also the final elements are usually created in an exceedingly single step while not the necessity for any further process. Earlier accumulative producing techniques were used for "Rapid Prototyping" within the domain of product style and development for the conception effective, pattern building, assembly verification and purposeful testing. Within the recent times, Additive producing techniques are being employed throughout the fabrication of actual end-use commodities for varied sectors. one in all the fascinating province with vital potential impact is that the medical field, wherever implants, scaffolds and pros-thesis are being factory-made victimization Additive producing Technique [1]. Traditional ways of scaffold fabrication embrace fiber bonding, solvent casting and particulate action [2], membrane lamination, gas foaming, refrigerant evoked section separation [2, 3] then on. However, all of those techniques are primarily supported manual work and lack of a corresponding de-signing process; thus further procedure was required to on-taking appropriate form and also the microstructure. These ancient techniques even have several disadvantages like long fabrication periods; poor repeatability and short property of pores [4]. To overcome the constraints of those typical techniques, automatic computer-controlled fabrication techniques, like Additive producing (AM), are being explored. Pre-production stage Finite element analysis

A. K. Abraham (✉)
Department of Mechanical Engineering, GISAT, Kottayam, India
e-mail: abyk.abraham2014@vit.ac.in

V. G. Sridhar
School of Mechanical and Building Sciences VIT Chennai, Chennai, India

 
K. Antony and J. P. Davim (eds.), *Advanced Manufacturing and Materials Science*, Lecture Notes on Multidisciplinary Industrial Engineering,
https://doi.org/10.1007/978-3-319-76276-0_16

(FEA) was carried out to designed cobalt chrome scaffolds for the mechanical properties evaluation.

The researchers present the patient-specific implant's fabrication and its property evaluation. Metallic materials like cobalt alloys have been successfully soiled in clinical applications for orthopedic implants for a long-time [1]. Besides good biocompatibility and non allergic tissue response, cobalt chrome has an elastic modulus lesser than of that of stainless steel [3]. There are still some complications associated with dense metallic implants such as sterile loosening being one major problem until the end prosthetic treatment leading to wavering of the implants and to obvious early revision [2, 4]. One reason in the loss of bone that has been identified is an insufficient and reduced load transfer into the adjacent bone stock due to the stiffness mismatch between implant and the surrounding tissue, known as stress shielding [5, 6]. In order to decrease the mass and implant stiffness, to increase the load shift through the bone up to tissue and to progress the raw boned fixation for enough long-term stability, open porous structures can be engaged for designing customized or typical implants [7–10]. There are different potential of incorporate porosity into dense materials, like space receptacle particles [11], direct 3D printing [12] or additive manufacturing [13, 14]. The final offers the advantage of fabricating capricious geometrical structures with a spacious range of involuntary properties [15], shapes [16, 12] and material designs with prearranged nonlinear properties or the use of multipart, cellular and purposeful mesh arrays [8].

By using additive manufacturing techniques open-porous scaffolds can be made-up in a wide range of design variations with specific properties satisfying both mechanical and biological prerequisites. It is vital to know that geometry deviation, the worn material as well as the engaged additive manufacturing techniques have an influence on scaffold properties. The fabrication of structures using additive manufacturing technologies is grave characterized by the belongings of the process; moreover the accuracy of the fabricated scaffolds is compulsory for the evaluation of the perfunctory properties of open-porous structures. In addition, the properties of fabricated scaffolds can be openly controlled by the geometrical shape and the structural opus of the scaffolds [17]. Thus, mechanical properties can be custom-made to the patient-precise requirements of the relevance area, e.g. an elastic modulus parallel to that of person cortical bone. Instead, persuade of the unit cell orientation has to be resolute. Furthermore, the functional correlation involving porosity and elastic modulus as well as the strut orientation and the loading angles play a critical role in the assessment of open-porous structures. Challis et al. investigated the directional hope of the Young's modulus on conventional encrusted grid structures and topology-optimized structures by via the finite element method. By using topology optimization techniques they generated a design with isotropic characters. i.e., that had no route of low stiffness. By scaling the size of the struts, any exact elastic modulus could be implemented; still they found a non-linear relationship between the porosity and the elastic modulus.

Amendment and drop of the scaffolds stiffness is essential in order to prevent stress shielding of the nearby bone tissue just about stiff implants [5, 6]. As we know the mechanical properties of bone depends on several factors and the properties vary about a broad range [17]. So it is extremely important that the mechanical properties of an additive manufactured construction for a medical application has the same values as the bone. This should be the way to defeat the mismatching of elastic properties stuck between the human bone and the metallic implant in the boundary and with that the troubles caused by stress shielding of the bone due to the extreme stiffness of the metallic implant. But the direction of the structural elements from end to end the production influences the mechanical properties obviously.

## 1.1 Design of the Open-Porous Scaffold

For this study structural can developed by CAD software, the design data exported to fabrication through SLM technology. In this study a twisted design of crossing struts had reated. Two different designs in varying dimensional parameters in width (a), depth (b) and height (c) strut diameter (d) are used and the designed properties of the structural composition are investigated.

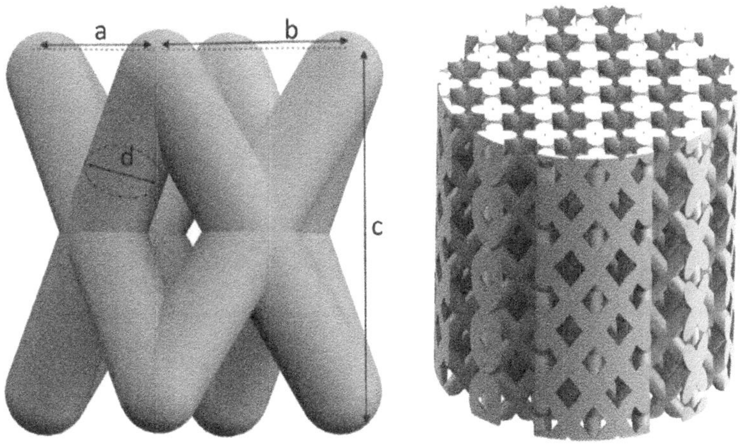


Based on structural unit, heavy cylindrical scaffolds modeled as a linear pattern of the sole unit. Based with DIN EN 50106 (Testing of Metallic Materials, Compression Test) the ratio between height (H) and outer diameter (D) of the model in between 1 and 2 in order to prevent buckling of the scaffolds under load compression. All parts were built with a height (H) of 18.0 mm and a diameter (D) of 17.0 mm. The design of the sole unit and the geometry for scaffold parts (twisted) is shown in Fig. 1 and the geometric components for the investigated models are listed in Table 1.

**Table 1** Overview of geometric parameters for the unit structure variation relating to the investigated designs of large-size scaffolds for mechanical testing. All values are derived from CAD data and are given in mm

| Configuration | Height c (mm) | Width a (mm) | Depth b (mm) | Strut diameter d (mm) |
|---|---|---|---|---|
| T1 | 3.00 | 2.12 | 2.12 | 0.90 |
| T2 | 4.00 | 2.83 | 2.83 | 1.10 |

**Table 2** Spatial orientation of the twisted designs T1 and T2

| Orientation type | T2-I | T2-II | T2-III | T2-IV | T2-V | T1-I | T1-IV | T1-V |
|---|---|---|---|---|---|---|---|---|
| Orientation | 0° | 45°–y-axis | 45°–y-axis 45°-x-axis | 90°-y-axis | 90°-y-axis 45°-x-axis | 0° | 90°–y-axis | 90°-y-axis 45°-x-axis |
| Unit cell (Z, X) | | | | | | | | |
| Scaffold (Z, X, Y) | | | | | | | | |
| Scaffold (Z, X) | | | | | | | | |
| Scaffold (Z, Y) | | | | | | | | |
| Scaffold (X, Y) | | | | | | | | |
| Number of unit cells in a scaffold | 64 | 64 | 72 | 60 | 60 | 192 | 182 | 210 |

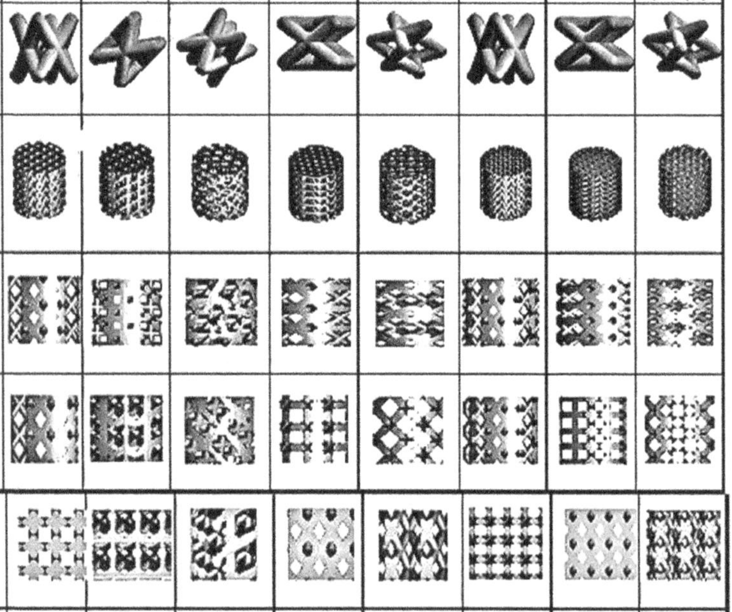

Design difference for the purpose of anisotropic material properties for the segmentation of anisotropic material properties the spatial direction of the twisted design has rotated within the external cylindrical shape of the scaffolds being perfunctorily tested (Table 2). The model parameters for the turning round of the twisted scaffolds are the same as scheduled in Table 1 with the listed strut diameter.

The sole organization has been rotated by 45° and 90° approximately the y-axis and in combination with the x- and the y-axis, in that order.

This direction categorizes that all specimens built in the same SLM process. The twisted design T2 manufactured in orientation types I, II, III, IV and V. The twisted design T1 prototyped in orientation types I, IV and V. The orientation of the unit cell and the consequential scaffold shapes are shown in the listed overview. The views lateral view and top view are total representations of the scaffold configurations. The illustrations allow for a better accepting of the modeling of the unit cell and the behavior struts. The final structure has been rotated by 90° around y-axis T-90-0 (IV) and rotated around both axes T-90-45 (V).

## *1.2 Porosity*

Even though the cell structure remains the same there are alters in the porosity of the specimens caused by a mix together of structural cell specimens. In order to confirm the accuracy and the reproducibility of the manufacturing process compared to the CAD data, porosity for the scaffolds with the twisted design under dissimilar loading orientations was calculated. Porosity for CAD scaffolds as well as for fabricated scaffolds was calculated according to the subsequent equations.

$$\text{Porosity}_{\text{CAD}} = (1 - \text{Vstr}/Vcyl) \times 100\%$$

where Vstr is the volume of the CAD scaffold struts and Vcyl is the overall volume enclosed by the outer periphery.

$$\text{Porosity}_{\text{SLM}} = (1 - \rho\text{sc}/\rho 0) \times 100\%$$

where 0 is the density of non-porous cobalt chrome (i.e. 8.5 g/cm$^3$) and sc is the density of the manufactured scaffolds, average bone mineral density is around 3.88 g/cm$^3$ in males and 2.90 g/cm$^3$ in females (Table 3).

The FEA is a tool that allows calculation of stress sharing in the contact area of implants with bone. The purpose of this work is to evaluate the things of biomechanical aspects of implant design with fascination conditions on the implant, the bone–implant interface. This paper focuses on various scaffolds models implants with fixation of implant to bone with dense stress shielding effect is simulated using FEA software to compare the results with the mechanical design variations.

**Table 3** Scaffolds cad porosities described in percentage

| Model | T1-1 | T1-4 | T1-5 | T2-1 | T2-2 | T2-3 | T2-4 | T2-5 |
|---|---|---|---|---|---|---|---|---|
| % CAD porosities | 55.5 | 55 | 54.5 | 60 | 58.5 | 56 | 59.5 | 60 |

## 1.3 Finite Element Analysis of Designed Scaffolds

Three dimensional cobalt chrome model is performed in solid modeling software, The model exported by ANSYS and subdivided into nodes and elements. Collection of elements is called mesh and it is make mesh optimization to get more accuracy results. Mesh optimization is occurred until the FEA results and analytical solutions are similar to each other. After the determination of mesh model, apply boundary conditions, initial conditions, and load stress are applied on the designed model. Scaffold bottom plate (A) side is fixed and the upper side scaffold with pressure plate is applied on the 1000 N force on upper plate surface. Then FEA analysis of seven models is performed and results are compared with the each other. It is aimed to determine the total von-mises stress and the elongation.

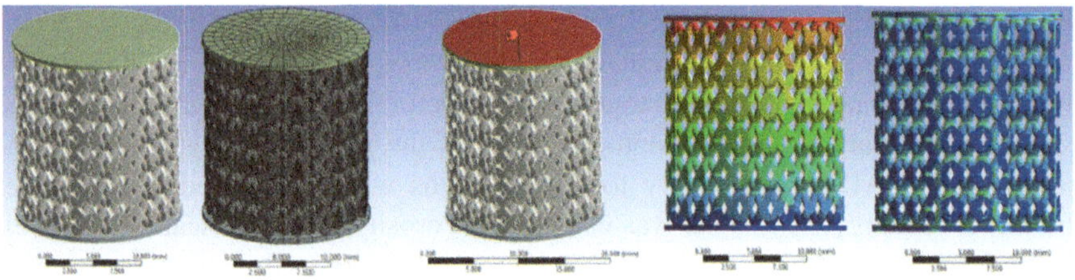

1.SCAFFOLD WITH FIXED AND MOVABLE PLATES FOR COMPRESSION ANALYSIS ,2.AFTER MESHING CONVERTED TO FEA MODEL ,3.DEFINING THE MATERIAL PROPERTIES AND BOUNDARY CONDITIONS APPLIED COMPRESSIVE FORCE 1000 NEWTON ON PLATE B, 4.STATIC STRUCTIURAL TOTAL DEFORMATION WHEN APLIED LOAD, 5. STATIC STRUCTURAL EQUILANT VON-MISES STRESS DISTRIBUTION

## 1.4 Results and Discussion

Analysis results are classified according to the different pore sized scaffold models listed below table.

**Table 4** Classified scaffold model pore size analysis results

| Model | T1-1 | T1-4 | T1-4 | T1-5 | T2-1 | T2-2 | T2-3 | T2-4 | T2-5 |
|---|---|---|---|---|---|---|---|---|---|
| Max. deformation in mm | 0.03692 | 0.17948 | 0.12019 | 0.050616 | 0.093846 | 0.27017 | 0.15981 | 0.2746 | 0.03692 |
| Max. equivalent von-mises stress in MPa | 75.501 | 464.8 | 256.2 | 78.749 | 427.14 | 565.46 | 260.42 | 171.35 | 75.501 |
| % CAD porosities | 55.5 | 54.5 | 55.5 | 60 | 58.5 | 56 | 59.5 | 60 | 55.5 |

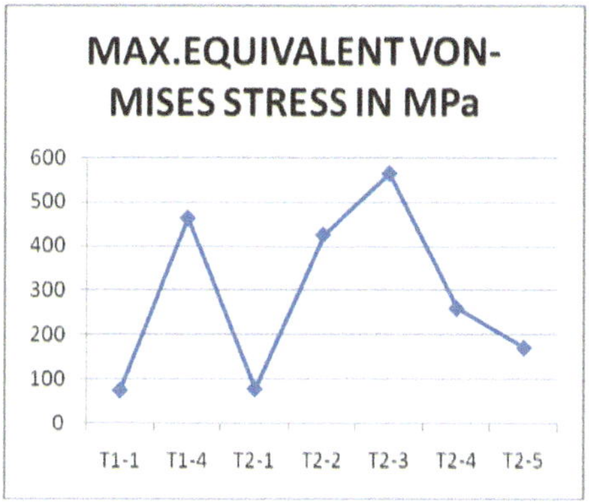


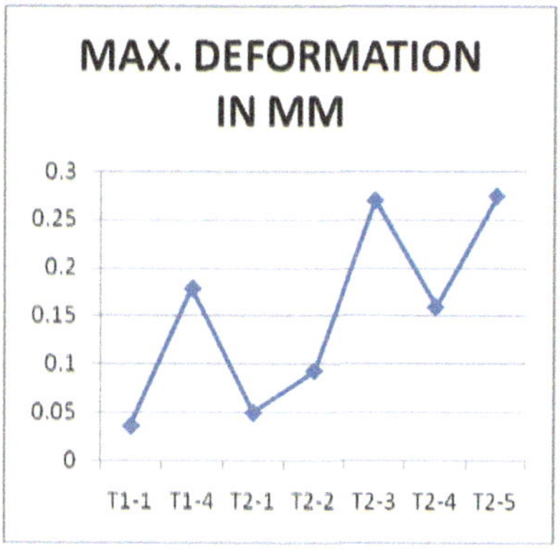


Finite Element Analysis was carried out using ANSYS 13 software, to compare the stress fields developed under uniaxial compression. It was found from the compression testing that the behavior of the material is linear at 1000 N. In finite element simulation the same compressive load was applied, and the finite element analysis was carried out using ANSYS 13, for the eight compression testing models. The details of the experimental stress and deformations are presented in Table 4. Porosities varied between 60.0 and 56.0% in model T2 and between 55.5 and 60% in model T1. The examination of the developed models shows divergences between CAD—porosity, all other values lie in between them. Models T1-IV and T1-V correspond to the CAD dimensions. Models T1-IV and T1-V correspond to the CAD dimensions. Porosity of the scaffolds also has an influence on the structural designs. In the case of the examined structure and dimensions, this connection is directly represented in a functional relation to the characteristic values obtained in the analysis. Finite element analysis of cobalt chrome specimen was completed and thereby compared the properties with the properties with various twisted designs. Thus SLM process has been able to produce the customized bone porous scaffold with properties similar to human bone.

## References

1. Weißmann V, Bader R, Hansmann H, Laufer N (2016) Influence of the structural orientation on the mechanical properties of selective laser melted Ti6Al4V open-porous scaffolds. Artic Mater Des
2. Long M, Rack HJ (1998) Titanium alloys in total joint replacement—a material science perspective. Biomaterials 19:1621–1639
3. Vannoort R (1987) Titanium—the implant material of today. J Mater Sci 22:3801–3811
4. Bahraminasab M, Sahari BB, Edwards KL, Frahmand F, Arumugam M (2013) Aseptic loosening of femoral components—materials engineering and design considerations. Mater Des 44:155–163
5. Tigges S, Stiles RG, Roberson JR (1994) Complications of hip-arthroplasty causing periprosthetic radiolucency on plain radiographs. Am J Roentgenol 162:1387–1391

6. Merle C, Streit MR, Volz C, Pritsch M, Gotterbarm T, Aldinger PR (2011) Bone remodeling around stable uncemented titanium stems during the second decade after total hip arthroplasty: a DXA study at 12 and 17 years. Osteoporos Int 22(11):2879–2886
7. Niinomi M, Nakai M (2011) Titanium-based biomaterials for preventing stress shielding between implant devices and bone. Int J Biomater
8. Harrysson OL, Cansizoglu O, Marcellin-Little DJ, Cormier DR, West HA (2008) Direct metal fabrication of titanium implants with tailored materials and mechanical properties using electron beam melting technology. Mater Sci Eng C Biomim Supramol Syst 366–373
9. Murr LE, Gaytan SM, Medina F, Lopez H, Martinez E, Machado BI, Hernandez DH, Martinez L, Lopez MI, Wicker RB, Bracke J (2010) Next-generation biomedical implants using additive manufacturing of complex, cellular and functional mesh arrays. Philos Trans R Soc A 368:1999–2032
10. Marin E, Fusi S, Pressacco M, Paussa L, Fedrizzi L (2010) Characterization of cellular solids in Ti6Al4V for orthopaedic implant applications: trabecular titanium. J Mech Behav Biomed Mater 3:373–381
11. Wegrzyn J, Kenton R, Hanssen AD, Lewallen DG (2015) Performance of porous tantalum versus titanium cup in total hip arthroplasty: randomized trial with minimum 10-year follow-up. J Arthroplast
12. Devika D, Arumaikkannu G (2011) Evaluation of mechanical behaviour of bone, implant and bone–implant interface by numerical simulation of two surgical fixation procedures using finite element analysis. Int J Comput Appl Technol 42(2/3)
13. Adachi T, Osako Y, Tanaka M, Hojo M, Hollister SJ (2006) Framework for optimal design of porous scaffold microstructure by computational simulation of bone regeneration. Biomaterials 27:3964–3972
14. Borden M, Attawia M, Khan Y, Laurencin CT (2002) Tissue engineering microsphere-based matrices for bone repair: design and evaluation. Biomaterials 23:551–559
15. Boschetti F, Raimondi MT, Migliavacca F, Dubini G (2006) Prediction of the micro-fluid dynamic environment imposed to three-dimensional engineered cell systems in bioreactors. J Biomech 39(3):418–425
16. Ahmed SE, Raheem E, Hossain S (2011) Absolute penalty estimation. In: Springer International Encyclopedia of statistical science. New York
17. Begum SR, Arumaikkannu G (2015) Computational fluid dynamic analysis and additive manufacturing of customised bone scaffolds. Prog Comput Fluid Dyn 15(3)

# Effect of Silicon Carbide as Filler Reinforcement on the Mechanical Properties of Glass Fiber Epoxy Composites

**Antara Bhattacharjee, S. Sujaykhosh and B. K. Nanda**

## 1 Introduction

Composites are structural materials made up of one or more constituents at the macroscopic level and are insoluble in each other. Depending on the type of reinforcements, composites are divided into fiber-reinforced, particulate-reinforced, flake-reinforced composites. Fiber reinforced (FR) polymer composites are widely used owing to high strength to weight ratio, higher stiffness, strength and fatigue properties. Among these FR composites, glass fiber reinforced epoxy (GFRE) composites are commonly used because these are cheap, readily available and can be easily manufactured. These find application in aerospace, automotive, marine, structural and machine tool industries as they have low densities, high strength, high stiffness, impact resistant and are non-corrosive. Further, the strength and stiffness of the FR composite can be increased by reinforcing it with particulates.

Raghavendra et al. [1] observed that calcium sulphate filled fiber reinforced epoxy composites exhibit high tensile strength and low impact strength and increased hardness compared to neat composites. The mechanical properties of the fiber reinforced polymer matrix primarily depend upon the strength and modulus of the fibers, the strength and chemical stability of the matrix and the effectiveness of the bonding between fibers and matrix in transferring stress across the interface. Nassar et al. [2] observed that unfilled carbon fiber reinforced epoxy composites exhibit homogeneous nature whereas CFRE composites with filler reinforcements exhibit heterogeneous nature due to the uneven distribution of the fine particles in

A. Bhattacharjee · S. Sujaykhosh (✉) · B. K. Nanda
Department of Mechanical Engineering, NIT Rourkela, Rourkela, Odisha, India
e-mail: sujaykhosh24@gmail.com



K. Antony and J. P. Davim (eds.), *Advanced Manufacturing and Materials Science*,
Lecture Notes on Multidisciplinary Industrial Engineering,
https://doi.org/10.1007/978-3-319-76276-0_17

the matrix. Agarwal et al. [3] have shown that addition of SiC particulates enhances the thermo-mechanical properties of GFRE composites. Suresha et al. [4] have demonstrated that filling glass epoxy composite with graphite particulates improves the mechanical properties such as hardness, tensile strength, tensile modulus, percentage elongation as well as wear resistance.

Therefore, the current research focuses on the evaluation of the effect of SiC particulate reinforcements on mechanical properties of GFRE composites. Mechanical properties such as tensile strength and flexural strength are estimated. Moreover, the best composite is identified which can be used as a substitute for structures in automotive and aerospace industries.

## 2 Material Selection and Methodology

Bi-directional woven E-glass fiber mat is used as the fiber reinforcement and epoxy under trade-name Bisphenol-A-Diglycidyl ether is used as the matrix material. The hardener (chemical name: tri-ethylene tetra-amine) is mixed with epoxy in the weight ratio of 1:10. Silicon Carbide particulate of average diameter 250 μm is used as filler reinforcement. 15 layers of glass fiber mats are used for a single specimen.

Hand lay-up technique is used for fabrication of the laminated composites. This is one of the most common methods used for the preparation of fiber reinforced composites in small amounts as it is simple and economic. It is an open molding process. The mold is coated with an anti-adhesive release agent. The filler material is mixed with epoxy and hardener according to requirement. This mixture is applied over the fiber mat placed in the mold using a brush. Similarly, 14 more layers of fiber mats and the epoxy hardener mixture are added one by one. Mold relieving spray and sheet is used to cover the mold surface. Adequate care is taken to ensure proper mixing of epoxy and hardener as well as matrix coating on the glass fibers. Hand roller is used frequently to remove any trapped gas or tangled fibers. The composite is cured under load of 400 N at room temperature for 24 h.

The combinations of weight percentage of the various constituents used for manufacturing composites are given in the Table 1.

**Table 1** Material designation and compositions

| Composite | Glass fiber wt% | Matrix wt% | SiC wt% |
|---|---|---|---|
| C1 | 46 | 50 | 4 |
| C2 | 44 | 50 | 6 |
| C3 | 42 | 50 | 8 |
| C4 | 40 | 50 | 10 |

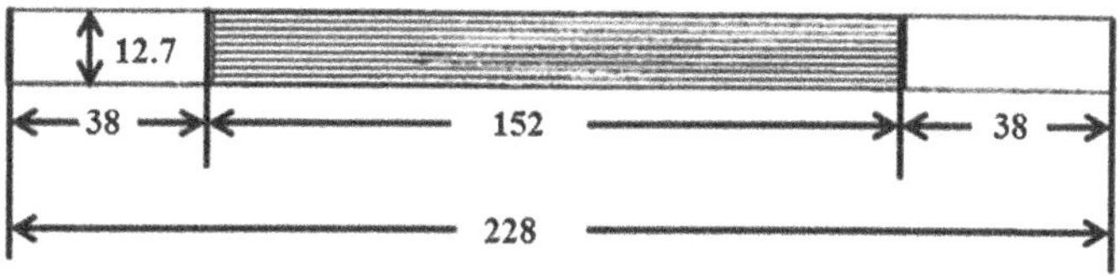


**Fig. 1** Specimen geometry for tensile test samples

# 3 Experimentation

Samples for tensile tests are made as per ASTM standard ASTM3039-76. The effective length of the specimens is maintained 152 mm with an average thickness of 5 mm. The schematic configuration of specimens for the tensile test is shown in Fig. 1. Samples for flexural tests are prepared as per ASTM standard ASTM D2344-84. Tensile and Flexural tests are done in the Universal testing machine (INSTRON UK/SATEC 600 kN). Microstructural analysis of the fractured surfaces of the broken tensile samples is done using scanning electron microscopy (SEM, Model: JEOL-JSM 6480LV).

# 4 Results and Discussions

## *4.1 Tensile Properties*

The engineering stress strain curves for the 4 composite samples are obtained from the tensile tests. The variation of tensile strength with SiC particles content is shown in Fig. 2. It can be seen that the tensile strength value increases with the increase in the SiC weight percentage. But the rate of increase of tensile strength decreases for higher SiC content. The high strength and aspect ratio of SiC particles as well as the good interfacial adhesion between epoxy and SiC particles attribute to the increased tensile strength. The tensile strain value decreases with increase in the SiC content. This shows that the presence of SiC particulates makes the composite more brittle.

The microstructural analysis of fracture surface of the broken tensile samples are given in Fig. 3. We can see from the images that more than one mode of failure is involved in the 4 cases. The main observable failure modes are matrix cracking, fiber matrix debonding and fiber failure with fiber elongation. The uneven distribution of the SiC fillers in the matrix material and the presence of cavities are the reasons for the variation in failure modes. Moreover, it can be observed that with increase in SiC content, average diameter of pores have increased and a number of pores have joined to form a bigger crack as evident from Fig. 3c, d.

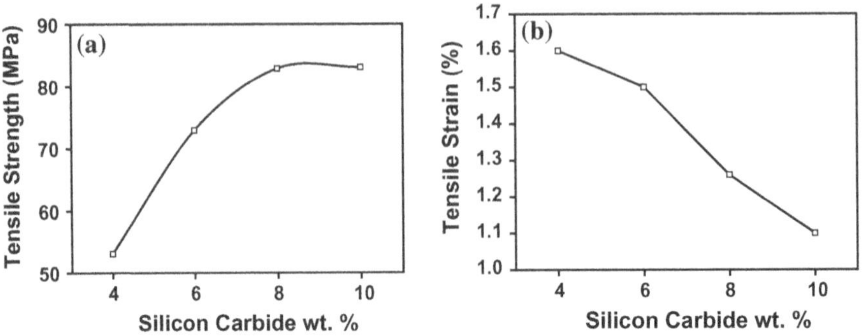


**Fig. 2** Effect of SiC particle content on tensile property of GFRE composites **a** tensile strength, **b** tensile strain

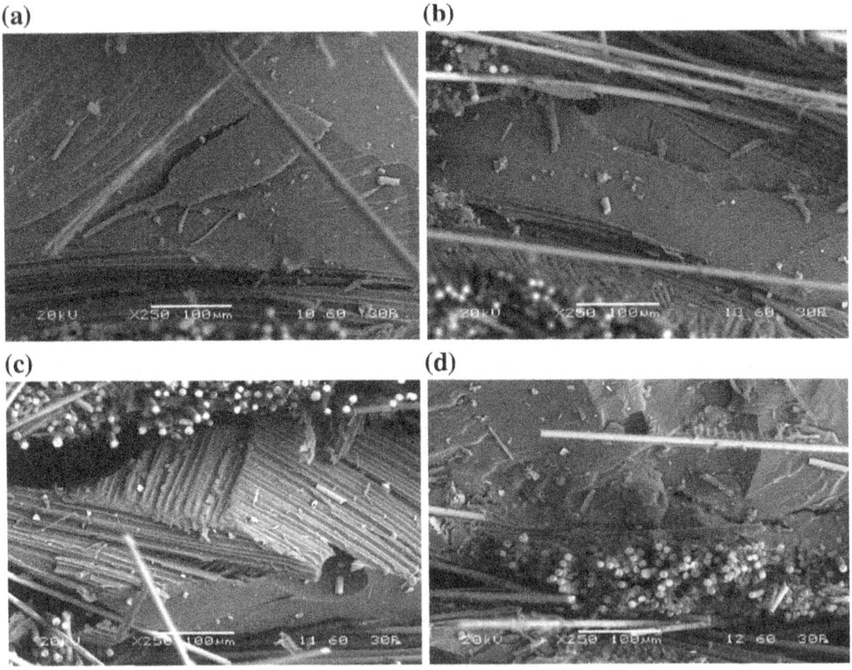


**Fig. 3** SEM images of the fractured GFRE composite samples with **a** 4 wt%, **b** 6 wt%, **c** 8 wt% and **d** 10 wt% SiC particle content

## 4.2 *Flexural Test*

The flexural strength of the GFRE composites reinforced with SiC fillers with different % filler loading is plotted in Fig. 4. From Fig. 4, it is observed that the

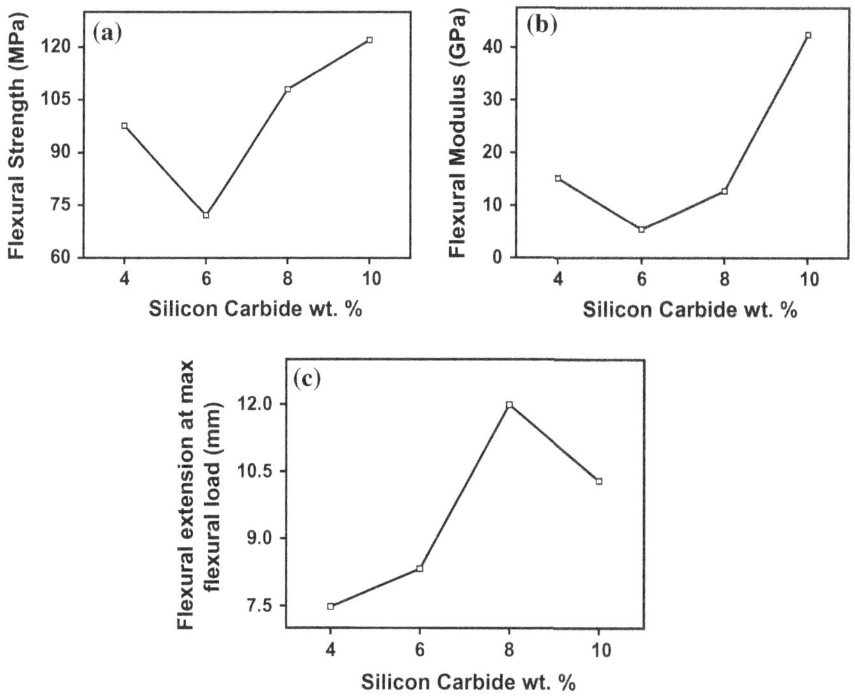


**Fig. 4** Effect of SiC particle content on **a** flexural strength, **b** flexural modulus and **c** flexural extension at maximum flexural load of GFRE composites

flexural strength of the composite material increases with increase in SiC content. The composite C2 shows an exception to this and have lower flexural strength than C1. This is due to an unexpected early failure of the composite by delamination since the matrix material between two fiber layers was unable to hold or transfer the load.

## 5 Conclusions

The study of the effect of SiC reinforcements on the mechanical properties of glass fiber epoxy composites has resulted into the following conclusions:

- The tensile strength of the glass fiber reinforced epoxy composite increases with the increase in SiC content. However, the percentage elongation of the samples is found to decrease with increase in SiC content.

- The microstructural analysis of the fracture surface is done. The different failure modes observed in the composites are matrix cracking, fiber matrix debonding and fiber failure.
- The flexural strength of the glass fiber reinforced epoxy composite is increased with the addition of SiC particulates.

## References

1. Raghavendra PN, Amaresh Kumar D (2015) Effect of silicon carbide and calcium sulphate on E-glass/epoxy composites. IJMET 6(7):08–15
2. Nassar A, Nassar E (2014) Thermo and mechanical properties of fine silicon carbide/chopped carbon fiber reinforced epoxy composites. Univers J Mech Eng 2(9):287–292
3. Agarwal G, Patnaik A, Sharma RK (2013) Thermo-mechanical properties of silicon carbide-filled chopped glass fiber-reinforced epoxy composites. Int J Adv Struct Eng 5 (1):21–28
4. Suresha B, Chandramohan G, Renukappa NM, Siddaramaiah (2006) Mechanical and tribological properties of glass-epoxy composites with and without graphite particulate filler. J Appl Polym Sci 103:2472–2480

# Optimization of Weld Bead Geometry of C263 Nimonic Alloy Electron Beam Weld Joints

**Shaik Rifshu Hussain, K. P. V. Krishna Varma, Sriniwash Anand and C. Lavanya**

## 1 Introduction

### *1.1 Electron Beam Welding*

Electron Beam Welding (EBW) is an integration welding process in which a high-velocity electrons beam is used to join two materials. The kinetic energy of the electrons is transformed into heat upon impact which melts the work pieces and let it flow. EBW can also be performed under vacuum conditions to stop dissipation of the ray. It had been developed by the German scientist Karl-Heinz Steigerwald, at the time working on various electron beam applications.

In fusion attachment the two surfaces of metals soften either directly or in conjunction with the filler and therefore the inter-mixed liquefied metal solidify to make a welded joint. A strong metallurgical bond is made giving adequate strength to the joint. Procedure to form a hole using EBM has been shown in Fig. 1 and the schematic of the set up used is as shown in the Fig. 2.

### *1.2 EBW Equipment*

The equipment includes Electron gun, generating the electron beam, Working chamber evacuated to "low" or "high" vacuum, Work piece manipulator (positioning mechanism), Supply and control/monitoring electronics.

S. R. Hussain (✉) · K. P. V.Krishna Varma · S. Anand
Mechanical Department, CMR College of Engineering and Technology, Hyderabad, India
e-mail: sunnyhussain140@gmail.com

C. Lavanya
Mechanical Department, GITAM University, Visakhapatnam, India



K. Antony and J. P. Davim (eds.), *Advanced Manufacturing and Materials Science*, Lecture Notes on Multidisciplinary Industrial Engineering,
https://doi.org/10.1007/978-3-319-76276-0_18

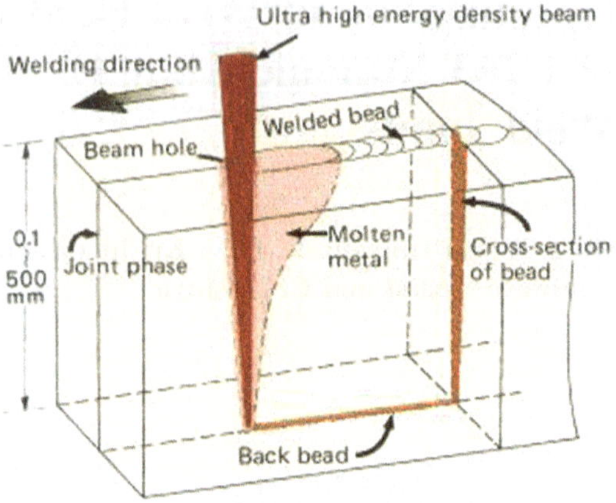


**Fig. 1** Key hole formation

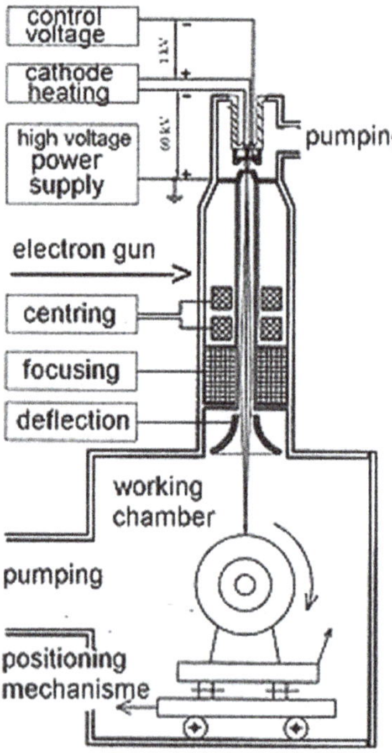


**Fig. 2** Line diagram of EBW equipment

## 1.3 Nimonic C263 Alloy

Nimonic C263 Alloy, an air thawed nickel-base alloy, was developed by Rolls-Royce Ltd. to provide a sheet material that can be fabricated readily and can show improved ductility in welded assemblies to replace Nimonic 80A alloy. It had been designed as sheet material to satisfy specific standard in terms of proof stress

**Table 1** Composition of Nimonic C263 alloy

| Element | C | Si | Mn | S | Ag | Al | B | Bi |
|---|---|---|---|---|---|---|---|---|
| % | 0.04 | 0.40 | 0.60 | 0.007 | 0.0005 | 0.60 | 0.005 | 0.0001 |
| Element | Co | Cr | Cu | Fe | Mo | Pb | Ti | Ni |
| % | 21.0 | 21.0 | 0.20 | 0.70 | 6.1 | 0.0020 | 2.4 | Balance |

and creep strength. It is now accessible in all standard forms. The techniques used for this alloy for welding are similar to those that are in general use for other age-hard enable nickel-base alloys. It's a precipitation-hard enable nickel-chromium-cobalt alloy with an addition of Mo for solid-solution strengthening. It's high strength and corrosion resistance in conjunction with smart formability and high-temperature plasticity in welded structures. The alloy is very appropriate for sheet applications. Utilized in gas turbines for rings, casings, and varied sheet fabrications. The composition of the nimonic alloy has been as shown in Table 1.

## 2 Literature Survey

### *2.1 Electron Beam Welding*

Nickel-base super alloys always had been a great interest in the materials research field for aerospace and industrial applications. The properties like creep and de-oxidation make them the core material for aero-engine components in hot sections, such as turbine discs, blades and casings. In such applications, the possibility to assemble different parts by welding is a fundamental task for the design and industrialization of these products. Thus, the study of the welding properties of this type of material is given most importance in order to obtain sound welds. Their hot cracking sensitivity, arising from a synergy of phenomena such as wide solidification temperature interval, chemical composition and thermal stresses, is the most important problem which has to be solved since it enormously compromises the low cycle fatigue resistance. For this reason, high power density welding technologies, with a selected attention to EBW, are the foremost favourite. In fact, they ensure a minimum size of the Fusion Zone (FZ) as well as of the Heat-Affected Zone (HAZ) and decrease in distortion and residual stresses. In particular, the high vacuum, which characterizes the EBW, avoids material to get contaminated and yields deep penetration. Experimental results allotted on a restricted number of nickel-base alloys, showed that it's attainable to get sound welds by optimizing the attachment speed, pre-heating, beam focus and attachment current [1].

### 2.1.1 Bead Geometry

The weld bead shape is an indication of bead geometry which affects the load carrying capability of the weldments and variety of passes required to fill the groove of a joint. Reinforcement is the crown height of the weld bead from the bottom plate. It affects the strength of the weld joint [2]. The weld bead geometry has been shown in Fig. 3 and the flow chart for RA has been shown in Fig. 4.

The process parameters are as shown in Table 2. The input and output parameters as shown in the Table. The input parameters being voltage, beam current, focus current, speed and standoff distance. The output parameters being the bead height, bead width, back bead width, weld penetration.

### 2.1.2 Input and Output Process Variables in EBW

Koleva [3, 4] used statistical regression analysis to develop the correlation of input parameters. Dey et al. [5] used nonlinear regression analysis to predict the bead profile.

## 2.2 *Full Factorial Experiment*

A full factorial experiment consists is used when there are two or more factors, with different attainable values or levels and whose experimental values consider taking into account all the possible combinations of these levels across all such factors. A full factorial design may also be called a fully crossed design.

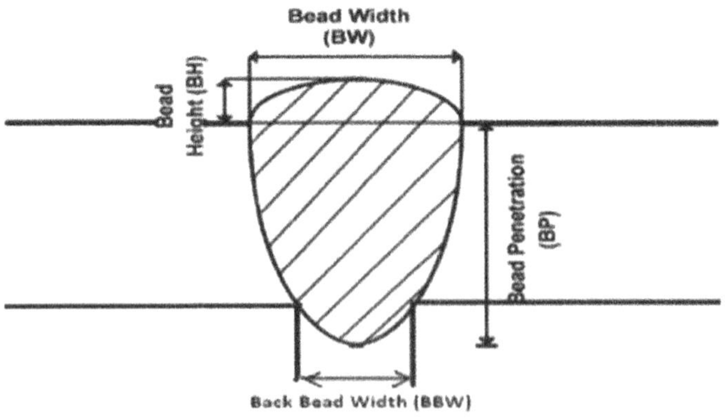


**Fig. 3** Weld bead geometry

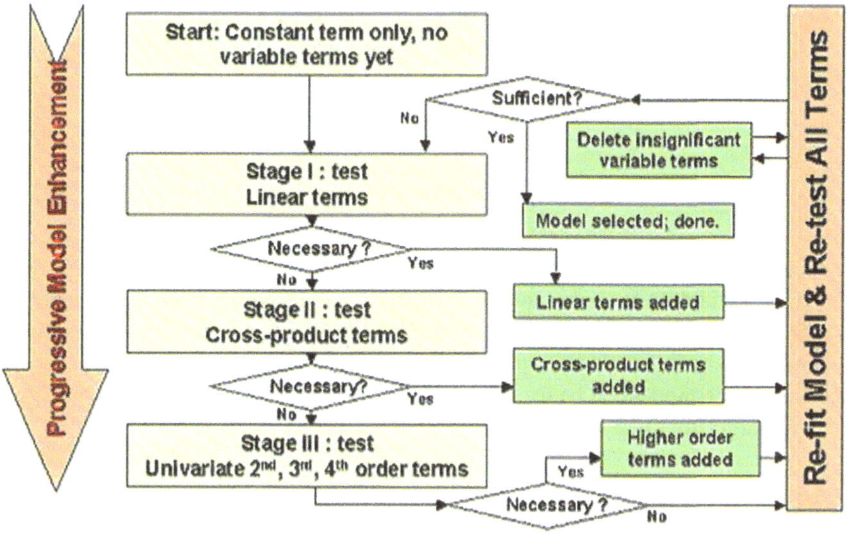


**Fig. 4** Flow chart of RA

**Table 2** Process variables

| Input parameters | Output parameters |
|---|---|
| 1. Voltage | 1. Bead height |
| 2. Beam current | 2. Bead width |
| 3. Focus current | 3. Back bead width |
| 4. Speed | 4. Weld penetration |
| 5. Standoff distance | |

## 2.3 *Regression Analysis*

It is used to determine the relationships among variables. For modelling and analysing several variables, there are several approaches. When the focus is on the relationship between a dependant and one or more independent variables it is widely used to estimate the relationship between variables.

## 2.4 *Genetic Algorithm*

The Genetic Algorithms (GA) is defined as a procedure search based on the mechanisms of genetics and natural selection. These types of algorithms solve difficult problems in a fast, suitable and reliable way. GA was developed by John

Holland. Regression analysis is carried out response-wise. Thus, it might not be able to capture the complete dynamics of a process. These problems could be solved using genetic algorithms [6].

### 2.5 Response Surface Methodology

It analyses the relationships between several explanatory variables and one or more response variables. In order to obtain an optimal response from a sequence of designed experiments, RSM is widely used. In factorial experiment or a fractional factorial design first-degree polynomial model can be used. However, the second-degree model can be used for optimization (maximize, minimize, or attain a specific target for) [7].

## 3 Experimental Setup

### 3.1 Bead on Plate Welds and Mounting

The experiments were carried on Mnemonic C263 Alloy plates of size 110 mm 95 mm × 1.6 mm. 81 bead-on-plate runs were done according to the process parameter combinations. The specimens were cut into 2 in. samples as marked in Fig. 6 using an abrasive cutter at a minimum distance of 30 mm from the edge to make mounts. Mounts of two inch diameter were made one for each run and each test with Bakelite powder on an Auto-mount machine. The samples are placed on the repository and taken inside, after which the mixture is heated to 120 °C and then cooled to 60 °C after ten minutes. The grinding and polishing of all the mounts was carried out on a polishing machine using the 400, 600, 800, 1000 and 1200 grade papers in the same order. The polished samples were etched using Ralph's reagent to reveal the micro-structure. All the etched samples were observed with an Olympus optical microscope having a magnification range of 50× to 2000× to measure the bead profile. The bead profiles in terms of output parameters, namely bead width (BW), back bead width (BBW), bead height (BH) and bead penetration (BP) for all the runs and test cases were measured and recorded.

### 3.2 Regression Analysis

Regression analysis was carried out using Minitab-15 software, on the experimental data collected as per the full factorial design. The parameters were expressed in a coded form as a function of process parameters, namely V, I, S and F. The Fitted

Plots indicate a relation between the input parameters (V, I, S, F) and the Response Variables. Quadratic form of input Model is seen to yield a better correlation for V, I and S as indicated by their respective R-Sq values in each graphs; so the Quadratic model is adopted for further analysis of Regression over the Linear model. The results of the regression analysis has been shown in Fig. 5 and the bead on the plate is as shown in Fig. 6.

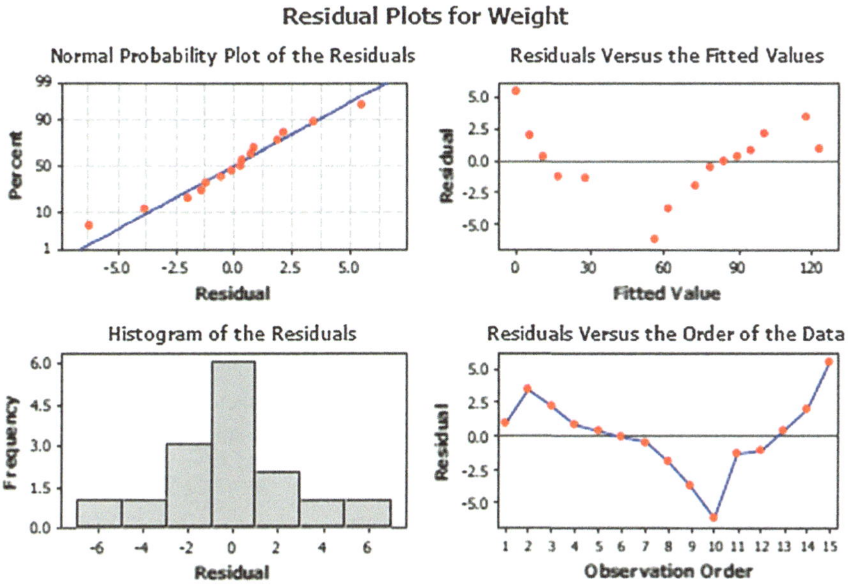


**Fig. 5** Interpretation of graphs

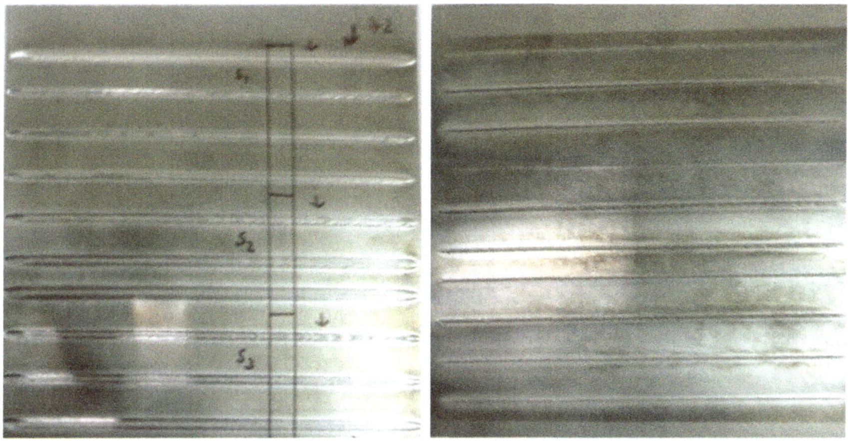

**Fig. 6** Bead on plate weld

The corresponding test statistics are 2.53, −1.52 and 2.97 indicating that these lie on the extreme ends of t-curve. Thus we reject the null hypothesis and conclude that the predictors play a significant role in the regression model. The standard deviation of the error terms is 11.9087.

## 3.3 *Genetic Algorithm*

A genetic algorithm (GA) is a method for solving both constrained and unconstrained optimization problems based on a natural selection process that mimics biological evolution. The algorithm repeatedly modifies a population of individual solutions. Over successive generations, the population "evolves" towards an optimal solution. The run time visualizations using genetic algorithm has been depicted in Fig. 7.

## 3.4 *Response Surface Methodology—Surface Plots*

These plots show how a response variable relates to two factors based on a model equation. Contour and surface plots are useful for establishing desirable response

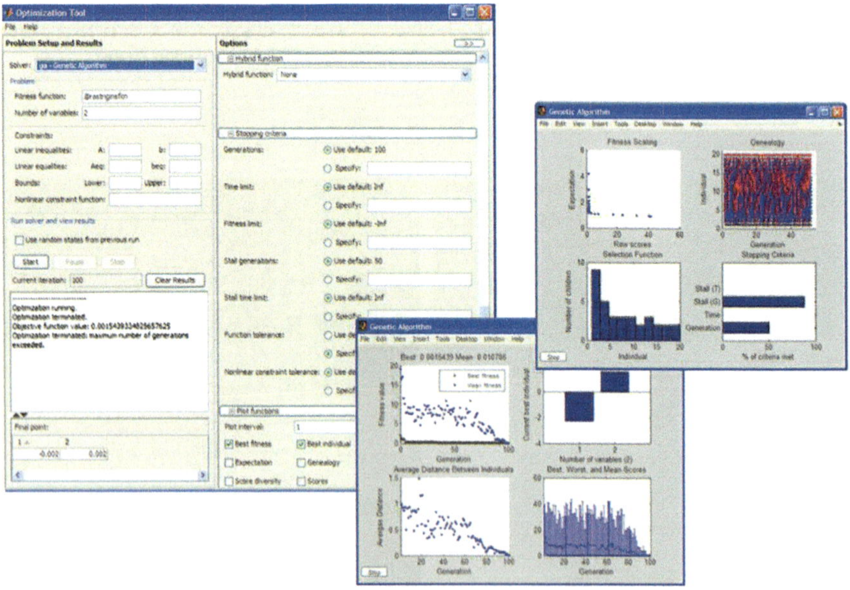

**Fig. 7** Run-time visualizations (right) generated while the function is being optimized using genetic algorithm plot functions selected in the optimization tool (left)

values and operating conditions. In a contour plot, the response surface is viewed as a two-dimensional plane where all points that have the same response are connected to produce contour lines of constant responses. A surface plot displays a three-dimensional view that may provide a clearer picture of the response surface.

# 4 Results and Discussions

## *4.1 Full Factorial Table for the Experiment*

**Material**: Nimonic C263 Alloy
**Thickness of Plate (T) mm**: 1.6 mm
**Dimensions mm**: 110 × 95
**No. of Factors**: 4
**No. of Levels**: 3
**Standoff Distance (D) mm**: 283 mm

The results of the full factorial experiments after a set of design of experiments are carried out and are tabulated and shown in Table 3 and are as follows.

## *4.2 Regression Analysis and RSM Surface Plots*

**Linearity check**:
The Fitted Plots indicate a linear relation between the input parameters (V, I, S, F) and the Response Variable. Quadratic form of input Model is seen to yield a better correlation for V, I, S and F as indicated by their respective R-Sq values in each graphs; so the Quadratic model is adopted for further analysis of Regression.

The Normal Probability Plot as seen in Figs. 8 and 9 for V, I, S, F showing their Goodness of Fit (Anderson-Darling statistic) with $<0.005$ P indicates that the observed data can be adequately fit into a statistical model as majority of the data lie within the control limits in the linear normality distribution.

## *4.3 Genetic Algorithm*

The optimal GA-parameters were obtained through a thorough and careful study.

The necessary conditions that were considered for this optimization are as follows:

| | |
|---|---|
| Initial population | : 100 |
| Cross over fraction | : 0.5 |

**Table 3** Full factorial table for the experiment

| Run order | Voltage (V) kV | Beam current (I) mA | Welding speed (S) m/min | Focusing current (J) A | Bead height (BH) mm | Bead width (BW) mm | Weld penetration (WP) mm | Back bead width (BBW) mm | Remarks |
|---|---|---|---|---|---|---|---|---|---|
| 1 | 45 | 12 | 0.6 | 103 | 0.069 | 2.5 | 1.262 | 0.5401 | [a] |
| 2 | 45 | 12 | 0.8 | 103 | 0.056 | 2.125 | 1.229 | 0.521 | [a] |
| 3 | 45 | 12 | 1.0 | 102 | 0.105 | 1.656 | 3.1 | 0.399 | [a] |
| 4 | 45 | 12 | 1.0 | 103 | 0.0678 | 1.829 | 1.08 | 0.717 | [a] |
| 5 | 45 | 14 | 0.8 | 103 | 0.09 | 2.09 | 1.58 | 0.44 | [a] |
| 6 | 45 | 14 | 1.0 | 103 | 0.07 | 1.76 | 1.26 | 0.48 | [a] |
| 7 | 45 | 16 | 1.0 | 103 | 0.07 | 1.86 | 1.751 | 0.32 | [b] |
| 8 | 50 | 12 | 0.6 | 109 | 0.066 | 2.652 | 1.49 | 0.56 | [a] |
| 9 | 50 | 12 | 0.8 | 109 | 0.06 | 1.857 | 1.47 | 0.454 | [a] |
| 10 | 50 | 14 | 0.6 | 107 | 0.0680 | 1.8185 | 1.349 | 0.48 | [a] |
| 11 | 50 | 14 | 1.0 | 109 | 0.0985 | 1.990 | 1.602 | 0.511 | [a] |
| 12 | 50 | 16 | 1.0 | 109 | 0.04 | 2.07 | 1.83 | 0.46 | [b] |
| 13 | 55 | 12 | 0.6 | 115 | 0.06 | 2.78 | 1.4 | 0.56 | [a] |
| 14 | 55 | 12 | 0.8 | 115 | 0.05 | 2.23 | 1.2 | 0.6 | [a] |
| 15 | 55 | 12 | 1.0 | 114 | 0.092 | 2.063 | 1.528 | 0.6731 | [a] |
| 16 | 55 | 16 | 0.6 | 113 | 0.0915 | 1.9191 | 1.171 | 0.662 | [a] |

[a]Best weld bead geometry condition
[b]Weld conditions of bead with partial penetration

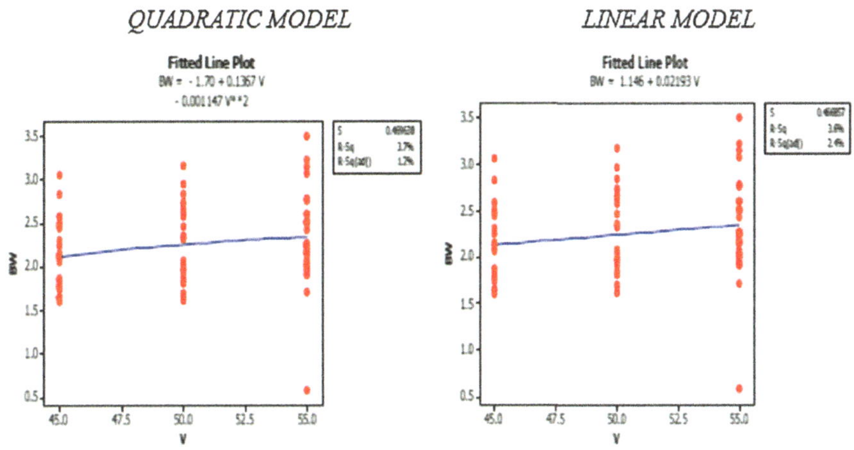


**Fig. 8** Fitted line plots

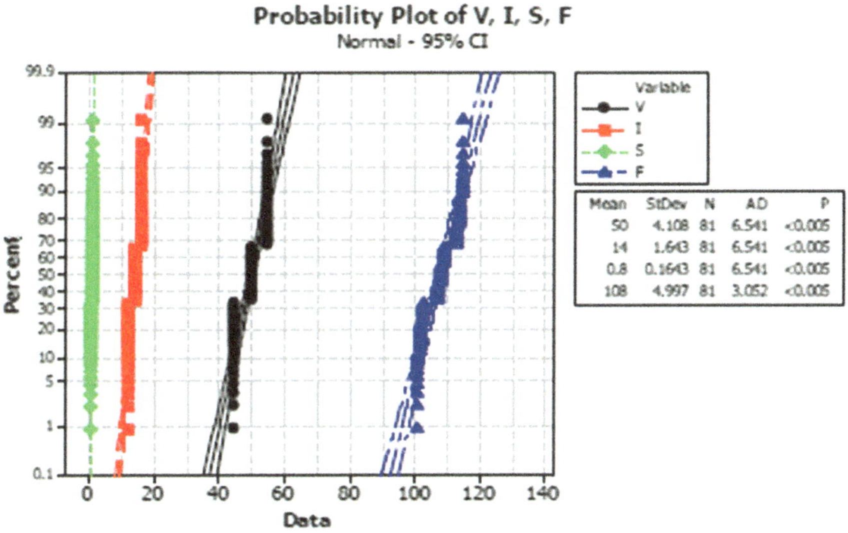


**Fig. 9** Probability plot

| | |
|---|---|
| Selection function | : roulette |
| Mutation function | : adaptive feasible |
| Lower bounds[V, I, S, F] | :[45, 12, 0.6101] |
| Upper bounds[V, I, S, F] | :[55, 16, 1115] |

With these conditions different optimized parameters were achieved i.e.

1. The optimized **Bead Width** was found to be 1.6275 **mm.**
   And the point at which this optimized value was obtained was:

$$V = 45.374\ \text{kV};\ I = 12.03\ \text{mA};\ S = 0.997\ \text{mm/min};\ F = 101.016\ \text{A}$$

2. The optimized **Bead Height** was found to be **−0.0806 mm.**
   And the point at which this optimized value was obtained was:

$$V = 45.511\ \text{kV};\ I = 12\ \text{mA};\ S = 1\ \text{mm/min};\ F = 101\ \text{A}$$

3. The optimized **Weld Penetration** was found to be 2.093 **mm.**
   And the point at which this optimized value was obtained was:

$$V = 46.397\ \text{kV};\ I = 12\ \text{mA};\ S = 0.998\ \text{mm/min};\ F = 101\ \text{A}$$

4. The optimized **Back Bead Width** was found to be 0.3856 **mm.**
   And the point at which this optimized value was obtained was:

$$V = 45.002\ \text{kV};\ I = 16\ \text{mA};\ S = 0.601\ \text{mm/min};\ F = 115\ \text{A}$$

5. To get the optimized area minimum back bead width equation was found to be:

i.e., maximum bead penetration, minimum bead width, and minimum bead height for a thickness of 1.6 mm, the equation was found to be:

$$\boldsymbol{AREA}_{nimonic}c263 = \left(\frac{(BW)^2 + 4(BH)^2}{8(BH)}\right)^2 cos^{-1}\left(\frac{(BW)^2 - 4(BH)^2}{(BW)^2 + 4(BH)^2}\right) - \left(\frac{(BW)^3}{16(BH)}\right) + \left(\frac{(BH)(BW)}{4}\right) + 0.8(BW) + \frac{(BP)(BBW)}{2}$$

Lower bounds [BH, BW, WP, BBW]: [−0.0806, 1.6274, 1.73, 0.03856]
Upper bounds [BH, BW, WP, BBW]: [0.315, 3.419, 2.09395, 1.407]
Thus the optimized value was found to be **1.7421 mm$^2$**.
And the point at which this optimized value was obtained was:
BH = 0.049 mm; BW = 1.627 mm; WP = 1.73 mm; F = 0.386 mm.

## 5 Conclusion

The weld bead geometry of the electron beam welds are optimized with respect to the variation in the welding parameters such as gun voltage, beam current, focus current and welding speed. Regression equations relating the welding parameters and the weld bead geometry parameters are developed and optimized using genetic algorithms. The optimized parameters are mentioned below.

- The input parameters for achieving an optimum bead width of 1.6275 mm were:

$$\mathbf{V = 45.374\,kV; I = 12.03\,mA;\ S = 0.997\,mm/min;\ F = 101.016\,A}$$

- The input parameters for achieving an optimum bead height of -0.0806 mm were:

$$\mathbf{V = 45.511\,kV;\ I = 12\,mA;\ S = 1\,mm/min; F = 101\,A}$$

- The input parameters for achieving an optimum weld penetration of 2.093 mm were:

$$\mathbf{V = 46.397\,kV;\ I = 12\,mA; S = 0.998\,mm/min; F = 101\,A}$$

- The input parameters for achieving an optimum back bead width of 0.3856 mm were:

$$\mathbf{V = 45.002\,kV;\ I = 16\,mA;\ S = 0.601\,mm/min;\ F = 115\,A}$$

- The optimum bead geometry to achieve a minimum bead area of 1.7421 $mm^2$ were:

$$\mathbf{BH = 0.049\,mm;\ BW = 1.627\,mm; WP = 1.73\,mm;\ F = 0.386\,mm.}$$

## References

1. Ferroa P, Zambonb A, Bonolloa F (2004) Investigation of electron-beam welding in wrought Inconel 706—experimental and numerical analysis
2. Effect of welding parameters on bead geometry and flux consumption—Shodhganga
3. Koleva E (2001) Statistical modeling and computer programs for optimization of the electron beam welding of stainless steel. Vacuum 62:151–157
4. Koleva E (2005) Electron beam weld parameters and thermal efficiency improvement. Vacuum 77:413–421
5. Dey V, Pratihar DK, Datta GL, Jha MN, Saha TK, Bapat AV (2009) Optimization of bead geometry in electron beam welding using a genetic algorithm. J Mater Process Technol 209:1151–1157

6. Jha MN, Pratihar DK, Dey V, Saha TK, Bapat AV (2011) Study on electron beam butt welding of austenitic stainless steel 304 plates and its input output modelling using neural networks. Proceed Inst Mech Eng Part B J Eng Manuf 225:2051 (originally published online 12 Sept)
7. Box GEP, Draper N (2007) Response surfaces, mixtures, and ridge analyses, 2nd edn. (Empirical model-building and response surfaces, 1987). Wiley

# Development of a System for Simultaneous Rotation of Multiple Stacks of Uranium Dioxide Pellets

M. Sriniwash Anand, Meesala Shivasai, G. Sai Sasank, Gorkala Madhu and Shaik Rifshu Hussain

## 1 Introduction

Our present system is consisting of a live roller tray designed in such a way where it can be moved upon the linear motion guide. The rods of the tray are supported by the miniature bearing at both ends. The rods fixed between the bearings are extended further so that the worm and worm wheel mechanism can be used for rotating the pellets. A 150 rpm motor is present at the left side of the tray to run the gear mechanism. To the left side of tray a linear motion guide is fixed. A cantilever beam is supported at the end of the linear motion guide for proper transferring of pellets.

## 2 Design Components and Their Specifications

The components used to design [1] are as follows

(1) Live roller tray.
(2) Gears: worm and worm wheel.
(3) Bearings: deep groove ball bearings.
(4) Synchronous motors.
(5) Linear motion guide.
(6) Cantilever beam.
(7) Pneumatic cylinders.
(8) Flexible coupling.

M. S. Anand (✉) · M. Shivasai · G. S. Sasank · G. Madhu · S. R. Hussain
Mechanical Department, CMR College of Engineering and Technology, Hyderabad, India
e-mail: anandsrinivash@gmail.com



K. Antony and J. P. Davim (eds.), *Advanced Manufacturing and Materials Science*,
Lecture Notes on Multidisciplinary Industrial Engineering,
https://doi.org/10.1007/978-3-319-76276-0_19

## 3 Live Roller Tray

Live roller tray is the development of rod tray in which the cylindrical rods are welded at both the ends. The rods of live roller tray are supported in the bearings and are rotated using the help of gear mechanism. The live tray on which the pellets are automatically rotated for visual inspection consists of 21 cylindrical rods each of diameter 12.5 mm placed equidistantly from each other at the fixed space of 19.5 mm. The tray ends which supports the cylindrical rods are separated by distance of 600 mm. The end support is provided with gap so that pellets can pass easily while transferring for inspection from rod tray to live tray and vice versa.

## 4 Gears

A **gear** is a rotating machine part having cut teeth, which mesh with another toothed part to transmit torque [2].

## 5 Worm Gear Design Parameter

Worm gear provides a normal single reduction range of 5:1 to 75-1. The pitch line velocity is ideally up to 30 m/s. The efficiency ranges from 98% for lowest ratio to 20% or highest ratio. The worm box is designed to disperse heat to the surroundings and lubrication is essential requirement. As the frictional heat generation is generally to high.

## 6 Gear Details

According the requirement of the design and limitations of various gears worm gear mechanism is perfectly suitable for the rotation of pellets.

**Table 1** Ground worm shaft

| Number | Axial module | Number of starts | Lead angle | Hand of thread | Total length | Face width | Pitch diameter | Outside diameter |
|---|---|---|---|---|---|---|---|---|
| KWG0.5-R1 | 0.5 | 1 | 3°11' | R | 410 | 12 | 9 | 10 |

## 7 Gear Specifications

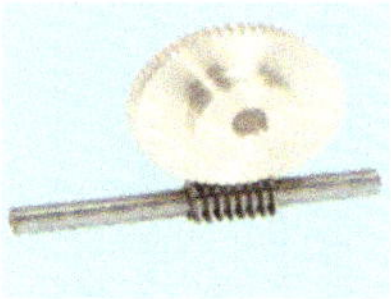

Worm gear model KWG0.5-R1

MODULE 0.5 GROUND WORM SHAFT [3] (Table 1).
MODULE 0.5 WORM WHEELS (Tables 2 and 3).

## 8 Bearings

Bearing is a device used for relative motion between machine members. The rods of the live roller tray are to be rotated for the automated rotation of pellets for which the rods are supported in the bearings at both ends of tray [4].

## 9 Radial Shaft Load Between Bearings

(a) Radial loads between bearings

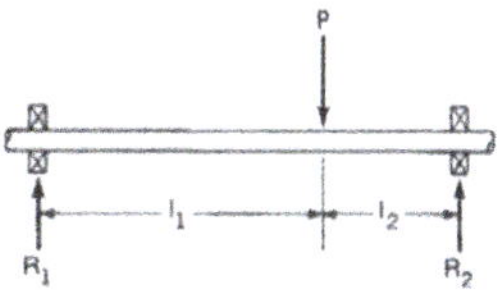


p= radial load
$R_1$, $R_2$= bearing loads
$l_1$, $l_2$= distance from radial load to bearing load
$R_1 = (l_2 P)/(l_1 + l_2)$
$R_2 = (l_1 P)/(l_1 + l_2)$

(b) Over hung radial load

P= radial load
$R_1$, $R_2$= bearing loads
$l_1$, $l_2$= distance from radial load to bearing load
$R_1 = (l_2 P)/(l_1 + l_2)$
$R_2 = (l_1 P)/(l_1 + l_2)$

**Table 2** Worm wheel

| Number | Transverse module | Reduction ratio | Num. of teeth | Num. of starts | Hand of thread | Bore | Hub diameter | Diameter of pitch | Outer diameter |
|---|---|---|---|---|---|---|---|---|---|
| AG0.5-20R2 | 0.5 | 1/10 | 20 | 1 | R | 4 | 9 | 10 | 11 |

| Number | Total length | Mounting distance | Hub length | Allowable torque (Nm) | Backlash (mm) | Weight (kgf) | Face width |
|---|---|---|---|---|---|---|---|
| AG0.5-20R2 | 12 | 9.5 | 7 | 0.27 | 0.06–0.13 | 0.01 | 5 |

**Table 3** Specifications of worm and worm wheel

| Catalogue number | KWG | AG |
|---|---|---|
| Precision | KHK W 001 grade 2 | KHK W 002 grade 2 |
| Reference angle | Axial | Rotating plane |
| Gear teeth | JIS 3 type standard full depth | Standard full depth |
| Normal pressure angle | 20° | 20° |
| Material | SCM640 | CAC702 |
| Heat treatment | Teeth induction hardening after thermal refining | – |
| Tooth hardness | 50–55 HRC | – |
| Tooth surface finish | Ground | Cut |
| Datum reference surface for gear cutting and grinding | Shaft (ground position) | Bore |
| Secondary operations | Possible except tooth area | Possible |

## 9.1 Determination of Basic Load Rating

The basic load rating C for a rating life of one million revolution for radial and angular contact ball bearing, except filling slot bearing, with balls which is not larger than one in diameter is given by equation:

$$C = F_c(i\cos\alpha)^{0.7}Z^{2/3}D^{1.8}N.$$

where

- $i$ number of rows of balls in the bearings.
- $\alpha$ nominal angle of contact.
- $Z$ number of balls per row.
- $D$ ball diameter = $(D\cos\alpha)/dm$.
- $d_m$ pitch diameter of ball races.

## 9.2 Bearing Details

After going through the bearing catalogues NTN 676AZZ deep groove shielded ball bearings are selected [5]. The bearing details and nomenclature are as follows-

676AZZ-first 6 represent deep groove ball bearing. 7 represent the diameter series seven, 6A represent nominal bore diameter 6 mm, and ZZ represent shield non-contact ZZ (Table 4).

**Table 4** Bearing specifications

| Inner diameter d | Outer diameter D | Width | $r_s$ (min) | Basic load rating | | Factor | Limiting speed | | Model |
|---|---|---|---|---|---|---|---|---|---|
| | | | | Static | Dynamic | | Grease | Oil | |
| 6 | 10 | 2.5 | 0.1 | 465 | 196 | 15.2 | 43,000 | 51,000 | 676AZZ |

# 10 Motor

A motor is the device that converts electrical energy to mechanical energy.

# 11 Motor Specification

The motor used are of synchronous type and specifications of selected motor are as follows. The motor number is 2SK4A-AULA [6].

**2SK4A-AULA** where 2 represent motor frame size, S represent motor type, K represent motor series, 4 represent output power, A represent motor shaft type and another A represent voltage, UL represent recognized and A represent inch size shaft.

## *11.1 Specification*

Table 5.

## *11.2 Motor Gear Head Specification*

Table 6.

# 12 Linear Motion Guide

## *12.1 LM Guide Details*

The detail of linear motion guide with model number KR55 20B 1380LP 00000 [7] are as follows

**Table 5** Motor specifications

| Model round shaft type | Power (W) | Voltage (VAC) | Frequency (Hz) | Current (A) | Starting torque (N-m) | Rated torque (N-m) | Rated speed (rpm) | Capacitor (μF) |
|---|---|---|---|---|---|---|---|---|
| 2SK4A-AULA | 4 | 115 | 60 | 0.24 | 22 | 22 | 1800 | 1 |

**Table 6** Gear head specification

| Model | Speed (rpm) | Gear ratio | Torque (N-m) | Weight (Ks) |
|---|---|---|---|---|
| 2GN12.5 KA | 114 | 12.5 | 0.22 | 0.7 |

Interpretation of model number

| KR55 | 20 | B | 1380L | P | 0 | 0 | 0 | 0 | 0 |
|---|---|---|---|---|---|---|---|---|---|
| Model no. | Ball screw lead | Type of nut | LM rail length | Accuracy grade | Presence/ absence of censor | Presence/ absence of cover | Presence/ absence of cover | Type of housing | Presence/ absence of intermediate housing |

And the last **0000** represents control number (Table 7)
**Bellows** (Table 8)

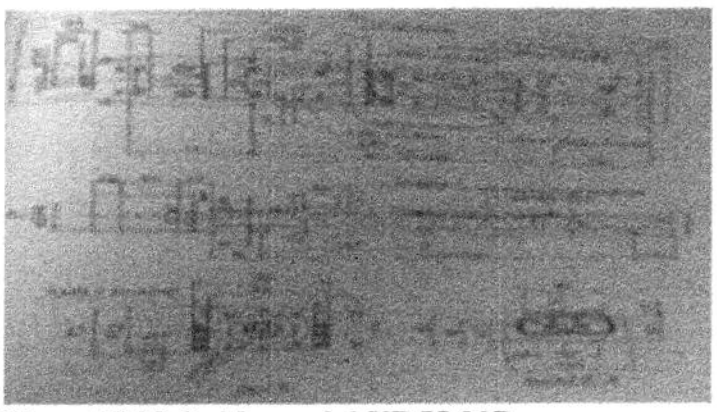

View of LM Guide model KR55 20B

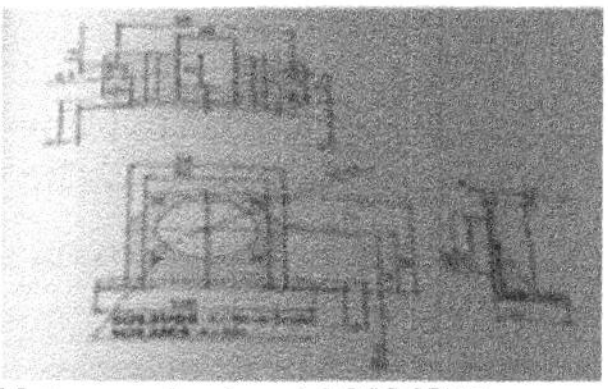

Motor gear head model-SOL2U08

## 13 Cantoliver beam

The dimensions of this cantilever beam are 500 mm in length, 30 mm in width and 12 mm in height. The rod will be obstructing the transfer of pellets is it's rectangular. So to avoid this and for easy motion and uninterrupted transferring of pellets

**Table 7** LM guide specifications

| LM rail length (mm) | Full length (mm) | Possible stroke (mm) | H (mm) | G (mm) | N | $n_1$ | Units total weight (kg) |
|---|---|---|---|---|---|---|---|
| 1380 | 1489 | 1098 | 90 | 15 | 10 | 7 | 28.6 |

**Table 8** Bellow specifications

| Model | LM rail length L | Stroke length S | Min/ max | Motor side A | Non-motor side B | $L_1$ | $L_2$ | W | T | $T_1$ |
|---|---|---|---|---|---|---|---|---|---|---|
| KR55 | 1380 | 998 | 77.4/0 75.4 | 127 | 127 | 223.1 | 128 | 154 | 77 | 42 |

the beam is provided with vertical projection of 15 mm which is controlled by pneumatic system [8].

## 14 Pneumatic Cylinders

Pneumatic cylinders are a mechanical device which produces forces, often in combination with movement, and are powered by compressed gases. To perform their function, the force is imparted by pneumatic cylinders by converting the potential energy of compressed gas to kinetic energy.

## 15 Flexible Coupling

Couplings are used in drive trains to transmit power and motion between two independent shafts that may not be properly aligned. Flexible coupling generally allow for some parallel and angular misalignment. Depending upon their design more misalignment may be accommodated, have higher torsional stiffness and load capacity, or be capable high speed [9].

## 16 Result and Calculation

### *16.1 Gear Selection Calculation*

Lead angle of worm = $\lambda$ = 12.5 mm
Pressure angle = $\phi$ = 20
Speed of pellet required = 10 rpm
Minimum number of teeth on worm gear = 21
Module = 0.5
Lead of worm = $l = P_a X n_W$, where $P_a$ = axial pitch of worm = $\pi X m$ and $n_W$ = number of starts = 1

Now lead of worm = $\pi * 0.5 * 1 = 1.571$
Velocity ratio = $N_1/N_2 = Z_2/Z_1$, $30/10 = 21/Z_1$
Number of threads in worm = 7

$$\text{Tan}\lambda = L/\pi d_1$$

Pitch diameter of worm = $d_1 = 2.255$
Pitch diameter of worm gear = $d_2 = Z_2 p/\pi = 21 * 1.571/\pi = 10.5$ mm
So the required gear should have pitch diameter of 10.5 mm.

## 16.2 Motor Power Calculation

Mass of each pellet = 32 gms
Rotation speed of pellet = 15 rpm
Diameter of pellet = 14 mm
Mass moment of inertia of cylindrical pellet = $I = Mr^2/2 = (0.032 * 0.007^2)/2 = 7.84 \times 10^{-7}$ kg $m^2$
Mass moment of inertia of stack of 30 pellets = $7.84 \times 10^2 * 30 = 2.352 \times 10^{-5}$ kg $m^2$
Let the acceleration to be 0.1 s
Angular acceleration = $\alpha = \dot{\omega}/0.1$
Where $\dot{\omega} = 2\pi N/60 = 2\pi * 15/60 = 15.7$ rad/$s^2$
Torque required to rotate pellets = $I * a = (2.352 \times 10^{-5}) * (15.7) = 7.39 \times 10^{-3}$ Nm.

## 16.3 ROD

Diameter of rod = 12.5 mm
Speed of the rod if pellet rotates at 15 rpm = $D_2/D_2 = N_1/N_2 = 12.5/14 = 15/N_2$, $N_2 = 16.8$ rpm = 17 (approx.)
Mass moment of inertia of rod = $I_r = Mr^2/2$ [8]
Mass pf cylindrical rod = density * volume = $(7 \times 10^3 * 7.36 \times 10^{-5}) = 0.515$ kg
$I_r = (0.515) * (6.25 \times 10^{-3})^2/2 = 1.005 \times 10^{-5}$ kg $m^2$
For 21 rods it would be = $I_r * 21 = 2.1 \times 10^{-4}$kg $m^2$
Angular acceleration of rod = $\alpha_r = \dot{\omega}_r/0.1 = 17.8$ rad/$s^2$
Torque required for rotation of rods = $I_r * \alpha_r = (2.1 \times 10^{-4}) * (17.8) = 3.75 \times 10^{-3}$ Nm
Bearing efficiency may be assumed as 95%

So due to frictional losses torque = $3.75 \times 10^{-3}/0.952 = 3.94 \times 10^{-3}$ Nm
Total torque = T = pellets torque + rods torque = $3.94 \times 10^{-3} + 7.38 \times 10^{-3}$ = 0.01132 Nm
Power required = $2\pi NT/60$ = 1.94 W.

## 16.4 Motor for LM Guide

Velocity of guide = v = 0.2 m/s
Lead = 20 mm = $20 \times 10^{-3}$ m
Length of tray = 600 mm = 0.6 m
Time required = 0.6/0.2 = 3 s
Required rpm of motor with lead 20 mm and time 3 s is, rpm = 30 * 60/3 = 600 rpm
Angular acceleration = $\dot{\omega}/0.1 = 2\pi N/60 * 1$ = 62.83 rad/s
Mass = density * volume = 7000 kg/m$^3$ * $\pi r^2 h$ = 7000 * $4.02 \times 10^{-4}$ = 2.81 kg
Mass moment of inertia of ball screw = $I = Mr^2/2 = 2.81 * 0.01^2/2 = 1.405 \times 10^{-4}$ kg m$^2$
Torque required = $T = I * \alpha = (1.405 \times 10^{-4} * 62.83)$ = 0.00882
Power = $2\pi * 600 * T/60$ = 0.55 W.

## 16.5 Calculation for Linear Motion Guide

LM guide section ($C$ = 31800 N; $C_o$ = 61900 N)
Ball screw section ($C_a$ = 3620 N; $C_{oa}$ = 9290 N)
Supporting bearing ($C_a$ = 7600 N; $C_{oa}$ = 3990 N)
Velocity = V = 200 mm/s
Acceleration = $\alpha$ = 1 m/s$^2$
Stroke = $l_s$ = 1100 mm, g = 9.81 m/s$^2$.

## 16.6 Imposed Load on Nut Block

$P_1 = mg + k_c mg * 40 = (25 * 9.81) + 2.83 \times 10^{-2} * 5 * 9.81 * 40 = 522.87$ N
During acceleration
$P_{1a} = P_1 + K_A * m\alpha * 193 = 596.69$ N
$P_{1at} = -K_B * m\alpha * 40 = 150$ N
During deceleration
$P_{1d} = P_1 - K_A * m\alpha * 193 = -201.32$ N
$P_{1Dt} = K_B * m\alpha * 40 = 150$ N.

## 16.7 Combined Radial and Thrust Loads

During uniform motion, $P_{1E} = P_1 = 522.87$ N
During acceleration, $P_{1aE} = P_{1a} + P_{1at} = 746.78$ N
During deceleration, $P_{1dE} = P_{1d} + P_{1dT} = 150$ N
Static safety factor, $F_s = C_0/p_{max} = 61900/522.87 = 118.38$
Rated life, $L = (c/F_w P_m)^3 * 50$.

## 16.8 Rated Life of Ball Bearing

During forward uniform motion, $F_{a1} = \mu mg + F = 0.005 * 25 * 9.81 + 2.5 = 3.726$ N
During forward acceleration, $F_{a2} = F_{a1} + m\alpha = 3.726 + 25 * 1 = 28.726$ N
During forward deceleration, $F_{a3} = F_{a1} - m\alpha = 3.726–25 = -21.27$ N
During reverse uniform motion, $F_{a4} = - F_{a1} = -3.726$ N
During reverse acceleration, $F_{a5} = F_{a4} - m\alpha = -3.726–25 = -28.726$ N
During reverse deceleration, $F_{a6} = F_{a4} + m\alpha = 21.274$ N
Static safety factor, $F_s = C_{oa}/F_{amax} = 9290/28.726 = 323.4$
Bucking load, $P_1 = \eta E i \pi^2/l_a^2 = (4 * \pi^2 * 2.06 \times 10^5 * 20^4 * \pi/4)/1489^2 = 28808.95$ N
Permissible tensile compressive load, $P_2 = \sigma\ (\pi/4)d_1^2 = 147 * (\pi/4) * 17.5^2 = 35357.64$ N.

## 16.9 Bearing Calculation

Load by pellets per mm on the rod = 1.77 g
Load by the self-weight of the rod per mm = 0.83 g
Combined load = 1.77 + 0.83 = 2.61 gm/mm
For UDL in midpoint of the rod the concentrated load will be acting i.e. at center of gravity as the combined load is less for calculation purpose the load is assumed as 10 gm/mm = $10 \times 10^3 * 600 = 6$ kg = 6 N
Let $R_A$ and $R_B$ be equivalent dynamic load at both the ends of rod
$W = R_A = R_B = 60/2 = 30$ N
Rating lift of ball bearing,
$L = (C/W)^K \times 10^6$, where L = rating life, C = Basic dynamic load = 465 N, W = equivalent dynamic load = 10 N (approx.), K = 3, for ball bearings
$L = (465/30)^3 \times 10^6 = 3723.875 \times 10^6$ revolution
Life in hours, $L_H = L/(60 * N) = (3723.875 \times 10^6)/(60 * 17) = 3.65 \times 10^6$ h.

## 17 Conclusion

The objective of the paper is to simultaneously rotate the multiple stacks of $UO_2$ pellets at adequate speed so that an operator can inspect pellets for any surface defects present on them. In this we have designed a system, where the live roller tray rods are rotated so that the pellets laced on it are rotated simultaneously.

The rotation can be interrupted by the operator as and when desired to pick out the defective pellets manually. After discussing various ideas/concepts, which were considered to be with minimum drawbacks, have been perused. All the input conditions such as drawing, rotation, operation and transferring have been successfully achieved. The only drawback in this system is to draw and transfer the pellets from rod-tray to the live roller tray without damaging them as they are made of soft material. The operator should understand the operation and work on it. The operator can reverse the rotation of pellets in between the rotation and can also stop it when he is interested during the inspection time. Hence the input and output conditions of drawing the pellets into live roller tray from rod tray and transferring back the pellets after inspection into the rod tray is achieved with on limitations occurring. In future, by further development this system can be upgraded to higher level where entire can be carried by using a remote control device.

## References

1. Venables M (2006) The IET Manuf Eng J 84(6)
2. https://en.wikipedia.org/wiki/Gear
3. KHK Stock Gears Catalogue
4. Kannaiah P (2006) Machine design. SciTech Publishers
5. NTN Small and Miniature Ball and Roller Bearings Catalogue
6. Synchronous Motors Catalogue
7. THK Integrated LM Guide and Ball Screw High-Rigidity/High-Precision Actuator Product Catalogue
8. Bhavikatti SS, Rajeshekarappa KG (2005) Engineering mechanics. New Age International Publishers
9. Suresh S (1998) Fatigue of materials. Cambridge University Press, Cambridge, England

# Investigation on the Influence of Nano Structured Zirconia Coating on the Corrosion Inhibition of SS 304 Stainless Steel

**A. V. Ajay, Sivakumar S. Nair, Sreejith Mohan and Y. Vaisakh**

## 1 Introduction

Corrosion resistance is an important material property. Reports indicate that 1/5th of the global energy and as much as 4.2% of gross national product (GNP) are depleted year after year due to corrosion [1]. Stainless steel is known for its superior resistance to corrosion in aggressive environments. However, it is additionally protected when used in chloride containing environments due to the susceptibility of localized corrosion [2]. Among the variety of stainless steels currently used in the industry, AISI 304 austenitic stainless steel is the most common. However, the low hardness, poor wear resistance, likelihood of pitting corrosion and stress corrosion cracking in chloride solution impose limitations to its applications. Hence, reduction in corrosion is among the priorities and challenges that scientists are devoted to achieve. Though it is difficult to completely eliminate corrosion, its intensity can be brought down by; (a) selecting appropriate materials for particular application, (b) using new alloys, (c) using corrosion inhibitors and (d) depositing protective films and coatings onto the metal surface. Generally, ceramics and ceramic coatings show good corrosion resistance in aggressive media on account of its superior passivity, insulating properties and tribological properties. Therefore, ceramic oxide films and coatings like $TiO_2$, $Al_2O_3$, $ZrO_2$, $SiO_2$, etc. can be achieved on metals to obtain enhanced surface properties [3–5]. Among these, $ZrO_2$ coatings are preferred due to its superior hardness, low thermal conductivity, good mechanical and chemical durability and high temperature properties. In addition to this, since the thermal expansion coefficient of $ZrO_2$ matches with many of the metals, the coatings are less prone to cracking

A. V. Ajay · S. S. Nair · S. Mohan (✉) · Y. Vaisakh
Sree Buddha College of Engineering, Nooranad, Kerala, India
e-mail: drsreejithmohan@gmail.com

A. V. Ajay
e-mail: ajayav007@gmail.com



K. Antony and J. P. Davim (eds.), *Advanced Manufacturing and Materials Science*, Lecture Notes on Multidisciplinary Industrial Engineering,
https://doi.org/10.1007/978-3-319-76276-0_20

or spalling. $ZrO_2$ films can be prepared by techniques such as chemical vapor deposition [6], plasma spraying [7], reactive sputtering [8], electron beam evaporation [9], pulsed laser deposition [10] and sol-gel synthesis [11]. Among these, the sol-gel dip coating method is an environmentally friendly technique of surface protection which avoids the need of traditional toxic pretreatments [12].

There are several works reporting the favorable influence of sol-gel derived $ZrO_2$ on the corrosion protection of steel. Li and co-workers [13] achieved sol-gel zirconia coating from zirconium-n-propoxide. In a separate work, acetyl acetone and ethyl acetoacetate were used to decrease the rate of hydrolysis of the precursor and improve the homogeneity of the coating [14]. Besides, several works on the favorable influence of sol-gel $ZrO_2$ coatings on the corrosion protection of stainless steel have been reported [15, 16].

However, the preparation of $ZrO_2$ nano films through the sol-gel dip coating technique involve several process parameters. Hence, a one parameter at-a-time approach fails to indicate the complex interaction of parameters and its effect on corrosion protection. In such situations, several statistical techniques like design of experiments, artificial neural networks etc. can be employed. Among the various DoE techniques, the taguchi method gives better estimate of parameter effects by conducting of minimal number of experiments. There are no reports on the application of taguchi method for identifying the effects of sol-gel $ZrO_2$ coating process parameters on the corrosion inhibition of AISI 304 austenitic stainless steel.

This work have manyfold objectives. Primarly, to deposit nano structured $ZrO_2$ on the surface of SS 304 austenitic stainless steel through the sol-gel dip coating technique. The second objective being the estimation of corrosion resistance of the modified surface using impedance and polarization tests and its comparison with the uncoated surface. Lastly, to identify the mechanism responsible for the variation in corrosion resistance for the modified surfaces.

# 2 Materials and Methods

## *2.1 Experimental Design*

Prior to the commencement of experiments, the coating process parameters which likely influence the properties of $ZrO_2$ coating have been distinguished through an extensive literature study. The molar concentration of the $ZrO_2$ sol, dipping time and annealing temperature were the selected process parameters. Through preliminary experimental trials in the laboratory a workable range of these parameters was selected as shown in Table 1.

**Table 1** Coating process parameters and their workable range

| | | | |
|---|---|---|---|
| Molar concentration of $ZrO_2$ sol (M) | 0.3 | 0.75 | 1.2 |
| Dipping time (min) | 30 | 60 | 90 |
| Annealing temperature (°C) | 500 | 600 | 700 |

**Table 2** Experimental runs and estimated $I_{Corr}$ values for nano $ZrO_2$ coated specimens

| Run | Concentration of $ZrO_2$ (M) | Time (min) | Annealing temp (°C) | $I_{corr}$ ($\mu A/cm^2$) |
|---|---|---|---|---|
| 1 | 0.3 | 30 | 500 | 4.85 |
| 2 | 0.3 | 60 | 600 | 5.32 |
| 3 | 0.3 | 90 | 700 | 1.97 |
| 4 | 0.75 | 30 | 700 | 3.15 |
| 5 | 0.75 | 60 | 500 | 3.72 |
| 6 | 0.75 | 90 | 600 | 4.54 |
| 7 | 1.2 | 90 | 600 | 2.89 |
| 8 | 1.2 | 30 | 700 | 2.49 |
| 9 | 1.2 | 60 | 500 | 9.03 |

Since there were three process parameters each at three levels, a full factorial design demands the conduct of $3^3$ experiments [17]. Nevertheless, since taking out these many experimental tests were not only pricey but also time consuming, the Taguchi methodology of DoE was employed in this study. Agreeing to this methodology, a set of nine experimental runs was to conducted as indicated in Table 2.

## 2.2 Nano $ZrO_2$ Coating

The methodology adopted for the deposition of nano $ZrO_2$ using the sol-gel dip coating technique is depicted in Fig. 1.

The precursor was Zirconium iso propoxide (98% pure, CDH chemicals). The solvent used was n-propanol (99.9% pure, Loba chemicals), and acetyl acetone (98% pure, Loba chemicals) was used as a chelating agent.

## 2.3 Coating Surface Characterization

The structural analysis of the coating was carried out using X-Ray Diffraction (XRD) Analysis (Rigaku Ultima III) and its morphological analysis was done using the Scanning Electron Microscope (SEM) (TESCAN VEGA SBH). From the XRD spectra, the average crystallite size of the coating was found using the Scherrer's equation as shown below:

$$T = \frac{0.9\lambda}{\beta \cos \theta} \tag{1}$$

**Fig. 1** Flow chart showing the preparation of nano $ZrO_2$ using sol-gel dip coating technique

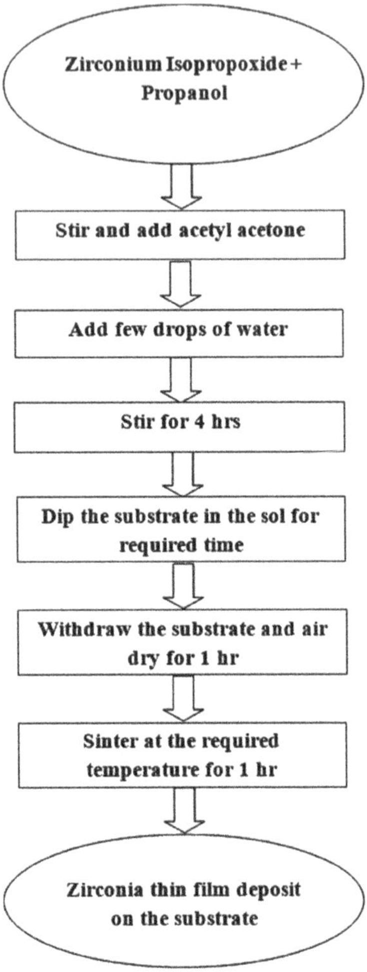


where T is the mean crystallite size, $\lambda$ is the X-ray wavelength; $\beta$ is the Full Width at Half Maxima (FWHM) of the peak corresponding to the diffraction angle, $\theta$.

The thickness of the coating was measured using a coordinate measuring machine (CMM, TESA MICRO HITE 3D).

## 2.4 Evaluation of Corrosion Resistance

To estimate the corrosion rate, the electrochemical impedance spectroscopy (EIS) was employed. The impedance experiments were conducted using the electrochemical system. In the three-electrode assembly, saturated calomel electrode (SCE) was employed as the reference electrode. A platinum electrode having an area of 1 $cm^2$ was taken as the counter electrode. Metal specimens having a cross section of 1 $cm^2$ was used as the working electrode, while 0.5 M NaCl solution was used as the electrolyte. The working area of the metal specimens were exposed to the electrolyte for 1 h. Following this, the polarization curves were scanned in narrow (±0.02 V vs. open circuit potential) and wide (±0.15 V vs. open circuit potential) potential window at a scan rate of 0.166 mV/s. The data obtained in the narrow potential window were used for the determination of the polarization resistance, i.e. linear polarization method, while the data obtained in wide potential window were analyzed by the TAFEL extrapolation method to get the corrosion current ($I_{corr}$) value. The corrosion resistance was estimated as a function of corrosion current. Lower values of the corrosion current implied higher corrosion resistance and vice versa.

## 2.5 Statistical Analysis

In the statistical analysis, the signal to noise (S/N) ratio was estimated. The equation for S/N ratio depends on the criterion of the performance parameter to be analyzed, corrosion current in this case. Since, a lower value of corrosion current was desired; the following equation for 'lower the better' type of S/N ratio was applied.

$$\eta = -10 \log \left[ \frac{1}{n} \sum_{i=1}^{n} y_i \right] \tag{2}$$

where η is the S/N ratio, $y_i$ the measured output value for the ith repetition and n the number of repetitions in a trial [18].

# 3 Results and Discussion

## 3.1 Structural Analysis of the Coating

As an example, the XRD spectrum of the $ZrO_2$ coated specimen corresponding to run no. 9 is shown in Fig. 2

The peaks obtained at 2θ values of 25.72° and 30.23° confirmed the presence of $ZrO_2$ [19]. The peak observed at 44.56° represents Fe in the base material (SS 304).

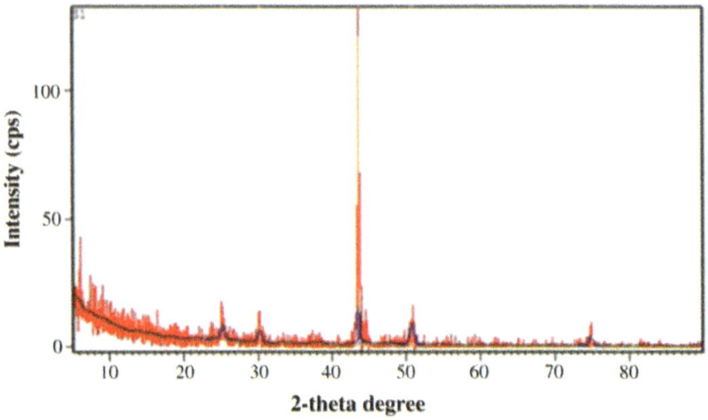


**Fig. 2** XRD spectra of nano $ZrO_2$ coated specimen

The average crystallite size of the coating calculated using the Scherrer's equation was found to be 25 nm confirming the nano structured nature of the coating.

## 3.2 *Electrochemical Measurements*

The potentiodynamic polarization curves for the coated and uncoated specimens are shown in Fig. 3

From the PDP curves, the $I_{corr}$ value was estimated by the TAFEL extrapolation method. The corrosion current for uncoated specimen was 12.45 μA/cm$^2$. Table 2 Shows the $I_{Corr}$ values obtained for the nano $ZrO_2$ coated specimens. A close examination of Table 2 interestingly reveals variation in $I_{corr}$ values with each

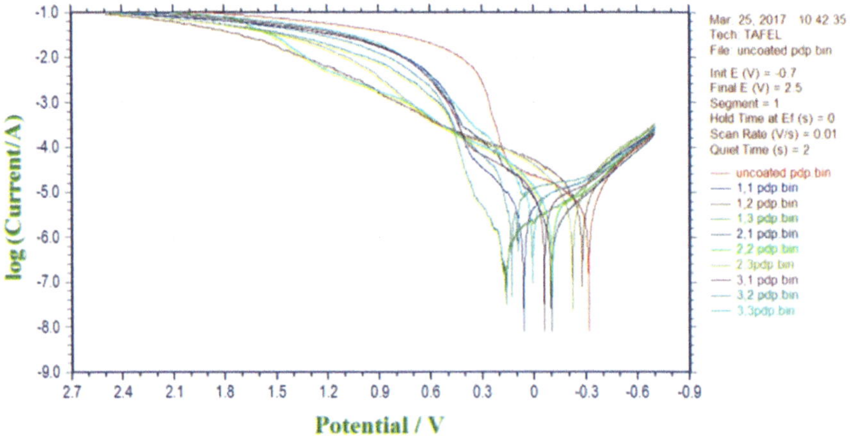


**Fig. 3** Potentiodynamic polarization curves

experimental runs. The lowest $I_{corr}$ value was noted for the experimental run 3 with 0.3 M molar concentration of $ZrO_2$, 90 min dipping time and 700 °C annealing temperature. Further, it could be observed that run 9 resulted in a higher $I_{corr}$ value of 9.03 μA/cm$^2$ and run 2 yielded a median $I_{corr}$ value of 5.32 μA/cm$^2$. The aforementioned three experimental runs were thenceforth referred to as critical runs for analysis. From the measured $I_{corr}$ value, the corrosion inhibition efficiency ($\eta_{pol}$) of the critical runs was calculated using the following relationship:

$$\eta_{pol} = \left[\left(I_{corr} - I'_{corr}\right)/I_{corr}\right] \times 100 \tag{3}$$

Where $I_{corr}$ and $I'_{corr}$ represents the corrosion current values of the uncoated and coated specimens respectively. Table 3 gives the $\eta_{pol}$ values of the three critical runs.

As evident from Table 3 the highest corrosion inhibition efficiency was exhibited by run no. 3 while the lowest was by run no. 9.

Further from the impedance measurements, the percentage of inhibition ($\eta_{EIS}$) was estimated from charge transfer resistance values by the following Eq. (4):

$$\eta_{EIS}\% = \frac{R_{ct} - R'_{ct}}{R_{ct}} \times 100 \tag{4}$$

where $R_{ct}$ and $R'_{ct}$ are the charge transfer resistances of working electrode with and without corrosion inhibition coatings.

The $\eta_{EIS}$ values of the critical runs are shown in Table 3. As evident, the highest $\eta_{EIS}$ value was noted for run no. 3 while the lowest belonged to run no. 9. To understand the reason behind the variation of corrosion inhibition behavior, the specimens corresponding to the three critical runs were subjected to morphological and thickness analysis. Figure 4 shows the SEM images of the three critical runs.

As apparent, the surface morphology of the coating corresponding to run no. 3 was more homogenous compared to the other two critical runs (2 and 9). This finding was in accordance with that of Tiwari et al. [20]. As the coating became more homogenous, the contact of electrolyte with the surface was prevented thereby enhancing the corrosion resistance.

The thickness of the coating for run nos. 3, 2 and 9 is shown in Table 3. It could be observed from Table 3 that, the experimental run with the lowest $I_{corr}$ value and

**Table 3** The measured $\eta_{pol}$ values for the experimental runs 3, 2 and 9

| Designation | $I_{corr}$ (μA/cm$^2$) | $R_{pore}$ (Ω) | $R_{ct}$ (Ω) | $C_{coat}$ (F) | $\eta_{Pol}$ (%) | $\eta_{EIS}$ (%) | Coating thickness (μm) |
|---|---|---|---|---|---|---|---|
| Run 2 | 5.32 | 40.76 | 3493 | 1.114 | 57.26 | 63.19 | 24 |
| Run 3 | 1.97 | 29.04 | 2091 | 1.188 | 84.17 | 70.02 | 39 |
| Run 9 | 9.03 | 68.87 | 2062 | 7.927 | 27.46 | 49.41 | 11 |
| Uncoated specimen | 12.45 | – | 1043 | 2.168 | – | – | |

Fig. 4 SEM images of **a** run 3, **b** run 2 and **c** run 9

the highest $\eta_{EIS}$ and $\eta_{pol}$ values also had the highest thickness and vice versa. As the coating thickness increases, it inhibits the passivation thereby the corrosion resistance increases [21].

## 3.3 Statistical Analysis

The statistical analysis of the corrosion current values was carried out to estimate the optimum process parameter combination yielding the highest corrosion resistance. Table 4 shows the S/N ratios for each process parameter. A lower S/N ratio implied better performance characteristic.

From Table 4, it is evident that the level 1 of molar concentration, level 3 of dipping time and level 3 of annealing temperature was influential in increasing the corrosion resistance.

**Table 4** S/N ratios of process parameters of nano $ZrO_2$ coating

| Level | Molar concentration of $ZrO_2$ sol (M) | Dipping time (min) | Annealing temperature (°C) |
|---|---|---|---|
| 1 | −9.266 | −17.579 | −14.534 |
| 2 | −14.352 | −11.730 | −13.593 |
| 3 | −16.264 | −10.573 | −11.755 |

## 4 Conclusions

This paper describes an experimental study conducted to analyze the influence of nano $ZrO_2$ coating on the corrosion inhibition of stainless steel SS 304. The coating was achieved using the sol-gel dip coating method and the corrosion resistance was estimated through electrochemical measurements in 0.5 M NaCl solution. The following are the major findings of this study:

a. The coating process parameters had a strong influence on the corrosion current. The lowest corrosion current value implying the highest corrosion resistance of as much as 84% was noted for 0.3 M molar concentration, 90 min dipping time and 700 °C annealing temperature.
b. An increase in homogeneity and thickness of the coating resulted in an increase in corrosion resistance and vice versa.
c. Statistical analysis revealed that, level 1 of molar concentration of sol, level 3 of dipping time and level 3 of annealing temperature was influential in reducing the corrosion current or increasing the corrosion resistance.

## References

1. Wang D, Bierwagen GP (2009) Sol–gel coatings on metals for corrosion protection. Prog Org Coat 64:327–338
2. Oldfield JW (1987) Test techniques for pitting and crevice corrosion resistance of stainless steels and nickel-base alloys in chloride-containing environments. Int Mater Rev 32:153–172
3. Shen GX, Chen JC, Lin CJ (2005) Corrosion protection of 316L stainless steel by a $TiO_2$ nanoparticle coating prepared by sol–gel method. Thin Solid Films 489:130–136
4. Ruhi G, Modi OP, Singh IB (2009) Pitting of AISI 304L stainless steel coated with nano structured sol–gel alumina coatings in chloride containing acidic environments. Corros Sci 51:3057–3063
5. Shan CX, Hou X, Choy KL (2008) Corrosion resistance of $TiO_2$ films grown on stainless steel by atomic layer deposition. Surf Coat Technol 202:2399–2402
6. Ito A, YouY Ichikawa T, Tsuda K, Goto T (2014) Preparation of $Al_2O_3$-$ZrO_2$ nanocomposite films by laser chemical vapour deposition. J Eur Ceram Soc 34(1):155–159
7. Yin B, Liu G, Zhou HD, Chen JM, Yan FY (2010) Microstructures and properties of plasma sprayed FeAl/$CeO_2$/$ZrO_2$ nano-composite coating. Appl Surf Sci 256(13):4176–4184
8. Li N, Suzuki M, Abe Y, Kawamura M, Sasaki K, Itoh H, Suzuki T (2012) Effects of substrate temperature on the ion conductivity of hydrated $ZrO_2$ sol. Energy Mater Sol Cells 99:160–165
9. Shen YM, Shao SY, Yu H, Fan ZX, He HB, Shao JD (2007) Influences of oxygen partial pressure on structure and related properties of $ZrO_2$ thin films prepared by electron beam evaporation deposition. Appl Surf Sci 254(2):552–556
10. Balakrishnan G, Sairam TN, Reddy VR, Kuppusami P, Song J (2013) Microstructure and optical properties of $Al_2O_3$/$ZrO_2$ nano multilayer thin films prepared by pulsed laser deposition. Mater Chem Phys 140:60–65
11. Liu WM, Chen YX, Ye CF, Zhang PY (2002) Preparation and characterization of doped sol-gel zirconia Films. Ceram Int 28:349–354
12. Dimitriya B, Helmuth M, Dmitry GS (2011) Mesoporous silica nanoparticles for active corrosion protection. ACS Nano 5:1939–1946

13. Jianye W, Garth LW (1996) Organic/inorganic hybrid network materials by the sol–gel approach. Chem Mater 8:1667–1681
14. Briois V, Belin S, Zucolotto C (2004) Solid-state and solution structural study of acetylacetone-modified tin (IV) chloride used as a precursor of $SnO_2$ nanoparticles prepared by a sol–gel route. Chem Mater 16:3885–3894
15. Gordon D, Bierwagen P (2009) Sol–gel coatings on metals for corrosion protection. Prog Org Coat 64:327–338
16. Valérie M (2006) Review on methods to deposit catalysts on structured surfaces. Appl Catal 23:1–17
17. Sreenivas R, Ganesh Kumar C, Shetty P (2008) The taguchi methodology as a statistical tool for biotechnological applications a critical appraisal. Biotechnol J 3:510–523
18. Sreejith M, Sivapirakasam SP (2016) Application of taguchi method in the optimization of process parameters for a sol–gel-derived nano-alumina film. Proc imechE Part L: J Mater Des Appl 230:574–585
19. Gazala R, Modi OP, Singh IB (2013) Hot corrosion behavior of sol-gel nano structured zirconia coated 9Cr1Mo ferritic steel in alkali metal. J Surf Eng Mater Adv Technol 3:55–60
20. Tiwari SK, Tripathi M, Singh R (2012) Electrochemical behavior of zirconia based coatings on mild steel prepared by sol–gel method. Corros Sci 63:334–341
21. Lidija C, Helena OC (2013) Enhancement of corrosion protection of AISI 304 stainless steel by nanostructured sol–gel $TiO_2$ films. Corros Sci 77:176–184

# Size Effect in Mechanical Behavior of Nb Microwires

K. S. Athira and Atul H. Chokshi

## 1 Introduction

In high-pulsed magnets, a large magnetic field is generated for a very short time interval. This application necessitates two important properties of the materials used. One is to have a large conductivity to reduce the Ohmic heating effects. The other is to have a high strength to withstand the stress due to the Lorentz force created by the large magnetic field. These two requirements can be met by using certain techniques to fabricate composites of two or more materials. It is essential to characterize the strengthening component in this composite. We need to also compare the processed material to the one before processing to understand the mechanisms by which the strengthening happens. One way to do this is to extract out a material from the composite and compare its properties to a commercially available material.

There could be other parameters that could change the property of the material apart from the processing. In case of microwires, there could be a change in the diameter of the wires between the commercially available one and the processed one. This change has been known to cause changes in properties and this is called size effect [1].

The aforesaid effect is referred to as an external size effect when the sample approaches the microstructural length scale that will affect the mechanical properties of the material. There is also an internal size effect, in which the strength of a bulk material depends on different microstructural length scales such as grain size, inter-particle spacing, and inter-dislocation spacing.

K. S. Athira (✉) · A. H. Chokshi
Department of Materials Engineering, Indian Institute of Science,
Bangalore 560012, India
e-mail: athiraiisc@gmail.com



K. Antony and J. P. Davim (eds.), *Advanced Manufacturing and Materials Science*,
Lecture Notes on Multidisciplinary Industrial Engineering,
https://doi.org/10.1007/978-3-319-76276-0_21

The existing research on size effect has been done largely on single crystals, which has shown a dependence of the yield strength on the size of the specimen [2–4]. Many models like weakest link theory [5], dislocation starvation [2, 3], and dislocation source truncation [4] have been proposed to explain the increase of strength with a decrease in the size of the material. There are very few studies on the size effect in polycrystalline materials. Some studies show an increase in strength with a decrease in the size of the specimen, while some others show an increase in strength.

There has been limited research on the mechanical behavior of Nb wires extracted from Cu–Nb composite. The present study is on comparing the properties of Nb wires extracted from Cu–Nb composite to that of commercially available Nb microwires and to observe the effect of external size and internal microstructure on their strength.

## 2 Background

### 2.1 *Size Effect in Polycrystals*

Fleck et al. [6] observed that there is no size effect in tensile tests of polycrystalline Cu microwires. But Greer et al. [7] show that the strength of nanoscale gold pillars increases with a decrease in the specimen size. Yaxin et al. [8] showed that the strength of polycrystalline Al wires decreases with a decrease in specimen size. This shows that the size effect in polycrystals is not well understood and that it varies from material to material.

Studies have been conducted on the effect of grain size and specimen diameter on the mechanical properties [9]. There are three areas of mechanical behavior which are based on the thickness and the number of grains across the thickness, that have been reported: polycrystalline, multicrystalline and quasi-single crystalline. Based on the ratio of specimen thickness (t) to that of grain size (d), the shift between these can be observed. The transition results in a change in the mechanism that governs size effect. When the t/d ratio is above a certain value, the Hall-Petch effect dominates and grain size determines strength. At lower t/d ratios, the specimen thickness has a strong effect on the strength.

### 2.2 *Cu–Nb Composite*

The Cu–Nb composites can be fabricated using a process called the accumulative drawing and bundling (ADB) process. The Cu matrix is composed of Oxygen Free High Conductivity (OFHC) Cu. The composite is produced by inserting a Nb rod in an OFHC Cu tube and then extruding and cold-drawing it. They are cut into 55 or

85 segments (the number comes from the number of hexagons that can be cut out from a circle). The segments are bundled in a can of OFHC Cu and reduced as before. This procedure is repeated n times (n = 4 or 5 usually) to get composites that have $55^n$ or $85^n$ Nb nanotubes embedded in Cu matrix [14].

Dubois et al. [10] examined the microstructure of the composite (Fig. 1). Nb fibers were observed (55 in this case), and the external Cu jacket is labelled as Cu-4. Cu matrix is distributed into the inter-filamentary channels, labelled as Cu-0 and super-filamentary channels, labelled as Cu-1, Cu-2 and Cu-3. The Cu–Nb interfaces are semicoherent with (111)Cu planes parallel to the (110)Nb planes and perpendicular to the wire axis.

# 3 Experimental Materials and Procedure

## 3.1 Materials

Nb microwires are obtained in two ways for the study.

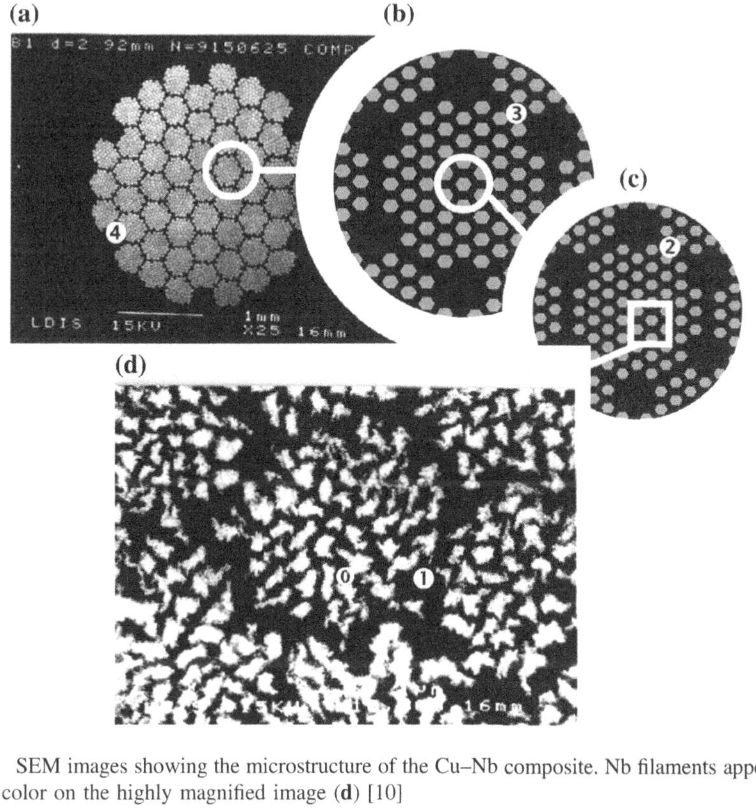


**Fig. 1** SEM images showing the microstructure of the Cu–Nb composite. Nb filaments appear in white color on the highly magnified image (**d**) [10]

A. Cu–Nb composite procured from Prof. L. Thilly (Université de Poitiers, France) which was fabricated using the ADB process with N = $85^2$ Nb nanotubes in a Cu matrix. From this, Nb microwires with a diameter of 85 μm were extracted from it by etching out the Cu matrix using 10 M HNO3 at room temperature. The whole process took about 30 min.
B. Nb microwire of diameter 250 μm with a purity level of 99.99% was purchased commercially from Alfa Aesar.

## 3.2 *X-Ray Diffraction*

XRD was done on 250 μm wires, 85 μm wires, and a standard Si sample using PANalytical's X'PERTPro machine (using a copper source) to determine the grain size.

## 3.3 *Tensile Testing*

The main difficulty with tensile testing was difficulty in gripping the sample and possibility of misalignment of the sample. The problem was solved using two techniques. Technique one is to make a groove in the center of an aluminum sample holder and to stick the wire to the groove so that the misalignment problem is solved. This is repeated at the other end of the wire. The aluminum holder is stuck to a paper template that has a rectangular cut in the region of gage length so that it is easy to handle (Fig. 2a). Another problem faced was slippage of the sample when tested under tension when normal glues like Fevi Kwick was used to stick the samples. So, Araldite was used instead of it, and after letting it dry for two days, it could hold the wire till its UTS. Technique two involved using a sample holder which was made that was bigger and easier to handle (Fig. 2b).Tensile tests were carried out on an ADMET machine at a crosshead velocity of 0.03 and 30 cm/min, on samples of gage length 5 mm, which translates to a strain rate of $10^{-3}$ $s^{-1}$ and 1 $s^{-1}$, respectively. A Futek load cell with a capacity of 667 N was used for load measurement and the elongation was calculated from the set crosshead velocity by the machine itself. In the technique two, the strain was calculated using Digital Image Correlation (DIC) by recording a video using a microscope. The test was terminated when the samples fractured.

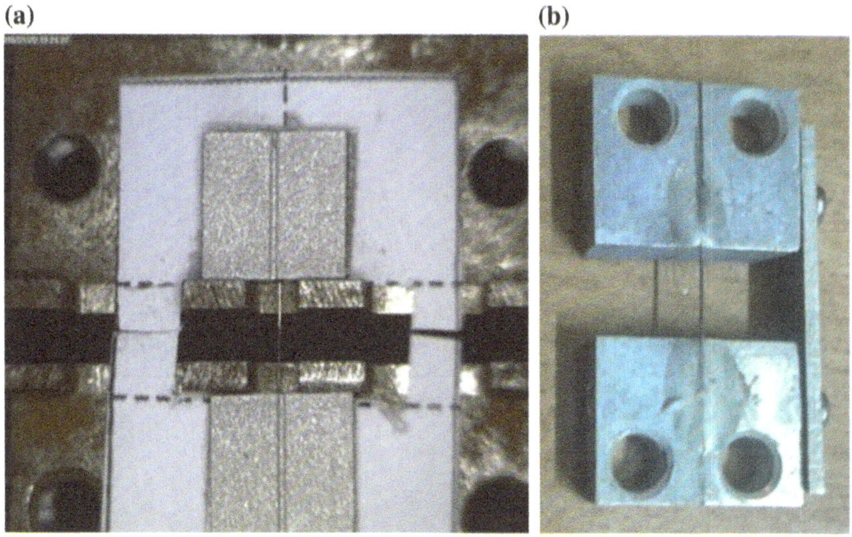


**Fig. 2** **a** Technique one using grooved Al sample holder. **b** Technique two using the machined sample holder **Fracture surface analysis**. The fractured samples were observed under SEM (ESEM Quanta 200). The images were compared to see the effect of size on the fracture

# 4 Results and Discussions

## *4.1 XRD*

The XRD data for 250 and 85 μm is plotted in Figs. 3 and 4, respectively. From the analysis of 110 peak using Scherrer equation, it was found that the 250 μm wires have a crystallite size of 12 nm and the 85 μm wires have a crystallite size of 10 nm.

**Fig. 3** XRD data for 250 μm wires

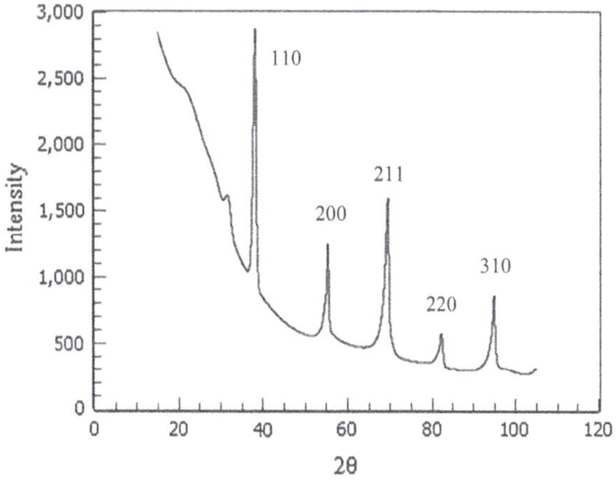

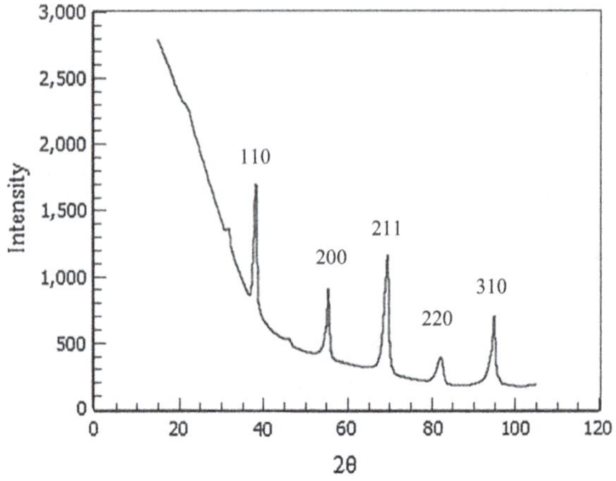


**Fig. 4** XRD data for 85 μm wires

## 4.2 *Tensile Test*

The tensile test results are given in Fig. 5.

The average UTS of 250 μm wires from five tests is ~940 MPa and that of 85 μm wires is ~530 MPa from four tests.

## 4.3 *Fracture Surface Analysis*

The fracture surface analysis of both the wires are given in Figs. 6 and 7.

From Fig. 6, a cup and cone mode of fracture is observed and thus it can be said that the 250 μm wire is plastic in nature. Many ductile dimples and voids could also

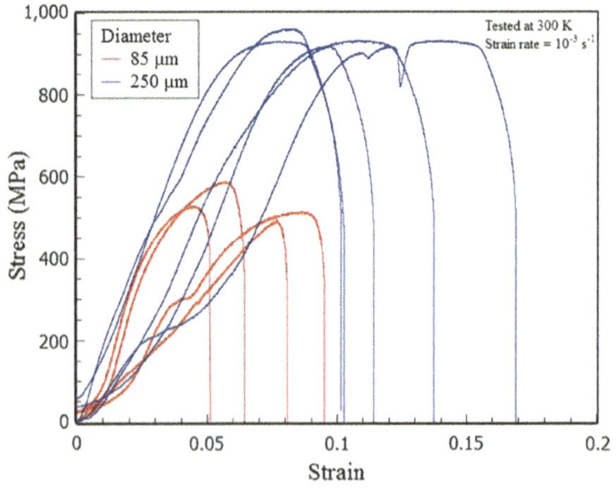


**Fig. 5** Tensile test data for both the microwires

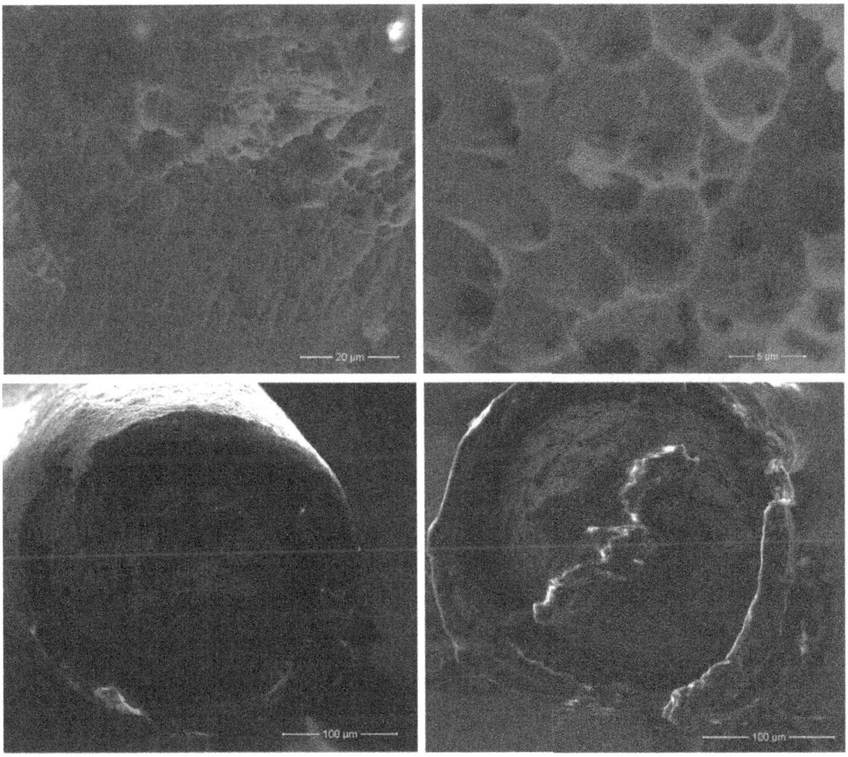


**Fig. 6** Fracture surface of 250 μm wires

**Fig. 7** Fracture surface of 85 μm wires

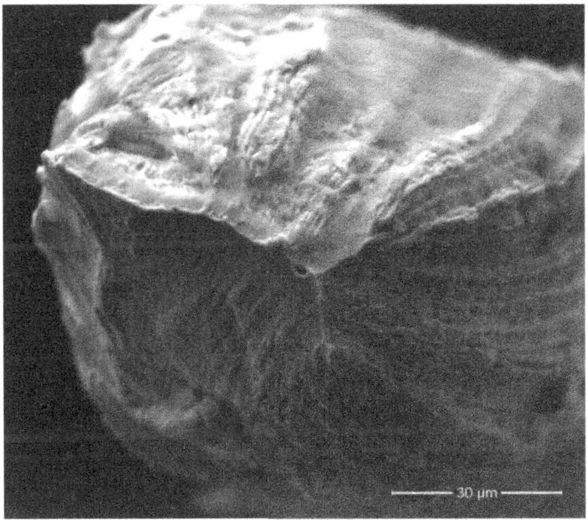

be observed. In case of the 85 μm wires, many facets could be observed in the micrograph. It can be inferred that the wires seem to be considerably plastic in nature, leading to a knife-edge fracture tip.

## 5 Conclusions

- From XRD, it was calculated that the 250 μm wires have a crystallite size of 12 nm and the 85 μm wires have a crystallite size of 10 nm. This verified that both of the wires have similar grain size and so the effect of size on strength can be due to size effect.
- From tensile tests, the average UTS of 250 μm wires is 940 MPa and that of 85 μm wires is 530 MPa. Thus "smaller is weaker" size effect is observed. But differences in processing technique make it difficult to compare the two microwires.
- The fracture surface analysis for 250 μm wires shows a cup and cone type of fracture with dimples and voids. Thus, it can be said that they are very plastic in nature. The 85 μm wires were also found to show some plasticity from the fractographs obtained.

## References


1. Chen XX, Ngan AHW (2011) Specimen size and grain size effects on tensile strength of Ag microwires. Scripta Mater 64:717–720
2. Uchic MD, Dimiduk DM, Florando JN, Nix WD (2004) Sample dimensions influence strength and crystal plasticity. Science 305:986–989
3. Greer JR, Nix WD (2006) Nanoscale gold pillars strengthened through dislocation starvation. Phys Rev B 73:245410
4. Parthasarathy TA, Rao SI, Dimiduk DM, Uchic MD, Trinkle DR (2007) Contribution to size effect of yield strength from the stochastics of dislocation source lengths in finite samples. Scripta Mater 56:313–316
5. El-Awady JA, Wen M, Ghoniem NM (2009) The role of the weakest-link mechanism in controlling the plasticity of micropillars. J Mech Phys Solids 57(1):32–50
6. Fleck NA, Muller GM, Ashby MF, Hutchinson JW (1994) Strain gradient plasticity: theory and experiment. Acta Metall Mater 42:475–487
7. Greer JR, Nix WD (2006) Nanoscale gold pillars strengthened through dislocation starvation. Phys Rev B 73:245410
8. Zhu Y, Li Z, Huang M (2013) Coupled effect of sample size and grain size in polycrystalline Al nanowires. Scripta Mater 68:663–666
9. Keller C, Hug E, Feaugas X (2011) Microstructural size effects on mechanical properties of high purity nickel. Int J Plast 27:635–654
10. Dubois JB, Thilly L, Lecouturier F, Olier P, Renault PO (2012) Cu/Nb nanocomposite wires processed by severe plastic deformation for applications in high pulsed magnets: effects of the multi-scale microstructure on the mechanical properties. IEEE Trans Appl Supercond 22:6900104

# Modeling and Predicting the Mechanical Properties of Nano Clay Dispersed Natural Fiber Hybrid Intra-ply Woven Fabric Composite by Using Response Surface Methodology

**K. Jayakrishna, M. Rajesh, A. Soundhar, K. Surendhar and M. T. H. Sultan**

## 1 Introduction

In the recent scenario there has been a significant increase in the usage of natural fibre because it is biodegradable and eco-friendly but does not provide much strength when compared to synthetic fibres. Natural fibres are preferred over synthetic fibres for their convincing properties such as low cost, low density, low weight, recyclability and biodegradability [1, 9]. They are also popular for its high strength, high stiffness and renewable characters. Natural fibre composites posses the emergent areas in material science. These are sustainable and recyclable but a fully bio-degradable composite can be expensive, sensitive to moisture and temperature, and also biodegradation might be difficult to control [2]. On hybridizing a natural fibre with a synthetic or a natural fibre, the mechanical properties of the natural fibre can be increased and it can overcome the limitations of natural fibre composites [3, 10]. After number of investigations on natural fibres as reinforcements it is concluded that it possesses good stiffness when compared to glass fiber composite [7, 8]. In this work, mechanical properties of the composite were determined using the Box-Behnken (BBD) experimental design of Response Surface Methodology (RSM). Based on the experimental design, experimental tests were conducted to develop a mathematical model and predict the mechanical properties of banan/jute reinforced with polyester hybrid composite. The model with predicted values was correlate with experimental values.

K. Jayakrishna (✉) · M. Rajesh · A. Soundhar · K. Surendhar
School of Mechanical Engineering, VIT University, Vellore, India
e-mail: mail2jaikrish@gmail.com

M. T. H. Sultan
Laboratory of Biocomposite Technology, Institute of Tropical Forestry and Forest Products (INTROP), Universiti Putra Malaysia, 43400 Serdang, Selangor, Malaysia

 
K. Antony and J. P. Davim (eds.), *Advanced Manufacturing and Materials Science*, Lecture Notes on Multidisciplinary Industrial Engineering,
https://doi.org/10.1007/978-3-319-76276-0_22

## *1.1 Experiment Details*

### 1.1.1 Materials and Methods

Jute/banana natural fibre intra-ply hybrid woven mat was used as primary reinforcement and montmorillonite (MMT) nanoclay was used as secondary reinforcement in a polyester resin matrix [6] (Fig. 1).

### 1.1.2 Preparation of Composite Specimens

Intra-ply woven composite laminates were prepared by compression moulding technique. The mould was prepared using a steel sheet of dimensions 300 mm 300 mm × 3 mm. The top and bottom parts were covered by EN steel parallel plates. At first based on experimental design data obtained from BB design of RSM, nano-clay mixture was prepared. Then, the intra-ply hybrid natural fibre was kept on it. To remove voids, the stacked laminates were rolled several times using a steel roller. Then, the mould was coat by two plates by utilizing 20 bar pressure using compression moulding machine [6]. The curing time and spindle speed were taken from the experimental design data values obtained from BB design of RSM.

### 1.1.3 Testing of Composite Specimens

Tensile test and flexural tests were done by universal testing machine (UTM) and impact test was complete using Izod impact test setup. Three point bending test was

**Fig. 1** Jute/banana hybrid composite

done as per ASTM standard D-790. Tensile test was done as per ASTM standard D-638. Izod test was done as per ASTM standard D-256 [6].

## *1.2 BB Parameters and Levels*

BB process control parameters were identified to be nano clay loading, speed and curing time. Each parameters is placed three equally spaced values generally coded as level −1, level 0, and level +1. The range of the input parameters Nano clay loading, speed and curing time were selected as 1–5 (wt%), 700–1000 rpm, 3–5 h respectively as shows Table 1.

## *1.3 Design of Experiments*

In this study the numbers of trials were designed with the help of Design Expert software. The factorial portion of the Box-Behnken Design(BBD) is a full factorial design with all the combination of the parameters at three levels (high +1 and low −1) and collected of twelve star points and five central points (coded level 0). The Box-behnken Designs contains 17 experimental values at three input variables and after conducting experiments the output responses are shown in Table 2.

# 2 Results and Discussion

## *2.1 Statistical Analysis*

Analysis of variance (ANOVA) was conducted with the aim of analyzing the influence of Nano clay loading, speed and curing time. Tables 3, 4 and 5 shows the results of ANOVA for tensile strength, flexural strength and impact strength. The analysis was carried out at 5% significance level and 95% confidence level [5, 11].

**Table 1** Parameters and levels

| Parameters | Level −1 | Level 0 | Level 1 |
|---|---|---|---|
| Nano clay loading (wt%) | 1 | 3 | 5 |
| Speed (rpm) | 700 | 850 | 1000 |
| Time (Hr) | 3 | 4 | 5 |

**Table 2** Experimental values

| Run | Nanoclay loading (wt) % | Speed (rpm) | Time (Hr) | Tensile strength (MPa) | Flexural strength (MPa) | Impact strength (J/m) |
|---|---|---|---|---|---|---|
| 1 | 3 | 1000 | 5 | 25.7 | 48.3 | 290 |
| 2 | 3 | 1000 | 3 | 24.5 | 46.5 | 295 |
| 3 | 1 | 700 | 4 | 18.5 | 34.2 | 132 |
| 4 | 1 | 1000 | 4 | 20.5 | 37.8 | 140 |
| 5 | 5 | 1000 | 4 | 19.6 | 36.7 | 156 |
| 6 | 3 | 850 | 4 | 23.4 | 45 | 269 |
| 7 | 3 | 850 | 4 | 22.5 | 45.8 | 274 |
| 8 | 3 | 850 | 4 | 23 | 46.2 | 276 |
| 9 | 3 | 700 | 3 | 24 | 44 | 280 |
| 10 | 3 | 700 | 5 | 24.5 | 45 | 271 |
| 11 | 5 | 850 | 5 | 20.5 | 37.8 | 156 |
| 12 | 5 | 700 | 4 | 21.2 | 38.2 | 153 |
| 13 | 1 | 850 | 5 | 17.9 | 35 | 140 |
| 14 | 3 | 850 | 4 | 23.7 | 46 | 260 |
| 15 | 1 | 850 | 3 | 19.5 | 35.8 | 132 |
| 16 | 5 | 850 | 3 | 21.7 | 36.2 | 150 |
| 17 | 3 | 850 | 4 | 25 | 46 | 279 |

**Table 3** Analysis of variance (ANOVA) for tensile strength

| Source | Sum of squares | df | Mean square | F value | *p*-value Prob > F | Remarks |
|---|---|---|---|---|---|---|
| Model | 84 | 9 | 9.33 | 8.81 | 0.0045 | Significant |
| A-nano clay loading | 5.45 | 1 | 5.45 | 5.14 | 0.0577 | |
| B-speed | 0.55 | 1 | 0.55 | 0.52 | 0.4941 | |
| C-time | 0.15 | 1 | 0.15 | 0.14 | 0.7167 | |
| AB | 3.24 | 1 | 3.24 | 3.06 | 0.1238 | |
| AC | 0.04 | 1 | 0.04 | 0.038 | 0.8514 | |
| BC | 0.12 | 1 | 0.12 | 0.12 | 0.7438 | |
| A^2 | 73.3 | 1 | 73.3 | 69.2 | <0.0001 | |
| B^2 | 1.53 | 1 | 1.53 | 1.44 | 0.2687 | |
| C^2 | 1.29 | 1 | 1.29 | 1.21 | 0.3071 | |
| Residual | 7.42 | 7 | 1.06 | | | |
| Lack of fit | 3.87 | 3 | 1.29 | 1.45 | 0.353 | Not significant |
| Pure error | 3.55 | 4 | 0.89 | | | |
| Cor total | 91.42 | 16 | | | | |

**Table 4** Analysis of variance (ANOVA) for flexural strength

| Source | Sum of squares | df | Mean square | F value | *p*-value Prob > F | Remarks |
|---|---|---|---|---|---|---|
| Model | 397.3387 | 9 | 44.14874 | 99.61038 | <0.0001 | Significant |
| A-nano clay loading | 4.65125 | 1 | 4.65125 | 10.49436 | 0.0143 | |
| B-speed | 7.80125 | 1 | 7.80125 | 17.60153 | 0.0041 | |
| C-time | 1.62 | 1 | 1.62 | 3.655117 | 0.0975 | |
| AB | 6.5025 | 1 | 6.5025 | 14.67123 | 0.0065 | |
| AC | 1.44 | 1 | 1.44 | 3.248993 | 0.1145 | |
| BC | 0.16 | 1 | 0.16 | 0.360999 | 0.5669 | |
| A^2 | 373.0322 | 1 | 373.0322 | 841.6521 | <0.0001 | |
| B^2 | 0.479605 | 1 | 0.479605 | 1.082107 | 0.3328 | |
| C^2 | 0.148026 | 1 | 0.148026 | 0.333984 | 0.5814 | |
| Residual | 3.1025 | 7 | 0.443214 | | | |
| Lack of fit | 2.2225 | 3 | 0.740833 | 3.367424 | 0.1357 | Not significant |
| Pure error | 0.88 | 4 | 0.22 | | | |
| Cor total | 400.4412 | 16 | | | | |

**Table 5** Analysis of variance (ANOVA) for impact strength

| Source | Sum of squares | df | Mean square | F value | *p*-value Prob > F | Remarks |
|---|---|---|---|---|---|---|
| Model | 75297.31 | 9 | 8366.368 | 151.545 | <0.0001 | Significant |
| A-nano clay loading | 630.125 | 1 | 630.125 | 11.41383 | 0.0118 | |
| B-speed | 253.125 | 1 | 253.125 | 4.585005 | 0.0695 | |
| C-time | 0 | 1 | 0 | 0 | 1.0000 | |
| AB | 6.25 | 1 | 6.25 | 0.11321 | 0.7464 | |
| AC | 1 | 1 | 1 | 0.018114 | 0.8967 | |
| BC | 4 | 1 | 4 | 0.072454 | 0.7956 | |
| A^2 | 74396.02 | 1 | 74396.02 | 1347.58 | <0.0001 | |
| B^2 | 182.0237 | 1 | 182.0237 | 3.297104 | 0.1123 | |
| C^2 | 142.8658 | 1 | 142.8658 | 2.587813 | 0.1517 | |
| Residual | 386.45 | 7 | 55.20714 | | | |
| Lack of fit | 165.25 | 3 | 55.08333 | 0.996082 | 0.4803 | Not significant |
| Pure error | 221.2 | 4 | 55.3 | | | |
| Cor total | 75683.76 | 16 | | | | |

## 2.2 Mathematical Model for Tensile, Flexural and Impact Strength

The quadratic models developed for output variables tensile strength, flexural strength and impact strength were evaluated by the F test. From the results the mathematical model is statistically significant to analyze the response variables. The adequacy of model is tested using ANOVA analysis. It was found that the model F-ratio of 8.81 was obtained for tensile strength. However, the F-ratio of the lack of fit is 1.45. It was found that the model F-ratio for flexural strength is 99.61 and the F-ratio of the lack of fit is 3.36. it was found that the model F-ratio for impact strength is 151.54 and the F-ratio of the lack of fit is 0.996. Hence the models are significant and their lack of fit is insignificant [4, 5].

$$\begin{aligned} \textit{Tensile strength} = {} & +37.06049 + 9.02125 \times A - 0.039439 \times B - 5.69917 \\ & \times C - 3.00000E - 003 \times A \times B + 0.050000 \times A \times C + 1.16667E \\ & - 003 \times B \times C - 1.04313 \times A^2 + 2.67778E - 005 \times B^2 + 0.55250 \\ & \times \mathrm{C}^2 \quad (\text{R-Squared value : 0.9189}) \end{aligned}$$

$$\begin{aligned} \textit{Flexural strength} = {} & +21.21562 + 16.91250 \times A - 0.011500 \times B - 0.083333 \\ & \times C - 4.25000E - 003 \times A \times B + 0.30000 \times A \times C + 1.33333E \\ & - 003 \times B \times C - 2.35313 \times A^2 + 1.50000E - 005 \times B^2 \\ & - 0.18750 \times C^2 \quad (\text{R-Squared value : 0.9923}) \end{aligned}$$

$$\begin{aligned} \textit{Impact strength} = {} & +240.70347 + 208.36667 \times A - 0.47344 \times C - 51.51667 \\ & \times C - 4.16667E - 003 \times A \times C - 0.25000 \times A \times C + 6.66667E \\ & - 003 \times C \times C - 33.23125 \times A^2 + 2.92222E - 004 \times C^2 \\ & + 5.82500 \times C^2 \quad (\text{R-Squared value : 0.9949}) \end{aligned}$$

where, A-Nano clay loading (wt%) B-Speed (rpm) C-Curing time (Hr).

## 2.3 BB Parameters and Their Influence

The illustration of the predicted models (3D response surface plot) are shown in Fig. 2 for the depth insight of the results. The tensile strength of intra ply woven composite is affected notably from the input process parameters, which is validated

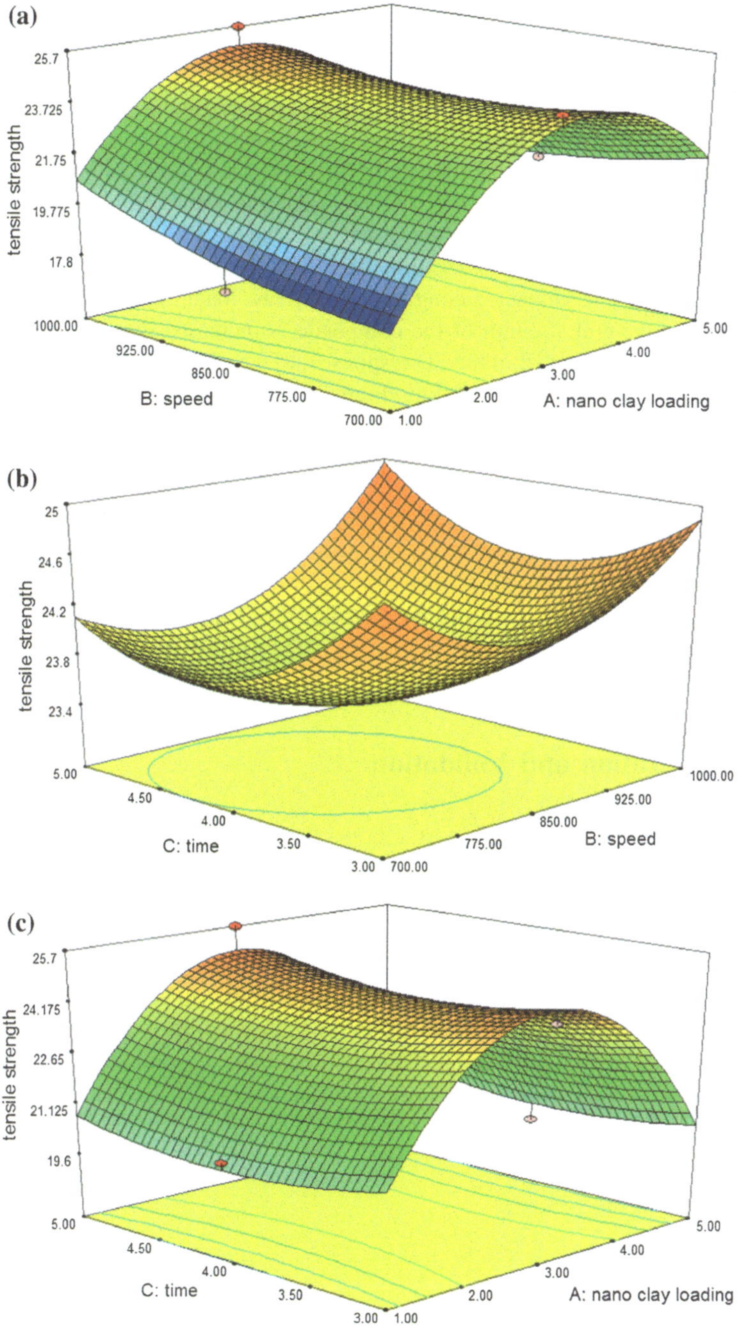


**Fig. 2** 3D response surface plots showing **a** the impact of speed and nano clay loading **b** the impact of speed and curing time and **c** the impact of curing time and nano clay loading for the tensile strength of jute/banana hybrid polymer composites

from the results. Figure 2a displays the impact of speed and Nano clay loading on the tensile strength of the intra ply woven composites. the tensile strength is increased by the inclusion of Nano clay weight percentage(3.02 wt%).the tensile strength value is decreased during addition of nanoclay (more than 3.02 wt%).the tensile strength is affected by the speed and curing time as shown in Fig. 2b and the tensile strength is influenced by the curing time and nano clay loading as shown in Fig. 2c. The nano clay loading 3.02 (wt%), speed 1000(rpm) and curing time five hours produce the maximum tensile strength value which is seen from the 3D response surface plot. The observation from Fig. 3 depict the flexural strength. Figure 3a depict the impact among the nano clay loading, speed and flexural strength. The flexural strength of the composite is increased during the increased speed which is observed from the 3D response surface plot. The flexural strength is also affected by the Nano clay loading. Figure 3b displays the impact of speed and curing time of the flexural strength. The flexural influenced by the curing time and nano clay loading shown in Fig. 3c. The observation from Fig. 4 depict the effect of impact strength under different input conditions. Figure 4a depict the impact of speed and nano clay loading for the impact strength. Figure 4b depict the impact of speed and curing time of the impact strength and Fig. 4c display the impact of curing time and nano clay loading of the impact strength. The nano clay loading 3.02 (wt%), speed 1000(rpm) and curing time five hours produce the maximum impact strength value which is seen from the 3D response surface plot.

## 3 Optimization and Validation

The combination has been evaluated with the help of Design Expert software. The desirability falls between zero and one shown in Table 6. The solution with the highest desirability and close to 1 is chosen as the optimal setting and the validation experiments are conducted. The optimum parameter setting is 3.02 (wt%), 1000 (rpm), 5(Hr) is chosen as the optimal solution. Validation experiments results shown in Table 7. Through confirmatory experiments, it is seen that the error between predictions and actual values fall within 5%. Thus the model can be effectively used to predict the mechanical properties. Responses observed were 24.974 (MPa), 47.587 (MPa) and 290.69 (J/m).

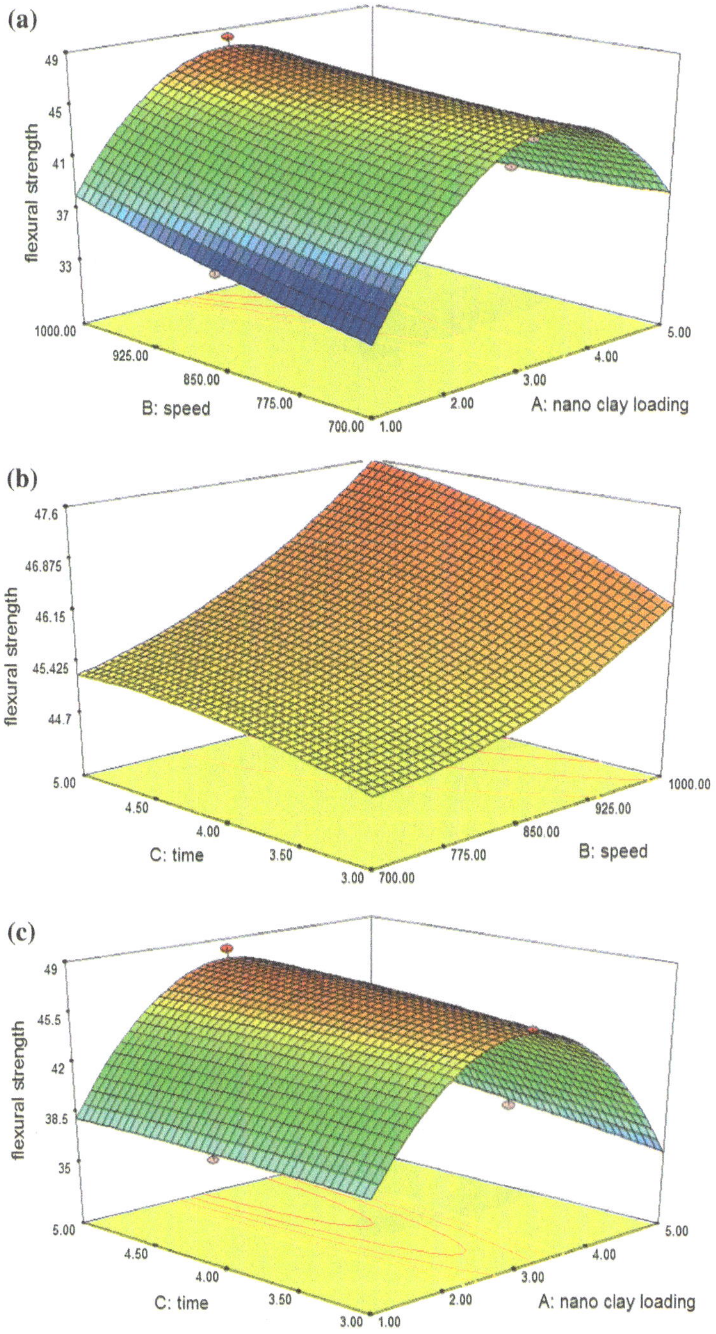


**Fig. 3** 3D response surface plots showing **a** the impact of Nano clay loading and speed, **b** the impact of speed and curing time and **c** the impact of Nano clay loading and curing time for the flexural strength of jute/banana hybrid polymer composites

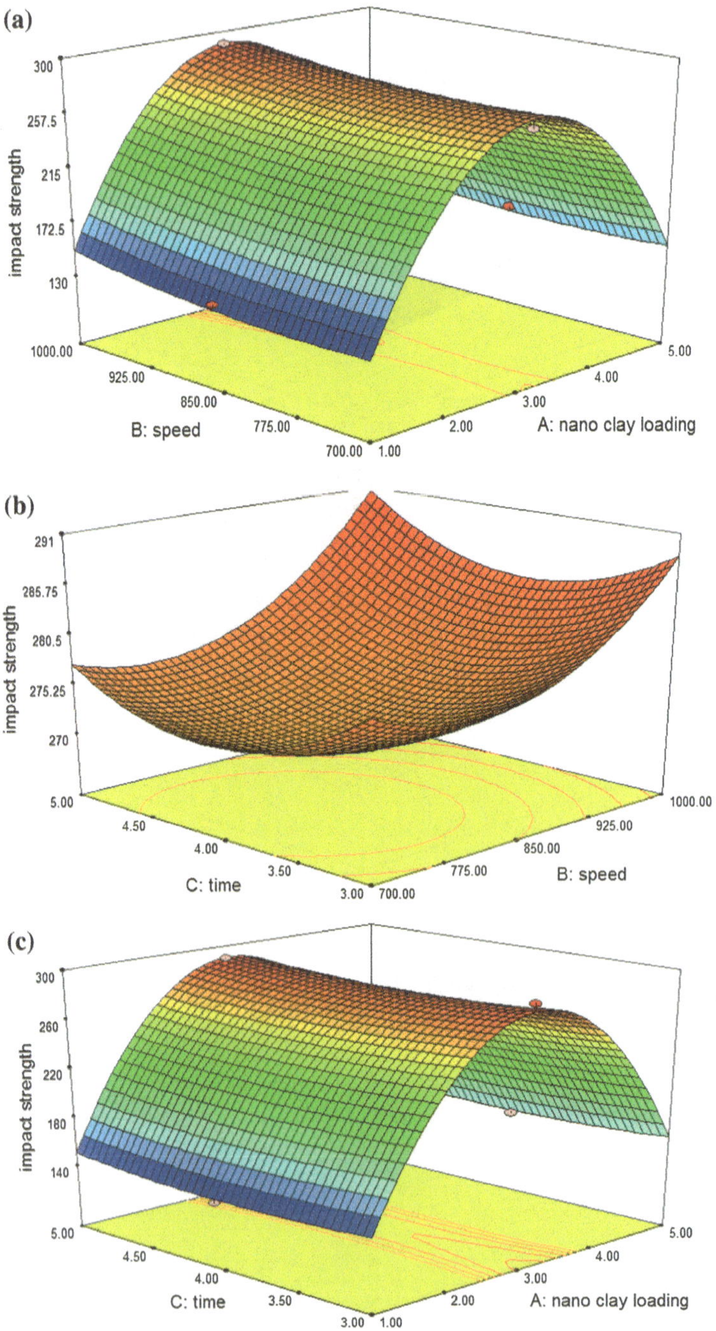


**Fig. 4** 3D response surface plots showing **a** the impact of Nano clay loading and speed, **b** the impact of speed and curing time and **c** the impact of Nano clay loading and curing time for the impact strength of jute/banana hybrid polymer composites

**Table 6** Optimal solution as obtained by design-expert

| Number | Nano clay loading | Speed | Time | Tensile strength | Flexural strength | Impact strength | Desirability | |
|---|---|---|---|---|---|---|---|---|
| 1 | 3.02 | 1000 | 5 | 24.974 | 47.587 | 290.69 | 0.9435 | Selected |
| 2 | 3.07 | 1000 | 5 | 24.970 | 47.577 | 290.70 | 0.9431 | |
| 3 | 3.09 | 1000 | 5 | 24.967 | 47.572 | 290.67 | 0.9428 | |
| 4 | 3.12 | 1000 | 5 | 24.961 | 47.559 | 290.58 | 0.9421 | |
| 5 | 2.92 | 1000 | 5 | 24.966 | 47.567 | 290.09 | 0.9416 | |
| 6 | 3.02 | 1000 | 5 | 24.945 | 47.554 | 290.32 | 0.9408 | |
| 7 | 3.04 | 1000 | 5 | 24.927 | 47.573 | 290.19 | 0.9404 | |
| 8 | 3.01 | 1000 | 4.5 | 24.858 | 47.555 | 289.36 | 0.9356 | |
| 9 | 2.9 | 1000 | 5 | 24.861 | 47.534 | 288.79 | 0.9342 | |
| 10 | 2.96 | 1000 | 3 | 24.901 | 46.305 | 288.42 | 0.8996 | |

## 4 Conclusion

Modeling and predicting the mechanical properties of jute/banana hybrid polyster composite investigation was done by using RSM, the following conclusions are drawn as follows:

(1) The analysis of each input parameter Nano clay loading (wt%), curing time and speed on tensile strength, flexural strength and impact strength was investigated in detail. The influential parameters were identified and their levels were fixed. Experiments were conducted and regression equation has been formed for tensile strength, flexural strength and impact strength.

(2) ANOVA showed that the model was statistically a reasonably good predictor of tensile, flexural and impact strength. Through confirmatory experiments, it is seen that the error between predictions and actual values fall within 5%. Thus the model can be effectively used to predict the mechanical properties of jute/banana Hybrid Intra-ply Woven Fabric Composite.

(3) The mechanical properties of the composite is increased by the inclusion of Nano clay weight percentage up to (3.02 wt%). Further inclusion of nano clay (more than 3.02 wt%) in the composite decrease the mechanical properties.

**Table 7** Confirmation test results

| Nanoclay loading | Speed | Curing time | Tensile strength | | Flexural strength | | Impact strength | | % of errors | | |
|---|---|---|---|---|---|---|---|---|---|---|---|
| | | | Predicted | Actual | Predicted | Actual | Predicted | Actual | Tensile strength | Flexural strength | Impact strength |
| 3.02 | 1000 | 5 | 24.974 | 23.88 | 47.587 | 46.25 | 290.69 | 286 | 4.581 | 2.890 | 1.639 |

## References


1. Sanjay MR, Yogesha B (2017) Studies on natural/glass fiber reinforced polymer hybrid composites: an evolution. Mater Today: Proc 4(2):2739–2747
2. Venkateshwaran N, ElayaPerumal A, Alavudeen A, Thiruchitrambalam M (2011) Mechanical and water absorption behaviour of banana/sisal reinforced hybrid composites. Mater Des 32 (7):4017–4021
3. Venkateshwaran N, Elayaperumal A, Sathiya GK (2012) Prediction of tensile properties of hybrid-natural fiber composites. Compos B Eng 43(2):793–796
4. Athijayamani A, Das MC, Sekar S, Ramanathan K (2017) Mechanical properties of phenol formaldehyde hybrid composites reinforced with natural cellulose fibers. BioResources 12 (1):1960–1967
5. Athijayamani A, Ganesamoorthy R, Loganathan KT, Sidhardhan S (2016) Modelling and analysis of the mechanical properties of Agave Sisalana Variegata fibre/vinyl ester composites using Box-Behnken design of response surface methodology. Strojniški vestnik-J Mech Eng 62(5):273–280
6. Rajesh M, Jeyaraj P, Rajini N (2016) Mechanical, dynamic mechanical and vibration behavior of nanoclay dispersed natural fiber hybrid intra-ply woven fabric composite. Nanoclay reinforced polymer composites. Springer, Singapore, pp 281–296
7. Gurunathan T, Mohanty S, Nayak SK (2015) A review of the recent developments in biocomposites based on natural fibres and their application perspectives. Compos A Appl Sci Manuf 77:1–25
8. Sanjay MR, Arpitha GR, Naik L, Gopalakrishna K, Yogesha B (2016) Applications of natural fibers and its composites: an overview. Nat Resour 7(03):108
9. Pickering KL, Efendy MA, Le TM (2016) A review of recent developments in natural fibre composites and their mechanical performance. Compos A Appl Sci Manuf 83:98–112
10. Sathishkumar TP, Naveen J, Navaneethakrishnan P, Satheeshkumar S, Rajini N (2016) Characterization of sisal/cotton fibre woven mat reinforced polymer hybrid composites. J Ind Text 1528083716648764
11. Aly M, Hashmi MSJ, Olabi AG, Benyounis KY, Messeiry M, Hussain AI, Abadir EF (2012) Optimization of alkaline treatment conditions of flax fiber using Box-Behnken method. J Nat Fibers 9(4):256–276

# Design and Evaluation of Al Hybrid Composite for Connecting Rod

M. S. Kiran Sankar, S. Akhil and O. Harikrishnan

## 1 Introduction

Engine and its components contribute the considerable portion of the weight of an automobile. By reducing the engine weight and its components, the automobile total weight can be curtailed and there by enhances the performance and fuel efficiency. The reduction in weight without compensating the performance is achieved by using composite materials. The matrix in the composite materials binds the fibres which contribute different properties to the overall composite. The selection of the matrix depends upon mainly on the composite application. The superior properties of AMCs are its greater strength, reduced density, improved stiffness, improved high temperature properties, improved abrasion and wear resistance and controlled thermal expansion coefficient [1].

The thesis aims on performance analysis of the automobile connecting rod by the developed hybrid composite material. The connecting rod is subjected to dynamic loading state mainly due to combustion pressure and inertia of the reciprocating parts. In shortly due to the alternate tensile and compressive stresses, the fatigue strength of the material is a key factor in the evaluation of component reliability. Automobile internal combustion engine connecting rod is a high volume production component subjected to complex loading. Proper optimization of this component, which is critical to the engine fuel efficiency and more vigorously pursued by the automotive industry in recent years, necessitates a detailed assimilation of the

M. S. Kiran Sankar (✉) · S. Akhil · O. Harikrishnan
Department of Mechanical Engineering, Sree Buddha College of Engineering,
Pattoor P.O, Alappuzha, Kerala 690529, India
e-mail: kiransankarms@gmail.com

S. Akhil
e-mail: akhils7314@gmail.com

O. Harikrishnan
e-mail: hkrishna3060@gmail.com

 
K. Antony and J. P. Davim (eds.), *Advanced Manufacturing and Materials Science*,
Lecture Notes on Multidisciplinary Industrial Engineering,
https://doi.org/10.1007/978-3-319-76276-0_23

applied loads and resulting stresses under in-service conditions. The real performance of the component is accurately predicted by the dynamic analysis. The design will be safe if the predicted fatigue life of the component is obtained in the range $1 \times 10^6$ to $1 \times 10^8$.

## 2 Design of Connecting Rod

Because of the complex loading conditions of connecting rod, the measurements of the connecting rod must meet the design criteria to ensure the safety.

The computation are based on the Engine data of Maruti 800 [2] Engine type: Four-stroke, Displacement: 796 (cc), Max Torque: 59 Nm @ 2500 rpm, Max Power: 37 BHP @ 5000 rpm, Bore: 68.5 mm, Stroke: 72 mm, Compression ratio: 9.2:1

Density of petrol at 288.855 K = $737.22 \times 10^{-9}$ kg/mm$^3$.

Molecular weight M = 114.228 g/mole and Ideal gas constant R = 8.314 JK$^{-1}$ mol$^{-1}$.

From gas equation

$$``PV = m \times R_{specific} T", \tag{1}$$

Because of the gas pressure minimum load on the piston is equal to the load on the connecting rod.

### 2.1 *Dimension of I Division*

At the centre portion thickness of rib and web section = t

Thickness of flange and web section = t

Width of the section B = 4t

Depth of portion H = 5t

Area of the portion, A = $2 \times 4t \times t + 3t \times t = 11t^2$

Moment of Inertia of portion about x-x axis and y-y axis

$$``I_{xx} = \frac{1}{12} \times \left[4t \times (5t)^3 - 3t \times (3t)^3\right]" = \frac{419t^4}{12}, \tag{2}$$

$$``I_{yy} = \frac{1}{12} \times \left[2t \times (4t)^3 - 3t \times (t)^3\right]" = \frac{131\, t^4}{12}. \tag{3}$$

$\frac{Ixx}{Iyy} = 3.2$, since the value of $\frac{Ixx}{Iyy}$ are in between 3 and 3.5, therefore I-section selected is quite acceptable.

a. Thickness of section

From Rankine-Gordan formula, stress due to axial load, $F = \frac{419t^4}{12} = F = \frac{F_c}{A} = \frac{f_c}{1+K\left(\frac{1}{k}\right)^2}$.

$F_c$ = Crippling load = Maximum gas force × Factor of safety, $f_c$ = Allowable unit stress = 400 $N/mm^2$,

K = Denominator constant = 1/7500, k = radius of gyration

Crippling load, $F_c = P \times \frac{\pi}{4} \times D^2 \times FS = 15.4 \times \frac{\pi}{4} \times 68.5^2 \times 3 = 170260.14$ N

$$``k = \left[\frac{bh^3 - b_1h_1^3}{12(bh - b_1h_1)}\right]^{\frac{1}{2}''} = \left[\frac{4t \times (5t)^3 - 3t \times (3t)^3}{12(4t \times 5t - 3t \times 3t)}\right]^{\frac{1}{2}} = \left[\frac{419\,t^4}{12 \times 11t^2}\right]^{\frac{1}{2}} = 1.78t \quad (4)$$

Solving the above equation we get $t^4 - 38.69\,t^2 - 37.13 = 0 \therefore t = 6.29 \approx 6$

Thickness of the rib and web t = 6 mm
Width of the section B = 4t = 24 mm
Depth or height of the section, H = 5t = 30 mm
Depth near big end $H_1$ = 1.1 H to 1.25 H = 1.2 H = 1.2 × 30 = 36 mm
Depth near small end, $H_2$ = 0.75 H to 0.90 = 0.85 H = 0.85 × 30 = 25.5 mm.

## 2.2 *Dimension of Crank Pin and Piston Pin*

Maximum gas pressure force (Fig. 1; Table 1),

$$F_g = F_L = \frac{\pi}{4} \times D^2 \times P = \frac{\pi}{4} \times (68.5)^2 \times 15. = 56753.38\ \text{N} \quad (5)$$

# 3 Experimental Details

The complete fabrication process has been done at Karunya University, Coimbatore. The fillers selected according to literature review are listed below:

1. Matrix—Aluminium alloy 6061 (Kaizen steels Kochi).
2. Reinforcements—Silicon Carbide (500 mesh) (Universal Steels Koratty), Flyash (Santiga enterprises, Thana) Graphite (Karunya University Coimbatore).
3. Wetting agent—Magnesium metal powder (Autocast Ltd., Alappuzha).

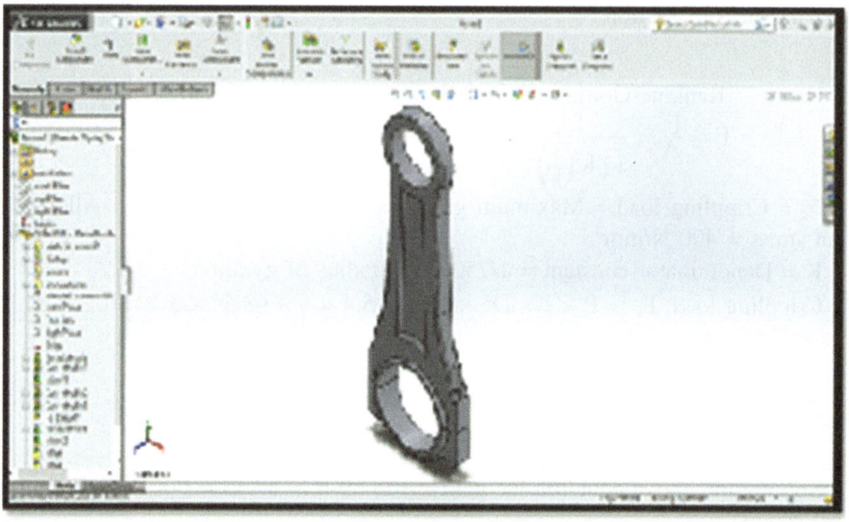

**Fig. 1** Model of the connecting rod

**Table 1** Load calculation of crank pin and piston pin

| Load on the crank pin $F_c = d_c \times l_c \times P_{bc}$ (6) | Load on the piston pin $F_p = d_p \times l_p \times P_{bp}$ (7) |
|---|---|
| $d_c$ = Diameter of crank pin<br>$l_c$ = Length of crank pin ($l_c = 1.25\ d_c$ to $1.5\ d_c$)<br>$P_{bc}$ = Bearing pressure<br>$Fc = d_c \times l_c \times P_{bc} = d_c \times 1.3$<br>$d_c \times 20 = 26\ d_c^2$<br>$F_g = Fc$, $56753.38 = 26\ d_c^2$<br>Diameter of crankpin,<br>$d_c = 46.72$ mm $\approx 47$ mm<br>Length of crankpin,<br>$l_c = 60.73$ mm $\approx 61$ mm | $l_p = 1.5\ d_p$ to $2\ d_p$<br>$Fp = d_p \times l_p \times P_{bp} = d_p \times 1.75$<br>$d_p \times 30 = 52.5\ d_p^2$<br>$F_g = F_p$, $56753.38 = 52.5\ d_p^2$<br>Diameter of piston pin,<br>$d_p = 32.87$ mm $\approx 33$ mm<br>Length of piston pin, $l_p = 57.53$ mm $\approx 58$ mm |

4. Degassing agent—Hexa-chloro ethane tablets (Karunya University Coimbatore).
5. Graphite crucible (Size No. 6) from Mannar.

The percentage composition of reinforcement chosen are Flyash-14%, Silcon-8%, Graphite-8% and Magnesium-5%. The weighted amount aluminium is melted in the furnace and its moved into a stir casting furnace and the temperature of the furnace was kept around 600 °C for complete melting of the Al6061. When the temperature reaches around 715 °C degassing was carried out by adding Hexa-chloro ethane tablets. The predetermined fillers are mixed thoroughly and preheated at 415 °C for one hour. Prior to the reinforcement addition, Magnesium metal powder (100 g) was added to the Al6061 matrix in order to increase weld ability. At 900 rpm mechanical sintering is carried out by using 3 bladed stainless steel sinter for 7 min, then its poured to the permanent mould of dimension 100 × 100 × 10 mm. After 2 min curing it is then removed [3–6]. To achieve improved properties heat treatment is also carried out. T6 heat treatment was completed after aging for three hours at 205 °C. T6 heat treatment was completed after aging for three hours at 205 °C. The specimens needed for mechanical testing were prepared from the cast samples using Wire cut Electric Discharge Machining (WEDM) for prepare the specimens needed for mechanical testing. ASTM E08 is used for tensile test.

The samples are prepared at Speed spark, Coimbatore using Electronic- Sprint Cut 734 Wire EDM [7].

# 4 Real Time Analysis and Its Influence on the System

While doing quasis dynamics of the connecting rod, external loads which are known from the load analysis are applied to both the crank and piston pin end of the con rod.

Dynamic analysis of connecting rod mainly consists of three steps data collection, validation and actual study. Data collection includes Geometrical data, Load conditions and material properties. The modelling is generally carried out considering the forces acting on the connecting rod [8]. The material data is customized by including properties of the hybrid composite to engineering data section. The model is built up in the Solid works 16 and the model analysis is done in ANSYS by applying the boundary condition. The solutions are critically examined for the dynamically developed stress.

Fatigue analysis is done in n-code design life interface of the Ansys software with the prepared model. For connecting rod fatigue life is. If the fatigue life of the connecting rod is obtain with in the value ($1 \times 10^6$ to $1 \times 10^8$), then the design will be safe under fatigue loading.

# 5 Result and Discussions

## *5.1 Evaluation of Mechanical Properties*

### 5.1.1 Tensile Strength

The Tensile strength, Yield strength and Ultimate strength of composite is 120, 83.50 and 186.10 MPa respectively and percentage elongation measured is 12.80. The high elastic modulus and hard SiC embedded into the Aluminium matrix are the root cause for the enhancement of the tensile strength [9–11]. The results show a sign of increase in the ductility behaviour of the composite. The graphite particles reduce the inter particle spacing by filling the void spaces between aluminium particles. Graphite escalates the grain boundary density which upsurge the resistant to flow. The hard flyash particle precludes the movement of dislocations, thereby strengthening the composite. Tensile strength of the composite up surges with boost in the content of flyash particles. Good bonding of smaller size fly ash particles with aluminium matrix is the root cause for this behaviour [12, 13]. But the augmentation of flyash beyond a certain limit causes reduction in tensile properties due to poor wettability.

### 5.1.2 Impact Strength

The average Impact strength of the specimen is found to be 6.67 J. Silicon carbide absorbs most of the energy because of high value of impact strength. Higher amount of flyash diminishes the impact strength because it is dispersed as precipitate in the aluminium matrix. The dispersion of the filler materials intensify the impact strength due to increase in density compared to other regions. Higher amount of graphite addition dwindled the impact strength because it acts as an effective solid lubricant which accelerates the movement of grains along the slip planes. Heat treatment increases the impact strength due to the grain refinement caused by cooling.

### 5.1.3 Hardness Test

The prepared composite has a high hardness value (52 BHN) in comparison to the virgin Al alloy. Hardness of the composite changes with addition of different fillers. Addition of silicon carbide upsurge the hardness of the composite because it prevents the dislocation movement in the composite. The addition of flyash obstacles the movement of dislocations and there by augmented the hardness of the composite. Increase in the flyash content helps to increase the hardness due to clustering of particles. Since graphite is a soft dispersed material, the inclusion of graphite curtails the hardness of the composite.

### 5.1.4 Density Test

Overall weight of the composite is the sum of the weights of both matrix and the various fillers added to it. The composite density calculated is 2.25 g/cc. The composite weight is much lower than that the aluminium matrix due to the addition of low density fillers. The silicon carbide addition results in the increment of the overall density due to the high density of Silicon Carbide. Addition of metered amount of magnesium also results in density reduction. The first specimen is not considered for testing because, in the first specimen, the fly ash content is more due to improper mixing, which reduces the density but cannot be casted to required shape since flow ability is very low. The overall engine weight (for a normal light vehicle engine) can be reduced to 22.4% by replacing the conventional Al alloy by proposed composite.

## *5.2 Metallographic Analysis*

### 5.2.1 SEM Analysis

The SEM image shows ductile portions coupled with tensile fracture at the SiC region. Silicon carbide added to the composite causes the entrapment of air bubbles which causes porosity in the composite. The imperfect blending of silicon carbide in composite is the reason for air entrapment. The porosity leads to the development of small cracks and advances on the application of load. Cup and cone type fracture is seen in some portions even though brittle fracture is seen in majority. The analysis confesses the uniform dispersion of flyash particles in the aluminium matrix.

### 5.2.2 Micro Examination

The microstructure examination remarks the flyash precipitation and distribution of SiC particles in the grain boundaries of aluminium. Precipitation of Fly ash occurs due to increased percentage of Fly ash and rapid cooling of the cast. The Fly ash particles are seen interconnected. The interconnectivity distribution of flyash harvest the augmentation of ductility and tensile strength. The highly dendrite structure of aluminium is refined by the SiC particles owing to the resistance offered to the freely flowing aluminium grain (Fig. 2).

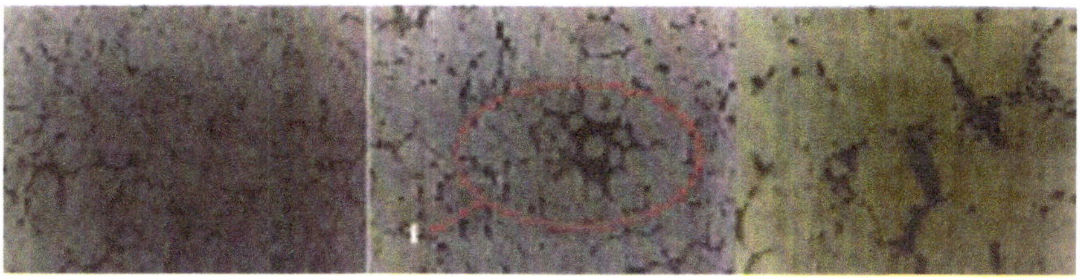

**Fig. 2** Micro structure of Al-SiC-flyash-graphite composite

**Table 2** Stresses and deformations in dynamic

| | General material | | | | Aluminium hybrid composite | | | |
|---|---|---|---|---|---|---|---|---|
| | Tension | | Compression | | Tension | | Compression | |
| | Max | Min | Max | Min | Max | Min | Max | Min |
| Stress (MPa) | 79.38 | 0.00067 | 23.30 | $3 \times 10^4$ | 79.375 | 0.00062 | 23.808 | $3.2 \times 10^4$ |
| Total deformation (mm) | 0.073 | 0.00019 | 0.004 | $4 \times 10^{13}$ | 0.078 | 0.00021 | 0.004 | $2.5 \times 10^{-12}$ |
| Fatigue life (cycle) | $1 \times 10^7$ | | | | $1 \times 10^7$ | | | |

### 5.3 *Dynamic Analysis Results*

The composite is designed safely on virtue of the lower value of stress in comparison with the yield stress for the selected application. The connecting rod had trivial deformation with in the safe zone. The tested fatigue life of the connecting rod made with the composite is $1 \times 10^7$, which is safe for connecting rod (Table 2).

## 6 Conclusion

The inclusions of desired percentages of SiC, flyash, graphite to the Aluminium 6061 matrix helps to curtail the density by 21.05% compared to the conventional aluminium alloy. The reinforcement materials accounted the brittle to ductile transition of Aluminium 6061. The excess slag formation had noticed with the fly ash percentage beyond 16%. The uniform dispersion of flyash and silicon carbide augmented the tensile strength and ductility. The silicon carbide inclusion results in the porosity by virtue diminish the composite property. The span of this project can be extended to the performance evaluation of Smart hybrid composite. The weight of the composite can be further abated by adding magnesium/magnesium matrix. The strength depreciation as a result of magnesium addition can be surmount by adding CNT, carbon fibre etc.

## References

1. Wahab VMN, Daud AR, Ghazali MJ (2009) Preparation and characterization of stir cast-aluminium nitride reinforced aluminium metal matrix composites. Int J Mech Mater Eng (IJMME) 4(2):115–117
2. Singha J, Chauhan A (2016) Characterization of hybrid aluminum matrix composites for advanced applications. A Rev J Mater Res Technol 5(2):159–169

3. Bhandare RG, Sonawane PM (2013) Preparation of aluminium matrix composite by using stir casting method. Int J Eng Adv Technol (IJEAT) 3(2). ISSN: 2249–8958
4. Hynes NRJ et al (2015) Production of aluminium metal matrix composites by liquid processing methods. In: International conference on condensed matter and applied physics (ICC 2015)
5. Balajia V, Sateeshb N, Manzoor Hussainc M (2015) Manufacture of aluminium metal matrix composite (Al7075-SiC) by stir casting technique. Mater Today: Proc 2:3403–3408
6. Sahu PS, Banchhor R (2016) Fabrication methods used to prepare Al metal matrix composites a review. Int Res J Eng Technol (IRJET) 3(10):123–132
7. Ozden S, Ekici R, Nair F (2007) Investigation of impact behaviour of aluminium based SiC particle reinforced metal–matrix composites. Compos Part A: Appl Sci Manuf 38:484–494
8. Irfan Sadaq S, Yunus SV (2017) Dynamic analysis of a connecting rod using FEA. IJMER 5 (5):2321–5747
9. Ozben T, Kilickap E, Cakir O (2008) Investigation of mechanical and machinability properties of SiC particle reinforced Al-MMC. J Mater Process Technol 198:220–225
10. Ramachandra M, Radhakrishna K (2004) Study of abrasive wear behaviour of Al-Si (12%)-SiC metal matrix composite synthesised using vortex method. In: Proceedings of the international symposium of research students on materials science and engineering, IIT Madras, India, 20–22 Dec 2004
11. VeereshKumar GB, Rao CSP, Selvaraj N, Bhagyashekar MS (2009) Studies on Al6061-SiC and Al7075-$Al_2O_3$ metal matrix composites. J Miner Mater Charact Eng (JMMCE), USA 9 (1):47–59
12. Malhotra S, Narayan R, Gupta RD (2013) Synthesis and characterization of aluminium 6061 alloy-fly ash & zirconia metal matrix composite. Int J Curr Eng Technol 3(5):1716–1719
13. Vipin K, Sharm RC, Chaudhary SR (2017) Effect of flyash particles with aluminium melt on the wear of aluminium metal matrix composites. Eng Sci Technol 20(4):1318–1323

# Literature Review of Hard Turning and It's Thermal Aspects

**Shahu Patil, Priya Jadhav and Satish Kumar**

## 1 Introduction

In current manufacturing, scenario manufacturers are trying to do manufacturing at the lowest cost and higher quality. Because of this trend hard turning is becoming popular as a pre-grinding process, because lower of low cost associated with it as compared to grinding. Because grinding wheels are costlier than tooling involved in hard turning. Hard turning can be performed on the same machine on which soft turning is done [1]. So it is important to understand hard turning process with all of its aspects. Material list which can be hard turned is too long, some of them are Inconel, Stellite and carburized and nitrided irons, Hastelloy, bearing steels, high-speed steels, hot and cold-work tool steels, die steels and case hardened steels [1].

## 2 What Is Hard Turning?

Hard turning is the process of turning workpieces having the hardness greater than 45 HRC. Generally, hard turned part are having the hardness of 58–68 HRC [2]. Hard turning can be an alternate option for those machining operations where very tight tolerances not required. But it cannot replace grinding operation. Material removal rates are 4–6 times than grinding. Well established hard turning process can achieve surface finishes below 0.0003 mm, roundness values of 0.00025 mm [1].

---

S. Patil (✉) · P. Jadhav · S. Kumar
Department of Mechanical Engineering, Symbiosis Institute of Technology,
Symbiosis International University, Pune, India
e-mail: patil.shahu@sitpune.edu.in



K. Antony and J. P. Davim (eds.), *Advanced Manufacturing and Materials Science*,
Lecture Notes on Multidisciplinary Industrial Engineering,
https://doi.org/10.1007/978-3-319-76276-0_24

## 3 Cutting Tool Materials

When we consider cutting tool material important factors to consider are application and finish requirements. Mainly there are following cutting tool materials (1) CBN (2) Ceramic (3) Cermet (4) Natural and Synthetic Diamonds.

1. **CBN**—CBN is preferred when workpiece hardness ranges between 50 and 68 Rc. CBN insert will give the surface finish of 11–15 micro-inches. CBN inserts are available with multiple grades, so important point is to select suitable grade according to different machining requirements. CBN is costlier than Ceramic inserts.
2. **Ceramic**—Ceramic is economical than another type of insert when machining hardened steel. But the achieved surface finish will not be as good as that achieved by CBN. Ceramic is not preferred in coolant cutting because it will not give a good performance because of the presence of thermal shocks.
   Ceramics are having same cost structure as that of carbide, which is 3–4 times less than CBN.
3. **Cermet**—Cermet will give more advantage than CBN while machining hardened steel turning through a hard case into a softcore. But wear resistance of cermet is less. Cost of cermet nearly equal to multi-layer coated carbides [3].

## 4 Importance of Temperature Study

Energy consumed in the cutting process mostly converted into heat. This heat is generated because of friction at the chip-tool interface and work-tool interfaces. Because of this temperature in cutting zone increases. If we consider the hard turning operation, then cutting zone temperature is approximately 1700°. Most of the heat from the cutting application should be carried away with the chip about 60–80% of the total heat, the workpiece carries away 10–20% of the heat, while the cutting tool carries approximately 10% of the heat. Excessive heat in tool and workpiece increases their temperatures. Increased temperature causes cutting tool to elongates which result in a dimensional error [4].

## 5 White Layer

Generally, in dry hard turning temperature in the cutting zone would be around 925 °C. Heat generated in machining heats the material ahead of the tooltip, as a result, the material gets soften. Hence localized heating improve ease of machining [1].

Griffiths BJ [5] observed in both wet and dry hard turning that material can lead to white layer formation, white layer can be formed because of any of following reasons: (i) Because of high-temperature products will be rapidly heated and quenched hence they get transformed (ii) material is also subjected to large plastic deformation and hence material structure will be changed to fine grain (iii) high-temperature surface also get easily reacted with environment. The White layer is not visible to naked eyes which is called as featureless. The thickness of the white layer is usually several micrometers. We can see the white layer only by destructive sectioning and then observed under a good microscope. Why people are concerned about white layer because white layer causes fatigue life of surface to reduce in cases where high contact stresses are present like in bearing races. The sharp tool may start forming white after becoming dull because after dull edge cutting zone temperature will be high, which is favorable to white layer formation.

As fatigue life is reduced because of white layer Kalam SA et al. [6] did an experiment to avoid it. To do this they used Shield gas coolant. They performed the experiments with CBN inserts in the atmosphere of Shield gas coolant. Shield gas (Argon with $CO_2$) keeps the cutting temperature low and hence atmospheric reaction will not happen which is the reason for white layer formation.

## 6 Temperature Measurement

1. **Experimental Methods**

Temperature measurement is most difficult in metal-cutting operations. As cutting edge is heated tool-workpiece interface temperature is very complex and hence it is difficult to develop a setup for measuring the temperature. Hence measurement of temperature with accuracy is still a big challenge [7].

We can use different types of equipment for measurement of the tool-chip interface temperature. Motorcu used an infrared thermometer (IR sensor) to measure the tool-workpiece interface temperature, a K-type thermocouple was used to measure tool temperature [4]. Nordgren A and Thakare A [8] used charge-coupled device (IR-CCD) technique. This imaging technique can be used for temperature measurement of 500–1000 °C, and with suitable calibration, good accuracy can be achieved. But the accuracy of the IR technique depends on the emissivity if both tool and workpiece.

Chinchanikar S et al. [9] investigated the tool-chip interface temperature while turning of AISI 4340 steel with two hardness values of 35 HRC and 45 HRC by CVD coated inserts. They develop a setup for temperature measurement based on tool-work thermocouple. This setup is shown in the Fig. 1.

They used workpiece as a cold end and the tip of inserts as a hot junction to develop thermocouple. Then emf was measured using a multimeter. For calibration of temperature, they developed another setup using a heating coil and K-type thermocouple. The results were recorded and based on regression analysis

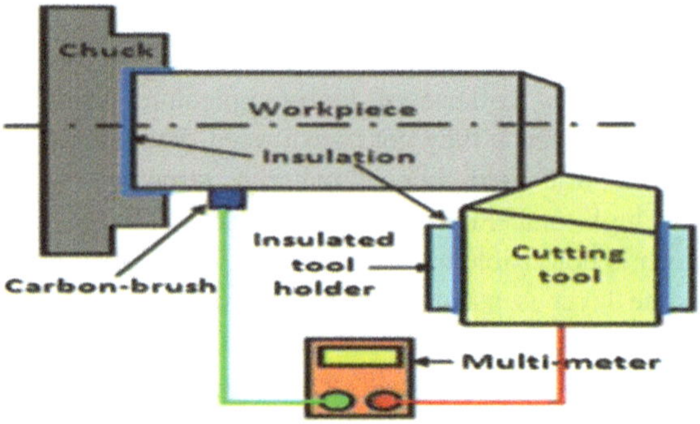


**Fig. 1** Experimental set-up referred from Chinchanikar S et al. [9]

regression model was developed for two hardness values. Predicted values of mathematical models were very close to experimental values. Interface temperature was more in case of hard work material. Interface temperature has more influence of cutting speed than the depth of cut.

Chen Lei et al. [10] studied the machined surface temperature with different experimental methods and 3d FEM method in order to avoid the problem of white layer formation in hard turning, which causes damage to the fatigue life of workpiece. They developed two experimental setups (1) tool-foil thermocouple (2) embedded thermocouple method. In tool foil thermocouple method lot of time is invested in setup preparation. Embedded thermocouple method requires additional FEM modeling. So both experimental methods were proved as feasible.

## 2. Analytical Method

Different analytical methods can be used to develop a model which can predict the temperature value for given set of variables. RSM (Response Surface Methodology) based Study of Cutting Temperature during Hard Turning With Multilayer Coated Carbide Insert did by Shihab SK et al. [11]. Gosaia Mehul and Bhavsarb Sanket N [12] used RSM for studying temperature and tool wear respectively in turning operation. Sahoo AK and Sahoo B [13] used Regression analysis for developing a numerical model which can predict temperature values for certain variable condition. Finite Element Method (FEM) is a good method for material removal simulation, tool wear study and also for thermal study. Once we model the whole setup of an experiment in FEM software then by varying different cutting parameters we can see changes in result in quick time. Kumar SS et al. [14] carried out heat transfer analysis of cutting tool in DEFORM3D software with tool wear estimations also. Then he observed that finite element method can predict tool life as well as wear profile.

Singh G et al. [15] find out Optimized setting of input parameters for the maximum Material removal rate while hard turning AISID3 Die steel using NiN/TiCN/Al2O3/TiN multi-layer coated inserts with different thickness using Taguchi method used for optimization. Silva GC et al. [16] Developed fuzzy control mechanism using an ARDUINO UNO module for data acquisition. Also, FEM thermal analysis was done in ABAQUS software to obtain temperature results then those results were compared with the experimental result. Using k-type thermocouple temperature was measured. Tool wear measured by machining the workpiece without FUZZY control. Then by using Fuzzy control cutting speed was varied to maintain cutting temperature conditions. So it is possible to control temperature using Fuzzy control mechanism.

## 7 Heat Sources in Machining

Major heat source while machining is shear plane heat source and flank face friction heat source as here concentration is on the machined surface temperature (Fig. 2).

For FEM modeling advection thermal model concept was used by Chen Lei et al. [10]. A heat source at a shear plane is applied as constant source and flank face heat source is moved along cutting tool path by discrete time steps. Experimental measurements were converted to machined surface temperature by inverse heat transfer analysis. The shear plane heat source was observed as a major heat source.

## 8 Effect of Different Cutting Parameters

As we discussed earlier temperature is the main factor for affecting tool life. Keeping tool-work interface temperature within certain temperature can increase tool life and also it would lead to better machining of the workpiece. But for keeping the temperature below certain there are many cutting parameters affecting the temperature which should be controlled. There are three aims of any machining operation at a time maximum production rate, the good surface integrity of workpiece, minimum cost (increased tool life). So while machining, deciding the optimum cutting parameters keeping above aims of machining is important.

Bapat PS et al. [17] developed temperature distribution model using ABAQUS software which was based on heat generation. A number of thermal simulations were carried out to find out the value of temperature and region in which it occurs.

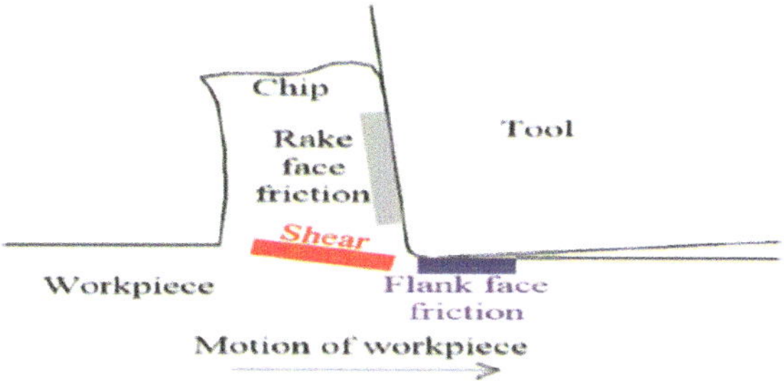


**Fig. 2** Heat sources in machining referred from Chen Lei et al. [10]

Also, temperature was directly proportional to cutting speed. The main reason was as the cutting speed increases friction between tool and workpiece also increases.

Lotfi M et al. [18] carried simulation of turning of Incone l625 using carbide and ceramic inserts(PVD-TiAlN coated) in software DEFORM3D. Then observation was tool wear more in areas where the concentration of temperature and stresses was more. The values of temperature increase as we increase the depth of cut. Chinchanikar S and Choudhury SK [19] measured temperature while turning workpiece of AISI 4340 Steel. They observed that the most affecting parameters on temperature are cutting speed and feed. Klockea F and Lunga D [20] did FEM modeling of turning AISI 1045 steel in software DEFORM™. The study focused on the effect of different parameters on temperature. The result shows that heating of workpiece mainly depends on uncut chip thickness. The cutting speed has a considerable effect on the rate of heat flow into the workpiece. Ravindran VN and Babu RJ [21] A practical explicit 3D FEA model has been developed using ANSYS to simulate turning of Al 7075-T6 heat treated alloy. Using Carbide Tipped Tool. They observed that there is a direct proportionality between temperature and cutting speed, as the speed increased temperature was also increased. Heat generated also increases gradually from room temperature as the point of contact progresses along the workpiece.

Melzi N and Temmar M [22] developed a numerical model and also temperature distribution model using ANSYS software. Different simulations were carried out to obtain the region of maximum temperature at various cutting conditions. The results of the simulation were having good agreement with the results available in the literature. The developed model which could be used to predict the temperature and according to which it is possible to choose correct process parameters during machining of XC48 steel. Shihab SK et al. [11] Studied the effect of coated carbide inserts on temperature while machining AISI 52,100 alloy steel. The experimental setup was prepared using a K-type. Effect of cutting parameters was studied using statistical analysis of variance (ANOVA). ANOVA result showed that all cutting parameters are having a significant influence on cutting temperature. Regression model also predicted for cutting temperature, predicted values of this equation are closer to experimental values.

## 9 Insert Coating Effects

Srithara A et al. [23] used coated carbide insert in hard turning of AISI D2(66HRC). Then observed that change in microstructure which increases strength and wear resistance. Chinchanikar S and Choudhury SK [19] measured temperature During turning of AISI 4340 Steel. Effect of different cutting parameters and the type of coating, namely, Physical vapour deposition (PVD) and Chemical Vapour Deposition (CVD) on tool-workpiece interface temperature was observed using mathematical models. PVD coated tool produced less interface temperature in comparison to CVD-coated tool. For both coated and uncoated depth of cut was a negligible parameter in terms of affecting interface temperature.

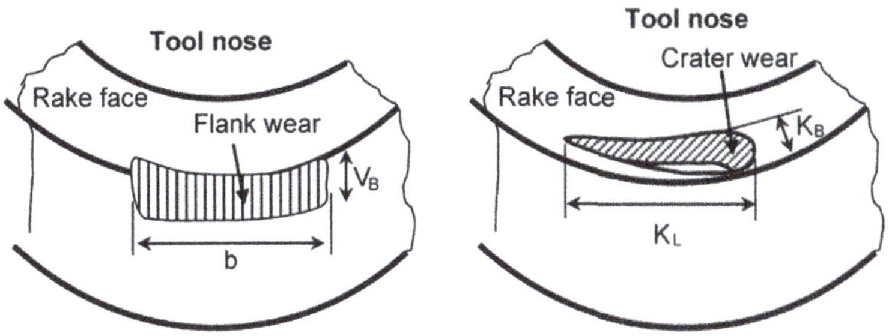


**Fig. 3** Crater wear and flank wear at tool nose referred from Aslantas K et al. [26]

Chetan et al. [24] used cryogenically treated inserts in turning of Nimonic 90 alloy. The observation was AlTiN coated carbide inserts are having more hardness than uncoated carbide inserts. Aneiro FM et al. [25] did the experimental procedure for turning of AISI 4340 (hardened to 48–50 HRC) using coated carbide inserts and PCBN insert. Then they observed tool life for the coated carbide tested, which was around 7000 m of cutting length and which was half of PCBN insert life. But PCBN prices are double to coated carbide insert price. Aslantas K et al. [26] investigated the wear behavior of coated and non-coated ceramic $Al_2O_3$–TiCN mixed inserts while machining of AISI 52,100 which was hardened. Crater wear is dominant wear type for both coated and uncoated cutting tools (Fig. 3). In Fig. 3 KB is crater width, KL is length of crater, bis flank wear length and VB is flank wear width.

Author also studied comparative tool life for inserts of both types then they are having following observations

According to Fig. 4, it appears that the because of TiN coating the life of the cutting tool is tripled. Uncoated tools are having low thermal conductivity because of which tool-chip interface temperature increases. Increased temperature leads to a bimetallic effect on the upper and lower side of chip and hence chip is formed with a smaller radius of curl (Fig. 5).

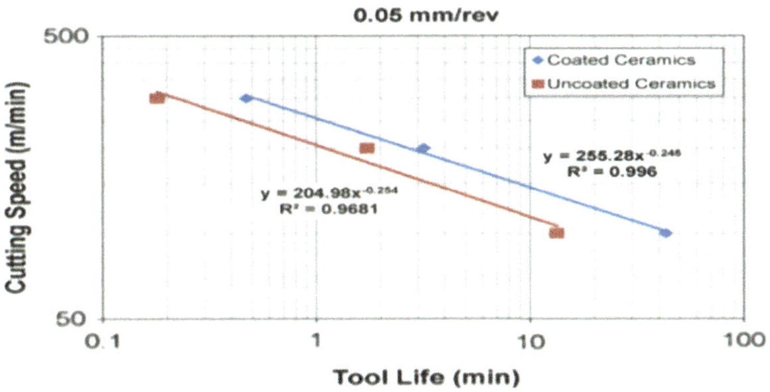


**Fig. 4** Comparison of tool life of coated and uncoated ceramic inserts for f = 0.05 mm/rev [26]

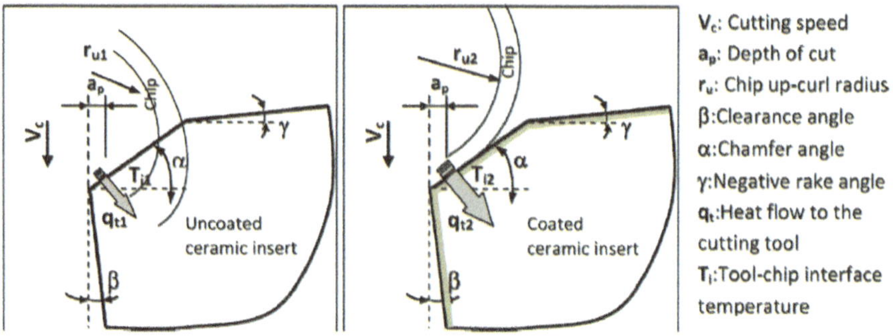


**Fig. 5** Chip flow direction in case of coated and uncoated ceramic inserts [26]

More AS et al. [27] studied Tool wear of CBN–TiN coated carbides inserts and were compared. It is found that the CBN–TiN coated carbide cutting edge was having average tool life of 18–20 min and PCBN tools was having average tool life of 32 min. Though the tool life of PCBN is better than CBN–TiN coated carbide letter will be a better choice if we consider cost as the main factor. Chinchanikar S and Choudhury SK [19] based on the previous study extended their work to study the effect of coating on turning AISI 4340 steel using PVD and CVD coated inserts. By using the previously developed experimental model to measure the temperature they regression model for temperature estimation in both cases of the coating. Then they found that interface temperature was low for PVD coated inserts than CVD coated inserts. Also, it was observed that cutting speed is having more influence on temperature than the depth of cut.

Fang Z and Obikawa T [28] Studied turning of Inconel 718 under high-pressure jet coolant of 13 MPa condition. Made a slot in cooling channels in the insert to enhance heat transfer from the tool. The CFD code PHOENICS was used to calculate. Thermocouple setup was used to measure the temperature. Because of channels heat transfer increased which causes less tool wear. Hence life of insert with channel was found twice than insert having the not channel.

Gill SS et al. [29] used cryogenically treated TiAlN coated tungsten carbide inserts to understand the wear behavior of insert while turning. The TiAlN coated tungsten carbide inserts were subjected to cryogenic treatment at two levels −110 °C (shallow treatment) and −196 °C (deep treatment) of temperature independently. The observed that performance of shallow cryogenically treated inserts was significantly better as compared with deep cryogenically treated and untreated inserts.

Colak O [30] investigated machining performance of Inconel 718 under high-pressure cooling conditions(at the tool-chip interface) with TiN coated carbide tool. They studied for three sets of cutting parameters. Then they observed that tool flank wear and cutting forces decreases because of high-pressure coolant. Also, they did ANOVA study which shows that high-pressure cooling increased cutting tool life.

# 10 Conclusion

Hard turning is a process which can be used as a pre-grinding process as finishing operation with low cost. Different insert materials like ceramic, cermet, carbide can be used for hard turning. Temperature is an important factor in hard turning as heat involved is more in hard turning. There are many methods of increasing tool life like CVD or PVD coating, cryogenic treatment. This broad literature review is more focused on temperature effect on hard turning. Sometimes because of the high-temperature workpiece can be subjected to white layer formation. The different experimental methods like IR sensor, a K-type thermocouple can be used for temperature measurement. Also, different analytical methods like RSM, ANOVA, FEM are used by authors successfully, validating with experimental results. Cutting speed and depth of cut are two main parameters affecting the tool-chip interface temperature. In most of the literature coated inserts are having better performance than uncoated inserts. Tool wear is also more in case of the uncoated insert as compared to coated insert.

# References


1. Hard Turning and the Machine Tool, Soroka DP, Hardinge Inc. One Hardinge Dr.Elmira, NY 14902
2. Huddle D (2001) New hard turning tools and techniques offer a cost-effective alternative to grinding. Tooling and Production Magazine 80:96–103
3. Gosiger Solution (2012) Fundamentals of hard turning. http://cdn2.hubspot.net/hub/139128/file-17761415-pdf/docs/gos_wp_hardturning_f.pdf
4. Motorcu AR, Isik Y, Kus A, Cakir MC (2016) Analysis of the cutting temperature and surface roughness during the orthogonal machining of AISI 4140 alloy steel via the taguchi method. Mater Technol 50(3):343–351
5. Griffiths BJ (1987) Mechanism of white layer generation with reference to machining processes and deformation processes. ASME Trans J Tribol 109(3):525–530
6. Kalam SA, Azad A, Omkumar M, Giri Sankar S, Begum RV (2015) Elimination of white layer formation during hard turning of AISI D3 steel to improve fatigue life. IOSR 12(3):07–14
7. Abukhshim NA, Mativenga PT, Sheikh MA (2005) Investigation of heat partition in high speed turning of high strength alloy steel. Int J Mach Tools Manuf 45:1687–1695. https://doi.org/10.1016/j.ijmachtools.2005.03.008
8. Nordgren A, Thakare A (2015) Experimental study and modeling of steady state temperature distributions in coated cemented carbide tools in turning. Procedia CIRP 31:234 23
9. Chinchanikar S, Choudhury SK, Kulkarni AP (2013) Investigation of chip-tool interface temperature during turning of hardened AISI 4340 alloy steel using multi-layer coated carbide inserts. Adv Mater Res 701:354–358
10. Chen Lei, Tai Bruce L, Chaudhari Rahul G, Song Xiaozhong, Shih Albert J (2017) Machined surface temperature in hard turning. Int J Mach Tools Manuf 121:10–21
11. Shihab SK, Khan ZA, Mohammad A, Siddiqueed AN (2014) RSM based study of cutting temperature during hard turning with multilayer coated carbide insert. Procedia Mater Sci 6:1233–1242

12. Gosaia Mehul, Bhavsarb Sanket N (2016) Experimental study on temperature measurement in turning operation of hardened steel (EN36). Procedia Technology 23:311–318
13. Sahoo AK, Sahoo B (2013) A comparative study on performance of multilayer coated and uncoated carbide inserts when turning AISI D2 steel under dry environment. Measurement 46 (8):2695–2704
14. Kumar SS, Reddy PS, Murthy AB (2017) Analysis of 3d tool wear for turning operation based on finite element method, IJESRT, June 2017
15. Singh G, Singh H, Kumar J (2017) Effect of coating thickness with carbide tool in hard turning of aisid3 cold work steel. Int Res J Eng Tech 04(04):436–440
16. Silva GC, Malveira BM, Carneiro JRG, Brito PP, Silva TA (2017) Modelling of machining operations wear and thermal analysis of WC inserts in turning operations by fuzzy modeling. Procedia CIRP 58:523–528
17. Bapat PS, Dhikale PD, Shinde SM, Kulkarni AP, Chinchanikar SS (2015) A numerical model to obtain temperature distribution during hard turning of AISI 52100 steel. Mater Today Proc 2:1907–1914
18. Lotfi M, Jahanbakhsh M, Farid AA (2016) Wear estimation of ceramic and coated carbide tools in turning of Incone l625:3DF Eanalysis. Tribol Int 99:107–116
19. Chinchanikar S, Choudhury SK (2014) Evaluation of chip-tool interface temperature: effect of tool coating and cutting parameters during turning hardened AISI 4340 steel. Procedia Mater Sci 6:996–1005
20. Klockea F, Lunga D (2013) FEM-Modelling of the thermal workpiece deformation in dry turning. Procedia CIRP 8:240–245
21. Ravindran VN, Babu RJ (2015) Thermal analysis of machining Al 7075–T6 using carbide tipped tool. IJIET 5(4):160–166
22. Melzi N, Temmar M (2016) Applying a numerical model to obtain the temperature distribution while machining, 1–3 June 2016
23. Srithara A, Palanikumarb K, Durgaprasadc B (2014) Experimental investigation and surface roughness analysis on hard turning of AISI D2 steel using coated carbide insert. Procedia Eng 97:72–77
24. Chetan, Ghosh S, Rao PV (2017) Performance evaluation of deep cryogenic processed carbide inserts during dry turning of Nimonic 90 aerospace grade alloy. Tribol Int 115:397–408
25. Aneiro FM, Reginaldo T, Lincoln C (2008) Turning hardened steel using coated carbide at high cutting speeds J Braz Soc Mech Sci Eng 30(2):104–109
26. Aslantas K, Ucunb İ, Cicekc A (2012) Tool life and wear mechanism of coated and uncoated $Al_2O_3$/TiCN mixed ceramic tools in turning hardened alloy steel. Wear 274–275:442–451
27. More AS, Jiang W, Brown WD, Malshe AP (2006) Tool wear and machining performance of CBN–TiN coated carbide inserts and PCBN compact inserts in turning AISI 4340 hardened steel. J Mater Process Technol 180:253–262
28. Fang Z, Obikawa T (2017) Turning of Inconel 718 using inserts with cooling channels under high pressure jet coolant assistance. J Mater Process Technol. https://doi.org/10.1016/j.jmatprotec.2017.03.032
29. Gill SS, Singh J, Singh H, Singh R (2011) Investigation on wear behaviour of cryogenically treated TiAlN coated tungsten carbide inserts in turning. Int J Mach Tools Manuf 51(1):25–33
30. Colak O (2012) Investigation on machining performance of inconel 718 under high pressure cooling conditions. J Mech Eng 58(11):683–690
31. Kumar S, Khedkar NK, Jagtap B, Singh TP (2017) The effects of cryogenic treatment on cutting tools. In: *IOP Conference Series: Materials Science and Engineering*, vol 225, IOP Publishing, p 012104

# Photocatalytic Degradation of Rhodamine-B by Advance Oxidative Process Using Electrochemically Synthesized ZnO–$V_2O_5$ Nanostructures

Jenice Jean Goveas, Naveen Praveen Mascarenhas and Richard Adolf Gonsalves

## 1 Introduction

Water is among the most important resources required to sustain life. Out of the existing vast hydrosphere on planet earth, only 0.7% is the form of fresh water and only 0.01% is accessible to living organisms [1]. It is known that 15% of the world's total production of dyes are lost into aquatic systems during the process of dyeing and is released as textile effluents that act as the major source of non-aesthetic pollution, eutrophication and perturbations to the aquatic ecosystem [2]. The production and applications of synthetic chemical products have become rampant during the last century and this poses a serious threat to the environment especially the water resource. Organic dyes are the major group of contaminants released into industrial effluents especially from textile industries and most of them are mutagenic to human health [3]. These chemicals are nonbiodegradable and hence novel techniques of water treatment must be developed to eliminate or degrade perilous contaminants present in waste water.

Semiconductor oxide materials such as ZnO, CuO, $TiO_2$, $Al_2O_3$, $SnO_2$, $ZrO_2$ and $WO_3$ are attractive photocatalysts as they are not only environmentally sustainable but also show high catalytic efficiency for the degradation of environmental pollutants which include pesticides, dyes, detergents and other volatile organic compounds [4]. Among the oxides of vanadium such as $V_2O_5$, $VO_2$, $V_2O_3$ and VO, vanadium pentoxide ($V_2O_5$) possesses exceptional properties and has wide applications. It may be used as a catalyst, as material for ceramics, solar cells, chemical sensors, electrical and optical devices, infra-red detectors and also as cathode

J. J. Goveas (✉) · R. A. Gonsalves
Department of Chemistry, St. Aloysius College Autonomous,
Mangalore 575003, India
e-mail: jenicegoveas@gmail.com

N. P. Mascarenhas
Department of Physics, St. Joseph Engineering College, Vamanjoor 575028, India

 
K. Antony and J. P. Davim (eds.), *Advanced Manufacturing and Materials Science*,
Lecture Notes on Multidisciplinary Industrial Engineering,
https://doi.org/10.1007/978-3-319-76276-0_25

material for rechargeable batteries [5]. The catalytic activity is attributed to the easy reduction and oxidation between the different oxidation states of vanadium oxides. Zinc oxide having wurtzite hexagonal phase, direct band gap of 3.37 eV and large exciton binding energy of 60 meV is a promising material for photocatalytic degradation of water pollutants [6]. There is thus a need for an efficient method for the bulk synthesis of these semiconductor metal oxide nanoparticles. Although electrodeposition has been one of the oldest techniques for deposition prevailing from the time of Alexander Volta (1800), this technique has not received the attention it requires [7]. This paper discusses a simple hybrid electrochemical thermal method to generate a large quantity of zinc–vanadium mixed metal oxide nanoparticles which have been studied for their role as photocatalysts to degrade a model pollutant Rhodamine B.

A number of methods including physical, chemical and biological techniques have been developed for the degradation of organic compounds. Among the new advanced oxidation processes (AOP), heterogeneous photocatalysis stands out as an emerging destructive technology which brings about complete mineralization of the organic pollutants [8]. Rhodamine-B (RhB) is a dye, with Mol. Wt with 479.02 g/mol and the formula $C_{28}H_{31}N_{23}Cl$ belonging to the class of xanthenes. It shows strong absorption in the visible region (λmax = 540 nm) and is stable at different pH values. It is quite stable in aqueous solutions and is used in textiles, cosmetics and as an analytical reagent but can lead to aesthetic pollution in aquatic environments. It possess high resistance to biological and chemical degradation. It is prohibited in the use as food colour as it is carcinogenic and can cause mutagenesis and teratogenesis in rats [9] (Fig. 1).

This model organic dye has been selected for this research due to its structure (functional groups) and its presence in the environment.

**Mechanism of degradation**: The degradation reaction of the dye by OH˙ radical generated using the synthesized photocatalysts by UV visible light is as follows.

$$ZnO - V_2O_5(OH.)_{(ads)} - D(ads) + ZnO - V_2O_5 - D(ads) \rightarrow \text{intermediates (P)} \rightarrow CO_2 + H_2O$$

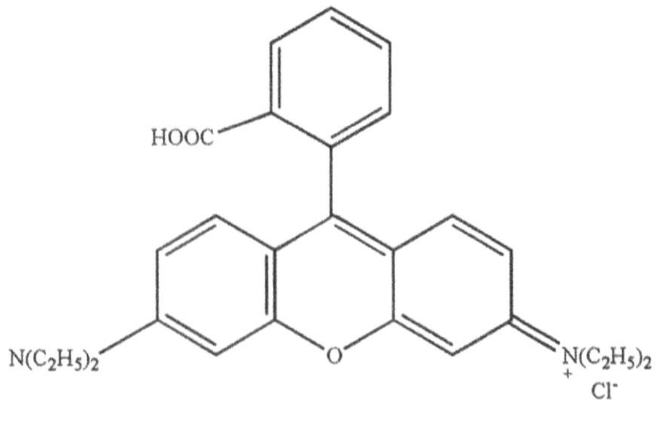


**Fig. 1** Structure of Rhodamine-B

where D is the model organic dye and P is an intermediate product of the reaction.

This process is very efficient due to high surface to volume ratio of the nanoparticles along with high surface energy and hence can prove to be a new green Chemistry technology to degrade pollutants.

## 2 Materials and Methods

All chemicals used in this research are of analytical grade and were used as received. High purity Zinc metal plates (99.99%), sodium bi-carbonate (AR grade: 99.5%), sodium vanadate (AR grade: 98.5%) were purchased from Sisco Research laboratories, Mumbai and Rhodamine-B from S.D. Fine Chemicals Ltd, India. Millipore water (specific resistance, 15 mΩ cm at 25 °C, Millipore Elix 3 water purification system, France) was used to prepare the electrolyte solution.

**Catalyst Synthesis**: ZVO nanoparticles were synthesized using a standard electrochemical technique [10]. Prior to electrolysis, the Zn plates were activated using dilute HCl followed by washing with Millipore water. The electrolyte was a 400 ml solution taken in a Pyrex vessel containing $NaHCO_3$ to increase the conductance and Sodium Vanadate for deposition. It was introduced into a rectangular undivided cell where pure Zn plates were used as both cathode and sacrificial anode. The electrolysis was carried out for about one hour under potentiostatic conditions; a constant voltage of 12 V was supplied by a DC power supply (model PS 618 potentiostat/galvanostat 302/2 A) with constant stirring at 500 rpm.

The pH of the electrolyte was recorded before and after electrolysis. A white precipitate was formed and filtered (by Whattman filter paper No.41) to be isolated from the solution. The electrolysis was also carried out in the presence of 3 different surfactants—cetrimide, PEG and SDS. The obtained powder was calcined at different temperatures from 60 °C to a maximum of 650 °C for an hour.

## 3 Characterization

The Fourier transform infrared spectra (FTIR) of the samples were recorded using Brucker-Alpha-P spectrometer. The XRD data was collected using mini Flex 600 tabletop powder diffractometer at a scan rate of 1°/min. SEM images were recorded using Hitachi TM3030 tabletop SEM with a magnification of ×10 K. Photocatalytic degradation of the pollutant was carried out using Shimadzu UV 1800 UV-Visible spectrophotometer. Fluorescence studies were conducted using Agilent fluorescence spectrometer.

## 4 Results and Discussion

An FTIR spectrum of mixed metal oxide is shown in Fig. 2. A peak at 638 $cm^{-1}$ is associated with the characteristic mode of Zn–O bond. IR absorption of V–O stretching band is observed at 829 $cm^{-1}$ which indicates the presence of $V_2O_5$. The absence of $VO_2$ can be confirmed by the absence of peaks at 508 and 714 $cm^{-1}$.

Figure 3 shows the XRD pattern of $ZnO–V_2O_5$ which was used to check the purity, crystallinity of the nanoparticles. A comparison with the standard card indicates that the diffraction peaks of the product can be indexed to hexagonal wurtzite structure. The crystallite size is calculated from the most intense peaks ($2\theta = 27.8°$) using Scherrer formula and it was found to be 30 nm. Lattice strain is 0.0088. ZnO nanoparticles also show wurtzite structure with most intense peak ($2\theta = 36.1°$). The diffraction peak observed at $2\theta$ values 20°, 30.6°, 42.4° and 48° indicate the presence of $V_2O_5$ nanoparticles in the lattice of ZnO nanocrystalline lattice. The shift of the most intense peak to a lower $2\theta$ value indicates the formation of mixed metal oxides. The diffraction peaks at $2\theta$ values 31.1°, 34.3°, 36.2° indicate the formation of nanocrystalline structures.

Scanning Electron Microscope images (SEM) of mixed metal oxides are shown in Fig. 4. It clearly shows that the particles are agglomerated to form spongy cave like structures with well-defined pores. The average size of the particles from SEM images is found to be 50–80 nm. The particles synthesized in the presence of surfactants as additives showed better porosity and lower particle size and thus were more effective as photocatalysts.

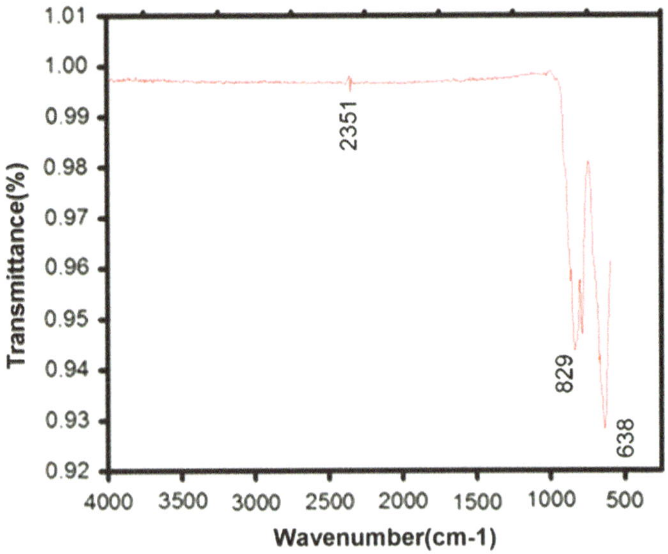


**Fig. 2** FTIR spectrum of ZVO nanoparticles

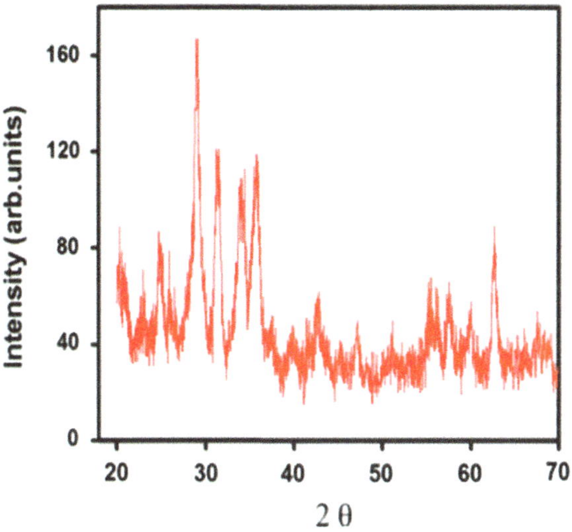


**Fig. 3** XRD pattern of ZVO nanoparticles

## 5 Photocatalytic Procedure

Rhodamine B is used as model pollutant to evaluate the photodegradation behaviour of the electrochemically synthesized of zinc–tin oxide microstructures under UV light degradation in the presence of $H_2O_2$. The photochemical reactor used in this study was made of a Pyrex glass jacketed quartz tube. A high pressure mercury vapour lamp (HPML) of 125 W was placed inside the jacketed quartz tube. To avoid fluctuations in the input light intensity, a supply ballast capacitor were connected in series with lamp. 50 ml of the solution ($Zn_2V_2O_5$ +RhB) was taken in the beaker and continuously stirred to ensure that the suspension of the catalyst was uniform. During the photocatalytic process the sample was at room temperature and ambient pressure. $H_2O_2$ was added to the sample batch to increase the oxygen source during the reaction avoiding the decrease of the catalytic activity during the reaction due to lack of oxygen [11]. Before irradiation, the catalyst was dispersed in the solution and kept in the dark under stirring for 10 min to reach adsorption-desorption equilibrium [12]. Irradiation time ranged from 0 to 60 min. The system was covered so that no other light reached the sample. Every 10 min an aliquot of 10 ml was taken out from the solution. The sample was filtered at vacuum to remove the particles of catalyst and the UV and fluorescence spectra were obtained. A decrease in the intensity of the absorption and fluorescence spectra was observed along the degradation process. 4 different nanoparticles were used for dye degradation—those synthesized without surfactants and those synthesized using cetrimide, PEG and SDS as additives. It was observed that the ZVO nanoparticles synthesized using SDS were most efficient for photodegradation of Rhodamine-B and gave the best results.

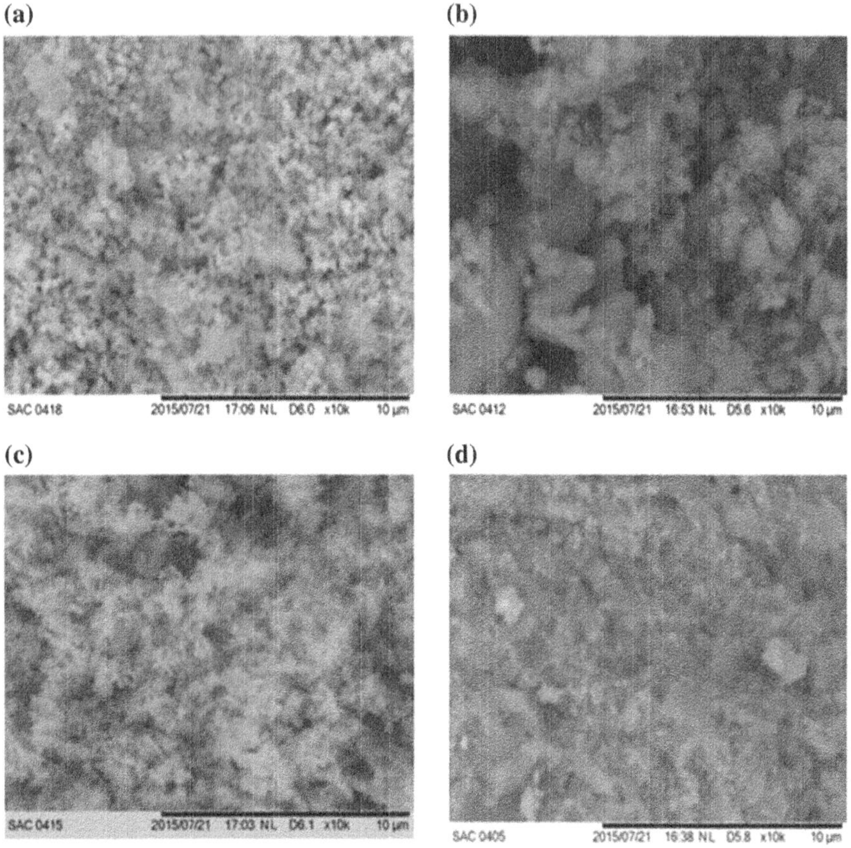


**Fig. 4** SEM images of ZVO nanoparticles **a** without additive **b** using SDS as additive **c** using PEG as additive **d** using Cetrimide as additive

Different intermediates have absorption peaks at different wavelengths than the original organic compounds and for this reason additional absorption peaks are observed during the degradation process. Maximum absorption peak also shifted from 540 to 555 nm after 30 min this proves that various intermediates are formed during the degradation process (Fig. 5).

The relevance of $H_2O_2$ and the photocatalyst for the degradation of Rhodamine B was determined at the beginning of the research. The photocatalytic degradation process of RhB was done in the absence of the catalyst and in the absence of hydrogen peroxide. In both cases no significant reduction in the pollutant concentration was observed during the irradiation process even after 60 min of irradiation. When the hydrogen peroxide and photocatalyst were added to the reaction mixture during the photocatalytic process a significant degradation was observed indicating the relevance of both the photocatalyst and the hydrogen peroxide for the efficiency of the process. Hydrogen peroxide is the source of OH.

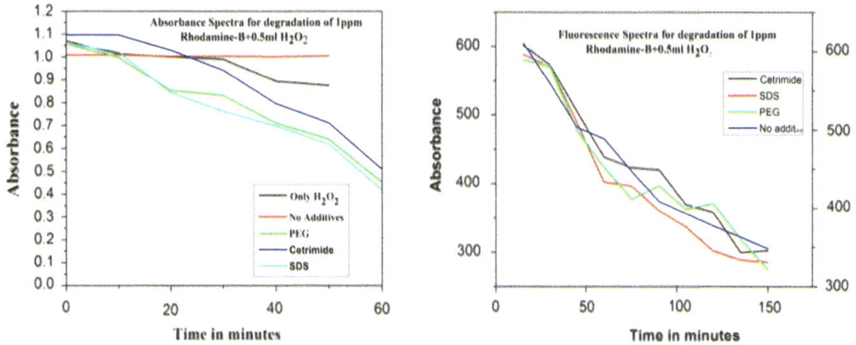


**Fig. 5** Photocatalytic degradation of 1 ppm Rhodamine-B

produced during the catalytic breakdown. It can accept the photogenerated electrons from the conduction band and forms hydroxy radicals preventing recombination of charges and thus increasing the efficiency of the photocatalyst [13]. The decomposition of $H_2O_2$ for the formation of radicals during the degradation process is given by [14]:

$$H_2O_2 + e^- \longrightarrow OH^- + OH^{\cdot}$$
$$H_2O_2 + O_2^{\cdot -} \longrightarrow OH^- + OH^{\cdot} + O_2$$

**Kinetics of dye degradation**: Kinetic studies of the photocatalytic degradation of the dyes reveal that it follows pseudo first order reaction according to Langmuir Hinshelwood mechanism $r = \frac{-dC}{dt} = \frac{kKC}{1+KC}$ where r is defined as the reaction rate, k the reaction rate constant, K the absorption coefficient and C is the concentration of the reactant. The semilogarithmic plots of concentration against time in minutes give almost straight lines where the slope represents the value of the reaction rate. This is confirmed by a plot of ln $C/C_0$ against time which appears as a straight line as shown in Fig. 6. The degradation leads to the conversion of organic carbon into harmless gaseous $CO_2$ and that of nitrogen heteroatoms into nitrate and ammonium ions [15].

## 6 Summary

AOPs are advantageous as they bring about the complete mineralization of the pollutants. They are nonselective processes and can be used efficiently in low concentration of contaminants. Heterogeneous photocatalysis using mixed metal oxide nanoparticles is one such AOP successfully implemented in this project. The semiconductor photocatalysts of ZVO generated were chemically inert, stable, inexpensive, easily synthesized and are therefore perfect photocatalysts for the

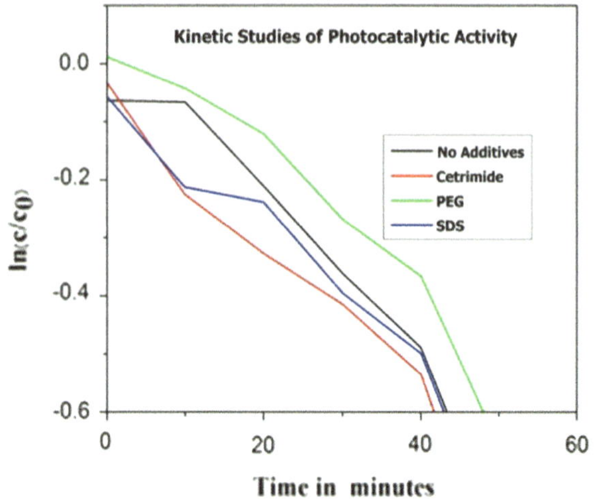


**Fig. 6** Kinetics study

degradation of organic pollutants. The electrochemically synthesized ZVO nanoparticles are highly efficient in degrading the model pollutant Rhodamine B. However, both $H_2O_2$ and the photocatalyst are needed for efficient degradation. Fluorescence spectra reveals a clear decrease in fluorescence indicating degradation pattern. Kinetic studies have confirmed the reaction to be of pseudo first order.

**Acknowledgements** The authors would like to thank the management of St. Aloysius College, Mangalore for providing laboratory facilities to undertake this project and the laboratory of Applied Biology, St. Aloysius College for providing SEM imaging facility.

## References

1. del María C, Teresa C, Eduardo E, Arancha G, Carmen M, Francisco M (2013) Am Chem Sci J 3(3):178–202
2. Zollinger H (1991) Color chemistry. Synthesis, properties and applications of organic dyes and pigments, 2nd edn. VCH, Weinheim
3. Han F, Kambala VSR, Srinivasan M, Rajarathnam D, Naidu R (2009) Appl Catal A 359:25W
4. Cun, Wang XM, Zhao JC, Mai BX, Sheng GY, Peng PA, Fu JM (2002) J Mater Sci 37:2989–2996
5. Pdtkov V, Trikalitis PN, Bozin ES, Billinge SJL, Vogt T, Kanatzidis MG (2002) Am Chern Soc 124:10157
6. Kazeminezhad I, Sadollahkhani A, Farbod M (2013) Mater Lett 92:29–32
7. Chandrappa KG, Venkatesha TV (2012) Nano Micro Lett 4:14–24
8. Herrmann JM (1999) Water treatment by heterogeneous photocatalysis. In: Jansen F, van Santen RA (eds) Environmental catalysis, catalytic science series, vol 1. Imperial College Press, London, pp 171–194
9. Sarma J, Sarma A, Bhattacharyya KG (2008) Biosorption of commercial dyes on *Azadirachta indica* leaf powder: a case study with a basic dye Rhodamine B. Ind Eng Chem Res 47:5433–5440

10. Chandrappa KG, Venkatesha TV, Vathsala K, Shivakumara C, Nanopart J (2010) Res. 12:2667–2678
11. Cotto-Maldonado MC (2012) Heterogeneous catalysis applied to advanced oxidation processes (AOPs) for degradation of organic pollutants. Dissertation, Universidad del Turabo
12. Sun M, Li D, Chen Y, Chen W, Li W, He Y, Fu X (2009) Synthesis and photocatalytic activity of calcium antimony oxide hydroxide for the degradation of dyes in water. J Phys Chem C 113(31):13825–13831
13. Arathi T, Madras G (2007) photocatalytic degradation of Rhodamine dyes with nano $TiO_2$. Ind Eng Chem Res 46(1):7–14
14. Yu K, Yang S, He H, Sun C, Gu C, Ju Y (2009) Visible light- driven photocatalytic degradation of Rhodamine B over $NaBiO_3$: pathways and mechanisms. J Phys Chem A 113 (37):10024–10032
15. Houas A, Lachheb H, Ksibi M, Elaloui E, Guillard C, Herrmann JM (2001) Appl Catal B Environ 31:145–157

# Facile Synthesis and Characterisation of Nanocomposite Doped Chitosan—Polystyrene Polymer Blends

Naveen Praveen Mascarenhas, Jenice Jean Goveas, Richard Adolf Gonsalves, T. Chandra Shekara Shetty and Vincent Crasta

## 1 Introduction

A polymer may be a natural or synthetic macromolecule comprised of repeating units of smaller molecules called monomers. The process of Polymerization involves formation of covalent bonds between the individual monomers giving rise to the polymer. Condensation polymerization occurs by loss of small molecules from the monomers which joint together. A polymer composite is a composite material containing polymers, or polymers along with other kinds of materials. The research dealing with polymer composites have reinforced the growing need of new materials with novel properties which can be modified for functional applications in, agriculture, industry, medicine, etc. Many polymers are known to be suitable matrices in the development of composite structures because of the ease of production and processing. They possess good adhesion with reinforcing elements, are resistant to corrosive environment, and are light weight. This results in the numerous applications viz in sensor technology, as super capacitors, corrosion inhibitors and for electromagnetic shielding. Polymers can also be used in and batteries and actuators [1–6]. An interesting property of polymer composites is the ability to alter the electrochemical, optical, chemical and mechanical properties of conducting polymers by changing the monomer or dopant incorporated into the

N. P. Mascarenhas (✉) · V. Crasta
Department of Physics, St Joseph Engineering College, Vamanjoor 575028, India
e-mail: naveenmascarenhas@gmail.com

J. J. Goveas · R. A. Gonsalves
Department of Chemistry, St Aloysius College Autonomous, Mangalore 575003, India

T. C. S. Shetty
Department of Post Graduate Studies in Physics, St Aloysius College Autonomous, Mangalore 575003, India



K. Antony and J. P. Davim (eds.), *Advanced Manufacturing and Materials Science*, Lecture Notes on Multidisciplinary Industrial Engineering,
https://doi.org/10.1007/978-3-319-76276-0_26

polymer. Organic-inorganic composites have attracted great attention in many applications, such as packing material, circuit board, interlayer dielectrics and passive protection. With the progress of the electronic industry, embedded capacitors exhibited special interest. The characteristics of high dielectric permittivity and better mechanical strength improve the electrical performance and reduce the size and cost of the polymer composites [7]. In addition the incorporation of ceramic fillers having high dielectric constant within a polymer matrix results in a composite with enhanced properties [8].

The process of blending is thus beneficial and involves mixing of two or more suitable polymers and copolymers. By blending the polymers [9] their properties can be enhanced for wide applications. Specific properties of natural polymers can be significantly improved by blending with synthetic polymers. The goal of synthesizing new polymer blends is not only to modify the properties of its components drastically, but also to extract the maximum possible performance of the blended materials.

In the last decade polymer blends have received much attention due to their ability to exhibit better physicochemical properties, which are largely observed when the individual components used for blending are compatible. Incorporation of inorganic fillers such as Transition metal oxides to form Composite Polymer materials is one of the ways to improve electrical, mechanical and photocatalytic properties of polymer blends [10].

PS and chitosan are immiscible, because PS is hydrophobic, whereas chitosan is a hydrophilic polysaccharide. Consequently, their blends suffer a drawback in thermal and mechanical properties which make them not competitive for many applications. A number of studies are focused on the improvement of compatibility between their two phases, either by addition of a cross linking agent like glutaraldehyde or by stimulation of a chemical reaction which could result in modification of their polymer interface.

The unique physical and chemical properties of transition metal oxides have greatly attracted the interest of researchers worldwide. Transition metal oxide photocatalysts have been gaining a lot of attention because of their promising applications in water disinfection, dye degradation and for the complete mineralization of toxic organic compounds which are present as impurities in industrial effluent water [11]. The electrochemical thermal route has the advantage of being simple, convenient, and swift facilitating the generation of pure, homogenous and larger quantities of nanoparticles. Mixed metal oxides have varying applications due to their acid-base and redox roperties. They play a vital role in organic synthesis. Zinc oxide nano particles are generally known for their unique properties which find applications in non-linear optics, photocatalysis, solar energy conversion, gas sensors and cosmetics [12]. $Ti^{+4}$ has a comparable ionic radius with that of $Zn^{2+}$. Hence it forms stable MMO of composition $ZnO/TiO_2$. In this paper an attempt is made to improve the quality of CS–PS polymer blend by incorporating mixed metal oxide nanomaterials of $ZnO–TiO_2$.

## 2 Materials and Methods

Chitosan from HiMedia laboratories Pvt Ltd, Polystyrene from Sigma Aldrich now Merck, High purity Zinc metal plates (99.99%), Sodium Bi-Carbonate (AR grade: 99.5%), Potassium Titanium Oxalate (ARgrade: 98.5%) purchased from Sisco Research laboratories, Mumbai, Millipore water (specific resistance, 15 mΩ cm at 25 °C. Millipore Elix 3 water purification system, France) was used for the preparation of electrolytic solution.

### 2.1 *Preparation of Polymer Bends*

CS being hydrophobic material, it is only soluble in some selected solvents. In this study, acetic acid was used as solvent. 1% CS solution was prepared using 2 N acetic acid and was agitated for 2 h using a magnetic stirrer in a 50 ml beaker. In another beaker 2.5% PS solution was prepared using research grade benzene as solvent and stirred well using a magnetic stirrer to get homogeneous solution. The two solutions were mixed in various compositions to make the blends of CS–PS. In all cases, the total weight of two components in solution was maintained as 1 g. CS solution was added to the PS solution in steps and the solutions were thoroughly mixed using a magnetic stirrer for 2–3 h to obtain homogeneous solution. The synthesized $ZnO–TiO_2$ nanopartcles weighing 0.01 g were introduced to the blend during the stirring process. The polymer blend solution doped with nanocomposites was sonicated using UV sonicator for 30 min. The blends were poured into Teflon coated plates and allowed to evaporate in dry air. After 2 days polymer films peeled from the plates (Fig. 2b).

The molecular structure of CS and PS are shown in the Fig. 1a, b respectively.

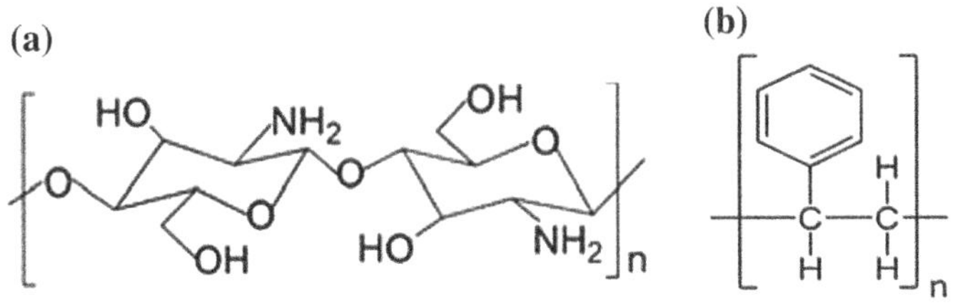


**Fig. 1** **a** and **b** Molecular structure of CS and PS

## 2.2 Synthesis of ZTiO Nanoparticles

ZTiO nanoparticles were synthesized using a standard electrochemical technique. Prior to electrolysis, the Zn plates were activated using dilute HCl followed by washing them with Millipore water. The electrolyte was a 400 ml solution taken in a Pyrex vessel containing $NaHCO_3$ to increase the conductance and Potassium Titanium Oxalate for deposition. It was introduced into a rectangular customized undivided cell where pure Zn plates played the role of both cathode and anode. The electrolysis was conducted for 60 min under potentiostatic conditions where a constant voltage of 12 V was supplied by a DC power supply (model PS 618 potentiostat/galvanostat 302/2 A) with continuous stirring at 500 rpm.

The white precipitate thus obtained was filtered (by Whattman filter paper No. 41) and isolated from the solution. Calcinations of these particles was performed at 600 °C for an hour. The obtained mixed metal oxide (MMO) nanoparticles were milled for half an hour (Fig. 2a).

The formation of $ZnO–TiO_2$ nanoparticles follows the following path

$$Zn \rightarrow Zn^{+2} + 2e. \tag{1}$$

$$2H^{+} + 2e \rightarrow H_2. \tag{2}$$

$$C_4K_2O_9Ti + 3H_2O \rightarrow 2K^{+} + Ti^{+4} + 4CO_2 + 6OH^{-}. \tag{3}$$

$$Zn^{+2} + Ti^{+4} + 6OH^{-} \rightarrow ZnTi(OH)_6 \rightarrow ZnTiO_3 + 3H_2O. \tag{4}$$

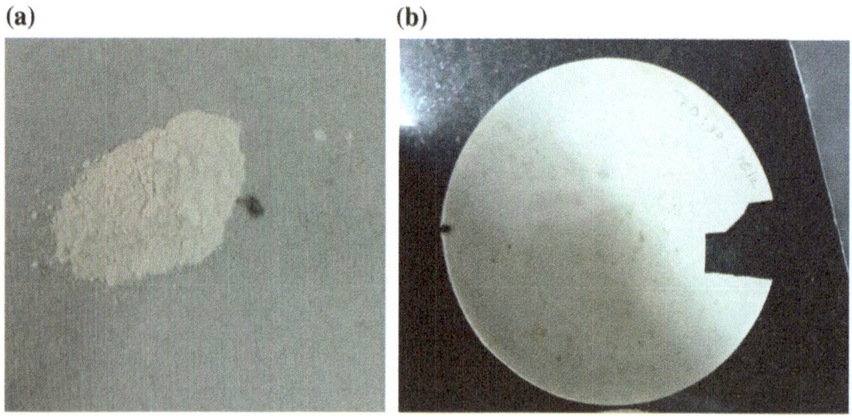

**Fig. 2** **a** and **b** Synthesised $ZnO\text{-}TiO_2$ nanoparticles and polymer film

# 3 Result and Discussion

## 3.1 FT-IR Analysis

The Fourier transform infrared spectrum of CS–PS blend without and with nanocomposites was recorded using KBr disc technique [13]. The FTIR spectra are obtained using JASCO FTIR 3006 spectrophotometer. The FTIR spectrum of CS–PS (20:80) blend with nanocomposites is shown in Fig. 3b. The characteristic transmission peaks are consistent with the functional groups present in the material. The assigned values of the absorptions peaks are presented in Table 1.

The shift in vibration frequencies is due to interactions between ZnO–$TiO_2$ nanoparticles with functional groups OH and $NH_2$ of chitosan. The small and broad

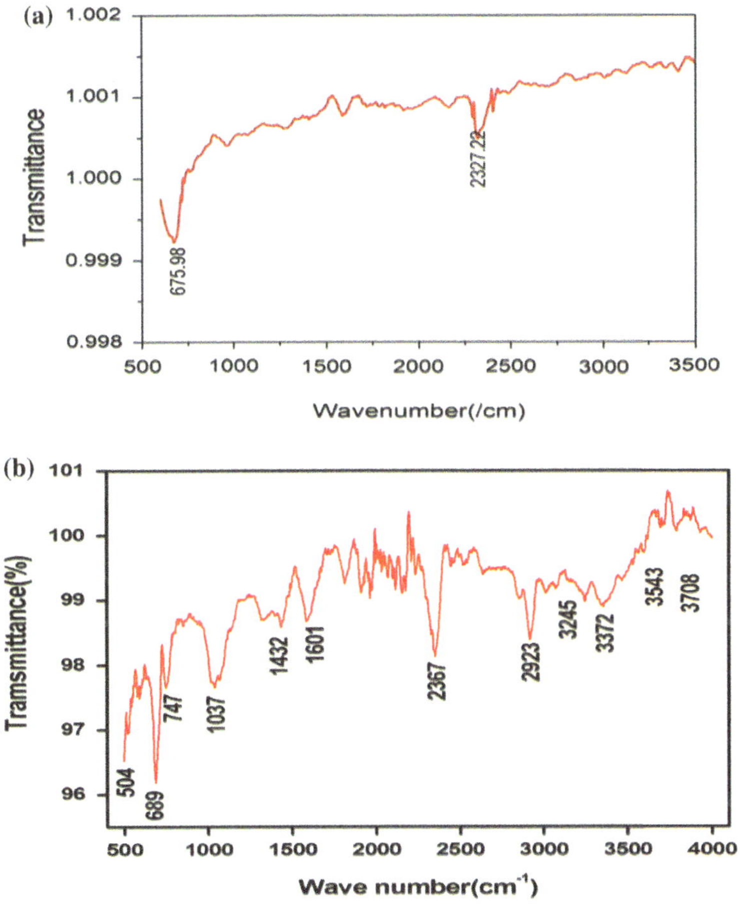


**Fig. 3** **a** and **b** FT-IR of ZnO-$TiO_2$ nano composite and polymer doped with ZnO-$TiO_2$

**Table 1** Functional group assignment for CS-PS blend doped with ZnO-$TiO_2$

| Wave number ($cm^{-1}$) | Mode of Assignments |
|---|---|
| 3708 | –OH stretch |
| 3543 | –NH bend |
| 3372 | |
| 3245 | |
| 2923 | Symmetric $CH_2$ stretching—vibration corresponding to pyranose ring, aromatic ring |
| 2367 | N–H stretch of $NH_3^+$ |
| 1601 | NH-bend |
| 1432 | $CH_2$ bend |
| 1037 | NH-wagging, C–N stretch peaks shifted due to the presence $TiO_2$ |
| 749 | |
| 689 | |
| 504 | |

bands at 3708 and 3372 $cm^{-1}$ are attributed to the stretching vibration of O–H and N–H primary amines (asymmetric stretch) [14]. This may be due to the immobilization of $Zn^{+2}$ onto O–H and N–H functional groups. The band at 2923 $cm^{-1}$ is exhibited due to asymmetric stretch of C–H group (alkanes). The absorbed frequency at 1601 $cm^{-1}$is due to N–H bend of secondary amines (symmetric stretch). The stretch at 1037 $cm^{-1}$ is due to carboxylic acids (C–O). A small peak at 504 $cm^{-1}$ was attributed to the stretching vibrations of N–Zn [15].

## *3.2 Powder Diffraction Study*

Data was collected using Miniflex 600 table top powder diffractometer at a scan rate of 1° per minute for a small portion of the CS–PS blend without and with nanocomposites. Wavelength of radiation used was 1.5418A°. Powder XRD pattern for a CS–PS blend in the ratio 20:80 are shown in the Fig. 4a, b respectively. With the increase in chitosan percentage it was observed that the amorphous nature of the film increases.

Figure 4 represents the XRD pattern of the blend without and with nanocomposites. In Fig. 4b the diffracted peaks appeared at 24.7°, 31.69°, 34.38°, 36.18°, 37.8°, 47.45°, 56.46°, 62.76°, 66.21°, 67.80°, 68.92°. The results demonstrated both characteristic peaks of the CS–PS blend and ZnO–$TiO_2$ nanoparticles which confirms the successful blending of the polymer composite [16].

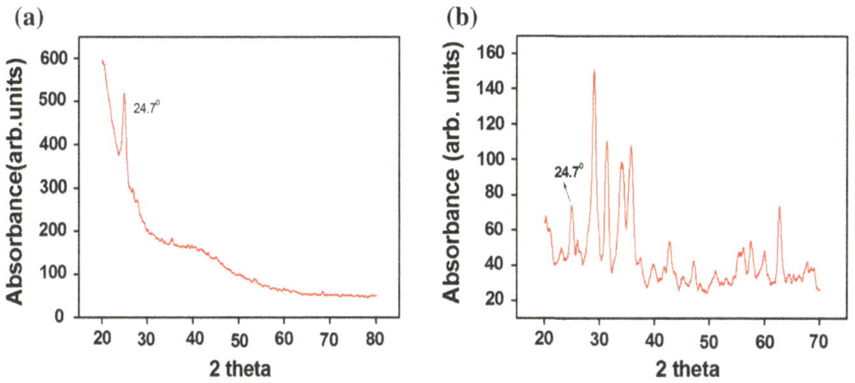


**Fig. 4** **a** and **b** Powder XRD patterns of polymer blend without and with ZnO-$TiO_2$ nanocomposites

## 3.3 *Surface Morphology Studies*

In order to study the surface morphology of the CS–PS blended films doped with nanocomposite, Scanning Electron Microscope (SEM) image was taken using a Hitachi TM3030 table top SEM with a magnification of 2.00 K X The image shows reinforcement of nanocomposites into CS–PS blend. Figure 5 represents SEM images of CS–PS blended film without and with nanocomposites.

SEM visual gives the approximate size and shape of nanoparticles. As presented in Fig. 5, the prepared CS–PS Blend doped with ZnO–$TiO_2$ exhibited globular morphology. Agglomeration observed in the SEM images is possibly due to the dispersion of nanoparticles in the powder form. The nanoparticles were successfully incorporated into the polymer matrix.

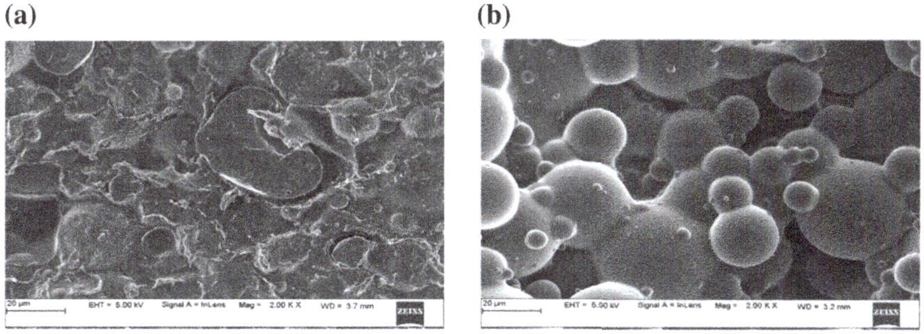


**Fig. 5** **a** and **b** SEM images of polymer blend without and with ZnO-$TiO_2$ nanocomposites

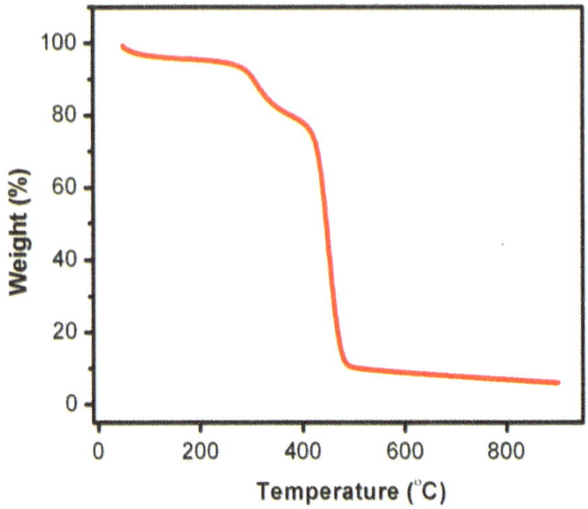


**Fig. 6** TGA of polymer blend with ZnO-$TiO_2$ nanocomposites

## 4 Thermal Analysis

To investigate the thermal stability of 20:80 CS–PS blend doped with nanocomposites, Thermo Gravimetric Analysis (TGA) and Differential Scanning Calorimetry (DSC) was carried out using PerkinElmer TGA/DSC analyzer. Taking a small part of the blend, the analysis was carried out under the nitrogen atmosphere at a heating rate of 20 °C/min. The TGA thermogram of CS–PS blend (20:80) is shown in Fig. 6.

The TG curve of the sample indicates its high stability up to 280 °C. The first weight loss by 14.539% was observed at 382.86 °C and the second weight loss by 79.65% was observed at 523.85 °C. 97.332% of Total Weight Loss was observed at 00 °C. The blended film shows the glass transition temperature at 94.61 °C.

## 5 Conclusion

Prepared blends are thermally stable and have good stability. From the SEM images it is concluded that the films have good surface morphology. The property of amorphous nature of the films observed from the XRD analysis finds application in adsorbent for dyes in the treatment of effluent water from textile industries.

**Acknowledgements** Authors thank the Principal, St Aloysius College (Autonomous), Mangalore for providing the Laboratory and experimental facilities for the research work carried out.

## References


1. Pfaendner R (2010) Photo-oxidation of polypropylene/montmorillonite nanocomposites. 1. Influence of nanoclay and compatibilizing agent. Polym Degrad Stab 95:369–373
2. Stejskal J, Sapurina I, Trchov′a M, Konyushenko EN (2008) oxidation of aniline: polyaniline granules, nanotubes, and oligoaniline microspheres. Macromolecules 41:3530
3. Wei Y (2001) Nonclassical or reactivation chain polymerization: a general scheme of polymerization J Chem Educ 78:551
4. Ciric-Marjanovic G, Trchova M, Stejskal J (2008) Theoretical study of the oxidative polymerization of aniline with peroxydisulfate: tetramer formation. Int J Quantum Chem 108:318
5. Tzou K, Gregory RV (1992) Kinetic study of the chemical polymerization of aniline in aqueous solutions. Synth Met 47:267
6. Wei Y, Sun Y, Tang X (1989) Effects of *p*-aminodiphenylamine on electrochemical polymerization of aniline. J Phys Chem 93:4878
7. Ortiz RP, Facchetti A, Marks TJ (2010) High-k organic, inorganic, and hybrid dielectrics for low-voltage organic field-effect transistors. Chem Rev 110:205
8. Chiang CK, Popielarz R (2002) Polymer composites with high dielectric constant. Ferroelectrics 1:275
9. Cao Y, Smith P, Heeger AJ (1993) Counter-ion induced processibility of conducting polyaniline. Synth Met 55–57:3514–3519
10. Winie T, Jamal A, Hanif NSM, Shahril NSM (2014) Hexanoyl chitosan-polystyrene blend based composite polymer electrolyte with surface treated $TiO_2$ fillers. Key Eng Mater 594–595:656–660
11. Yu CL et al (2012) Sci China Chem 55:1802–1810
12. Sang CK et al (2003) Physical 103:130
13. Waghmare VG, Kariya KP, Paliwal LJ (2011) J Mater Sci Eng B1:485
14. Munnawar et al (2017) Synergistic effect of chitosan-zinc oxide hybrid nanoparticles on antibiofouling and water disinfection of mixed matrix polyethersulfone nanocomposite membranes. Carbohydr Polym. http://dx.doi.org/10.1016/j.carbpol.2017.08.036
15. Dhanavel et al (2014) Photocatalytic activity of Chitosan/ZnO nanocomposites for degrading methylene blue. Int J chemtech 1880–1882
16. Hanif NSM, Shahril NSM, Azmar A, Winie T (2015) Studies on the effect of acid treated $TiO_2$ on the electrical and tensile properties of hexanoyl chitosan-polystyrene-$LiCF_3SO_3$ composite polymer electrolytes. In: American Institute of Physics Conference Proceedings 1674, 020028-1-020028-7

# Influence of EDM Control Factors for Aluminium Hybrid Composites

**Kamalkishor Ghisulal Maniyar and Dilip Sahebrao Ingole**

## 1 Introduction

The mechanical and tribological properties of hybrid composites can be highly enhanced by adding different reinforcements such as silicon carbide, alumina, graphite etc. Machining of Hybrid composites materials are new challanges in the field of conventional machining process. It is reduced by employing Non-conventional machining processes specially Electrical discharge machining process. Ahamed et al. [1], Gopalkannan et al. [2], Prakash et al. [3], Puhan et al. [4], Roy et al. [5], found the impact of control factors on the machining characteristics during the machining of Aluminium Composites. George et al. [6] investigated the parametric combination of control factors for boosting the MRR and limit the electrode wear. Hourmand et al. [7] used RSM to analyse EDM of Al-$Mg_2$-Si composite. They have discovered that machining parameters have good impact on the profile and microstructure of machined surface. Kumar et al. [8], Pecas et al. [9], Talla et al. [10] showed the better experimental results of powder added EDM as compared to conventional EDM. Karthikeyan et al. [11] found rise in weight fraction of SiC gives negative impact on MRR while it gives the positive impact on TWR and SR. Lin et al. [12], Singh [13] applied Taguchi based grey relational technique and found the optimal parametric combination of machining parameters. Liu et al. [14], Sing and Yeh [15] developed a high abrasive EDM process which assisted to enhance the efficiency and surface finish as compared to the conventional EDM process. Mohan et al. [16] analyzed the impact of input

K. G. Maniyar (✉)
Department of Mechanical Engineering, MGI's College of Engineering
and Technology, Shegaon, Maharashtra, India
e-mail: kkmaniyar1313@gmail.com

D. S. Ingole
Department of Mechanical Engineering, Prof. Ram Meghe Institute
of Technology and Research, Badnera, Maharashtra, India

 
K. Antony and J. P. Davim (eds.), *Advanced Manufacturing and Materials Science*,
Lecture Notes on Multidisciplinary Industrial Engineering,
https://doi.org/10.1007/978-3-319-76276-0_27

parameters on machining of Al-SiC composite and also obtained the good output results compared to stationary electrode. Rajkumar et al. [17] proved the microwave heat treatment an operative method compared to conventional heat treatment. Radhika et al. [18], Singh et al. [19] employed Taguchi DOE methodology to determine optimal parametric combination of control factors. Singh et al. [20] revealed the MRR found to be higher for larger current and pulse on time setting at the expenses of tapercity, radial over cut and surface finish. The objective of present research work is to examine the influence of control factors on machining characteristics of Aluminium hybrid composites.

## 2 Methodology

### *2.1 Manufacture of Hybrid Composite*

The liquid stir casting method is employed to manufacture three samples of hybrid composites wt of 70–90% Aluminium alloy reinforced with wt of 5–15% silicon carbide and wt of 5–15% graphite. The reinforcement materials are heated separately in graphite crucible and added into molten state of metal matrix material

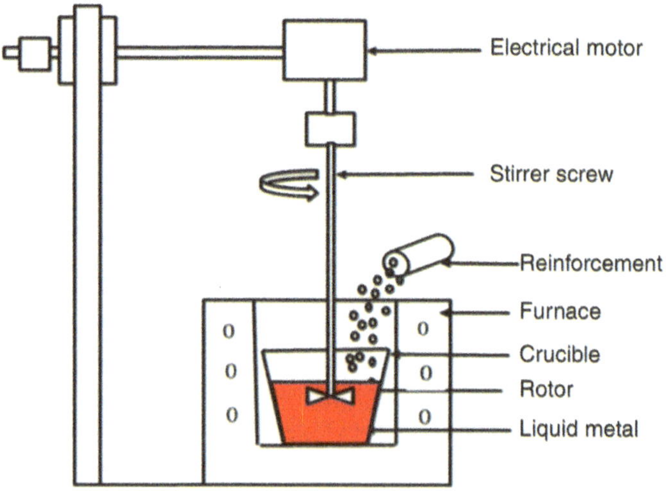


**Fig. 1** Set up of liquid stir casting method [1]

**Table 1** Chemical composition of AC2B Al alloy

| Al | Si | Fe | Cu | Mn | Ti | Other |
|---|---|---|---|---|---|---|
| 88.7 | 6.89 | 0.46 | 3.33 | 0.06 | 0.32 | 0.24 |

(Fig. 1). The mechanical stirrer is used to obtain homogenous mixture. The chemical composition of Aluminum alloy is as given in Table 1.

## 2.2 Control Factors and Machining Characteristics

The control factors are considered such as combined equal % volume of silicon carbide and graphite, current and pulse on time with three levels such as low, medium and high in each. The control factors and their level is as shown in Table 2. The machining characteristics are evaluated in terms of material removal rate, surface roughness and tool wear rate. Material removal rate and tool wear rate are evaluated by the proportion of weight difference of its material prior and after machining to machining time. Surface tester is used to measure reading of surface roughness of hybrid composites. The transverse length of each measurement is set to 5 mm and the average roughness value is noted.

## 2.3 Conduction of Experiments

The experimental trails are performed by non conventional machining process namely EDM process. Electrode is made up of copper which has good mechanical and electrical properties. It is most suitable tool material used for machining of hybrid composites. The experiments are conducted by Taguchi technique utilizing the machining set up. The experimental trials are planned by orthogonal array L27 and weight of the work piece and electrode are taken into account for evaluating the MRR and TWR respectively. Each experiment is repeated twice for obtaining better results and average values are considered for examination work.

**Table 2** Control factors and their level

| Control factors | Unit | Level 1 (low) | Level 2 (medium) | Level 3 (high) |
|---|---|---|---|---|
| combined equal % volume of SiC–Gr | % | 5 | 10 | 15 |
| Current | A | 4 | 8 | 12 |
| Pulse on time | μs | 200 | 400 | 500 |

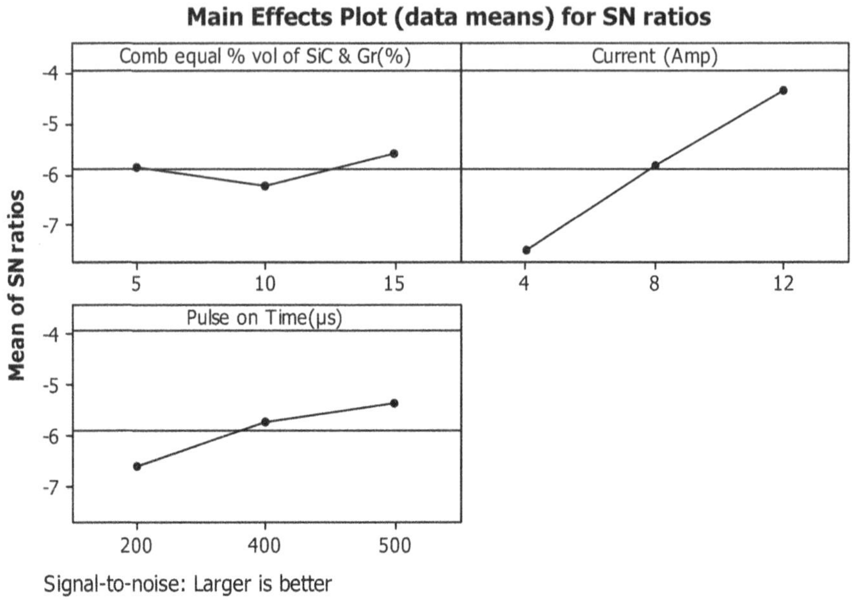


Fig. 2 Signal to noise ratio for MRR

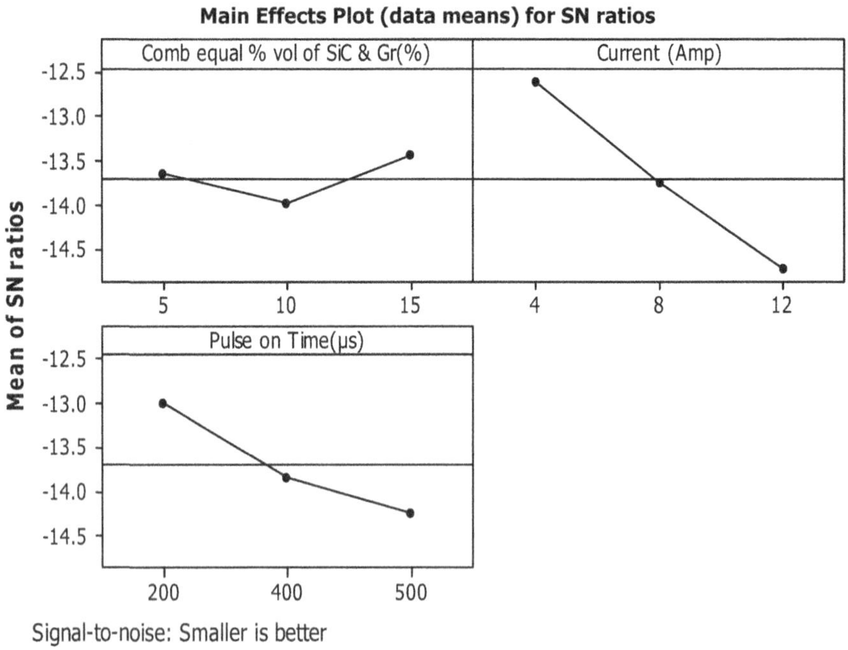


Fig. 3 Signal to noise ratio for SR

# 3 Result and Discussion

## 3.1 SN Ratio for Machining Characteristics

Signal to Noise analysis is applied to determine optimal setting of control factors for obtaining better results of machining characteristics. According to Fig. 2, the optimal parametric combination of control factors for higher material removal rate are combined equal % volume of SiC–Gr 15%, Current 12 A and Pulse on time 500 μs. To acquire lower surface roughness the ideal parametric mix of control factors are combined equal % volume of SiC–Gr 15%, Current 4 A and Pulse on time 200 μs is as shown in Fig. 3. The mean S/N proportion diagram for TWR is displayed in Fig. 4, it is noted the ideal parametric mix of control factors are combined equal % volume of SiC–Gr 15%, Current 4 A and Pulse on time 500 μs.

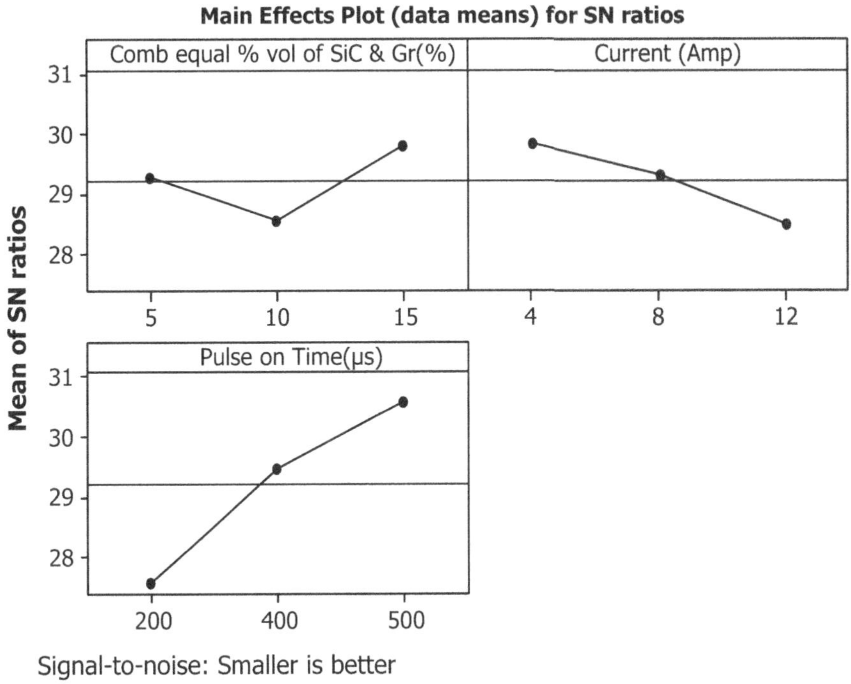


**Fig. 4** Signal to noise ratio for TWR

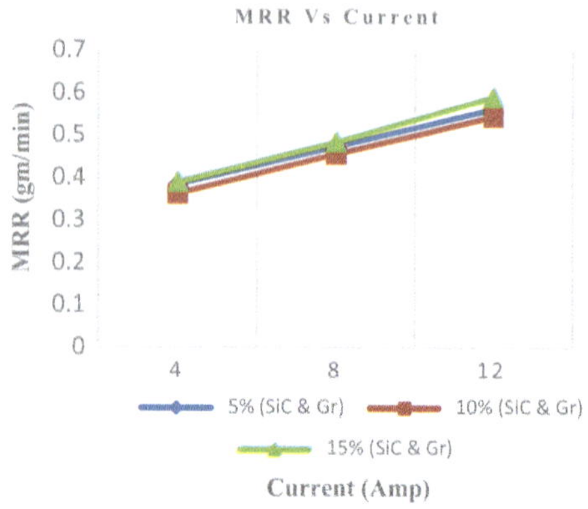


**Fig. 5** Correlation between MRR and current

## 3.2 *Effect of Control Factors on Machining Characteristics*

According to Figs. 5, 6 and 7 shows that current is directly proportional to the material removal rate, tool wear rate and surface roughness of hybrid composites. MRR, TWR and SR are greatly increased when the current increased from 4 to 12 A. From Figs. 8, 9 and 10 it is noted that the pulse on time is inversely proportional to tool wear rate whereas it is directly proportional to material removal rate and surface roughness of hybrid composites. MRR and SR values are higher for a higher pulse on time whereas TWR values are lower for a higher pulse on time. From Figs. 11, 12 and 13 it is observed the sample of hybrid composites Aluminium Alloy reinforced with 15% wt of silicon carbide and 15% wt of graphite produces higher metal removal rate, lower surface roughness and tool wear rate as measured to other samples. When the percentage of graphite is increases from 10 to 15% then it gives better machinability and also it increases metal removal rate. The tribological properties of the hybrid composite material is also increases when graphite percentage crosses from 10 to 15% and it helps to obtain the lower the surface roughness and tool wear rate.

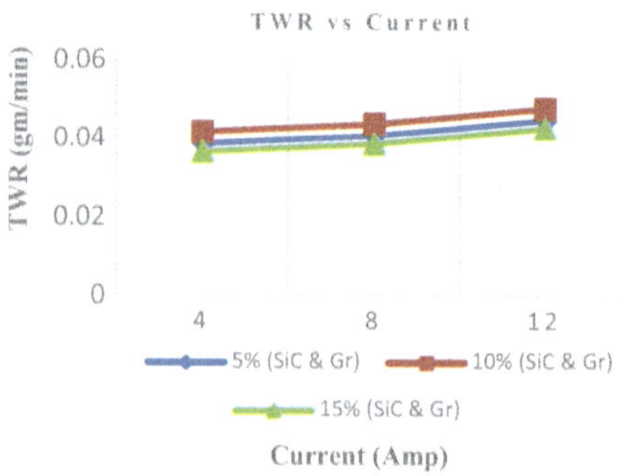


**Fig. 6** Correlation between TWR and current

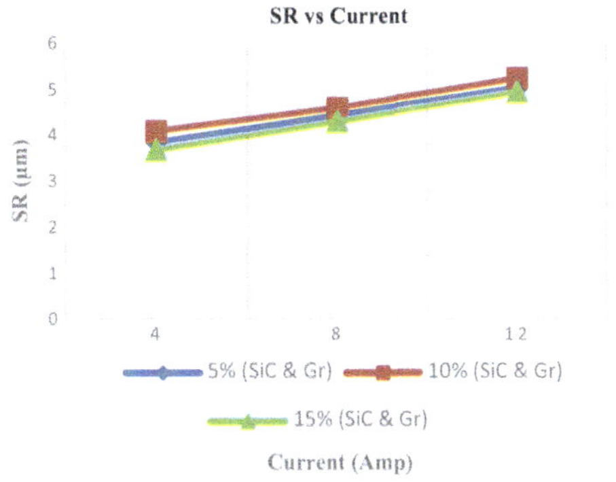


**Fig. 7** Correlation between SR and current

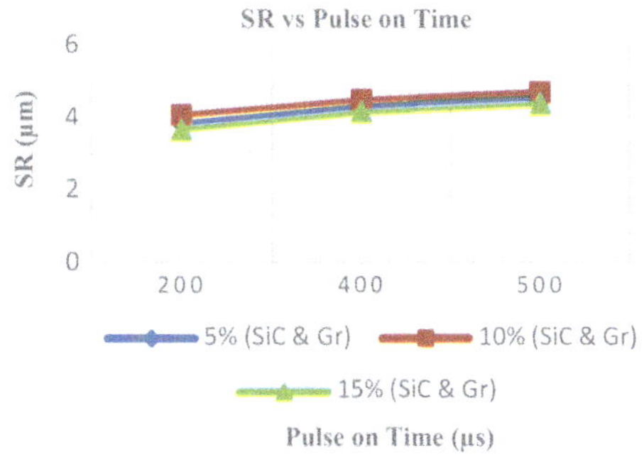


**Fig. 8** Correlation between SR and pulse on time

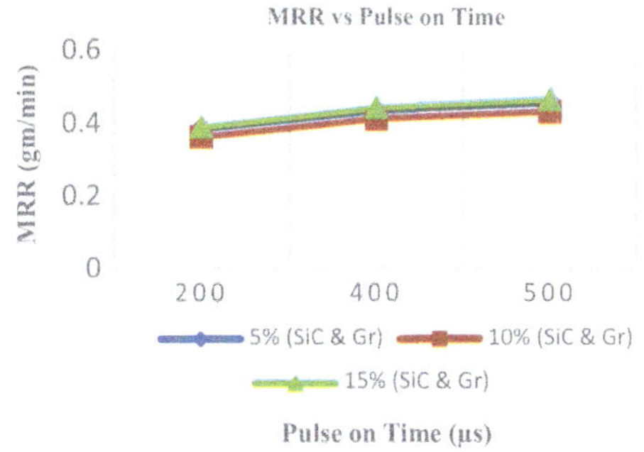


**Fig. 9** Correlation between MRR and pulse on time

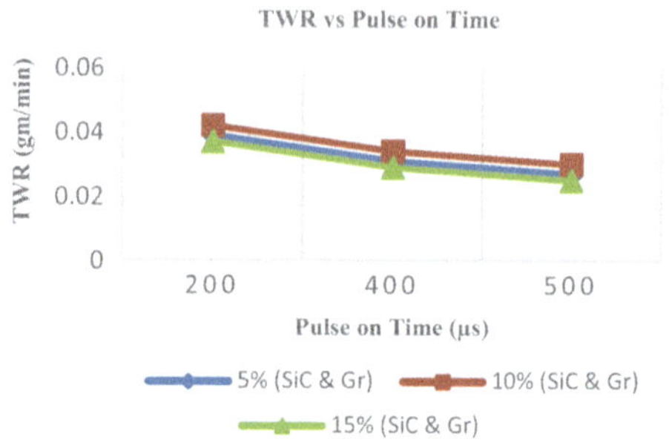


**Fig. 10** Correlation between TWR and pulse on time

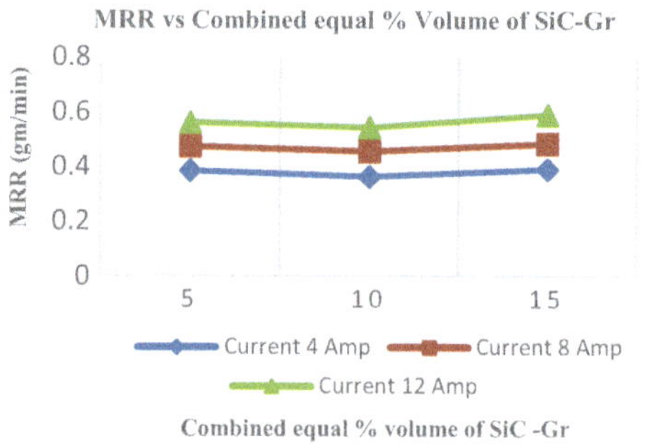


**Fig. 11** Correlation between MRR and combined equal % volume of SiC–Gr

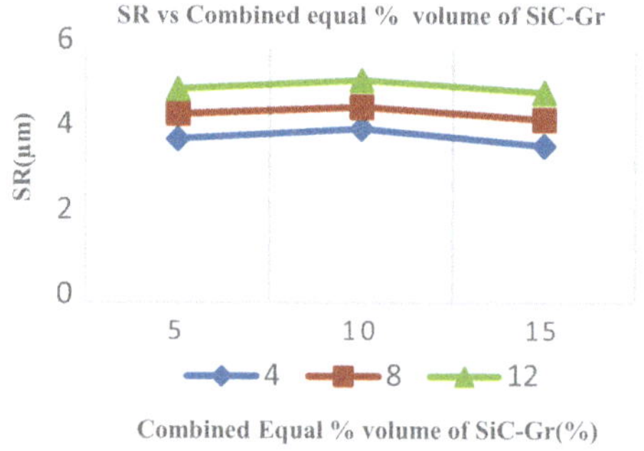


**Fig. 12** Correlation between SR and combined equal % volume of SiC–Gr

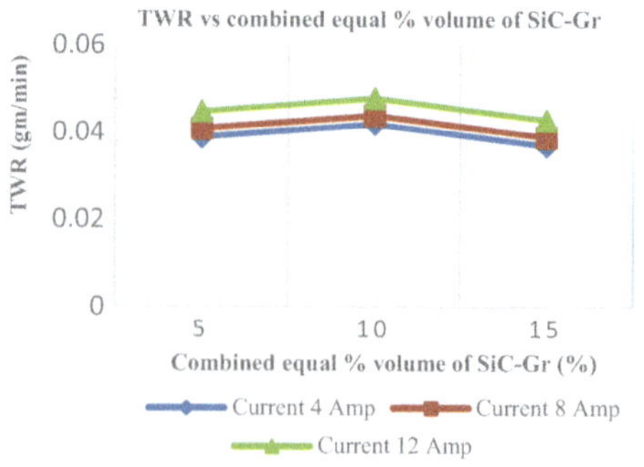


**Fig. 13** Correlation between TWR and combined equal % volume of SiC–Gr

## 4 Conclusions

The conclusion can be summarized into following points

1. The sample of hybrid composite Aluminium Alloy reinforced with 15% wt of silicon carbide and 15% wt of graphite produces higher metal removal rate, lower surface roughness and tool wear rate as measured to other samples.
2. Maximum material removal rate can be acquired by applying optimal setting of control factors such as current 12 A, pulse on time 500 μs and combined equal % volume of SiC and Gr (15%).
3. Minimum surface roughness can be given by employing optimal setting of control factors such as current 4 A, pulse on time 200 μs and combined equal % volume of SiC and Gr (15%).
4. Minimum tool wear rate can be obtained by utilizing optimal setting of control factors such as current 4 A, pulse on time 500 μs and combined equal % volume of SiC and Gr (15%).
5. It was examined that all three control factors have significant effect on machining characteristics considered in present study.

**Acknowledgements** Authors are gratitude to higher authorities, management and staff members of Dekson Casting Company Ltd. Aurangabad for manufacturing of Hybrid Composites. Authors are also greatful to faculty members, HOD and higher authorities of Indo German Tool Room, Aurangabad for experimentation.

## References

1. Ahamed AR, Asokan P, Aravindan S (2009) EDM of hybrid Al-SiCp–$B_4$Cp and Al-SiCp–glassp MMCs. Int J Adv Manuf Technol 44:520–528
2. Gopalkannan S, Senthilvelan T, Ranganathan S (2012) Modeling and optimization of EDM control factors on Machining of Al 7075-$B_4C$ MMC using RSM. Procedia Eng 38:685–690
3. Prakash JU, Moorthy TV, Peter JM (2013) Experimental investigation on machinability of aluminium alloy (A413)/fly ash/$B_4C$ hybrid composites using wire EDM. Procedia Eng 64:1344–1353
4. Puhan D, Mahapatra SS, Sahu J, Das L (2013) A hybrid approach for multi-response optimization of non-conventional machining on AlSiCp MMC. Measurement 46:3581–3592
5. Roy C, Syed K, Kuppan P (2016) Machinability of Al/10% SiC/2.5% $TiB_2$ metal matrix composite with powder-mixed electrical discharge machining. Procedia Technol 25:1056–1063
6. George PM, Raghunath BK, ManochaL M, Warrier AM (2004) EDM machining of carbon–carbon composite—a Taguchi approach. J Mater Process Technol 145:66–71
7. Hourmand M, Farahany S, Sarhan AD, Noordin MY (2014) Investigating the electrical discharge machining (EDM) parameters effect on Al-$Mg_2$-Si metal matrix composite (MMC) for high material removal rate (MRR) and less EWR-RSM approach. Int J Adv Manuf Technol
8. Kumar H, Davim JP (2011) Role of powder in the machining of Al-10% SiCp metal matrix composites by powder mixed electric discharge machining. J Compos Mater 45(2):133–151

9. Pecas P, Henriques E (2008) Electrical discharge machining using simple and powder-mixed dielectric: the effect of the electrode area in the surface roughness and topography. J Mater Process Technol 200:250–258
10. Talla G, Sahoo D, Gangopadhyay S, Biswas CK (2015) Modelling and multi-objective optimization of powder mixed electric discharge machining process of aluminum/alumina metal matrix composite. Eng Sci Technol Int J 18:369–373
11. Karthikeyan R, Lakshmi Narayanan PR, Naagrazaan RS (1997) Mathematical modelling for electric discharge machining of aluminium-silicon carbide particulate composites. J Mater Process Technol 87:59–63
12. Lin JL, Lin CL (2002) The use of the orthogonal array with gray relational analysis to optimize the electrical discharge machining process with multiple performance characteristics. Int J Mach Tools and Manuf 42:237–244
13. Singh S (2012) Optimization of machining characteristics in electric discharge machining of 6061 Al/$Al_2O_3$p/20P composites by grey relational analysis. Int J Adv Manuf Technol 63:1191–1202
14. Liu J, Wu YZ, Yue T (2015) High speed abrasive electrical discharge machining of particulate reinforced metal matrix composites. Int J Precis Eng Manuf 16(7):1399–1404
15. Singh S, Yeh MF (2010) Optimization of abrasive powder mixed EDM of aluminum matrix composites with multiple responses using gray relational analysis. J Mater Eng Perform 21 (4):481–491
16. Mohan B, Rajadurai A, Satyanarayana KG (2002) Effect of SiC and rotation of electrode on electric discharge machining of Al-SiC composite. J Mater Process Technol 124:297–304
17. Rajkumar K, Santosh S, Ibrahim S, Gnanavelbabu A (2014) Effect of electrical discharge machining parameters on microwave heat treated aluminium-boron carbide-graphite composites. Procedia Eng 97:1543–1550
18. Radhika N, Sudhamshu AR, Chandran GK (2014) Optimization of electrical discharge machining parameters of aluminum hybrid composites using Taguchi method. J Eng Sci Technol 9:502–512
19. Singh S, Maheshwari S, Pandey PC (2007) Optimization of multi performance characteristics in electric discharge machining of aluminium matrix composites (AMCs) using Taguchi DOE methodology. Int J Manuf Res 2:140–158
20. Singh PN, Raghukandan K, Rathinasabapathi M, Pai BC (2004) Electric discharge machining of Al-10% SiCp as-cast metal matrix composites. J Mater Process Technol 155–156:1653–1657

# Thermal Conductivity of Silicon Nitride Reinforced Al6061 Based Composites

**R. Keshavamurthy, N. P. Muthuraju and H. Govindaraju**

## 1 Introduction

Aluminum based alloys are often considered for most of the engineering applications owing to their excellent ductility, formability, malleability and mechanical properties. The mechanical, friction and wear characteristics aluminum composites can be largely improved by secondary manufacturing methods such as cold working, hot working and severe plastic deformation processing. The end properties of aluminum composites depend on various parameters such as morphology, size, content and dispersion of these particulates in the matrix and union of reinforced phase with matrix. In addition to this, processing techniques are very important in dispersion of these particulates. From the review of literature, it is higher the quantity of particulate in the alloy, mechanical, friction and wear characteristics increases. However, this may not same for fracture toughness, ductility and/or thermal properties where with higher percentage of reinforcement these properties generally decrease. Unlike very few researchers have proved increment in thermal conductivity with higher quantity of reinforcements [1–3].

R. Keshavamurthy (✉) · H. Govindaraju
Department of Mechanical Engineering, Dayananda Sagar College of Engineering, Bangalore 560078, India
e-mail: Keshavamurthy.r@gmail.com

H. Govindaraju
e-mail: govi_h@rediffmail.com

N. P. Muthuraju
Department of Mechanical Engineering, Vidyavardhaka College of Engineering, Mysore 570002, India
e-mail: Muthuraju.np@gmail.com



K. Antony and J. P. Davim (eds.), *Advanced Manufacturing and Materials Science*, Lecture Notes on Multidisciplinary Industrial Engineering,
https://doi.org/10.1007/978-3-319-76276-0_28

Molina et al. [4] have presented the thermal properties of SiC reinforced Al composites where two different sizes of SiC particulates were adopted for fabrication of composites. The composites were synthesized liquid infiltration route. The conductivity of the composite specimens was measured using steady state technique in which the temperature gradient between the reference and specimen was measured. The conductivity of composites with bimodal SiC particles showed a maximum of 250 W/mK can be achieved with the proposed technique. The composites with smaller size SiC particles showed very low thermal conductivity values when evaluated with alloy. It was credited to the fact that the occurrence of SiC will lead to dissolution of silicon into the matrix. Wang et al. [5] have reported the influence of aging on thermal conductivity of Al6061-SiC composites. The composites with 5 and 3% SiC were heat treated at T6 conditions. The thermal characteristics were measured using thermal diffusivity instrument which uses laser flash technique. The results showed that the addition of SiC to Al6061 led to fall in conductivity values. For all aging time conditions there was significant drop in conductivity of composites when evaluated with alloy. The fall in thermal conductivity was attributed to electron scattering due to presence of flaws. Most of these defects are point defects, larger defects like precipitates and presence of dislocations. The existence of coherent precipitates results in transformation induced strain results in dispersion of electrons which leads to fall in conductivity of aged composites. Guo et al. [6] have presented the influence of varying silicon content to Al and its influence on thermal conductivity of Al/diamond composites. Here pure Al and Al with varying silicon content was used as matrix while diamond were used as particulate material. The composites was developed using pressure infiltration technique in which the diamond preforms were infiltrated with liquid aluminium. In order to evaluate the thermal conductivity a cylindrical shaped composite samples of 10 mm dia and 3 mm thickness was used and subjected to laser flash technique. The thermal conductivity of composites enhanced with addition of silicon content up to 12.2 wt %. The thermal conductivity with 7 wt% Si showed value of 489 W/mK which was increased to 532 W/mK when the silicon content was increased to 12.2 wt%. However, further increase in silicon content resulted in severe decrease in conductivity. It was seen that for silicon content up to 20 and 40 wt%, the thermal conductivity dropped to 372 and 314 W/mK respectively. Out of all the composites with mixed size of diamond particles showed highest thermal conductivity of 558 W/mK when evaluated with single diamond particle composites.

On the other hand, limited information available as regards influence of forging on thermal conductivity of primarily processed aluminum based metal matrix composites. In the light of the above, this work reports on effect of forging and temperature on thermal conductivity of Aluminum based composites reinforced with silicon Nitride particles.

# 2 Experimental Details

## 2.1 Synthesis of Composite and Forging

Al6061 alloy was used as matrix material. Silicon Nitride powder having particle size 10–20 microns were procured as particulates. Stir casting technique was adapted to synthesis composites. Composites were prepared with 2, 4, 6, and 8 wt% $Si_3N_4$ particulates. $Si_3N_4$ particulates were metallic coated to enhance its wettability and interfacial bond with aluminum matrix. Both alloy and composites were hot forged at a temperature of 500 °C using hydraulic hammer at a constant strain rate. Both alloy and composites were characterized for its optical micrograph and thermal conductivity. More details on composite synthesis and forging are accessible in our earlier works [7–9].

## 2.2 Thermal Conductivity Test

Thermal conductivity of alloy and its composites were measured using parallel conductance method. The thermal conductivity was measured using an apparatus based on the parallel heat-flow technique under steady-state. The thermal conductivity of composites under steady state conditions is defined as, $K = (Qs/A)/(\Delta T/\Delta L)$, where Qs is the amount of heat passing per unit time, A is cross-sectional area and $\Delta T$ is temperature difference over a distance $\Delta L$. Specimens of size 50 mm length and 5 mm diameter were used as test samples in this study (Fig. 1). Thermal conductivity was calculated using parallel conductance method. More details on test procedure and test apparatus are reported elsewhere [10, 11]. The schematic representation of thermal conductivity facility employed in this study is shown in Fig. 1. The details of the test facility are given below. The specimen (1) is mounted inside a copper sample holder assembly. The sample holder includes two copper 10 mm cylindrical blocks (2 & 3). The crucial heat source copper block (2) is a 10 mm diameter solid cylinder on which heater is wound. The bottom heat-sink block (3) is fixed to the copper plate. The thermocouples [4 Nos] (4) are mounted at the source, one at the middle of the specimen, one at the sink, and one at the furnace surface respectively, which are attached to a 4 channel digital temperature indicator. The heater block is supported by a thin walled tube (5). This is provided with a load (6) mounted on a stainless steel support frame (7). Spring loading ensures the proper contact between the sample and the holder assembly. The central sample holder is mounted inside a furnace (9). Vacuum tight stainless steel chambers (10) encapsulate the whole arrangement to minimize convective heat exchange. The thermal contact between heat source block and the sample is made or broken by a lifting mechanism (11).

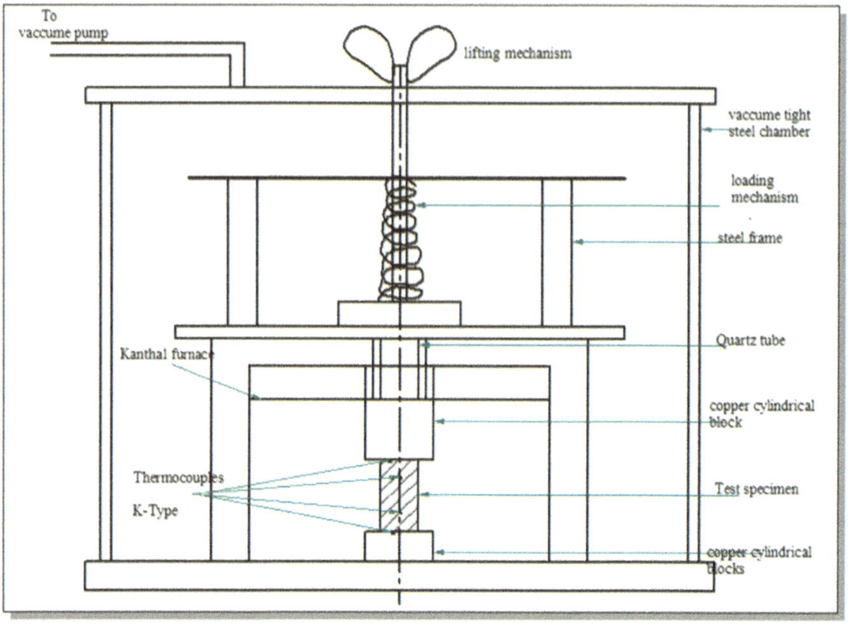


**Fig. 1** Constructional details of thermal conductivity test setup

## 3 Results and Discussions

### *3.1 Microstructure*

Optical Micrographs of Al6061-10 wt% $Si_3N_4$ composites are presented in Fig. 2. It can be seen that dispersion of $Si_3N_4$ is uniform in both the processing conditions. The reinforcements are observed to be present both in and around grain boundaries of Al6061 matrix. The electroless nickel coating and stirring action helps in dispersing the reinforcements uniformly in the Al6061 molten melt and help them in retaining in the melt. The second important factor is hot forging, which breaks the bunch of reinforcements if any clusters present in the cast composites. For a given content of $Si_3N_4$ the clustering is hardly visible in forged composite. The large pressure applied not only breaks the clusters but also helps in enhancing union of reinforcement with alloy.

### *3.2 Thermal Conductivity*

Figure 3 shows the thermal conductivity of alloy and its composites with the varying $Si_3N_4$ content. It is noticed that the thermal conductivity of composites

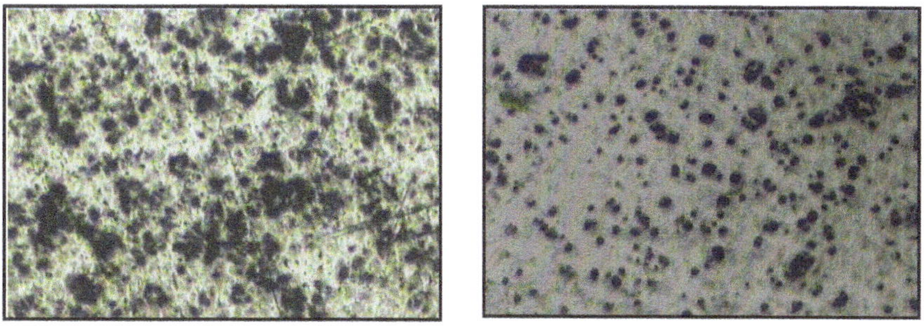

(a) Cast Al6061-10wt%$Si_3N_4$ composite (b) Forged Al6061-10wt%$Si_3N_4$ composite

**Fig. 2** Optical micrographs of cast and forged Al6061-10 wt% $Si_3N_4$ composite

were lower than that of unreinforced Al6061 alloy in both the processing techniques. Thermal conductivity of aluminum composites is highly dependent on reinforcing materials, processing techniques, microstructure and bond between the matrix and reinforcement. In the present case Silicon Nitride reinforcement possess lower thermal conductivity (30 W/mK). However, under all the conditions studied, forged system exhibits slightly higher thermal conductivity. The processing condition is a important factor that influences the composites. Though the composites synthesized possess homogenous dispersion and excellent bond, microstructure also plays a vital role. The dislocations produced around the reinforcement in the aluminum matrix due to difference in CTE can root the scattering of electrons and phonons between Al6061 and $Si_3N_4$. It is well known that the about 90% of heat carried in matrix material is by electrons while in case of reinforcements they conduct heat by phonons. With increasing $Si_3N_4$, the increase in the dislocation density is possible. So these dislocation causes dispersion of electrons and phonons on the elastic strain field [12–14]. Figure 4 displays the influence of temperature on thermal conductivity alloy and its composites. The thermal conductivity was calculated for different temperatures ranging from 50–200 °C. Under all the temperatures studied the thermal conductivity values have been decreased in both the processing techniques. However, forged system demonstrated significantly a superior thermal conductivity when evaluated with cast alloy and its composites. The enhanced conductivity of forged composites over cast ones may be attributed to reduced casting/structural defects, improved dispersion of reinforcement and improved bond constituents. The thermal conductivity depends on residual component which is due to structural defects like grain boundaries, precipitates and dislocations and along with this the existence of second phase particles in case of composites. So at high temperatures the jumping probability of electrons over such structural imperfections is not possible. Further, a large number of phonons cause dispersion of electrons and due to which the relaxation time between the collusions is decreased. Thus, with the increase in temperature from 50 to 200 °C, the percentage of distribution of phonons increases and small relaxation time leads to reduction in thermal conductivity values [15].

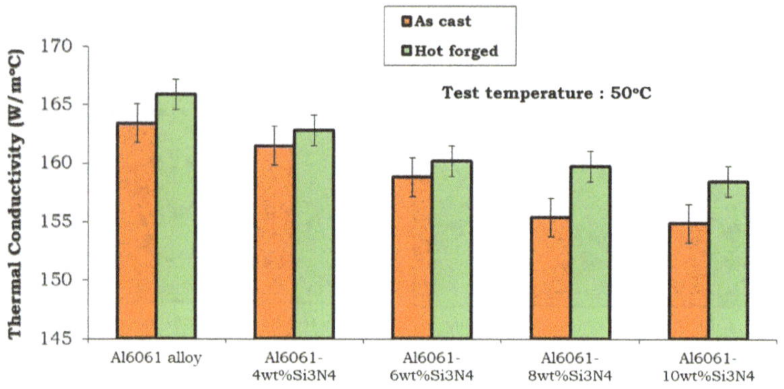


**Fig. 3** Effect of silicon nitride on thermal conductivity

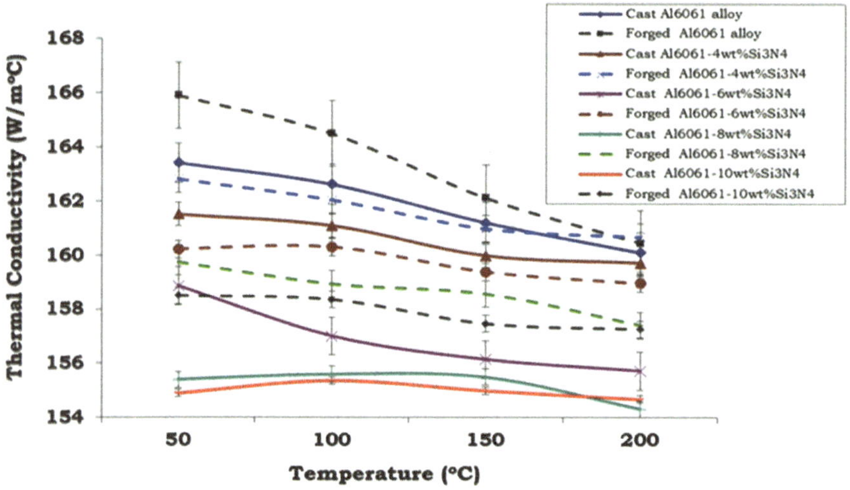


**Fig. 4** Effect of temperature on thermal conductivity

# 4 Conclusion

Al6061 composite reinforced with silicon nitride was synthesized by liquid metallurgy route and successfully hot forged. Optical micrograph shows uniformity in dispersion of $Si_3N_4$ particles. The thermal conductivity of composites decreased with increase in reinforcements mainly because of lower conductivity of reinforcement. However, Thermal conductivity of forged system was higher compared to cast ones. Thermal conductivity of alloy and their composites decreases with raise in the temperature.

## References


1. Shalaby EAM, Churyumov AY, Besisa DHA, Daoud A, Abou El-khair MT (2017) A Comparative Study of Thermal Conductivity and Tribological Behavior of Squeeze Cast A359/AlN and A359/SiC Composites. J Mater Eng Perform 26(7):3079–3089
2. Pradeep Kumar GS, Koppad PG, Keshavamurthy R, Alipour M (2017) Microstructure and mechanical behavior of in-situ fabricated AA6061-TiC in-situ composite. Arch Civil Mech Eng. 17(3):535–544
3. Pradeep Kumar GS, Keshavamurthy R, PrachiKumari (2016) Influence of hot forging on tribological behavior of Al6061-$TiB_2$ in-situ composites. In: IOP conference series: materials science and engineering, 2016, vol 149, p 012087
4. Molina JM, Narciso J, Weber L, Mortensen A, Louis E (2008) Thermal conductivity of Al–SiC composites with monomodal and bimodal particle size distribution. Mater Sci Eng A 480:483–488
5. Wang H, Lo SHJ (1996) Effects of ageing on the thermal conductivity of a silicon carbide particulate reinforced 6061 aluminium composite. J Mater Sci Lett 15:369–371
6. Guo C, He X, Ren S, Qu X (2016) Effect of (0–40) wt% Si addition to Al on the thermal conductivity and thermal expansion of diamond/Al composites by pressure infiltration. J Alloy Compd 664:777–783
7. Ramesh CS, Keshavamurthy R, Channabasappa BH, Ahamed A (2009) Microstructure and mechanical properties of Ni-P coated $Si_3N_4$ reinforced Al6061 composites. Mater Sci Eng A 502(1–2):99–106
8. Ramesh CS, Keshavamurthy R (2012) Influence of forging on mechanical properties of Ni–P coated $Si_3N_4$ reinforced Al6061 composites. Mater Sci Eng A 551:59–66
9. Keshavamurthy R, Madhusudhan J, Krishna A (2016) Effect of thermo-mechanical processing and heat treatment on the tribological characteristics of Al based MMC's. In: IOP conference series: materials science and engineering, 2016, vol 149, p 012118
10. Dasgupta T, Umarji AM (2005) Apparatus to measure high-temperature thermal conductivity and thermoelectric power of small specimens. Rev Sci Instrum 76:094901
11. Shivananda Murthy KV, Girish DP, Keshavamurthy R, Varol T, Koppad PG (2017) Mechanical and thermal properties of AA7075/$TiO_2$/Fly ash hybrid composites obtained by hot forging. Prog in Nat Sci: Mater Int 27:474–481
12. Xu Y, Tanaka Y, Goto M, Zhou Y, Yagi K (2004) Thermal conductivity of SiC fine particles reinforced Al alloy matrix composite with dispersed particle size. J Appl Phys 95:722–726
13. Kida M, Weber L, Monachon C, Mortensen A (2011) Thermal conductivity and interfacial conductance of AlN particle reinforced metal matrix composites. J Appl Phys 109:064907
14. Pal H, Sharma V (2015) Thermal conductivity of carbon nanotube–silver composite. Trans Nonferrous Met Soc China 25:154–161
15. Kim Y, Oh N, Oh S, Moon I (2001) Thermal conductivity of W-Cu composites at various temperatures. Mater Lett 51:420–424

# Grinding of Inconel 718 Using Soap Water Jet and Liquid Carbon Dioxide

**Arnab Kundu, Sirsendu Mahata, Manish Mukhopadhyay, Ayan Banerjee, Bijoy Mandal and Santanu Das**

## 1 Introduction

Grinding is a material removal process, frequently used to finish or semi-finish Ni based alloys, among which Inconel 718 is the most widely used. It contributes to almost 45% of wrought nickel based alloy products and as much as one-fourth cast nickel based production [1]. Inconel is extensively used in aerospace, petrochemical, medical and nuclear industries due to its enhanced mechanical and chemical properties viz. oxidation and corrosion resistance, suitability in extreme working conditions of heat (−423° to 1300 °C) and pressure. It can also be easily fabricated. Ni-based alloys constitute almost 50% of the total material required for the manufacture of an air-craft engine, and therefore, they are popularly known as aerospace materials. These alloys also find application in cryogenic storage tanks. These are high temperature applications where fatigue and creep become critical, that generally depend on the surface quality of the alloy. Grinding is the only conventional and economic process to achieve high surface finish, so grinding of Inconel alloys is a topic of general interest around the world.

However, grinding of this alloy is problematic due to its poor thermal conductivity, high coefficient of thermal expansion, chemical affinity towards abrasives, high hardness at elevated temperature. This leads to high temperature generation at the grinding zone, high wheel wear, and rapid wheel loading. Proper selection of grinding parameters can reduce these harmful effects to some extent. Choice of grinding wheel, either conventional, or super-abrasive, is quite important. Super abrasive grinding wheels exhibit lower grinding forces and longer wheel life, but they are quite costly. However, conventional abrasive wheels are still preferred for grinding based industries due to lower abrasive cost and their ease of use [2]. One

A. Kundu · S. Mahata · M. Mukhopadhyay · A. Banerjee · B. Mandal · S. Das (✉)
Department of Mechanical Engineering, Kalyani Government Engineering College, Kalyani, Nadia, West Bengal 741235, India
e-mail: sdas.me@gmail.com



K. Antony and J. P. Davim (eds.), *Advanced Manufacturing and Materials Science*, Lecture Notes on Multidisciplinary Industrial Engineering,
https://doi.org/10.1007/978-3-319-76276-0_29

other technique to control grinding temperature is by exploring an efficient cooling and lubrication technique. Addressing this issue, several researchers used different techniques to deliver fluid at the grinding zone, such as multi-nozzle cooling [3], Minimum Quantity Lubrication (MQL) [4], jet cooling [5], cryogenic cooling [6], use of solid lubricant [7], use of soap water [8], etc. However, use of grinding fluid is being discouraged due to a large number of reasons:

i. high procurement and disposal costs (costs associated with lubrication reaches almost 20–30% in machining difficult-to-machine materials [9],
ii. harmful effects of mists, gases in causing lung disease, allergy and even skin cancer [10],
iii. increasing bacterial growth on shop floor,
iv. restricted penetration of lubricants at the grinding zone, due to presence of a stiff layer of air [11], although it can be suppressed by employing compound nozzle and/or pneumatic barrier setup [12].

In contrast to conventional grinding/cutting fluids, gas based ones in liquid form are less harmful and quite environment friendly. Two common gases used in machining and grinding are liquid nitrogen and liquid carbon dioxide. Liquid nitrogen was first used in 1953 to machine steel. Since then, several researchers tried this coolant in their works. It was seen that cryogenic cooling by liquid nitrogen during machining AISI 1040 and AISI 4320 mainly resulted in reduction of cutting forces and cutting temperature, as tool maintained sharpness for a longer time [13]. An experimental study to enumerate benefits using liquid nitrogen during drilling of titanium with coated inserts was done by Ahmed et al. [14]. Reduction in cutting temperature of up to 9% and lesser tool damage was observed. However, the workpiece was hardened, with the maximum hardness observed at the highest feed rate.

In comparison, liquid carbon dioxide is easily available and environmentally neutral. Cryogenic milling of high strength stainless steel using liquid $CO_2$ was investigated by Cordes et al. [15]. Compared to dry milling, material removal rate was higher and tool wear was reduced considerably. Moreover, chip adherence was visibly suppressed, resulting in better surface quality. Another experimental investigation was conducted by Jerold et al. [16] to observe influence of cryogenic cooling when turning AISI stainless steel. It was seen that $CO_2$ gas could reduce cutting temperature up to 35% and surface roughness up to 52%, when compared to wet machining. Tool wear was also less. A hybrid machining system consisting of PAM (Plasma Arc Machining) to preheat Inconel 718 workpiece and application of liquid nitrogen as the grinding fluid was used by Wang et al. [17]. Remarkable results of 250% improvement in surface finish and 156% increase in tool life were achieved. Cutting forces were also reduced by 30–50%.

Soap water has been used in the past to machine certain metals. Teicheret al. [18] used alkaline soap water to improve grindability of Ti-6Al-4V alloy using brazed cBN grinding wheel. Lower forces and better chip formation were reported using soap water. Also, chip redeposition was suppressed, giving rise to smooth workpiece surface. Similar results were obtained by Babic et al. [19] when grinding AISI steel using soap water mist jet. Significant reductions in grinding temperature and forces were obtained. Soap water was applied in a drop by drop fashion during grinding of titanium grade I alloy with an alumina wheel [8]. It was seen that alkaline soap water is effective in reducing grinding forces, producing favourable chips and increasing G-ratio. The method used was simple and environmentally clean.

In this experimental study, surface grinding of Inconel 718 is performed using an alumina wheel with vitrified bond. Grinding ratio, surface hardness and condition of wheel loading are observed under each condition to assess grindability of Inconel 718 under given conditions.

## 2 Experimental Procedure

In the present work, Inconel 718 is used as the workpiece, in the form of rectangular flats. Alumina wheel with vitrified bond is used as the grinding wheel. Each experiment under each condition consists of 20 passes, with a depth of cut of 10 micron in the up-grinding direction. The table speed and wheel speed is kept constant at 14 m/min and 30 m/s respectively. Initially, dry grinding is conducted followed by grinding under soap water jet application, and finally under a liquid $CO_2$ assisted soap water application. The experimental details are listed in Table 1.

**Table 1** Complete list of experiments and equipment

| | |
|---|---|
| Surface grinding machine | Make: HMT Praga Division, India, Model: 452 P<br>Main motor power: 1.5 kW, Infeed resolution: 1 μm, maximum spindle speed: 2800 rpm |
| Grinding wheel | Make: Carborundum Universal Limited, Specification: AA60K5V8<br>Type: disc-type, dimension: 200 mm × 20 mm thick × 31.75 mm bore |
| Workpiece | Material: Inconel 718, Hardness: 340 HV, dimension: 120 mm × 60 mm × 6 mm |
| Force dynamometer | Make: Sushma Grinding Dynamometer, Bengaluru, India, Model: SA 116, Range: 0.1–100 kg, Resolution: 0.1 kg |
| Dressing details | Make: Solar, India, specification: 0.5 carat single point diamond tip<br>Dressing depth: 20 μm, dressing speed: 1.8 m/min |
| Process parameters | Infeed: 10 μm, grinding wheel velocity: 30 m/s, table speed: 14 m/min, mode: up-grinding |
| Environmental conditions | (a) Dry<br>(b) Wet (jet of soap water solution)<br>[Nozzle diameter: 1.25 mm (plastic), coolant: soap water, composition: solution of Clinic Plus shampoo and water, mixing ratio: 1:20 (by vol.), coolant flow rate: 210 ml/min]<br>(c) Liquid $CO_2$ assisted soap water jet<br>[$CO_2$ delivery nozzle diameter: 1.23 mm (stainless steel), fluid flow rate: 210 ml/min, $CO_2$ exit pressure: 3 kg/cm$^2$] |

Fig. 1 Setup for soap water jet application, and liquid $CO_2$ assisted soap water jet application

Soap water jet is delivered into the grinding zone by the fluid delivery arrangement as shown in Fig. 1. A specially designed nozzle of diameter 1.23 mm is used to supply soap-water solution (mixture of water and 'Clinic Plus' shampoo, 1:20 by vol.). A constant flow rate of 210 ml/min is maintained. The striking feature of this arrangement is the use of a pump of only 10 W power rating, ensuring the cost-effectiveness of this indigenous process. The fluid is recycled using a filter. Figure 2 shows the arrangement for Liquid $CO_2$ assisted soap water jet application. Liquid $CO_2$, at a temperature of −26° C, was used to cool the grinding zone, in addition to the jet of soap water.

## 3 Results and Discussions

Grinding ratio: Grinding ratio, or G-ratio, is the ratio of material removal rate to wheel removal rate. G-ratio values can be as high as 500, and for difficult-to-grind materials such as titanium and Inconel, they can be as low as 1. Grinding ratios

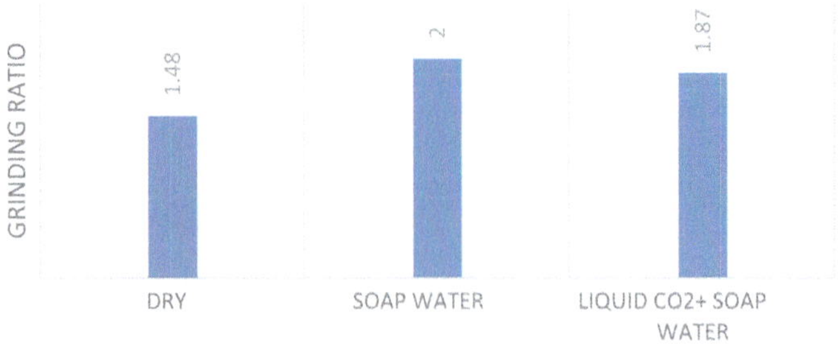


Fig. 2 Grinding ratios under different environmental conditions

have been calculated, at the end of each experiment, by measuring the initial and final wheel diameters and workpiece height. The ratios are presented in Fig. 2.

The lowest grinding ratio in this experiment is seen in case of dry grinding. Due to intense wheel loading, chips get embedded in inter-grit spaces of the wheel cutting face, thereby reducing the material removal rate, due to less availability of sharp cutting edges. Also, wheel grit wear is more, due to absence of any coolant, thereby resulting in low G-ratio. Highest G-ratio is obtained when grinding Inconel 718 using soap water jet. This is probably due to efficient material removal from the workpiece, due to penetration of the jet of soap water into the grinding zone that acts as a good lubricant and decreases grinding forces. Quite possibly, micro-fracture of alumina grits may have occurred to restore cutting ability of the grinding wheel, by removing the dull, worn out grits [20]. Such a phenomenon increases the grinding efficiency, and thus the grinding ratio by 35%, as compared to the G-ratio obtained in dry grinding. Addition of liquid $CO_2$ to the jet stream of soap water increases the grinding ratio as compared to that under dry grinding, but the percentage increase is less (26%). At a low temperature of −26 °C, liquid $CO_2$ may have increased the surface hardness of the alloy.

Surface hardness: Ground surface hardness of Inconel 718 workpieces was measured after every experiment. Vickers hardness test was conducted with a square base diamond indenter subjected to a load of 500 gm for a dwell time of 10 s. Vickers hardness is calculated by measuring the lengths of the two diagonals of an indentation. For each grinding environment, hardness was measured at five different locations on the ground surface. Hardness differs from grain to grain, as a result of which variations are shown in the form of error bars projected on the average hardness values, shown in Fig. 3.

It is seen from Fig. 3 that Inconel 718 is softened at the ground surface, after dry grinding. Due to very low thermal conductivity of the material, the heat generated during plastic deformation during grinding cannot dissipate quickly, leading to a rapid rise in localized temperature. This thermal softening effect, coupled with recrystallization of the grains in alloy, leads to a decrease in hardness. Lowest surface hardness is reported in case of grinding with soap water jet, with a decrease

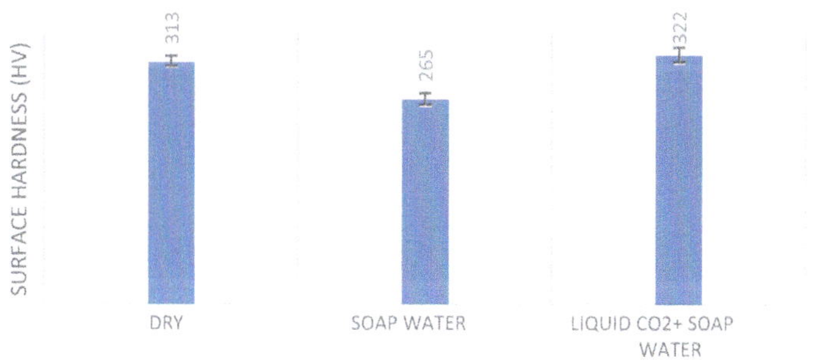


**Fig. 3** Variation of average surface hardness of Inconel 718 samples, for different environmental conditions

of 15%, as compared to that obtained in dry grinding. The primary reason of hardening in machining is rapid heating and subsequent cooling cycles, whereas mechanical deformation is secondary [21]. In case of using liquid $CO_2$ assisted soap water jet, the workpiece and wheel are further cooled, before the actual cut. This 'pre-cooling' technique reduces the temperature at the wheel-workpiece contact region, thus resulting in increased hardness and higher compressive residual stresses in the material. When actual grinding starts, the ground surface is subjected to rapid increase and decrease of temperatures, which may have led to higher quenching rates and subsequently, higher hardness. This is also backed by relatively higher friction coefficient values when under this condition. However, grinding forces simultaneously increase and thus,a balance needs to be obtained between forces and surface hardness of the alloy.

Wheel loading: Wheel loading is a phenomenon in which microscopic chips get embedded in the inter-grit spaces, thus reducing the cutting area and leading to thermal problems. This increases with the increase in grinding passes. Pictorial views of loaded wheels, taken after 20 passes using a digital camera of Nikon make, are shown in Fig. 4

It is evident from Fig. 4a that grinding wheel bears traces of intense wheel loading, which produces large cutting forces and reduced grinding ratios. This is due to absence of lubrication and increased ductility of Inconel 718 at higher grinding temperature. At higher grinding temperature, Inconel reacts with alumina grits via a complex chemical reaction in which nickel adheres and sticks to the inter-grit spaces, causing the wheel to get loaded. The wheel loading in the next two conditions are lower and not much different. Amount of wheel loading is seen to expectedly reduce with proper lubrication, as seen from Fig. 4b. This is due to application of the high velocity soap water jet, that may have helped in removing

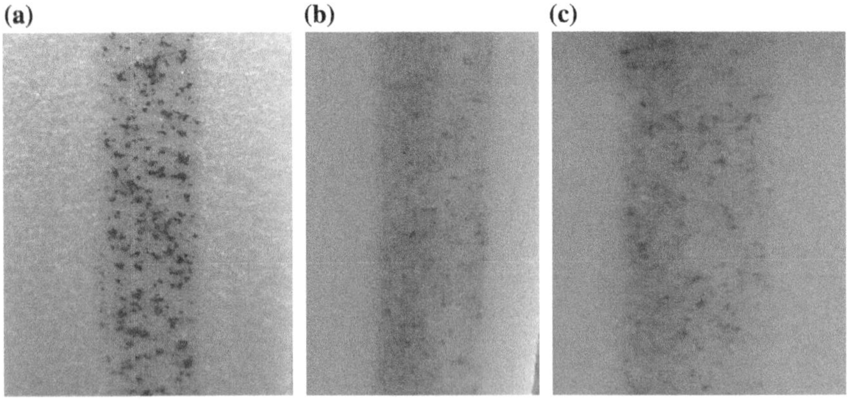


**Fig. 4** Images of the grinding wheel being loaded under **a** dry grinding, **b** using soap water jet, and **c** using liquid $CO_2$ assisted soap water jet

some of the chips from the wheel face, thereby reducing intensity of wheel loading. Using low temperature liquid $CO_2$ in addition to soap water jet might have helped in retention of grit sharpness, ensuring low wheel loading, as can be seen from Fig. 4c.

## 4 Conclusion

Grinding of Inconel 718 was conducted with an alumina wheel under three conditions: dry, using soap water jet and using liquid $CO_2$ assisted soap water jet. Using the soap water jet, increase in 35% grinding ratio and 15% surface hardness increase can be obtained, compared to those under dry conditions. Wheel loading is also reduced. However, addition of liquid $CO_2$ to the soap water jet stream does not aid in increasing G-ratio further. Moreover, the sample got hardened and strengthened along the ground surface that might lead to increased forces and increased energy consumption. Within this experimental domain, better grindability of Inconel 718 can be achieved by employing a jet of soap water, indicating its suitability in industry.

## References


1. Ezugwu E, Bonney J, Yamane Y (1989) An overview of the machinability of aeroengine alloys. J Mater Process Technol 134(2):233–253
2. Sinha MK, Setti D, Ghosh S, Rao PV (2016) An investigation on surface burn during grinding of Inconel 718. J Manuf Process 21:124–133
3. Mahata S, Mandal B, Mistri J, Das S (2013) A comparative study of grinding performance using different fluid delivery techniques. J Assoc Eng 83(3–4):63–70
4. Sinha MK, Setti D, Ghosh S, Rao PV (2016) An alternate method for optimisation of minimum quantity lubrication parameters in surface grinding. Int J Mach Mach Mater 18(5–6):586–605
5. Zhang D, Zhang Y, Changhe L, Jia D, Zhang X (2015) Experimental research on the energy ratio coefficient and specific grinding energy in nanoparticle jet MQL grinding. Int J Adv Manuf Technol 78(5–8):1275–1288
6. Venugopal KA, Paul S, Chattopadhyay AB (2007) Growth of tool wear in turning of Ti-6Al-4V alloy under cryogenic cooling. Wear 262(9–10):1071–1078
7. Goriparthi BK, Ravuri BP, Revuru RS, Anne VG (2016) Performance evaluation of grinding wheels impregnated with graphene nanoplatelets. The Int J Adv Manuf Technol 85(9–12):2235–2245
8. Kundu A, Mahata S, Mukhopadhyay M, Banerjee A, Mandal B, Das S (2017) Investigating the use of soap water in SQL mode during grinding titanium alloy. In: Proceedings of the national conference on advanced functional materials processing & manufacturing, CMERI, Durgapur, West Bengal, 2–3 Feb 2017
9. Pusavec F, Kramar D, Krajnik P (2010) Kopac, J Transitioning to sustainable production—part II: evaluation of sustainable machining technologies. J Clean Prod 18:1211–1221

10. Dhar NR, Islam S, Kamruzzaman M, Paul S (2006) Wear behavior of uncoated carbide inserts under dry, wet and cryogenic cooling conditions in turning C-60 steel. J Brazilian Soc Mech Sci Eng 28(2):146–152
11. Mandal B, Singh R, Das S, Banerjee S (2012) Development of a grinding fluid delivery technique and its performance evaluation. Mater Manuf Process 27(4):436–442
12. Mandal B, Biswas D, Sarkar A, Das S, Banerjee S (2013) Improving grindability of Titanium grade I using a pneumatic barrier. Reason-A Tech J 12:37–45
13. Dhar NR, Kishore NSV, Paul S, Chattopadhyay AB (2002) The effects of cryogenic cooling on chips and cutting forces in turning AISI 1040 and AISI 4320 steels. Proc Inst Mech Eng Part B: J Eng Manuf 216(5):713–724
14. Ahmed LS, Govindaraju N, Kumar MP (2016) Experimental investigations on cryogenic cooling in the drilling of Titanium alloy. Mater Manuf Process 31(5):603–607
15. Cordes S, Hübner F, Schaarschmidt T (2014) Next generation high performance cutting by use of carbon dioxide as cryogenics. Procedia CIRP 14:401–405
16. Jerold BD, Kumar MP (2012) Machining of AISI 316 stainless steel under carbon-di-oxide cooling. Mater Manuf Process 27(10):1059–1065
17. Wang ZY, Rajurkar KP, Fan J, Lei S, Shin YC, Petrescu G (2003) Hybrid machining of Inconel 718. Int J Mach Tools Manuf 43:1391–1396
18. Teicher U, Ghosh A, Chattopadhyay AB, Kunanz K (2006) On the grindability of Titanium alloy by brazed type monolayered superabrasive grinding wheels. Int J Mach Tools Manuf 46:620–622
19. Babic DM, Torrance AA, Murray DB (2005) Soap mist jet cooling of grinding processes. Key Eng Mater 291–292:239–244
20. Jackson MJ, Davim JP (2011) Chapter 1 Abrasive tools and bonding systems, machining with abrasives, 1st edn. Springer, New York, USA
21. Chou YK (2002) Surface hardening of AISI 4340 steel by machining: a preliminary investigation. J Mater Process Technol 124(1–2):171–177

# Mechanical Properties of Al7075-SiC-$TiO_2$ Hybrid Metal Matrix Composite

**G. S. Pradeep Kumar, R. Keshavamurthy, Vijay Tambrallimath, Riya Biswas and Yogesh Sahebrao Andhale**

## 1 Introduction

Metal matrix composite is gaining a large momentum in attaining light weight and huge utilization in various engineering applications. Monolithic metals and their alloys cannot always meet the demands of today's modern engineering requirements. With addition of more than one material it is possible to achieve tailor made properties and meet the performance requirements for specific application. Materials are mainly added to change their properties in order to fulfil the requirements of design [1, 2].

Few researchers have incorporated fly ash in AA7075 to provide a better wear resistant, strength and hardness composite. Microstructural characteristics of AA7075 and its composite were compared and analyzed using Optical Microscope (OM) and SEM.

It has been noticed that the fly ash particles have been dispersed uniformly in the composite. The composite has been prepared by liquid metallurgy technique. The hardness has been increased due to good union between flyash and Al7075 alloy and homogenous dispersion of fly ash [3]. Review on various combinations with Al7075 being the matrix and various components like SiC, $TiO_2$, $TiB_2$, $B_4C$, fly ash etc. being the reinforcements have contributed in improved mechanical characteristics. This paper emphasises on the composite prepared by liquid state processing techniques like stir casting methods. This process appears to be relatively simple and cost effective and can be made semi continuous or continuous by stirring [4].

G. S. Pradeep Kumar · Y. S. Andhale
D Y Patil School of Engineering Academy, Ambi, India

R. Keshavamurthy · V. Tambrallimath (✉) · R. Biswas
Dayananda Sagar College of Engineering, Bangalore, India
e-mail: vijay.tambrallimath@gmail.com

 
K. Antony and J. P. Davim (eds.), *Advanced Manufacturing and Materials Science*, Lecture Notes on Multidisciplinary Industrial Engineering,
https://doi.org/10.1007/978-3-319-76276-0_30

Reinforced basalt fibres into Al7075 has seen improvements in mechanical properties of the material. Microstructural evaluation of composite specimens discloses the consistent distribution of basalt fibres and other precipitates in the Aluminium alloy. The strength value comparisons with various theoretical studies recommends that the experimentally calculated values are in good agreement with theoretical studies considering the random dispersion of basalt fibres in the Al (7075) matrix [5]. Preparation of a cenosphere composite with Al7075. It has enhanced properties which include strength, toughness, hardness and wear resistance evaluated with alloys. Its applicable in aircraft for the reason that of its lower weight to strength ratio. It's been prepared by liquid metallurgy technique [6]. Some researchers have studied influence of $TiO_2$ and fly ash were studied by synthesizing composite using casting and hot forging techniques. The hybrid composites exhibited improvement in compressive strength by using two fillers and further increase in the content of $TiO_2$ particles. The coefficient of thermal expansion of hybrid composites decreased with increase in the content of $TiO_2$ and fly ash. On the other hand, thermal conductivity of composites decreases with hybrid reinforcements when evaluated with Al7075 alloy [7].

While designing the materials, hybrid metal matrix composites generally have more flexibility of design owing to presence of two or more reinforcements.

Thus, current study was undertaken to aluminium based hybrid metal matrix composites with low cost reinforcements like SiC and $TiO_2$ by stir casting. SiC and $TiO_2$ were chosen as hybrid reinforcements owing to their superior mechanical, Tribological and physical properties. From the extensive review of literature, it is clear that Silicon Carbide and Titanium dioxide can be considered as efficient reinforcements with multiple benefits for synthesis of hybrid metal matrix composites. Further, very limited information exist with reference to mechanical properties with the combination of titanium dioxide and silicon carbide reinforced Al7075 based metal matrix composites. In the light of the above, this work focuses on synthesis of Al7075-$TiO_2$-SiC hybrid metal matrix composites by liquid metallurgy technique and characterization of its mechanical properties.

## 2 Experimental Procedure

Composites were synthesized by mixing molten Al7075 alloy, Silicon Carbide and Titanium Carbide as reinforcements. Stir casting technique was employed to synthesize the hybrid composites with 2 wt% of each Silicon Carbide and Titanium Carbide particles. Al7075 alloy has been used in automotives, aerospace and marine industries due to their high strength to density ratio and formability. Table 1 shows the composition of matrix material used in present study.

The particle size of silicon carbide was in the range of 80–70 μm and the particle size of titanium dioxide was in the range of 40–50 μm. The percentage of silicon carbide and titanium dioxide used in the casting process was 2% each. Aluminum 7075 alloy was melted in an electrical resistance furnace. Molten Al7075 alloy was

**Table 1** Composition of Al7075

Chemical composition of A16061 alloy

| Element | Si | Fe | Cu | Mn | Ni | Pb | Zn | Ti | Sn | Mg | Cr | A1 |
|---|---|---|---|---|---|---|---|---|---|---|---|---|
| Percentage | 0.43 | 0.43 | 0.24 | 0.139 | <0.05 | 0.024 | 0.006 | 0.022 | 0.001 | 0.802 | 0.184 | Balance |

stirred at a speed of 300 rpm. Reinforcement particles were then slowly incorporated into the molten metal while maintaining a constant stirring speed. The composite melt at a temperature of 750 °C, was then transfered into preheated cast iron moulds. Developed hybrid composites were characterized for its microstructure, hardness and tensile strength. The test samples for tensile tests were machined from the castings as per ASTM A370 standards.

## 3 Results and Discussions

Figure 1 shows optical micrographs of Al7075-$TiO_2$-SiC hybrid aluminium composites. It is seen from the microphotograph that SiC and $TiO_2$ particles have been dispersed uniformly throughout the matrix material. The microstructure of composites can strongly influence strength, toughness, ductility of composites. Further, the developed composite do not show any visible defect in the micrograph indicating good quality of the casting process.

## 4 Hardness

Figure 2 shows comparison of hardness between as cast Al7075 alloy and Al7075-2 wt% $TiO_2$-2 wt% SiC hybrid composite. The enhancement in the hardness value is 24% over the unreinforced AA7075 alloy. A simple reason for increment in the hardness was the presence of hard reinforcing phase in the aluminium matrix. In general the presence of hard reinforcing phase present in the ductile aluminium matrix can significantly improve the hardness of overall

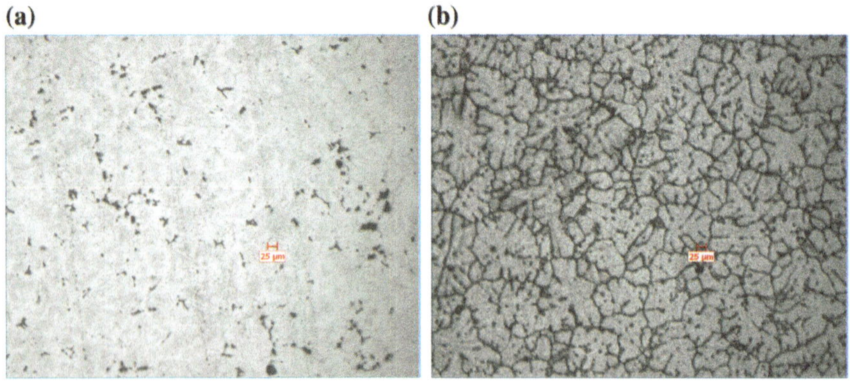


**Fig. 1** **a** Al7075-2 wt% $TiO_2$-2 wt% SiC composite **b** Al7075-2 wt% $TiO_2$-2 wt% SiC composite (etched)

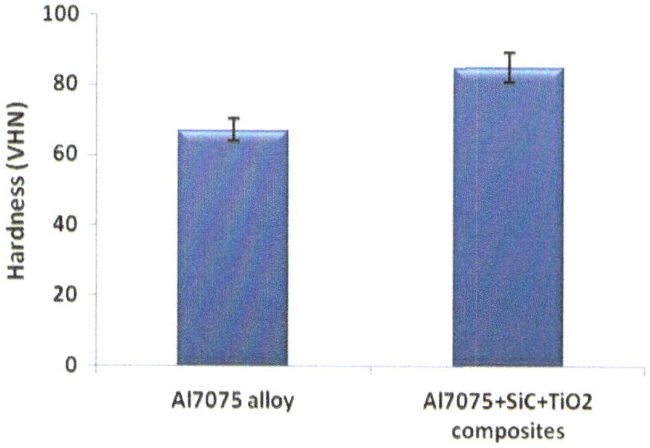


**Fig. 2** Hardness result of Al7075 alloy and Al7075-$TiO_2$-SiC composite

composite [8]. Increase in the hardness of the hybrid composites when compared with that of AA7075 alloy may be attributed to following reasons [9, 10].

1. Higher hardness of reinforcing phases. Addition of hard reinforcements (SiC and Titanium dioxide) in the soft ductile matrix contributes to improvement in the hardness of the composites.
2. Due to large variation in coefficient of thermal expansion between AA7075 alloy and reinforced particles, dislocation increases and acts as barrier for plastic deformation.
3. Uniform dispersion of particulates, excellent bond between matrix and reinforcing particles (SiC and $TiO_2$), and extensive grain refinement has significantly improved the hardness of composites.

## 5 Tensile Test

Figure 3 shows variation of ultimate tensile strength of Al7075 alloy and Al7075-$TiO_2$-SiC hybrid composites. It is observed that an improvement of 12% in the tensile strength of the composite is noticed when compared to the alloy Al7075. The reinforcements, SiC and $TiO_2$ are contributing to ultimate tensile strength of AA7075 matrix. The strengthening of AA7075 is mainly due to the consistent dispersion and superior bond of titanium dioxide and Silicon Carbide with the Aluminium 7075 alloy. The regular dispersion of reinforcements as seen in microstructure is reflected in the increment in ultimate strength values.

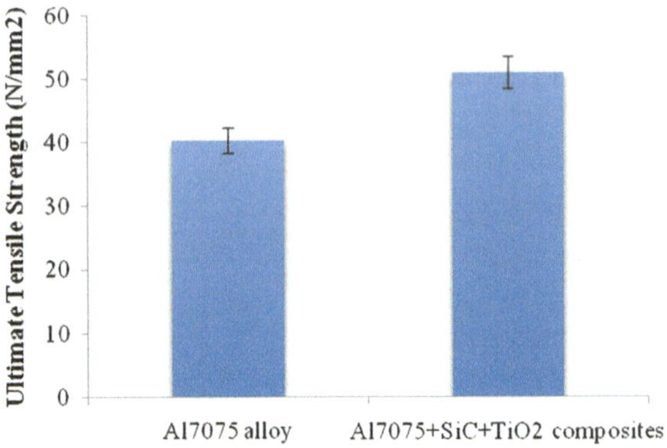


**Fig. 3** Tensile strength of Al7075 alloy andAl7075-$TiO_2$-SiC composite

## 6 Conclusions

Al7075-2 wt% $TiO_2$-2 wt% SiC hybrid composite has been successfully synthesised by liquid metallurgy technique. Optical microstructure studies clearly reveal uniform dispersion of particles in matrix material. Hybrid composite demonstrate considerable enhancement in hardness and ultimate tensile strength when evaluated with matrix alloy.

## References

1. Keshavamurthy R, Pavan M, Abdul Tazamul Haq M, Zinge D, Mohammed Y (2015) Hardness and tensile behavior of forged Al7075-SiC composite. Int J Appl Eng Res 10 (55):3769–3772
2. Ramesh CS, Keshavamurthy R, Koppad PG, Kashyap KT (2013) Role of particle stimulated nucleation in recrystallization of hot extruded Al 6061/SiCp composites. Trans Nonferrous Met Soc China 23(1):53–58
3. Mohanavel V, Suresh Kumar S, Srinivasan RV, Ganesha P, Anand KT (2017) Mechanical and metallurgical characterization of AA7075-fly ash composites produced by liquid state method. J Chem Pharm Sci (JCHPS) Special Issue (2):217–220
4. Nimbalkar SR, Vijay M, Satpute M (2015) Aluminium alloy Al7075 reinforcement and Stir casting–a review. IPASJ Inter J Mech Eng (IIJME) 3(11):038–046. ISSN 2321-6441
5. Karthigeyan R, Ranganath G, Sankaranarayanan S (2012) Mechanical properties and microstructure studies of aluminium (7075) alloy matrix composite reinforced with short basalt fibre. Eur J Sci Res 68(4):606–615
6. Chandel V, Bhatia OS, Sethi MS (2015) Fabrication and characterization of Al 7075-cenosphere composite & its comparison with pure Al 7075: a review. Int J Res Stud Sci, Eng Technol. ISSN: 2349–4751 (Print) & ISSN: 2349–476X (Online)
7. Murthy KVS, Girish DP, Keshavamurthy R, Varol T, Koppad PG (2017) Mechanical and thermal properties of AA7075/$TiO_2$/fly ash hybrid composites obtained by hot forging. Prog Nat Sci: Mater Int 27(4):474–481

8. Ramesh CS, Keshavamurthy R, Channabasappa BH, Ahmed A (2009) Microstructure and mechanical properties of Ni–P coated Si 3 N 4 reinforced Al6061 composites. Mater Sci Eng, A 502(1):99–106
9. Shivananda Murthy KV, Keshavamurthy R, Girish DP (2017) Mechanical characteristics of hot forged Al6061-Al2O3 composite. Appl Mech Mater 787:598–601
10. Keshavamurthy R, Sudhan JM, Gowda N, Krishna RA (2016) Effect of thermo-mechanical processing and heat treatment on the tribological characteristics of Al based MMC's. IOP Conf Ser: Mater Sci Eng 149(1):012118

# Study on Machinability of Aluminum Alumina Metal Matrix Composites

M. Sucharitha and B. Ravi Sankar

## 1 Introduction

The applications of metal matrix composites (MMCs) are being growing daily in each the aerospace and vehicle industries, due to their progressed properties as compared to monolithic metals [1–3]. Now a days, among various metal matrix composites (MMCs) synthesized, aluminum metal matrix composites in well known and discontinuously reinforced aluminum metal matrix composites, which include Al- $Al_2O_3$/SiCp specially, are emerged because the forerunner for a spread of fashionable and special packages [2, 3]. This trend has been attributed to their advanced specific strength and precise stiffness, excessive temperature capability, lower coefficient of thermal growth, higher wear resistance, stepped forward dimensional balance, and amenability to standard steel forming techniques [4, 5]. Further, improvement of stir casting path for synthesis has delivered down their price to an acceptable level in comparison to the ones processed by powder metallurgy and spray casting [3]. However, the presence of discontinuous second segment debris gives superior mechanical and bodily homes however on the other hand, is significantly have an impact on their "machinability" because the presence of tough reinforcement particles in different weight ratio makes them extraordinarily tough to machine as they result in speedy device put on [6, 7]. Even though attempts had been made to cast off machining operation, including close to internet shape forming and stir casting, they may be confined and therefore machining is still a quit essential a part of component manufacture. In addition, for lots additives, the production of true surface end is crucial. Research on machinability of mild

M. Sucharitha (✉)
Department of Mechanical Engineering, Bharath University, Chennai, India
e-mail: saisuchi2002@gmail.com

B. Ravi Sankar
Department of Mechanical Engineering, Bapatla Engineering College, Bapatla, India



K. Antony and J. P. Davim (eds.), *Advanced Manufacturing and Materials Science*, Lecture Notes on Multidisciplinary Industrial Engineering,
https://doi.org/10.1007/978-3-319-76276-0_31

alloy composites bolstered with $Al_2O_3$/SiC fibers/debris imply bad machinability because of abrasive wear of equipment. Furthermore, fine of the machined surface additionally deteriorates with device wear [8, 9].

Published literature at the machinability of particulate bolstered MMCs indicates that best CNMG gear offer a beneficial device lifestyles while machining those materials with CNMG tools, that's more difficult than alumina and silicon carbide and it additionally does no longer have a chemical tendency to react with the work piece material.

The prevailing observe has been carried out to look at the machinability of the AA6061/$Al_2O_3$ metal matrix composites at specific weight fraction of $Al_2O_3$ with the help of CNMG tool.

## 2 Experimental Setup

The cylindrical $Al_2O_3$ reinforced AA6061 MMCs are cut and the required samples (shown in Fig. 1) are prepared for machining purpose. Machining tests of the specimen is carried out in a conventional universal lathe machine (Padmini Lathes Pvt. Ltd.). The cutting tool is fitted in a rigid tool holder. The selected cutting tool material is CNMG. Chip breaker was not used during the experimental study and the machining tests are conducted under dry cutting processes (Figs. 2, 3 and 4).

The cutting forces (Ft, Ff and Fr) are measured at different cutting speed and depth of cut, at constant feed rate by using SYSCON instrument made lathe tool dynamometer. The selected machining parameters and the details of cutting tool used for experimentation are shown in Tables 1, 2, 3 and 4.

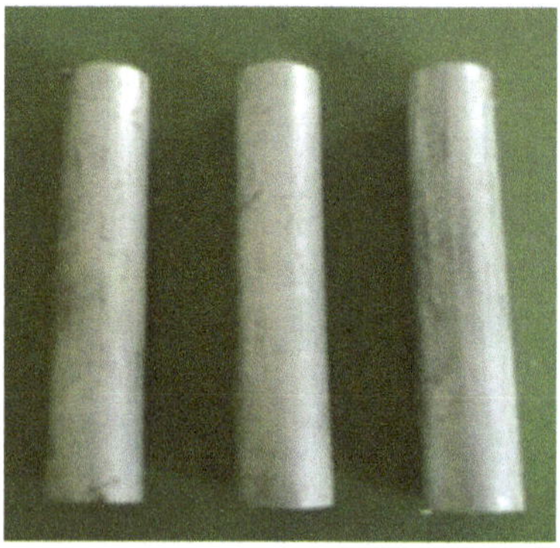

**Fig. 1** Fabricated AMMC rods

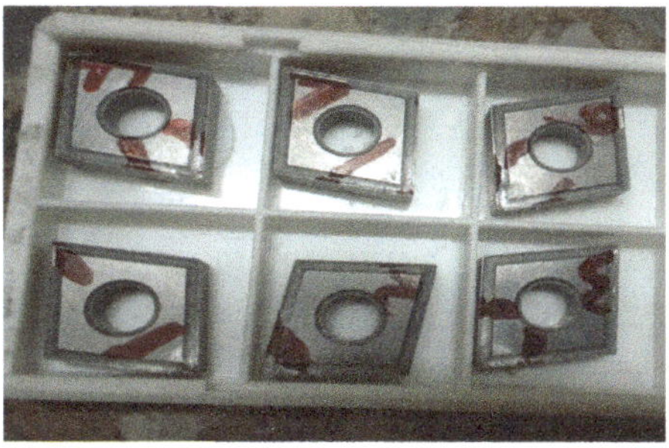

**Fig. 2** Cutting tools (CNMG)

**Fig. 3** Chip formation

**Fig. 4** AMMC rods after machining

**Table 1** Cutting tool details

| Tool material and grade | CNMG |
|---|---|
| Rake angle (°) | 5 |
| Clearance angle (°) | 7 |
| Cutting edge angle (°) | 80 |
| Nose radius (mm) | 4 |

**Table 2** Experimental values for 2% specimen

| Sl. no. | Speed (rpm) | Feed (mm/rev) | Depth (mm) | Wattmeter reading | | Cutting speed (M/Min) | Cutting force ($N/mm^2$) |
|---|---|---|---|---|---|---|---|
| | | | | Initial | Final | | |
| 1 | 500 | 0.04 | 0.4 | 0.10 | 0.12 | 43.19 | 1389.21 |
| 2 | 775 | 0.04 | 0.4 | 0.11 | 0.13 | 66.95 | 896.19 |
| 3 | 1200 | 0.04 | 0.4 | 0.14 | 0.17 | 103.67 | 868.13 |
| 4 | 500 | 0.04 | 0.8 | 0.10 | 0.12 | 43.19 | 694.60 |
| 5 | 775 | 0.04 | 0.8 | 0.11 | 0.13 | 66.95 | 448.09 |
| 6 | 1200 | 0.04 | 0.8 | 0.14 | 0.18 | 103.67 | 578.75 |
| 7 | 500 | 0.04 | 1.2 | 0.10 | 0.13 | 43.19 | 694.60 |
| 8 | 775 | 0.04 | 1.2 | 0.11 | 0.14 | 66.95 | 448.09 |
| 9 | 1200 | 0.04 | 1.2 | 0.14 | 0.19 | 103.67 | 482.30 |

**Table 3** Experimental values for 4% specimen

| Sl. no. | Speed (rpm) | Feed (mm/rev) | Depth (mm) | Wattmeter reading | | Cutting speed (M/Min) | Cutting force ($N/mm^2$) |
|---|---|---|---|---|---|---|---|
| | | | | Initial | Final | | |
| 1 | 500 | 0.04 | 0.4 | 0.10 | 0.12 | 43.19 | 1389.21 |
| 2 | 775 | 0.04 | 0.4 | 0.11 | 0.14 | 66.95 | 1344.29 |
| 3 | 1200 | 0.04 | 0.4 | 0.14 | 0.19 | 103.67 | 1446.90 |
| 4 | 500 | 0.04 | 0.8 | 0.10 | 0.13 | 43.19 | 1041.91 |
| 5 | 775 | 0.04 | 0.8 | 0.11 | 0.16 | 66.95 | 1120.24 |
| 6 | 1200 | 0.04 | 0.8 | 0.14 | 0.20 | 103.67 | 868.14 |
| 7 | 500 | 0.04 | 1.2 | 0.10 | 0.14 | 43.19 | 926.14 |
| 8 | 775 | 0.04 | 1.2 | 0.11 | 0.16 | 66.19 | 746.83 |
| 9 | 1200 | 0.04 | 1.2 | 0.14 | 0.22 | 103.67 | 771.68 |

**Table 4** Experimental values for 6% specimen

| Sl. no. | Speed (rpm) | Feed (mm/rev) | Depth (mm) | Wattmeter reading | | Cutting speed (M/Min) | Cutting force ($N/mm^2$) |
|---|---|---|---|---|---|---|---|
| | | | | Initial | Final | | |
| 1 | 500 | 0.04 | 0.4 | 0.10 | 0.13 | 43.19 | 2082.81 |
| 2 | 775 | 0.04 | 0.4 | 0.11 | 0.15 | 66.95 | 1792.38 |
| 3 | 1200 | 0.04 | 0.4 | 0.14 | 0.19 | 103.67 | 1446.90 |
| 4 | 500 | 0.04 | 0.8 | 0.10 | 0.13 | 43.19 | 1041.90 |
| 5 | 775 | 0.04 | 0.8 | 0.11 | 0.16 | 66.95 | 1120.24 |
| 6 | 1200 | 0.04 | 0.8 | 0.14 | 0.20 | 103.67 | 868.14 |
| 7 | 500 | 0.04 | 1.2 | 0.10 | 0.16 | 43.19 | 1389.21 |
| 8 | 775 | 0.04 | 1.2 | 0.11 | 0.17 | 66.95 | 896.19 |
| 9 | 1200 | 0.04 | 1.2 | 0.14 | 0.23 | 103.67 | 868.14 |

# 3 Result and Discussion

## 3.1 Cutting Forces

In the present investigation, dry turning operations have performed to evaluate the cutting forces. The cutting force 'Ft' (tangential component), feed force 'Ff' (thrust component) and 'Fr' the radial component are measured for analyzing the machinability characteristics of AA6061/$Al_2O_3$—MMCs at different weight fraction of $Al_2O_3$. The tangential component Ft, acts in the direction of cutting velocity vector is the main cutting force (Ft), feed force (Ff) and radial force (Fr) at constant feed rate i.e. 0.04 mm/rev and different intensity of cut i.e. 0.4, 0.8 & 1.2 mm, during machining of AA6061 reinforced with 2, 4 & 6% of $Al_2O_3$ respectively. The end result shows that the cutting force additives toes, Ft, Ff and Fr were decreases on increasing the cutting speed of the composites i.e. reinforced with 2, 4 & 6% of $Al_2O_3$. The determine additionally well-known shows that on growing the burden fraction of $Al_2O_3$ inside the matrix alloy the reducing forces are increases at the equal cutting conditions, as on growing the proportion of $Al_2O_3$ in solid MMCs, the hardness of the composites will increase linearly.

Figure 3 shows the effect of depth of cut on the cutting force (Ft), feed force (Ff) and radial force (Fr) at constant 0.04 mm/rev feed and different cutting speed. The experimental results represent that for all the cast composites the cutting force components Ft, Ff and Fr were increases on increasing the depth of cut. The cutting force components Ft, Ff and Fr are higher at depth of cut 1.2 mm when compared to 0.4 & 0.8 mm at constant cutting speed and feed rate. This figure also shows that the cutting force components are increases on increasing weight fraction of $Al_2O_3$ in the cast MMCs at the required cutting forces are increased during machining of cast MMCs at dry condition.

## 3.2 Chip Formation

The chip formation in the course of machining has observed by using very excessive plastic deformation at the shear region and owing to the dearth of enough ductility of the paintings material. The addition of $Al_2O_3$ particle reinforcement into the aluminum alloy matrix has induced a reduction in its ductility and produced a semi non-stop sort of chip in the course of machining of these MMCs without chip breaker. It isn't always only improves the machinability of this composite, but also complements its applicability in diverse Industries. The size of chips additionally laid low with the share of reinforcement particles in forged MMCs. it is found that the sizes of chips are decreases on increasing the load percentage of $Al_2O_3$ in cast mmcs.

The scale of chips produced at some stage in machining of AA6061 metal matrix composites reinforced with 2 wt% of $Al_2O_3$ are longer in comparison to the scale of

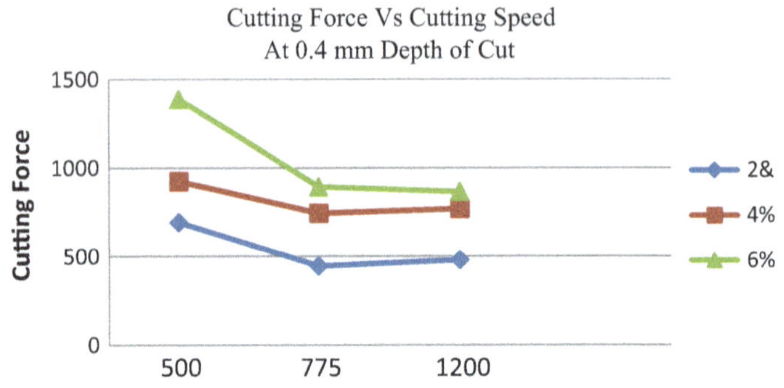


**Fig. 5** Graph between cutting force cutting speed at 0.4 mm depth of cut

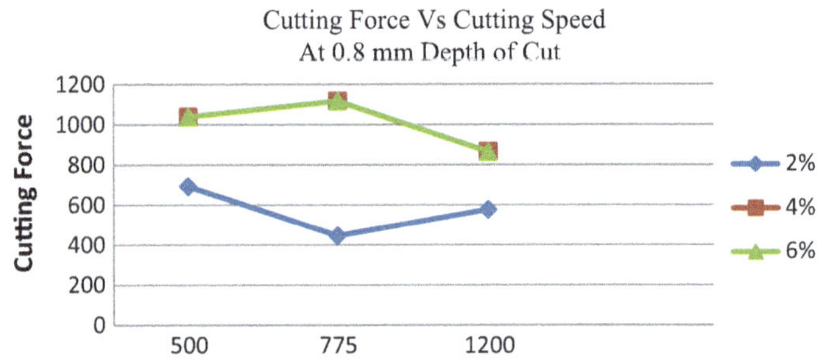


**Fig. 6** Graph between cutting force cutting speed at 0.8 mm depth of cut

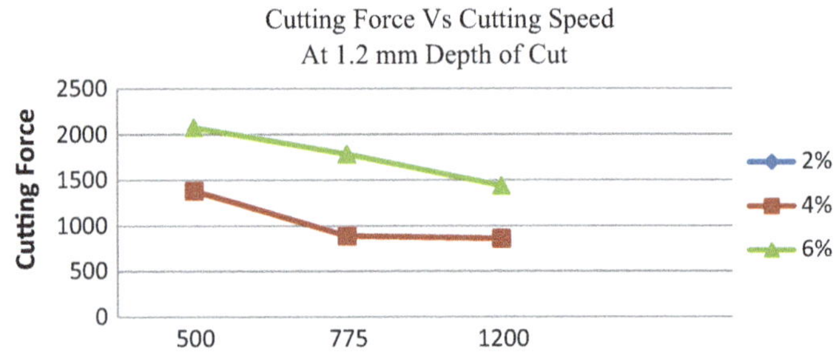


**Fig. 7** Graph between cutting force cutting speed at 1.2 mm depth of cut

chips produced in case of 4 and 6 wt% bolstered MMCs. As the presence of $Al_2O_3$ in solid MMCs will increase the brittleness of the Cloth, so the sizes of chips are progressively decreases. The consequences bringing out the reality that the forged MMCs strengthened with 2 wt% of $Al_2O_3$ is having true ductility assets in comparison to MMCs bolstered with 4 and 6 wt% of $Al_2O_3$ (Figs. 5, 6 and 7).

# 4 Conclusions

The experimental study on the machining parameters such as cutting forces of the cast composites at different weight fraction of reinforcements, during dry machining of MMCs by using tools concludes the following points:

- Aluminum matrix composites has been successfully fabricated by stir casting technique with fairly uniform distribution of $Al_2O_3$ particles with different percentages (2, 4, 6%).
- By increasing the cutting speed of the composites i.e., reinforced with increasing wt. percentage of $Al_2O_3$, the cutting force components were decreased.
- The result also reveals that, on increasing the percentage of $Al_2O_3$ in the matrix alloy the cutting forces are increases at the same cutting conditions.
- This paper is stating that by adding $Al_2O_3$ to AA6061, the cutting force values are increased. And the Strength of this metal matrix is directly proportional to the percentage of $Al_2O_3$.
- The sizes of chips are decreases on increasing the weight fraction of $Al_2O_3$ in the matrix metal.

## References

1. Dinwoodie J (1987) Automotive applications for MMCs based on short staple alumina fibres, SAE technical paper series, International Con Exp Detroit, MI, February, p 23–27
2. Chadwich GA, Heath PJ (1990) Machining of metal matrix composites. Met Mater 2–6:73–76
3. Dhadsanadhep C, Luangvaranunt T, Umeda J, Kondoh K (2008) Fabrication of Al/AL2O3 composite by powder metallurgy method from aluminium and rice husk ash. J Met, Mater Miner 18:99–102
4. Groover MP (2002) Fundamentals of modern manufacturing. Wiley
5. Fata A, Nikuei B (2010) The effect of the tool geometry and cutting conditions on the tool deflection and cutting forces. World Acad Sci Eng Technol (45):161–165
6. Luangvaranunt T, Dhadsanadhep C, Umeda J, Nisaratanaporn E, Kondoh K (2010) Aluminium-4 mass% copper/alumina composites produced from aluminium copper and rice husk ash silica powder by powder forging. Mater Trans 51:756–761
7. Mahboob H, Sajjadi SA, Zebarjad SM (2011) Influence of nanosized Al2O3 weight percentage on microstructure and mechanical properties of Al-matrix composite. Inst Mater Miner Mining 54:148–152
8. Radhika N, Subramanian S, Venkat Prasat S, Anandavel B (2012) Dry sliding wear behaviour of aluminium/alumina/graphite hybrid metal matrix composites. Ind Lubr Technol 64:359–366
9. Behera R (2013) Study on machinability of aluminium silicon carbide metal matrix composites. In: Transactions of 61st Indian foundry congress 2013

# Performance Evaluation of Cutting Parameters for Surface Roughness & Power Consumption in Turning of 904L Stainless Steel Using Vegetable Oil Based Cutting Fluids

**Rushikesh P. Waydande and Dayanand A. Ghatge**

## 1 Introduction

In recent years, new extremely corrosion-resistant austenitic stainless steels have entered the market place. They are commonly used to fabricate chemical and food processing equipment [1]. 904L stainless steel finds its application in oil refinery components, pulp and paper processing industries, seawater cooling devices as well as in gas scrubbing plants.It is also amongst the "difficult to cut" material and the difficulties such as poor surface finish and high tool wear are common. [2] In order to overcome these difficulties, castor oil and soyabean oil are used [3]. It is found that soyabean oil extended the surface finish for machining at taken cutting speed. In this Paper, this study becomes necessary to understand the theory behind the performance of vegetable oils during the machining of 904L stainless steel material [4].

### *1.1 Cutting Fluids*

Cutting fluids are employed in the machining method with the aim to boost the characteristics of the work piece-tool-chip system. Cutting fluids improve the potency of machining in terms of increased tool life, improved surface finish, reduced cutting force [5]. For applications where a metalworking fluid with good lubricating properties is required, a non-water-miscible fluid could also be counseled. The use of conventional petroleum-based cutting fluid is doubtless

R. P. Waydande (✉) · D. A. Ghatge
Department of Mechanical Engineering, Karmaveer Bhaurao Patil College of Engineering, Satara, India
e-mail: rushikeshwaydande@gmail.com

D. A. Ghatge
e-mail: dayanand.ghatge@kbpcoes.edu.in



K. Antony and J. P. Davim (eds.), *Advanced Manufacturing and Materials Science*, Lecture Notes on Multidisciplinary Industrial Engineering,
https://doi.org/10.1007/978-3-319-76276-0_32

dangerous. The effects of Specific cutting fluid on humankind, operating surroundings, the work piece and machine tool as well as usually on living surroundings as an entire are typically expressed by their ecological parameters. Machine operators are affected by contact with numerous substances at intervals the cutting fluids. The advantage of cutting fluid is that cools the work surface and tool. Another purpose of cutting fluids is that it lubricates the interface between the work surface and tool [6].

## 2 Experimentation Details

### *2.1 Selection of Workpiece Material & Tool*

It was reported that austenitic stainless steels are difficult to machine. Very few research work has been observed on 904L grade stainless steel. It has applications in piping systems, pollution management instrumentation, heat exchanges, and bleaching systems. This special grade of stainless steel additionally finds its application wherever robust reducing acids are used [7] (Tables 1 and 2).

**Table 1** Typical chemical composition for the 904L stainless steel

| | |
|---|---|
| Carbon (C) | 0.015 |
| Silicon (Si) | 0.254 |
| Manganese (Mn) | 1.592 |
| Sulphur (S) | 0.012 |
| Phosphor (P) | 0.032 |
| Chromium (Cr) | 19.856 |
| Nickel (Ni) | 24.745 |
| Molybdenum (Mo) | 4.152 |
| Copper (Cu) | 1.326 |
| Cobalt (Co) | 0.023 |
| Titanium (Ti) | 0.009 |
| Ferrous (Fe) | 47.923 |

**Table 2** Typical mechanical properties of 904L stainless steel

| | |
|---|---|
| Tensile Strength (MPa) min. | 490 |
| Yield Strength 0.2% Proof (MPa) min. | 220 |
| Hardness | Rockwell 70–90 typical |
| | Brinell 150 |
| Density (kg/m$^3$) | 7900 |
| Elastic Modulus (GPa) | 190 |
| Thermal Conductivity (W/m.K) | 11.5 at 20 °C |

**Fig. 1** 904 L stainless steel

**Table 3** Process parameters and their levels

| Symbol | **Factors** | **Level 1** | **Level 2** | **Level 3** |
|---|---|---|---|---|
| Vc | Speed (rpm) | 1000 | 1500 | 1700 |
| F | Feed rate (mm/rev) | 0.12 | 0.18 | 0.24 |
| D | Depth of cut (mm) | 0.6 | 1.0 | 1.5 |
| C, S, D | Type of cutting fluid | D | C | S |

Where, *D* Dry condition, *C* Castor oil, *S* Soyabean oil

In this study, 30 mm diameter 904L stainless steel rod is used as a work piece material. CNC Lathe machine is used for experiments. The Maximum RPM of the lathe machine is 3200 rpm, Photographic view of machine is shown in the Fig. 1. The experimentation is performed with coated carbide, TNMG 1604087N grade (Table 3).

## 2.2 *Vegetable-Based Cutting Fluids*

The present trend towards new kinds of cutting fluids supported Vegetable Oils in machining is clearly justified by their higher biodegradability and lower environmental impact. Many years, vegetable oils and fats have been used and maintained their importance as metalworking lubricants. The use of Vegetable oil in metal working applications may lighten problems faced by workers, such as Skin cancer and inhalation of toxic mist in the work environments [1]. The main advantage of vegetable oil based cutting fluids is low pollution of the environment. The other advantages are compatibility with additives, low production cost, wide production possibilities [8] (Table 4).

**Table 4** Experimental condition

| | |
|---|---|
| Work piece material | 904L stainless steel |
| Machining operation | Turning |
| Insert used | Coated Carbide tool insert |
| Insert designation | TNMG 1604087N (WIDIA Made) |
| Environment | Dry and wet |

Surface Roughness is measured with Instrument MITUTOYO surf test SJ-210. The work piece is attached to the detector unit of the SJ-210 P which traces the minute irregularities of the work piece surface. The vertical stylus displacement throughout the trace is processed and digitally showed on the liquid display of the SJ-210 P. The surf test has are solution varying from 0.01 to 0.4 μm depending on the measurement range [7] (Fig. 2).

The experimentation for this work was based on Taguchi's design of experiments (DOE) and orthogonal array.an outsized range of experiments got to be administrated once the amount of the Process parameters will increase. In this work, four cutting Parameters namely cutting speed, depth of cut, feed rate and types of cutting fluids were considered for experimentation.

For a four factors, three level experiments, Taguchi had specified L27 orthogonal array for experimentation. The response obtained from the trials conducted as per L27 array. Results were recorded and further analyzed (Table 5).

**Fig. 2** Photographic view of CNC lathe machine

**Table 5** Experimentation and observations

| Sr.No. | Speed (rpm) | Feed | Doc (mm) | Fluids/condition | Ra (μm) | Pc (W) |
|---|---|---|---|---|---|---|
| 1 | 1000 | 0.12 | 0.6 | Dry | 0.711 | 971.00 |
| 2 | 1000 | 0.12 | 0.6 | Castor oil | 1.207 | 969.00 |
| 3 | 1000 | 0.12 | 0.6 | Soyabean oil | 0.589 | 972.00 |
| 4 | 1000 | 0.18 | 1 | Dry | 2.409 | 1317.00 |
| 5 | 1000 | 0.18 | 1 | Castor oil | 2.612 | 1319.00 |
| 6 | 1000 | 0.18 | 1 | Soyabean oil | 1.592 | 1315.00 |
| 7 | 1000 | 0.24 | 1.5 | Dry | 3.689 | 1453.00 |
| 8 | 1000 | 0.24 | 1.5 | Castor oil | 4.989 | 1454.00 |
| 9 | 1000 | 0.24 | 1.5 | Soyabean oil | 3.879 | 1457.00 |
| 10 | 1500 | 0.12 | 1 | Castor oil | 1.211 | 1386.00 |
| 11 | 1500 | 0.12 | 1 | Soyabean oil | 0.503 | 1384.00 |
| 12 | 1500 | 0.12 | 1 | Dry | 0.954 | 1385.00 |
| 13 | 1500 | 0.18 | 1.5 | Castor oil | 3.487 | 1523.77 |
| 14 | 1500 | 0.18 | 1.5 | Soyabean oil | 1.917 | 1524.00 |
| 15 | 1500 | 0.18 | 1.5 | Dry | 1.999 | 1523.22 |
| 16 | 1500 | 0.24 | 0.6 | Castor oil | 4.517 | 1248.00 |
| 17 | 1500 | 0.24 | 0.6 | Soyabean oil | 2.968 | 1248.00 |
| 18 | 1500 | 0.24 | 0.6 | Dry | 2.640 | 1247.00 |
| 19 | 1700 | 0.12 | 1.5 | Soyabean oil | 0.986 | 1593.00 |
| 20 | 1700 | 0.12 | 1.5 | Dry | 0.790 | 1592.00 |
| 21 | 1700 | 0.12 | 1.5 | Castor oil | 2.096 | 1503.20 |
| 22 | 1700 | 0.18 | 0.6 | Soyabean oil | 2.084 | 1039.00 |
| 23 | 1700 | 0.18 | 0.6 | Dry | 2.337 | 1041.00 |
| 24 | 1700 | 0.18 | 0.6 | Castor oil | 2.567 | 1038.00 |
| 25 | 1700 | 0.24 | 1 | Soyabean oil | 3.654 | 1392.00 |
| 26 | 1700 | 0.24 | 1 | Dry | 2.819 | 1389.00 |
| 27 | 1700 | 0.24 | 1 | Castor oil | 4.284 | 1391.00 |

## 3 Analysis of Variance (ANOVA)

The observed values of surface roughness (*Ra*) were used for determining the significant factor influencing the machining process. The significant parameter influencing the surface roughness was found using the ANOVA procedure. Tables 6 show the ANOVA for Surface Roughness. From the calculations its being inferred that feed rate has more influence on Surface Roughness. Further its conjointly being inferred that cutting fluid has substantial influence on the process parameters (Fig. 3).

**Table 6** ANOVA for Surface Roughness

| Sr. No | Factor | D.O.F | Sum of squares | Mean squares | Variance | Contribution (%) |
|---|---|---|---|---|---|---|
| 1 | Cutting speed | 2 | 0.1562 | 0.0781 | 0.5948 | 3.684 |
| 2 | Feed rate | 2 | 33.0581 | 16.5291 | 125.888 | 77.97 |
| 3 | Depth of cut | 2 | 1.1967 | 0.5983 | 4.5567 | 2.8223 |
| 4 | Type of fluid | 2 | 5.6213 | 2.8106 | 21.4059 | 13.259 |
| 5 | Error | 18 | 2.3637 | 0.1313 | | |
| 6 | Total | 26 | 42.3959 | | | |

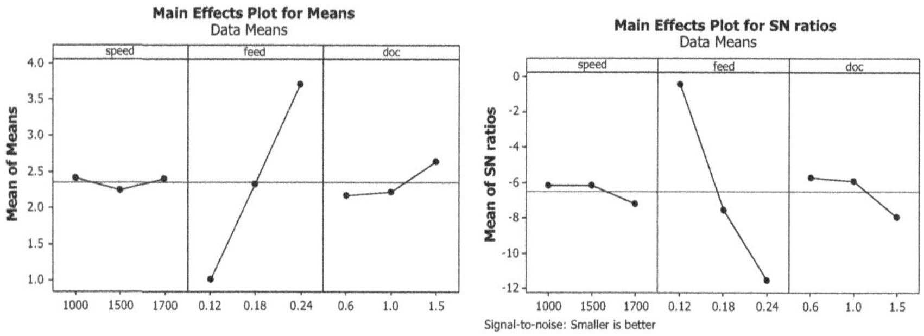


**Fig. 3** Main effect plots for surface roughness

## *3.1 ANOVA for Surface Roughness*

The above Table 6 shows the analysis of variance of surface roughness. It was found that percentage contribution of feed rate is high in surface roughness compare to other parameters. Feed rate has 77.97% contribution which is highly for surface roughness.

## *3.2 ANOVA for Power Consumption*

The above Table 7 shows the analysis of variance of Power Consumption. It was found that percentage contribution of Depth of cut is high in Power Consumption compare to other parameters. Depth of cut has 84.28% contribution which is highly for Power Consumption [8] (Fig. 4).

**Table 7** ANOVA for surface roughness

| Sr. No | Factor | D.O.F | Sum of squares | Mean squares | Variance | Contribution (%) |
|---|---|---|---|---|---|---|
| 1 | Cutting speed | 2 | 86,952 | 43,476 | 17.4182 | 8.6372 |
| 2 | Feed rate | 2 | 25,777 | 12,888 | 5.1634 | 2.5605 |
| 3 | Depth of cut | 2 | 848,464 | 424,232 | 169.96 | 84.2808 |
| 4 | Type of fluid | 2 | 590 | 295 | 0.1181 | 0.0586 |
| 5 | Error | 18 | 44,926 | 2496 | – | – |
| 6 | Total | 26 | 1,006,710 | | | |

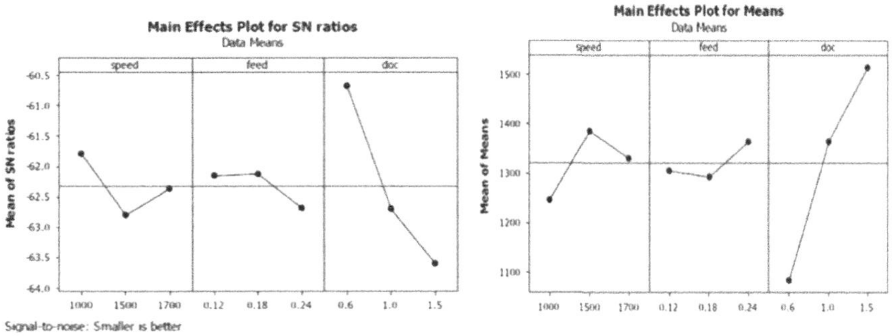


**Fig. 4** Residuals plots for power consumption

# 4 Result & Discussion

## *4.1 Surface Roughness*

The above plot is of main effect plot for surface roughness. From above plot it is seen that how speed, feed and depth of cut influence on surface roughness. The main effects plot used to examine variations between level suggests that for one or a lot of factors. There is a main effect when different levels of a factor affect the response differently. The signal-to-noise ratio of the quality characteristics is divided into the Larger the better, nominal the better, and the smaller the better. The above plot describes the smaller is better value. The smaller the surface roughness, the better the turning quality will be. In this case, the smaller the better is used to compute the signal-to-noise ratio of the surface roughness, as shown. L27 orthogonal array was used for the experimentation.

### 4.2 Power Consumption

The above plot is of main effect plot for Power Consumption. From above plot it is seen that how speed, feed and depth of cut influence on Power Consumption. The main effects plot used to examine differences between level means for one or more factors. There is a main effect when different levels of a factor affect the response differently. A main effects plot graphs the response mean for every factor level connected by a line. The above plot describes the smaller is better value. The smaller the Power Consumption, the better the turning quality will be. In this case, the smaller the better is used to compute the signal-to-noise ratio of the Power Consumption, as shown. L27 orthogonal array was used for the experimentation.

## 5 Conclusion

- Experiment involving coated carbide tool insert and AISI 904L stainless steel work material under varying machining parameters and with two different cutting fluids were performed. Cutting fluids were considered as important parameters in the machining process along with cutting speed, feed rate and depth of cut.
- The gradual increase of feed rate and depth of cut increases the surface roughness in machining of 904L Stainless steel by coated carbide insert.
- The result shows that the feed rate is highly controlling parameter, which influences the surface roughness in machining of 904L steel.
- An analysis of variance (ANOVA) was made and it was found that feed rate has greater influence i.e. 77.97% contribution on surface roughnessand depth of cut has greater influence on power consumption (84.28% contribution).
- It is observed that Soyabean oil was a better cutting fluid than the Castor oil in improving the tool life and reducing the surface roughness.

## References

1. Elmunafi MHS, Kurniawan D, Noordin MY (2015) Use of castor oil as cutting fluid in machining of hardened stainless steel with minimum quantity of lubricant. Procedia CIRP 26 408–411
2. Johnson RDJ, Wins KLD, Raj A, Beatrice BA (2014) Optimization of cutting parameters and fluid application parameters during turning of OHNS Steel. In: 12th global congress on manufacturing and management, GCMM 2014
3. Anthony Xavior ZM, Adithan M (2009) Determining the influence of cutting fluids on tool wear & surface roughness during turning of AISI 304 austenitic stainless steel. J Mater Process Technol 209:900–909

4. de Lacalle LNL, Angulo C, Lamikiz A, S′anchez JA (2006) Experimental and numerical investigation of the effect of spray cutting fluids in high speed milling. J Mater Process Technol 172:11–15
5. Isik Y (2010) An experimental investigation on effect of cutting fluids in turning with coated carbides tool. J Mech Eng 56(3), 195–201
6. Adegbuyi PAO, Lawal G, Oluseye O, Odunaiya G Analysing the effect of cutting fluids on the mechanical properties of mild steel in a turning operation. https://doi.org/10.5251/ajsir.2011.2.1.1.10
7. Avila RF, Abrao AM (2001) The effect of cutting fluids on the machining of hardened AISI 4340 steel. J Mater Process Technol 119:21–26
8. Hamdan A, Fadzil M, Abou-El-Hossein KA, Hamdi M Performance evaluation of different types of cutting fluid in the machining of AISI 01 hardened steel using pulsed jet MQL system
9. Ozcelika B, Kuram E, Demirbas E, ‚Sik E (2013) Effects of vegetable-based cutting fluids on the wear in drilling. In: Sadhana, vol 38. Part 4, Indian Academy of Sciences, pp 687–706

# Experimental Study on the Effects of Bio Ethanol Petrol Blends in a Spark Ignition Engine: Performance Analysis

**P. Anu Nair, S. Sarath, Agresh S. Kumar, Akshay Jegy George and Anoop G. Das**

## 1 Introduction

At present, virtually all the world's transportation and industries needs are supplied by the fuel derived from petroleum. Twenty first century with its prosperity will bring intricate problems to mankind. One such problem is the fast depletion of fossil fuels and rapid increase in atmospheric pollution. There will be a short fall of approximately 25% in crude oil supplies relative to demand by the end of last century. Hence, the sharply increasing crude oil prices and demand throughout the world triggered off a frantic search for alternative fuels. An alternative fuel is defined here as one whose production doesn't depend on natural crude oil from wells. Fortunately we have some alternate renewable and non-renewable fuels. Among them Bio ethanol has the best chance of enduring as an alternative fuel.

Bio ethanol can be used in two ways. The first is blending it with gasoline, with either small or no alteration in the base engine. The second way of using Bio ethanol in S.I engines is operation on pure Bio ethanol. This requires major modification to the engine and the fuel system. The merits of Bio ethanol are as follows. Bio ethanol can be obtained from renewable resources and can be produced from a wide range of materials such as molasses, a by product of sugar cane industry, sugar beet, starch containing materials like milo, corn, algae, cassava etc., cellulose materials such as bagasse, biomass, wood, agriculture, forest residue and saw dust.

Bio ethanol contains 35% oxygen in its molecule structure, hence it ensures complete burning, resulting in a substantial reduction in CO emission. Bio ethanol use in low level blends with gasoline results in an overall decrease in ground ozone formation which causes smog. The emissions produced by burning Bio ethanol are

P. Anu Nair (✉) · S. Sarath · A. S. Kumar · A. J. George · A. G. Das
Department of Mechanical Engineering, Gurudeva Institute of Science and Technology, Payyapady, Kottayam 686516, Kerala, India
e-mail: anunair67@gmail.com



K. Antony and J. P. Davim (eds.), *Advanced Manufacturing and Materials Science*, Lecture Notes on Multidisciplinary Industrial Engineering,
https://doi.org/10.1007/978-3-319-76276-0_33

less reactive with sun light than those produced by burning gasoline resulting in a lower potential for ground level ozone formation.

Alcohols create higher cooling effects on intake charge than gasoline, which persists throughout the engine cycle resulting in lower $NO_X$ emission levels. Bio ethanol works as an octane enhancer. High octane quality of alcohols enables higher compression ratios to be used than are possible with gasoline fueled engines and this improves both fuel economy and power. Alcohols burn faster and allow more efficient torque development.

Brazil, the number one, produces 16.1 billion liters/year of Bio ethanol, out of which 60% is used as a fuel. It has promulgated a federal law which make it compulsory to use 26% of anhydrous alcohol as a mixture with gasoline in all spark ignition engine powered automobiles. U.S.A is next with 5.75 billion liters and is emerging as an important user of fuel Bio ethanol for automobile application. In United States, Bio ethanol made from corn has become an increasingly important fuel additive over the past decade.

Basically India is an agricultural country. In India alcohol is manufactured from molasses, a viscous semi solid material left after the crystallization of sugar from concentrated sugar cane juice, and it produces 1.3 billion liters of alcohol/annum and our installed capacity is 3.2 billion liters. At present more than 70% of our petroleum requirement is met through import. Current gasoline consumption in India is 7–7.5 billion liters/annum. 5% Bio ethanol blending will reduce 300–325 million liters of petrol consumption.

Abdel Rahman and Osman [1] conducted investigation on varying the compression ratio of S.I engine working under different ethanol-gasoline fuel blends. They reported that 10% ethanol gasoline fuel blends increased the maximum pressure over that of pure unleaded gasoline. A higher compression ratio, for fuel blends above 20% Bio ethanol, improved the indicated thermal efficiency [2].

Palmer [3] from British Petroleum Company ran a wide range of vehicleperformance tests on oxygenated fuel blends. It was found that 10% ethanol in gasoline improves the engine power by 5%. The addition of Bio ethanol to lead free gasoline resulted in an increase of fuel research octane number by 5 units for each 10% ethanol addition. It was further reported that the exhaust emission of carbon monoxide was reduced considerably (by about 30%).

Hamdan and Jubran [4, 5] have studied the effect of ethanol addition on the performance of gasoline and diesel engines. The performance tests were carried out using different fuel blends of Bio ethanol and gasoline. The maximum percentage of ethanol used was 15%. The best performance was achieved when 5% ethanol—gasoline blend was used, with a thermal efficiency increase of 4 to 21%. However all their test were carried out under part load condition.

Yacoub and Bata [6, 7] conducted tests on a single cylinder spark ignition engine with unleaded test gasoline and high purity straight chain alcohols. Alcohols with carbon number ranging from $C_1$ to $C_5$ were individually blended with unleaded test gasoline. They found that adding lower alcohols ($C_1$, $C_2$ and $C_3$) to unleaded test gasoline improved its knock resistance. They found that Bio ethanol—gasoline blends provides the highest knock resistance improvement among all tested blends, and operate at higher efficiency compared with neat gasoline owing to its higher compression ratio. Similarly most of alcohol—gasoline blends showed reduction in CO and $CO_2$ emissions.

Cha-Lee Myung and Co authors [8] from Hyundai motor company worked on the Research and Development of Hyundai Flexible Fuel Vehicles. In their investigation, one of the suggestions was a V-grooved type spark plug. This spark plug with the gap size of 1 mm showed good ignitability and excellent anti-wetting characteristic at cold start. The cold start temperature limit with the V-grooved spark plus was lowered by nearly 5 °C compared to the conventional spark plug.

A literature released by SAE, titled "Alternative fuels for spark ignition engines" [9] gave some results about alcohol fueled engines. The lower Calorific Value of alcohol lower their stoichiometric A/F ratios, requiring more fuel for a given air volume. This and their much higher latent heats, make the fuel cooling effects on the intake charge much greater than with gasoline. Charge temperatures at full load are reduced, resulting in increased volumetric efficiencies and more torque and power from the engine.

## 2 Materials and Methods (Preparation of Bio Ethanol from Rice Water Waste)

### *2.1 Preparation of Rice Water Starch*

Take 250 gm jaggery and mix with 500 ml of water and heat the jaggery solution in an open vessel for 20 min at 100 °C. Then allow the solution to cool down at room temperature. Take 50 gm of yeast and dilute in 100 ml of water and mix it with the jaggery solution. Take 1 litre of cooking rice water waste mix it with the jaggery solution. Keep air tight tank for 10–15 days. After that remove the excess water above the starch solution (up to 500 ml).

### *2.2 Distallation of Rice Water Starch*

Distillation unit consist of 1 litre round bottom flask (RB flask) = 2 nos., Adapter, Waterpump, Condenser, Thermometer Electric heater. Taking the 1 litre mixed rice water solution in the RB flask. Keep it in the electric heater in perpendicular to heater. Connect the condenser unit to the RB flask with the help of adapters and place the thermometer in an oil contained tube for measuring the temperature of the vapour particles that develop in the flask. Keep the condenser unit to a 20°–30° with the help of a stand for the smooth flow. Connect the 1 litre RB flask to the condenser other end for collecting the liquid Bio ethanol.

**Table 1** Properties of Bio ethanol compared with gasoline

| Characteristics | Bio ethanol | Gasoline |
|---|---|---|
| Chemical formula | $C_2H_5OH$ | $C_4$–$C_{12}$ |
| Molecular weight | 46.07 | 100–105 |
| Octane number | | |
| Research | 108 | 90–100 |
| Motor | 92 | 81–90 |
| Cetane number | 8 | Below 15 |
| Calorific value kJ/kg | 29773 | 43729 |
| Heat of vaporization kJ/kg | 921.1 | 350 |
| Boiling point °C | 78.3 | 30–225 |
| Flash point °C | 12.8 | −43 |
| Auto ignition temperature °C | 422.8 | 257.2 |
| Freezing point °C | −114 | −43 |
| Stoichiometric A/F | 9 | 14.7 |
| Specific Gravity @ 60 F | 0.79 | 0.72–0.75 |
| Density kg/m$^3$ | 792 | 779 |
| Vapour Pressure @ 37.8 °C | 15.84 | 55–103.3 |

Check the whole setup of distilling unit and make sure that electrical heater is off and water supply to the condenser unit is connected properly. Take an amount of mixed rise water solution in a round bottom flask and place it in the electric heater. Switch on the power supply and gradually increase the temperature range of electric heater by adjusting the knob of electric heater. Switch on the water pump to supply water to the condenser unit to condense the Bio ethanol vapour generate in the round bottom flask. Condensed Bio ethanol is collected in the round bottom flask placed in the other end of distilling unit. This procedure is continued four hours for collect the Bio ethanol (Table 1).

# 3 Experimental Procedure and Data Acquisition

The experiments were carried out on a single cylinder, four stroke petrol engine named Villiers Enfield engine, model MK-25, manufactured by M/s Greaves Ltd. The engine was loaded by an alternator. The alternator was connected in series with a panel board consisting of a number of bulbs. By switching on the bulbs, the alternator was loaded in steps of 0.4, 0.8, 1.2, 1.6 and 1.8 KW. Different fuel blends of pure Bio ethanol and pure gasoline were prepared with 0% Bio ethanol up to 45% Bio ethanol in increments of 5%. A fuel tank with a measuring burette enables the engine fuel consumption to be measured.

## 3.1 Test on Engine Without Inlet Air Heating

The layout of the engine is shown in Fig. 1. The engine was made to run at 3000 rpm. The time taken for 10 cc fuel consumption was noted down with the help of a stopwatch. First sets of readings were taken with the base engine fuelled by pure gasoline to access its performance. Further different blending ratios were fuelled to the engine and readings were taken. When the quantity of Bio ethanol in the blended fuel was increased, the engine was not able to run satisfactorily beyond certain blending ratio. The latent heat of vaporization for Bio ethanol is nearly two times higher than gasoline. When the quantity of Bio ethanol is increased in gasoline blends, vaporization problem may occur in the carburetor.

Ice formation may take place at venturi throat and fuel nozzle when using alcohol. To avoid this problem, the intake air is to be heated, so that we can achieve maximum replacement of Bio ethanol in gasoline. Because of the above factors, in our investigation we found that the engine could run satisfactorily only up to 25% blending. When the engine was run at 30% blending it experienced starting difficulties and it was not able to take up the load. Hence it was decided to modify the base engine.

## 3.2 Test on Engine with Inlet Air Heating

A heat exchanger to heat the inlet air was fabricated and attached to the engine. In the heat exchanger, the part of the heat available with exhaust gases is used to preheat the inlet air. The heat exchanger is single pass counter flow type provided

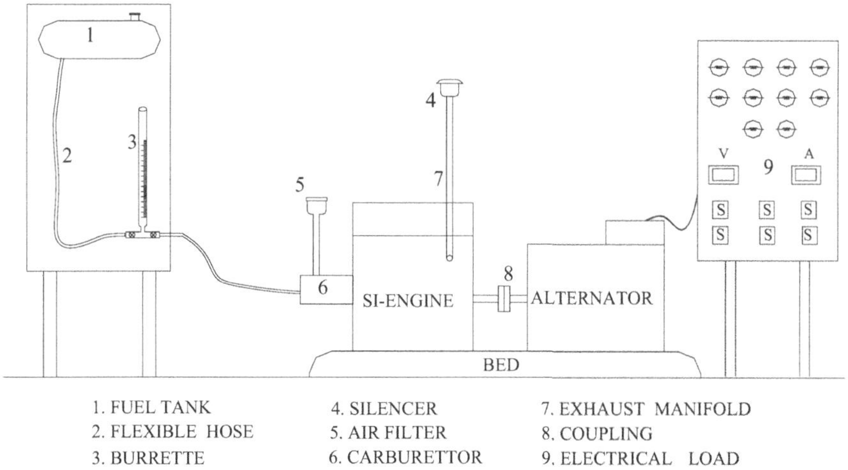


**Fig. 1** Experimental setup without air heating

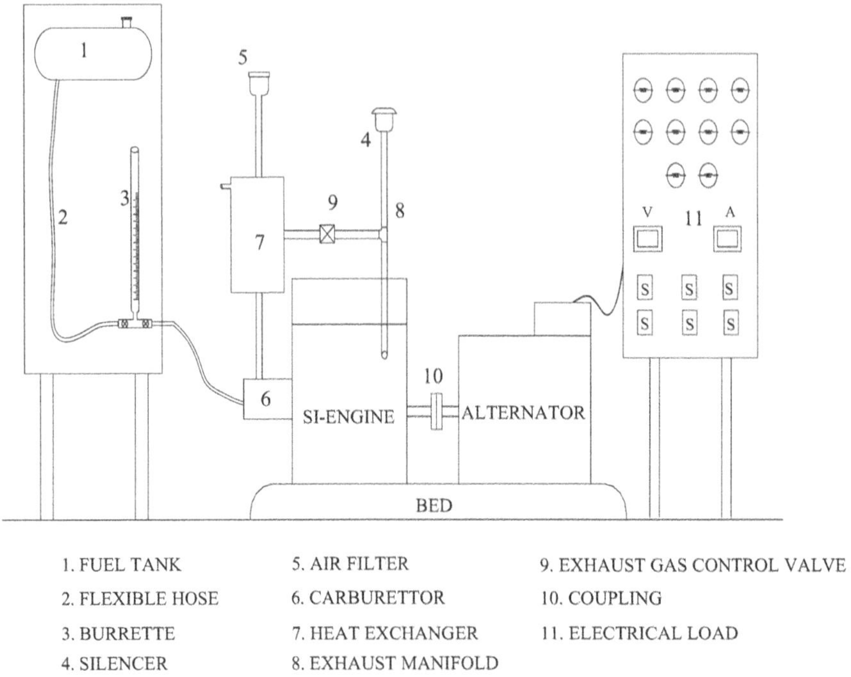


**Fig. 2** Experimental setup with inlet air heating

with the arrangement of measuring inlet air temperature by a thermometer. A control valve is provided in the path of exhaust gas to regulate the quantity of exhaust, so that desired inlet air temperature can be achieved. The following Fig. 2 shows the modified setup.

Under normal conditions, the engine inlet temperature was measured as 37 °C. Hence it was decided to heat the inlet air to 50 and 60 °C. The same procedure was followed as in the previous case and different readings were taken. With the readings taken the following graphs are drawn which show the effect of Bio ethanol addition in gasoline engine.

## 4 Results & Discussion

Figure 3 shows the comparison of brake thermal efficiency of the conventional gasoline engine, fuelled with neat gasoline and Bio ethanol blended gasoline ranging from 5 to 25% at different outputs. The engine attained a maximum efficiency of 13.49% at 25% of Bio ethanol blending, which is higher by 10% over neat gasoline fuelled engine. Next to 25%, 10% of Bio ethanol blending yielded 13.3% which is also higher by 8.48% when compared with neat gasoline fuelled engine. In the conventional engine, other blending proportions gave a slight reduction in thermal efficiency. Figure 4 shows the improvement in brake thermal efficiency at 25% blending which seems to the optimum without inlet air heating. In this blending proportion, we noted that the engine ran smoothly at different loads and did not have any starting difficulty. High latent heat of vaporization of Bio ethanol provides increase in volumetric efficiency, which results in increased charge density. This can result in higher power output and efficiency. The variation of specific fuel consumption is shown in Fig. 5. With 10% Bio ethanol blended fuel, the specific fuel consumption was decreased by 4.84% when compared with neat gasoline engine. The specific fuel consumption values for 25% blended and neat gasoline engine are very close.

**Fig. 3** Variations of brake thermal efficiency at different blending ratios without inlet air heating

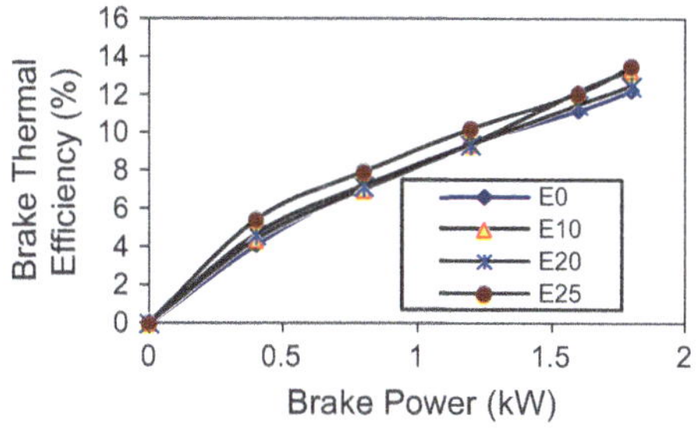


**Fig. 4** Variations of brake thermal efficiency at different blending ratios with inlet air heating

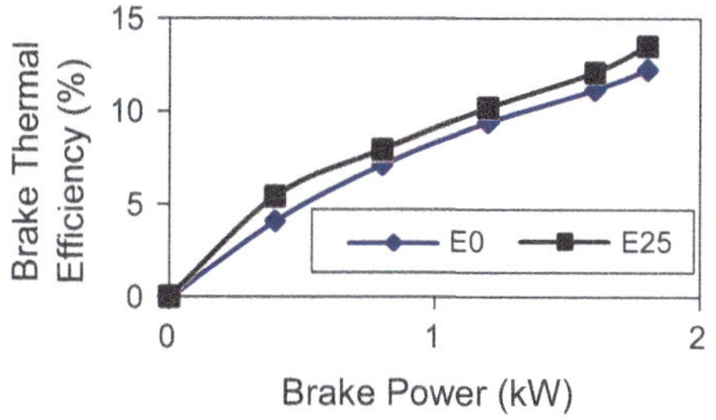


**Fig. 5** Variations of specific fuel consumption (SFC) for different blending ratios without inlet air heating

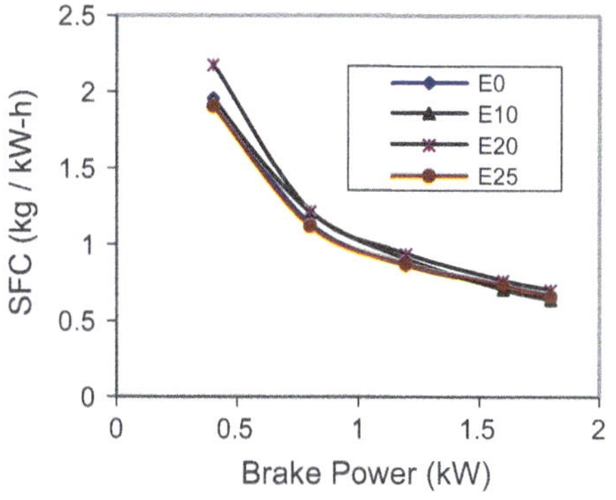

**Fig. 6** Variations of brake thermal efficiency for different blending ratios with inlet air heating to 50 °C

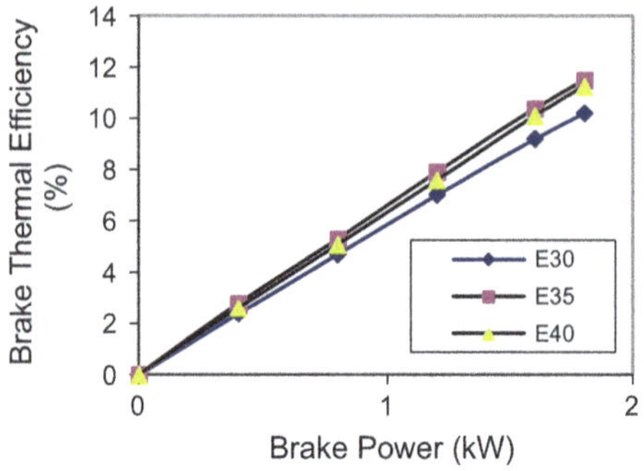


The engine ran smoothly up to 40% of blending when the inlet air temperature was raised to 50 °C in the modified setup. In this stage 35% was found to be optimum as brake thermal efficiency decreased when the blending ratio was increased to 40% which is shown in Fig. 6. Figure 7 shows that the most

**Fig. 7** Variation of specific fuel consumption (SFC) for different blending ratios at 50 °C with inlet air heating

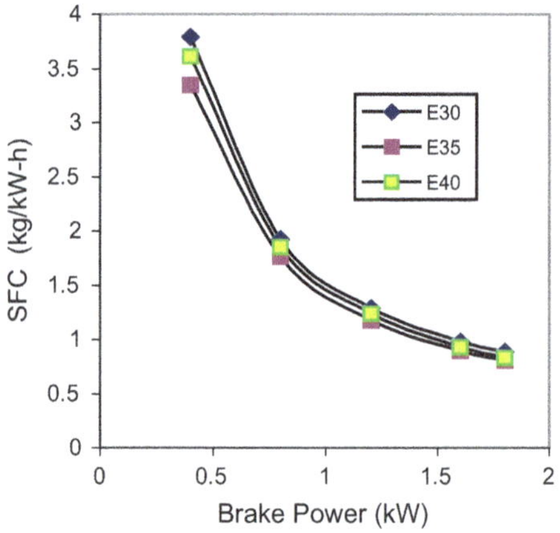


**Fig. 8** Variations of brake thermal efficiency for different blending ratios at 60 °C with inlet air heating

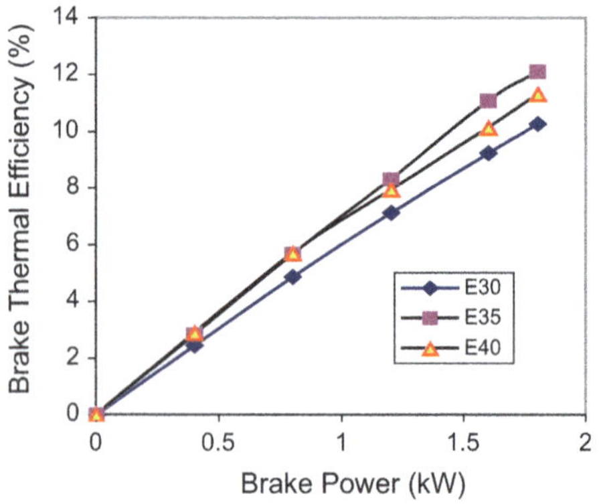

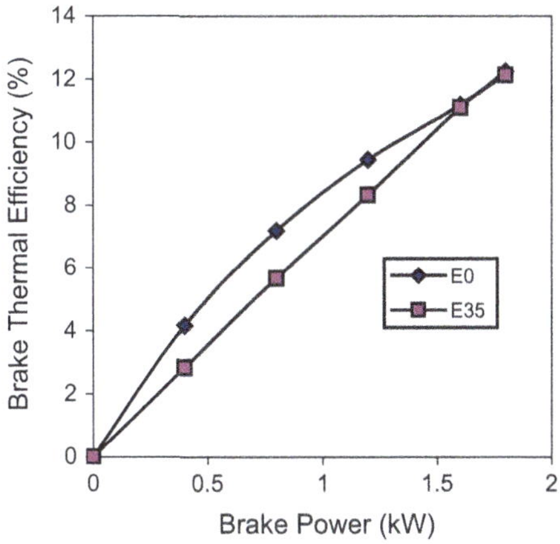


**Fig. 9** Compartion of brake thermal efficiency with neat gasoline fuelled engine and 35% blending ratio with 60 ° C air preheating

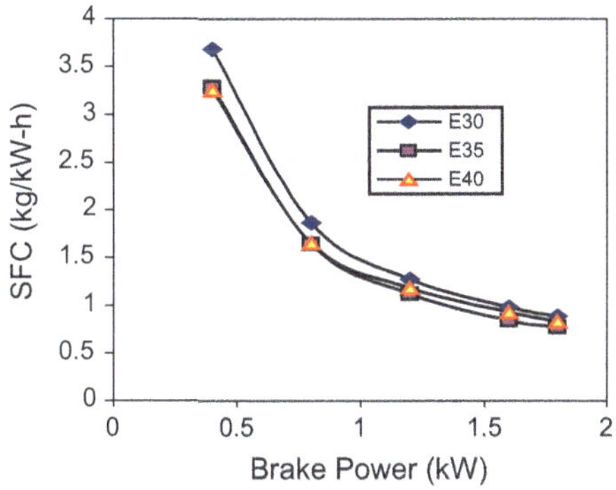


**Fig. 10** Variations of specific fuel consumption (sfc) for different blending ratios at 60 °C with inlet air heating

economical fuel consumption is attained in the same 35%. At the final stage, the temperature was raised to 60 °C and the performance of the engine was analysed. In this case also, 35% blending proved to be better than other blending proportions as shown in Fig. 8. But compared with neat gasoline, efficiency of E35 fuel was reduced by 1% which is shown in the following Fig. 9.

Regarding specific fuel consumption, almost all blending ratios have the closer value which is represented in Fig. 10. It is generally noticed that as the percentage of Bio ethanol increases beyond 25, the thermal efficiency decreases with respect to the base engine. This may due to the fact that as the percentage of Bio ethanol increases the heating value of fuel decreases, which may in turn result in lower thermal efficiency.

## 5 Conclusion

Based on the present study, the following conclusions are drawn. In a conventional engine, gasoline can be replaced by means of Bio ethanol up to 25 vol%. In 25% Bio ethanol blending, the thermal efficiency of the engine is increased with 10.37%, than when it runs with pure gasoline alone. By raising the intake air temperature to 60 °C, 40% of Bio ethanol replacement in gasoline can be achieved and the thermal efficiency of the engine is reduced with 7.59%, when compared with pure gasoline fuel. At this raised Temperature, 35% of blending gives better performance, even though it is less compared with pure gasoline by 1.06%, but the efficiency is higher compared with other blending ratios. From the above data, we can conclude that when the intake air temperature is increased, the replacement of Bio ethanol in gasoline is increased.

## References

1. Abdel Rahman AA, Osman MM (1997) Experimental investigation on varying the compression Ratio of S.I. engine working under different Ethanol-Gasoline fuel blends. J Energy Res 21:31–40
2. Park GY, Kang SH, Lee SW (2016) An experimental study on spray and combustion characteristics based on fuel temperature of direct injection bio-ethanol-gasoline blending fuel. J Mech Sci Technol 30:5239
3. Palmer FH (2013) Vehicle performance of gasoline containing oxygenates—International conference on petroleum based and automotive applications. Institutions of mechanical engineers conference publications (MEP), London, UK
4. Hamdan MAS and Jubran (1986) The effect of ethanol addition on the performance of diesel and gasoline engines. Dirasat 10:110–113
5. Saravanan S, Pitchandi K, Suresh G (2015) An experimental study on premixed charge compression ignition-direct ignition engine fueled with ethanol and gasohol. Alexandria Eng J 54:897–904
6. Yacoab Y, Bata R (1998) The performance and emission characteristics of C1–C5 alcohol—gasoline blends with matched oxygen content in a single—cylinder spark ignition engine. J Power Energy 212:363–379
7. Elfasakhany Ashraf (2015) Investigations on the effects of ethanol–methanol–gasoline blends in a spark-ignition engine: performance and emissions analysis. Eng Sci Techol Int J 18(4):363–379
8. Myung C-L, Lee S-H (1993) Research and development of hyundai flexible fuel vehicles. SAE paper 930330, pp 279–285
9. Alternative fuels for spark ignition engines. Automot Eng (2006) 91:30–33

# An Experimental and Theoretical Study on the Effect of Microstructure on the Tensile Behavior of Polylactide Nano Composites

**Anupam Glorious Lobo, B. K. Sarojini, Sabu Thomas, Adeeb Aman, Mahammad Nawaz and Abdul Mujeeb**

## 1 Introduction

Despite having applications in the packaging industry, 3D printing, biomedical, industrial applications [1], Polylactide is still not extensively used in electrical applications [2]. An ample number of researchers have shown interest on PLA-silica/PLA-clay composites mainly because of enhanced bonding between PLA-silica/PLA-clay. These studies have usually resulted in improving the mechanical attributes and biodegradability of PLA [3, 4]. However, not many works exist on polylactide-nano metal oxide composites [5, 6]. Nano $Al_2O_3$ and nano CuO have been used in other applications considerably enhancing the mechanical and thermal properties [7–9]. The current study focuses on preparing polylactide nano composites in the form of thin films using ultra sonication and solution casting. Nano $Al_2O_3$ and nano CuO have been used as dopants in low concentrations of 1 and 3 mg. The morphological and mechanical properties of PLA nano composites have been studied in order to extend the uses of PLA in

A. G. Lobo (✉) · S. Thomas
School of Chemical Sciences, Mahatma Gandhi University,
Kottayam 686560, Kerala, India
e-mail: anupamlobo5@gmail.com

B. K. Sarojini
Department of Industrial Chemistry, Mangalore University,
Mangaluru 574199, Karnataka, India

A. Aman · M. Nawaz
Department of Mechanical Engineering, P.A. College of Engineering,
Mangaluru 574199, Karnataka, India

A. Mujeeb
Department of Mechanical Engineering, Sahyadri College of Engineering
and Management, Mangaluru, Karnataka, India

 
K. Antony and J. P. Davim (eds.), *Advanced Manufacturing and Materials Science*,
Lecture Notes on Multidisciplinary Industrial Engineering,
https://doi.org/10.1007/978-3-319-76276-0_34

electrical applications. The finite element analysis has also been conducted to compare the tensile and electrical behavior of the nanocomposites in the presence of agglomeration of the nano particles.

## 2 Characterization Techniques

The SEM analysis was performed on an SEM of Carl Zeiss make, Neon 40 Crossbeam, Germany of resolution up to 1.1 nm.

Tensile properties were tested using ASTM D882-02 on Zwick Roell Z020 Universal Testing Machine having a load cell of 100 N, the preload applied was 0.2 N maintaining a test speed of 12.5 mm/min for all the samples. As per ASTM D 882-02, the gauge length of the specimen was 50 mm and at least 5 samples of each type were tested.

Density of the nano composites was determined using the simple principle of rule of mixtures given in Eq. 1.

$$\rho_c = V_{fN}\rho_N + (1 - V_{fN})\rho_p \tag{1}$$

where

$\rho_c$ Density of the nanocomposite (kg/m$^3$)
$\rho_p$ Density of the polymer matrix (kg/m$^3$)
$\rho_N$ Density of the nanoparticle (kg/m$^3$)
$V_{fN}$ Weight fraction of the nano particles

The finite element simulation was conducted on ANSYS® Academic Research Workbench 14.5 using the Transient Structural component system. The thin films were generated as Surface from Sketches using the Young's Modulus of the respective samples. The dimensions of the samples were 50 mm × 50 mm × 3 μm with step end time 19.75 s, velocity of pull 2.083 × 10$^{-4}$ m/s and initial strain

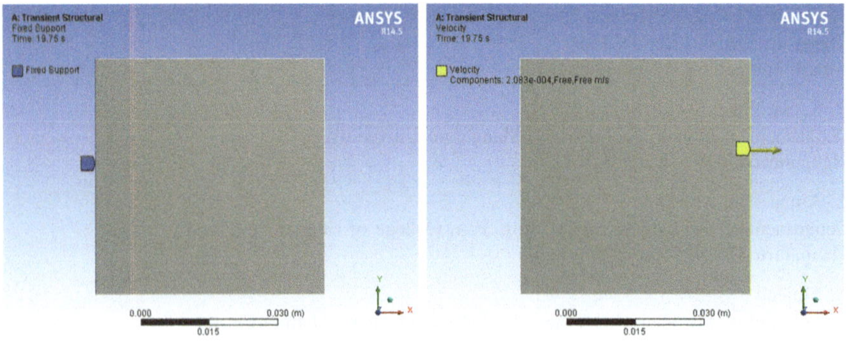


**Fig. 1** Simulation setup of tensile properties of PLA thin films

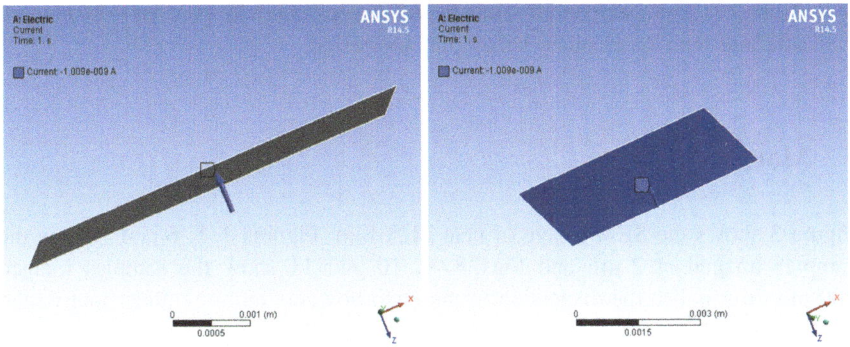


**Fig. 2** Simulation setup of I-V characteristics showing the sample and the direction of applied current

of 0.2 was applied using ANSYS® APDL commands. The initial setup of the simulation is as shown in Fig. 1.

The I-V characteristics were simulated using ANSYS® Academic Research Workbench 14.5 using the Electric component system. The material was generated using the Extrude method the resitivity values [10] of the respective nanocomposite thin films. The values for current were the input and the voltage values were the obtained output. The simulation setup is shown in Fig. 2.

## 3 Materials

Polylactide (PLA) was supplied by Shanghai Guangue Bio-Tech Pvt. Ltd. Shanghai, China. Nano $Al_2O_3$ and nano CuO manufactured by Sigma Aldrich was procured from Durga Labs Mangalore, India.

## 4 Nanocomposite Preparation

To prepare neat films, 15 ml Chloroform was taken along with 0.375 g of PLA and sonicated for about 20 min. This solution was poured into a petridish and placed aside for a day for the film formation. The formed film was then carefully peeled out.

The doped polymer films were made by dispersing PLA in chloroform as mentioned above. The required concentration (1 mg/3 mg) nano powder (nano $Al_2O_3$ or nano CuO) was added in minute amounts at various time intervals. The respective solutions were poured in separate petridishes and kept aside for film formation [6]. The peeled out films were dimensioned as per ASTM standard D882-02 and their mechanical behaviour was tested.

## 5 Morphology

Figure 3 shows the SEM image of neat PLA film. Figures 4, 5, 6 and 7 show the samples imaged at 2 μm and Figs. 8, 9, 10 and 11 show the samples imaged 200 nm to visualize the dispersion of the nano powders in the polymer matrix and to view the agglomeration of the nano powders in the polymer matrix respectively. Figures 8, 9, 10 and 11 reveal that the nano powders have agglomerated. Films doped with 3 mg nano CuO reveal that the nano powders have largely

**Fig. 3** Neat PLA film (100 nm)

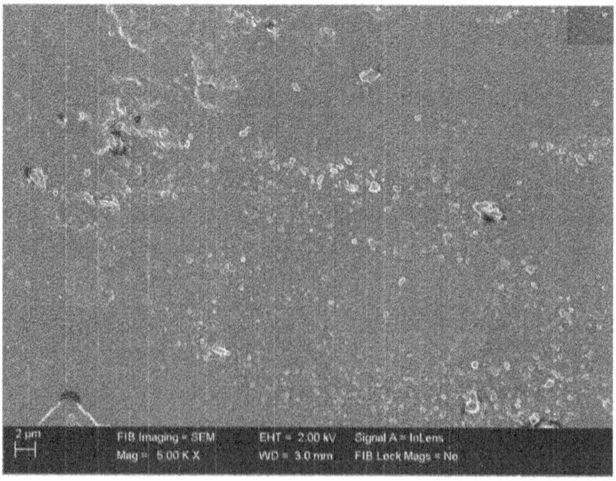


**Fig. 4** PLA+1 mg nano $Al_2O_3$ (2 μm)

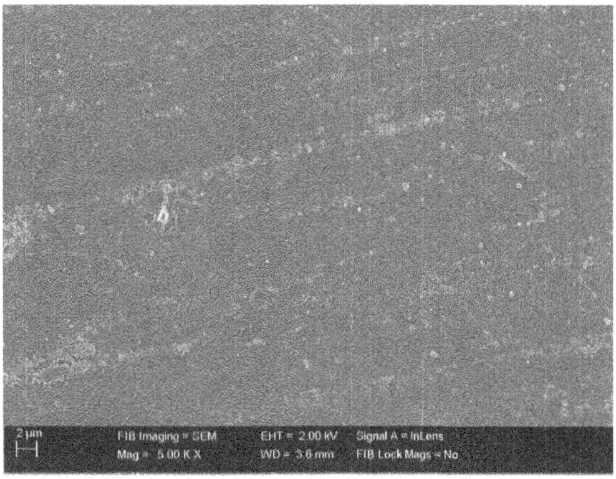


**Fig. 5** PLA+1 mg CuO (2 μm)

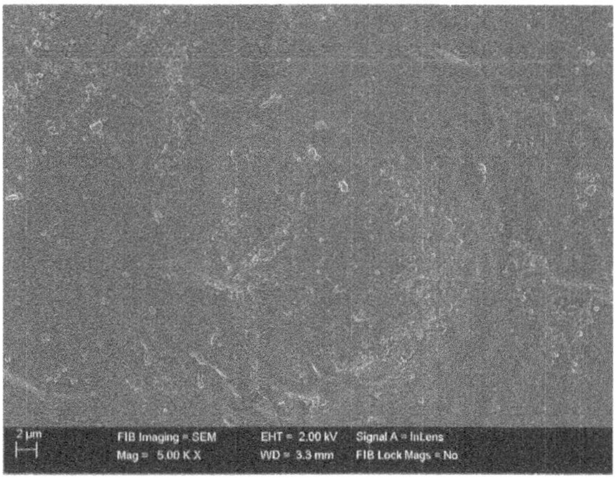


**Fig. 6** PLA+3 mg $Al_2O_3$ (2 μm)

**Fig. 7** PLA+3 mg CuO (2 μm)

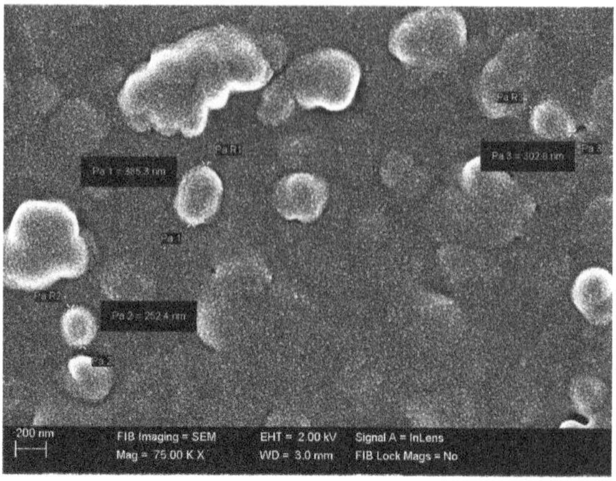


**Fig. 8** PLA+1 mg Nano $Al_2O_3$ (200 nm)

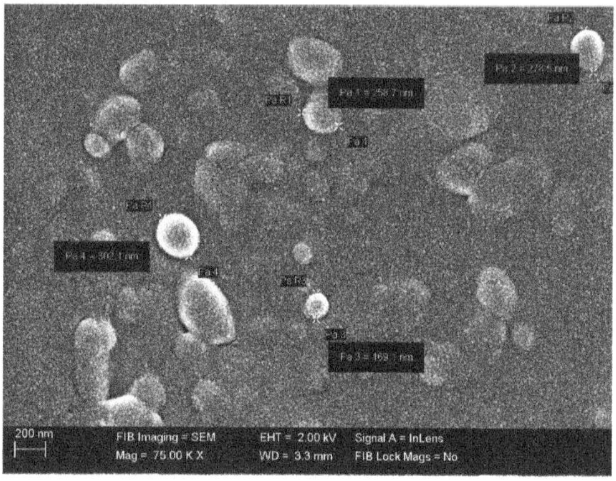


**Fig. 9** PLA+1 mg CuO (200 nm)

agglomerated and appear to be stacked up as in Fig. 9. However, spherical agglomerates are seen in many regions in all doped films. Better distribution of the nano powders is seen in films doped with 3 mg nano $Al_2O_3$.

## 6 Density

Due to very low doping concentrations, the density of the PLA nano composites did not change significantly. Table 1 shows the theoretical densities of the nano composites as calculated from rule of mixtures.

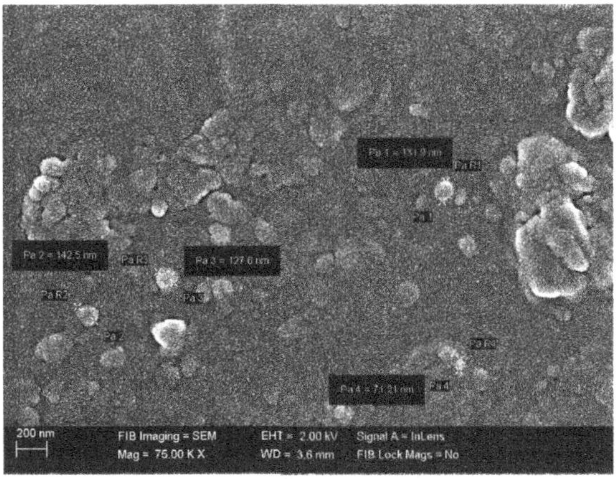


**Fig. 10** PLA+3 mg $Al_2O_3$ (200 nm)

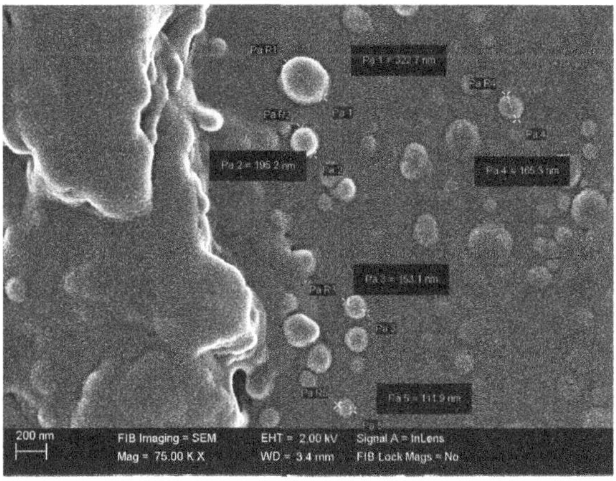


**Fig. 11** PLA+3 mg CuO (200 nm)

**Table 1** Density of neat and doped PLA samples

| Sample | Density (kg/m$^3$) |
|---|---|
| Neat PLA | 1270 |
| PLA+1 mg $Al_2O_3$ | 1272 |
| PLA+3 mg $Al_2O_3$ | 1276 |
| PLA+1 mg CuO | 1273 |
| PLA+3 mg CuO | 1278 |

# 7 Mechanical Properties

From this study it was observed that the mechanical properties of the PLA nanocomposites increased at low doping concentration and with very small variation in density of the nanocomposites in comparison to the neat films.

**Tensile Strength**. The Ultimate Tensile Strength (UTS) and the Tensile Strength at Break (TSB) follow similar trend for nano $Al_2O_3$ and nano CuO doped samples as shown in Fig. 12. The graph shows improvement of tensile strength for all doping concentrations in comparison with the neat PLA film despite the occurrence of agglomeration of nano powders. The tensile strength of PLA doped with nano $Al_2O_3$ and nano CuO is higher than PLA doped with nano titanium dioxide ($TiO_2$) [11]. It was found that the tensile strength of the samples in this study were considerably higher than most of the polymers used as electrical wire insulators [2]. From Fig. 13 it is clear that the finite element simuation values agree with the experimental values.

**Young's Modulus**. The variation of Young's modulus which is shown in Fig. 14. These values are in comparison with the Young's Modulus of PLA-epoxidized palmolien blends [12]. Since the samples doped with 1 mg nano $Al_2O_3$ and are showing highest values for tensile strength and Young's modulus, we may predict that the samples doped with 1 mg nano $Al_2O_3$ may be tougher and harder than the other samples in this study. PLA samples doped with nano $Al_2O_3$ and nano CuO were found to be considerably less stiff than PLA doped with nano $TiO_2$ [11].

**Percentage Elongation (PE)/ Ductility**. The variation of percentage elongation at break (PEB) and at tensile strength (PETS) is as shown in Fig. 15. The highest elongation is exhibited by PLA doped with 1 mg of nano CuO and the least is shown by sample doped with 3 mg of nano CuO. From the FTIR study of poly-lactide doped with nano $Al_2O_3$ and nano CuO it was observed that as the C-O stretch in the region 1000–1150 $cm^{-1}$ increased, the percentage elongation of the

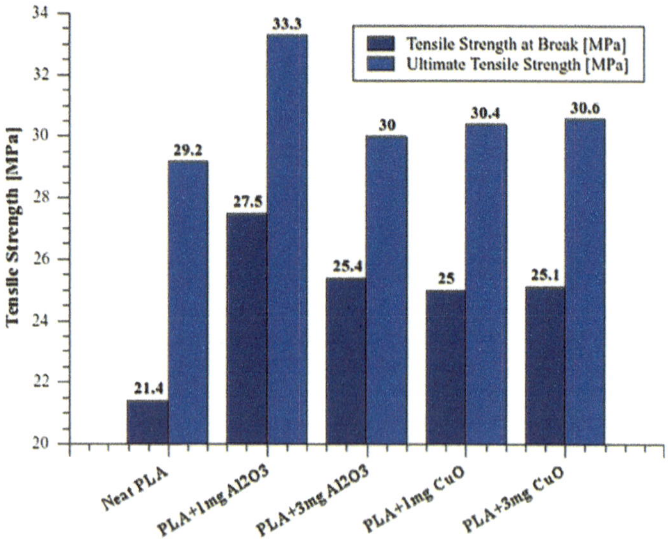


**Fig. 12** Variation of tensile strength

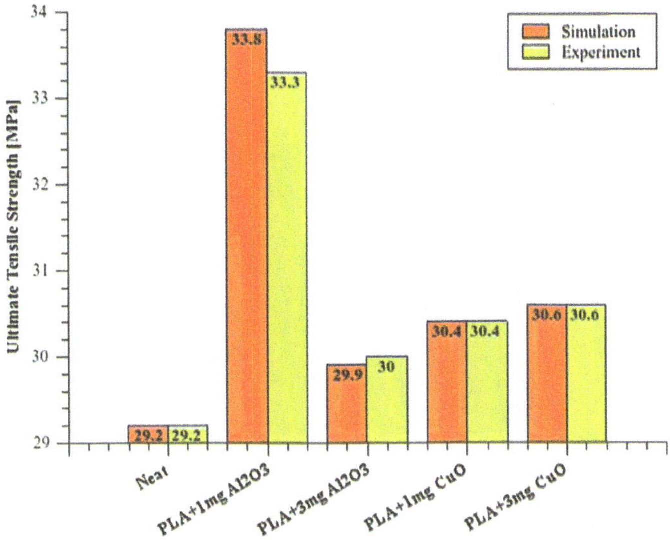


**Fig. 13** Comparsion of simulated and experimental values of tensile strength

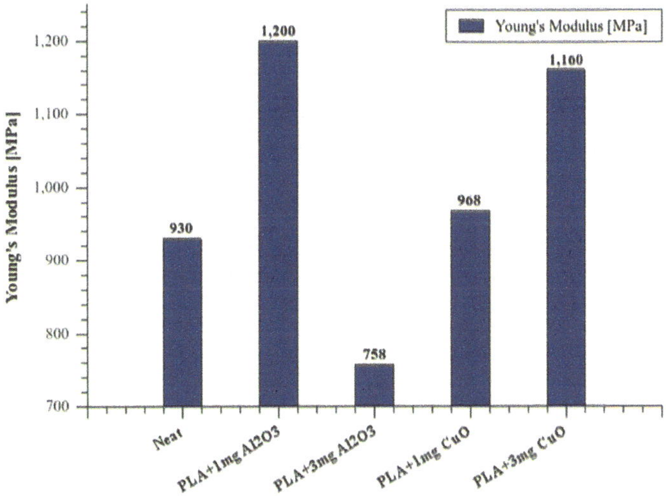


**Fig. 14** Variation of young's modulus

respective samples decreased [7]. The polymer samples in this study have found to exhibit better ductility in contrast to the polylactide grade used as electrical insulator [2]. Figure 16 shows the result of the finite element simulation where it is observed that the values for deformation of the samples show good congruence with the experimental results for most samples.

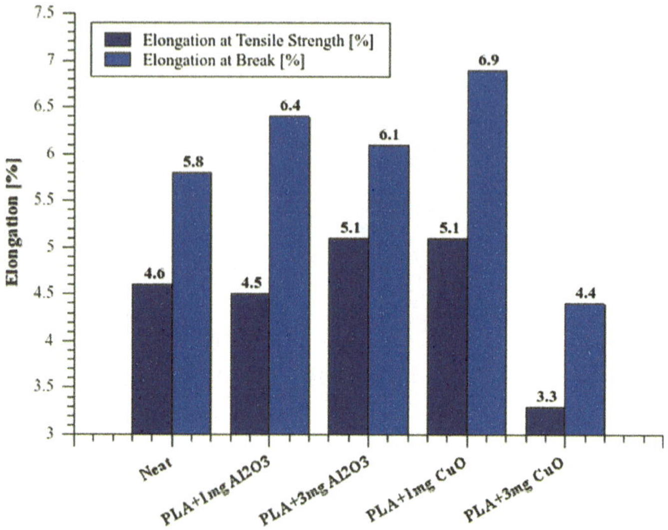


**Fig. 15** Variation of percentage elongation

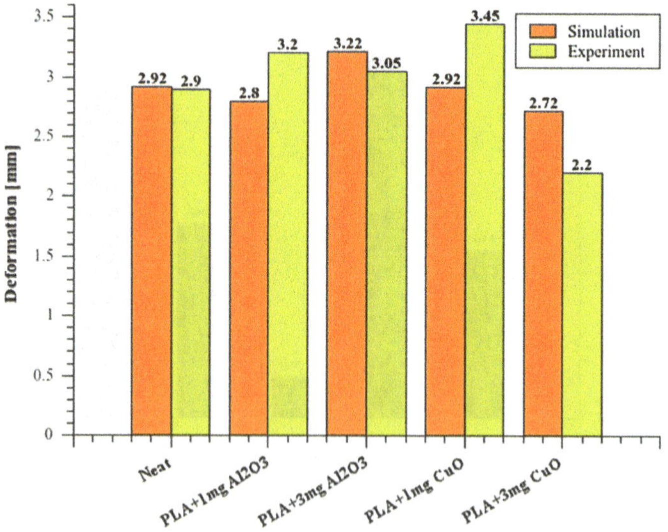


**Fig. 16** Comparsion of simulated and experimental values of deformation

Comparing the variation of the properties of all doped samples with neat polylactide film, it is seen that the values of tensile strength and Young's modulus are the highest for samples doped with 1 mg nano $Al_2O_3$ but at the same time are least for samples doped with 3 mg of the same dopant. Similarly, the values of ductility are highest of 1 mg doping of nano CuO and is least for samples doped

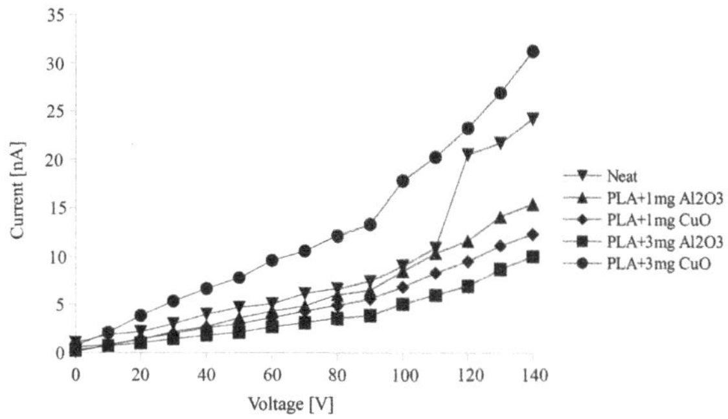


**Fig. 17** I-V characteristics graph obtained from finite element simulation

with 3 mg nano CuO. The sample doped with 1 mg nano $Al_2O_3$ is however having the best values for tensile strength, Young's modulus and ductility. Despite being thermally stable, the sample doped with 1 mg nano $Al_2O_3$ has the lowest value for electrical resistance in comparison with the other samples and hence it can be proposed to be used as a material for semi conducting polymer applications [10].

## 8 Electrical Simulation

The I-V characteristics obtained from the finite element simuation is shown in Fig. 17. Comparing this with the experimental graph [10], the trend in the variation is slightly comparable. However all the values of current and voltage are not comparable with the experimental data. The variations in trend for the experimental and simulated curves of the I-V characteristics were almost comparable but the corresponding experimental and simulated I-V values were distinct at few points.

## 9 Conclusions

1. The PLA nano composites showed 16.1% increase in tensile strength, 28.5% in Young's modulus, 10.9% in ductility even at low concentration doping of 1 and 3 mg and also with very small variation in density.
2. The values of the mechanical properties improved despite the presence of agglomeration of nano powders in the polymer matrix.
3. Enhancement of the tensile strength has been seen for all the doped samples in comparison to the neat samples, 1 mg nano $Al_2O_3$ doped sample displaying highest value of tensile strength and Young's Modulus and relatively high value for ductility.

4. The polymer sample doped with 1 mg nano $Al_2O_3$ proved to be a promising material for semi-conducting polymer applications.
5. The values obtained by simulating the tensile behaviour of the polymer nano composite thin films were in close agreement with the experimental values indicating that ANSYS® can be used to simulate the tensile behavior of polymer nanocomposite thin films.
6. The simulated I-V characteristics curves showed almost similar trend as that of the experimental data.

**Acknowledgements** The authors would like to thank Mr. Rajesh, Centre Head, CADD Centre, Bendoorwell, Mangaluru, Karnataka for providing the facility for simulation on ANSYS® Workbench.

## References

1. Jamshidian M, Tehrany EA, Imran M, Jacquot M, Desobry S (2010) Poly-lactic acid: production, applications, nanocomposites, and release studies. Compr Rev Food Sci Food Saf 9:552–571
2. Nakatsuka T (2011) Polylactic acid-coated cable. Fujikura Rev 40:39–46
3. Ray SS, Yamada K, Okamoto M, Ueda K (2003) New polylactide-layered silicate Nano composites. 2. Concurrent improvements of material properties, biodegradability and melt rheology. Polymer 44(3):857–866
4. Fukushima K, Fina A, Geobaldo F, Venturello GC (2012) Properties of poly(lactic acid) Nano composites based on montmorillonite, sepiolite and zirconium phosphonate. eXPRESS Polym Lett 6(11):914–926
5. Zubair M, Najeeb F, Mohamed M, Saeed A, Ramis MK, Abdul Mujeeb MS, Sarojini Bk (2014) Analysis of chitosan polymer doped with nano $Al_2O_3$ and nano CuO. Am J Polym Sci 4(2):40–45
6. Lobo AG, Mujeeb A, Nawaz M, Adeeb Aman AP, Basith A (2016) A Study on the FTIR spectroscopy of polylactide doped with nano aluminium oxide and nano cupric oxide. Polym Plast Technol Eng 55(16):1713–1719
7. Zheng L-Y, Lau K-T, Zhao L-X, Zhang Y-Q, Hui D (2009) Mechanical and thermal properties of nano-Al2O3/nylon 6 composites. Chem Eng Commun 197(3):343–351
8. Brüggemann D (2013) Nanoporous aluminium oxide membranes as cell interfaces. J Nanomaterials 1–18
9. Suleiman M, Mousa M, Hussein A, Hammouti B, Hadd TB, Warad I (2013) Copper(II)-oxide nanostructures: synthesis, characterizations and their applications-review. J Mater Environ Sci 4(5):792–797
10. Mujeeb A, Lobo AG, Antony AJ, Ramis MK (2017) An experimental study on the thermal properties and electrical properties of polylactide doped with nano aluminium oxide and nano cupric oxide. INAE Lett 2(4):145–151
11. Farhoodi M, Dadashi S, Mousavi SMA, Sotudeh-Gharebagh R, Emam-Djomeh Z, Oromiehie A, Hemmati F (2012) Influence of TiO2 nanoparticle filler on the properties of PET and PLA nanocomposites. Polymer (Korea) 36(6):745–755
12. Giita Silverajah VS, Ibrahim NA, Zainuddin N, Zin WM, Yunus W, Hassan HA (2012) Mechanical, thermal and morphological properties of poly(lactic acid)/epoxidized palm olein blend. Molecules 17(10):11729–11747

# Oxidation and Hot Corrosion Performance of $Al_2O_3$-40%$TiO_2$ Coating on Nickel Based Superalloys at 800 °C

**Naveen Kumar, Vivek Kumar and S. B. Mishra**

## 1 Introduction

The heat transfer pipes in coal-fired boilers, such as tubes for super heaters, economizer, blades in aero-engines and industrial gas turbine are exposed to an aggressive environment [1–3]. Due to hot corrosion many important structured engineering systems may fail in high temperature and hence forth main cause of failure modes of structured components in the high temp section of boilers and gas turbine. Which is the main failure modes of components in the hot sections of gas turbines, boilers etc. [4–6]. As per the properties of superalloys, superalloys have sufficient mechanical strength stability, creep and fatigue resistance [6, 7]. They find largest application in the turbine industry constituting over 40% of the gas turbine weight [7–10]. Although superalloys have adequate mechanical strength but they lack wear, oxidation and hot corrosion resistance [11]. One of the major life limiting factors of material at high temperature is oxidation [12, 13].

While keeping the aim of obtaining required specific or modified properties of the material, the layer of material formed naturally or synthetically or arranged artificially on the surface of an object made of another material and such type of process is known as coating [14]. Thermal spraying is an effective and low cost method to apply thick coatings to change surface properties of the component. Many different types of coatings are using in different aspect of applications widely in industry like including automotive systems, chemical process equipment, boiler components, power generation equipment and, aircraft engines, pulp, paper processing equipment, bridges, rollers, concrete reinforcement, orthopedics and dental,

N. Kumar (✉)
United College of Engineering Research, 211004 Naini Allahabad, India
e-mail: naveenkumar.me06@gmail.com

V. Kumar · S. B. Mishra
Motilal Nehru National Institute of Technology, Allahabad 211004, India
e-mail: 07.viveksrivastava@gmail.com

 
K. Antony and J. P. Davim (eds.), *Advanced Manufacturing and Materials Science*, Lecture Notes on Multidisciplinary Industrial Engineering,
https://doi.org/10.1007/978-3-319-76276-0_35

land based and ship, marine turbine, [15]. As per the desired specification, Detonation Gun (D-Gun) spray one the best suitable choices to get hard, wear resistive & dense coating, Gun spray & High velocity oxy fuel (HVOF) is the best option, among all the commercially available thermal spray [16]. To investigate the role of Detonation Gun spray coating in enriching the properties of surface of substrate to counter the problem of oxidation & hot corrosion etc. is the main objective of the present research work [17]. One of the thermal spray procedure is the Detonation Gun (D-Gun) spraying, which provides an excellent low porosity, good adhesive strength and more compressive residual stress of coating surface [18]. For the sprayed powder, Detonation gun (D-Gun) offers maximum velocity (800–1200 $ms^{-1}$) that is unattainable by the plasma and HVOF condition. The velocity of higher particle during deposition of coating results in desirable characteristics such as low porosity and higher hardness of the coating [19].

In thermal spray industry $Al_2O_3$ alloy with $TiO_2$ ceramics are popular due to easiness of spray along with environment adaptability [20]. In comparison to individual oxides, $Al_2O_3$ and $TiO_2$ have better performance. For lowering porosity in coating, $Al_2O_3$-$TiO_2$ powder contribution helps [21, 22]. The micro hardness of $Al_2O_3$-$TiO_2$ composite coating depends essentially on their composition. It observed that either increase of $TiO_2$ content decrement of $Al_2O_3$ content, the hardness decrement was observed linearly simultaneously [23].

Using the detonation gun spray process (D-Gun), $Al_2O_3$-40$TiO_2$ coating is deposited on superni-600 and superni-718 superalloys. Specimens test for hot corrosion test were taken out in the molten salt environment $Na_2SO_4$+82%$Fe_2(SO_4)_3$ at 800 °C for 50 cycles. Henceforth, the present work has been focused to study $Al_2O_3$-40$TiO_2$ coating on hot corrosion behavior and oxidation of Ni based superalloys. To study $Al_2O_3$-40$TiO_2$ coating and bare superalloys substrates, we used thermo gravimetric technique. To characterizing the oxidation and hot corrosion product of the coated and bare samples we use SEM.

# 2 Experimental Procedure

## *2.1 Development Coatings*

### 2.1.1 Substrate Material, Coating Powders

Superalloys Ni based namely as Supeni 600 and Superni 718 are selected. The superalloys Superni718 and Superni 600 were purchased through M/S Mishra Dhatu Nigam Ltd., Hyderabad, India in the rolled sheet form. In table, the compositions (chemical) of Superni 718 & Superni 600 stainless steels are shown. Each specimen measured approximate (20 × 15 × 5) mm were cut from the rolled sheet and polished by using 180, 220, 320, 400, and 600 grade SiC emery papers (Table 1).

**Table 1** Chemical compositions of Superni718 and Superni600 superalloys

| Midhani grade | | | | | | Composition (Chemical -wt%) | | | | | |
|---|---|---|---|---|---|---|---|---|---|---|---|
| Materials | C | Mn | Si | S | P | Cu | Cr | Fe | Ti | Al | Ni |
| Superni718 | 0.0047 | 0.01 | 0.02 | 0.002 | – | 0.010 | 17.78 | 19.56 | – | – | Bal |
| Superni600 | 0.08 | 0.03 | 0.03 | >0.005 | >0.005 | >0.01 | 20.19 | 0.10 | 0.30 | 0.1 | Bal |

**Table 2** Characteristics of $Al_2O_3$-$40TiO_2$ coating powders

| Powder | Make and commercial code | Chemical composition (wt%) | Morphology | Particle shape | Powder size |
|---|---|---|---|---|---|
| $Al_2O_3$-$40TiO_2$ Coating | H.C. StarckAmperit 740.001 | $SiO_2$ max. 0.1%<br>$Fe_2O_3$ max. 0.3%<br>$Na_2O$ max. 0.3%<br>$Al_2O_3$ balance | Fused | Regular and angular block | 45/22 μm |

The samples were blasted using alumina powder (grit 20) prior to coating. Commercially available $Al_2O_3$-$40TiO_2$ powder (H.C.Starck, Amperit 740.001) with its grit (particle size) 22–45 μm was used coat on the superalloys by D-Gun spray process. In Table 2, the Characteristics of $Al_2O_3$-$40TiO_2$ coating powders are given.

### 2.1.2 Detonation-Gun Technique

Detonation gun spray was ready to apply $Al_2O_3$-$40TiO_2$ coating on the superalloys substrates at SVX Powder M Surface Engg. (P.) Ltd, New Delhi (India). In Table 3, all the process parameters, including the spray distance, were kept constant throughout coating process. The process parameters are shown in.

## 2.2 *Oxidation Test*

The oxidation test is cyclic which is studied & performed in air environment for up to 50 cycles. In a Kanthol wire tube furnace, each cycle is consisting of 1 h heating at 800 °C followed by 20 min cooling at room temperature. As the alumina boat is kept for specimens. Alumina boats were preheated for 8 h at 800 °C so as to remove manufacturing defect present in it. It is preassumed that the weight of the alumina boats will remain constant during experiment all cyclic process throughout the experiment.

**Table 3** Parameter used in detonation gun spray

| Parameters | $Al_2O_3$-40$TiO_2$ coating |
|---|---|
| Nitrogen flow rate ($N_2$) | 800 SLPH |
| Pressure | 0.4 MPa |
| Acetylene flow rate ($C_2H_2$) | 1920 SLPH |
| Pressure | 0.14 MPa |
| Oxygen flow rate ($O_2$) | 4800 SLPH |
| Pressure | 0.2 MPa |
| Power | 450 VA |
| Fire rate | 1–10 Hz |
| Spray angle | 90° |
| Spray distance | 200 mm |
| Coating thickness (average) | 250 µm |

### *2.3 Hot Corrosion Test*

For 50 cycles under cyclic conditions, it is performed the Hot corrosion studies in a molten salt $Na_2SO_4$-82%$Fe_2$ $(SO_4)_3$ through experiment. For removal of dirt the specimen goes for wash in acetone and a salt of $Na_2SO_4$-82%$Fe_2$ $(SO_4)_3$is thoroughly mixed with distilled water and then specimen is heated in an oven at about 250 °C. It is essential requirement for proper adhesion of the salt layer through heating of the specimen, as the salt of $Na_2SO_4$-82%$Fe_2$ $(SO_4)_3$is properly mixed with distilled water and therefore a layer of this salt mixture with the help of a camel hair brush was applied uniformly on the warm polished specimens. The amount of salt coating was kept in the range of 4.0–5.0 mg/cm$^2$. Both alumina boats & salt coated specimen were then kept in the oven at 100 °C for at least 3–4 h and then they were again weighed before exposing to hot corrosion tests in the Kanthol wire tube furnace. With the help of Electronic Weight Balance Machine (with accuracy in 1 mg), both the weight of alumina boats and specimens was weighed together at end of each cycle during hot carrion process. During the process the spalled scale was also included during the measurements of weight change to find out the total rate of corrosion. Through weight change measurements of the bare and coated specimen, the rate of corrosion was calculated. Using the SEM for surface, the samples after corrosion tests were analyzed.

## 3 Result

### *3.1 Visual Observations for Oxidation*

In case of Superni718 superalloy Fig. 1a, the colour of Superni718 was blackish colour up to 5th and after that gradually converted to dark black colour and in case of coated Superni718 superalloys Fig. 1b, the colour of Superni718 was blackish

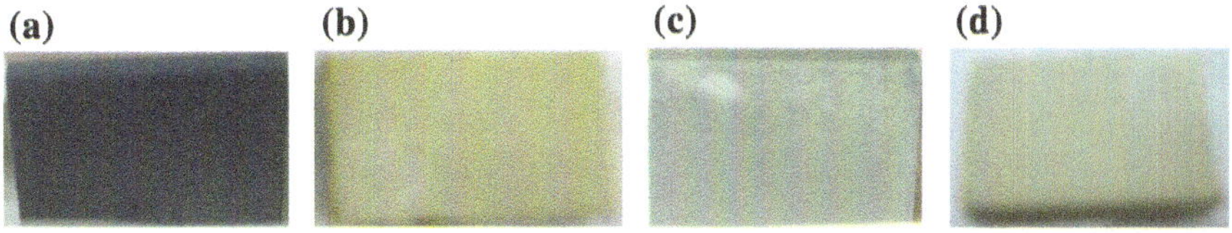


**Fig. 1** Surface macrographs for **a** Superni718 **b** Coated Superni718 and **c** Superni600 **d** Coated Superni600, oxidation in environment at 800 ° after 50 cycles

colour up to 8th and after that gradually converted to yellowish colour. The visual observations of the specimens were made of the end of each cycle. The colour of Superni600 super alloy was light blackish colour up to 10th cycles and after that gradually converts in dark greenish colour. The colour of coated Superni-600was light blackish colour up to 12th cycle, after that gradually convert in dark light yellowish.

## 3.2 *Cyclic Oxidation in Air*

Figure 2 shows the weight gain/unit area for the both the alumina boat as well as coated superalloys under cyclic conditions in the presence of environment. Figure 2 indicates that the weight gain in compare to bare superalloys in environment is higher than the weight gain of the coated superalloys. $Al_2O_3$-40$TiO_2$ Superni 600 coated superalloy has shown a slightly lower weight gain as compared to $Al_2O_3$-40$TiO_2$ Superni 718 coated superalloy.

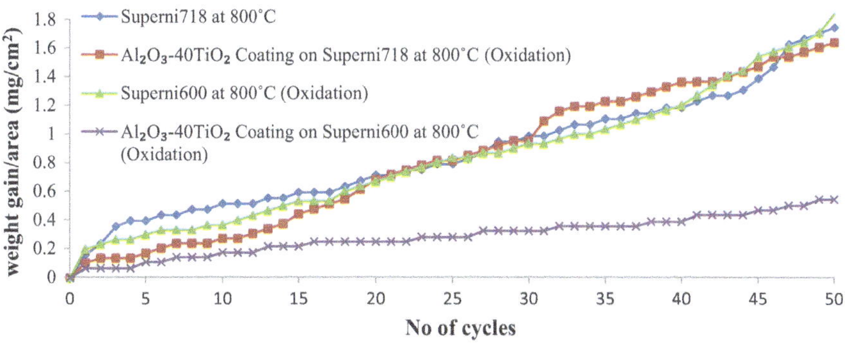


**Fig. 2** Weight gain/area versus number of cycles plot for $Al_2O_3$-40$TiO_2$ coated and bare superalloys at 800 °C for 50 cycles

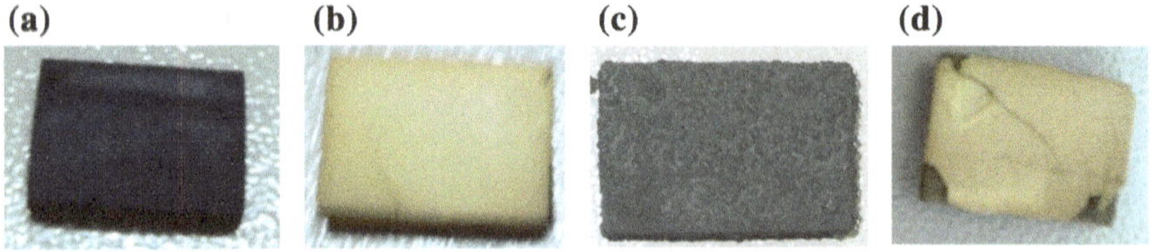


**Fig. 3** Surface macrographs for **a** Superni718 **b** Coated Superni718 and **c** Superni600 **d** Coated Superni600 after 50 cycles exposure to molten salt $Na_2SO_4$+82%$Fe_2(SO_4)_3$ environment at 800 °C

## *3.3 Visual Observations for Hot Corrosion*

In case of Superni718 superalloy Fig. 1a, spalling was observed after the completion of 6th cycle. The colour of Superni718 was blackish colour up to 8th and after that gradually converted to dark black colour and in case of coated Superni718 superalloys Fig. 1b, the colour of Superni718 was blackish colour up to 10th and after that gradually converted to yellowish colour. At end of each cycle, the visual observations of the specimen were made. The colour of Superni600 super alloy was light blackish colour up to 7th cycles and after that gradually converts in dark blackish colour. The colour of coated Superni600 was light blackish colour up to 9th cycle, after that gradually convert in dark yellowish (Fig. 3).

## *3.4 Cyclic Hot Corrosion in Molten Salt*

Figure 4 shows the weight gain/unit area for the bare as well as coated superalloys in the presence of molten salt $Na_2SO_4$-82%$Fe_2(SO_4)_3$ environment under cyclic conditions Fig. 4 indicate that the weight gain of the coated superalloys are lower as compared to those of bare superalloys in the molten salt environment. $Al_2O_3$-40$TiO_2$ Superni600 has shown a slightly higher weight gain as compared to $Al_2O_3$-40$TiO_2$ coated Superni718 superalloy.

Figure 5 shows the total weight gain/unit area in both cases of bare as well as coated superalloys.

Figure 6 shows the relationship between the weight gain square ($mg^2/cm^4$) versus time (number of cycles) plots are plotted to establish the rate law for the hot corrosion. Graph shows (from observation) that the coating follows a nearly parabolic rate of flow. The parabolic rate (Fig. 6) constant *Kp* was calculated by a linear least square algorithm to a function in the form of $(W/A)^2 = Kp\ t$, where W/A is the weight gain per unit surface area ($mg/cm^2$) and 't' indicates the number of cycles, represents the time of exposure [24].

The parabolic rate constant *Kp* for both superalloys is given in Table 4.

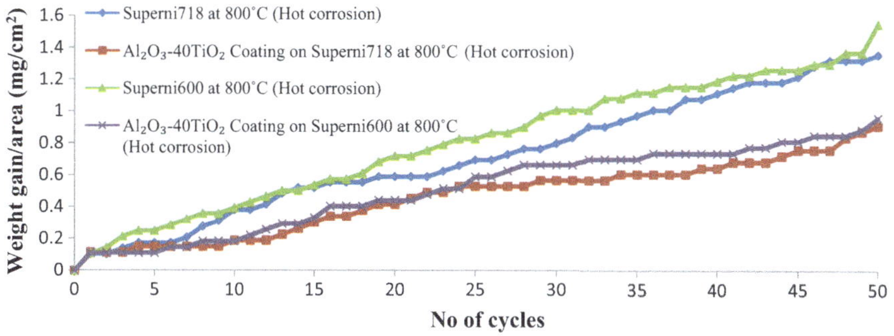


**Fig. 4** Weight gain/area versus number of cycles plot for $Al_2O_3$-40$TiO_2$ coated and bare superalloys subjected to $Na_2SO_4$-82%$Fe_2(SO_4)_3$ at 800 °C for 50 cycles

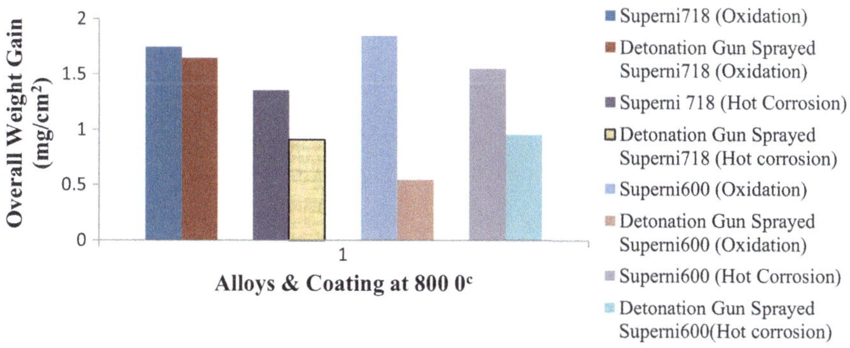


**Fig. 5** Bar chart showing overall weight gain per unit area for bare and $Al_2O_3$-40$TiO_2$ coated superalloys subjected to $Na_2SO_4$-82%$Fe_2(SO_4)_3$ at 800 °C for 50 cycles

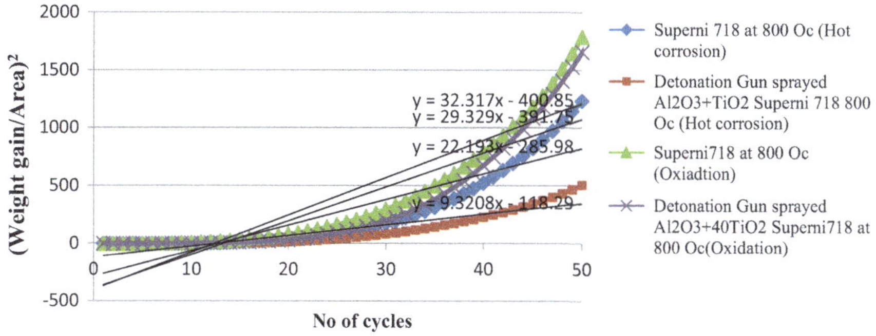


**Fig. 6** (Cumulative weight gain/area)$^2$ versus number of cycles for bare and $Al_2O_3$-40$TiO_2$ coated superalloys subjected to $Na_2SO_4$-82%$Fe_2(SO_4)_3$ at 800 °C for 50 cycles

**Table 4** The parabolic rate constant *Kp* for both superalloys

| S. No | Substrate | *Kp* Values ($\times 10^{-10}$ $g^2$ $cm^{-4}$ $s^{-1}$) |
|---|---|---|
| 1 | Superni718 | 66.376 |
| 2 | CoatedSuperni718 | 39.448 |
| 3 | Superni600 | 28.995 |
| 4 | Coated Superni600 | 6.724 |

## 3.5 SEM Analysis

The SEM micrographs of $Al_2O_3$-40$TiO_3$ coatings and uncoated superalloys after exposure to hot corrosion environment as given in Fig. 7. The microstructure consists of uneven surface. Almost each particle have deformed significantly. Fig (a) reveals that the microstructure is non-uniform and some unmelted particles can

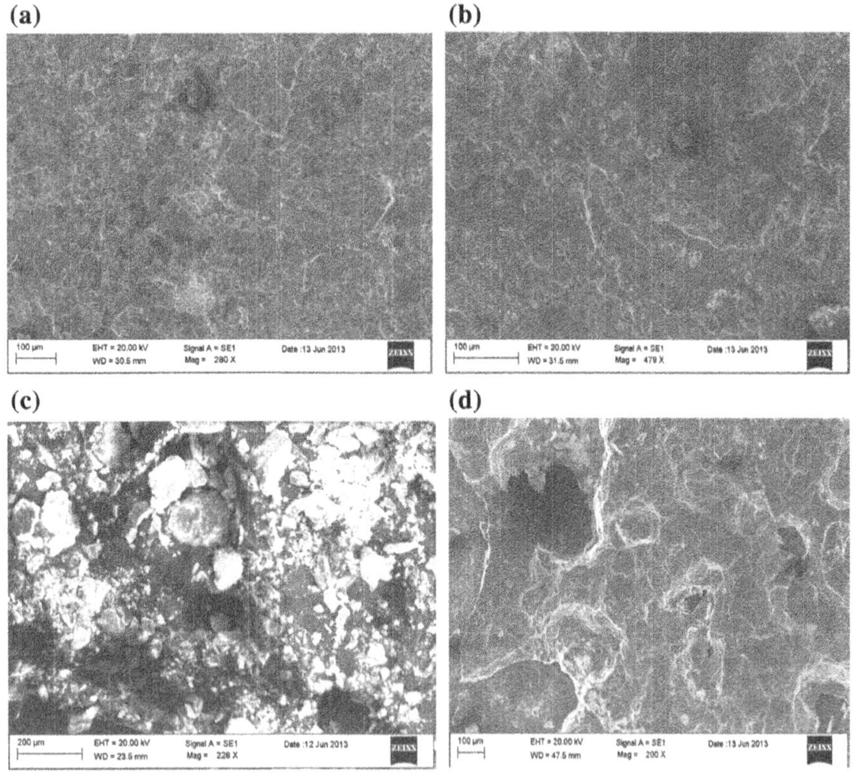


**Fig. 7** SEM micrographs of **a** $Al_2O_3$-40$TiO_2$ coated Superni 718 **b** $Al_2O_3$-40$TiO_2$ coated Superni 600 **c** Superni 718 **d** Superni 600 superalloys subjected to $Na_2SO_4$-82%$Fe_2(SO_4)_3$ at 800 °C for 50 cycles

be seen over the surface, in figure (b) the surface is uniform and some unmelted particles are present in the surface. The grain boundaries are distinctly seen in both the figures. Voids can be seen in figure (c) & (d) both.

## 4 Conclusions

In the present work, hot corrosion behaviour of Detonation Gun Sprayed $Al_2O_3$-40$TiO_2$ coatings on Superni 718 and Superni 600 superalloys in aggressive environment of $Na_2SO_4$+82%$Fe_2(SO_4)_3$ at 800 °C has been investigated and the following conclusions are made.

1. By experiment; it is shown some spalling and peeling of scale and mass gain was relatively more in bare superalloys.
2. The parabolic rate constants of bare superalloys are higher than the $Al_2O_3$-40$TiO_2$ coated superalloys subjected to cyclic hot corrosion in molten salt environment indicating the protective behaviour of the Detonation Gun (D-Gun) sprayed $Al_2O_3$-40$TiO_2$ coating. The values of parabolic rate constant *Kp* for coated Superni 718 were obtained to be 39.448 and for coated Superni 600 were obtained to be 6.724.
3. The hot corrosion resistance of $Al_2O_3$-40$TiO_2$ coating on Superni 718 was inferior as compared to that of Superni 600 superalloys in the given molten salt environment.
4. The Detonation Gun Spray process could be used successfully to deposit $Al_2O_3$-40$TiO_2$ coating on different superalloys, the Detonation Gun Sprayed $Al_2O_3$-40$TiO_2$ coating improved the hot corrosion behaviour in aggressive environment of $Na_2SO_4$-82%$Fe_2(SO_4)_3$ at 800 °C as Compared to bare superalloys.

## References

1. Ananthapadmanabhan PV, Thiyagarajan TK, Sreekumar KP, Satpute RU, Venkatramani N, Ramachandran K (2003) Co-sprayingofalumina–titania: correlation of coating composition and properties with particle behavior in the plasma jet. Surf Coat Technol 168(2–3):231–240
2. Pieraggi B, Dabosi F (1987) Mater Corros 38:584–590
3. Burakowski T, Wierzchon T (1999) Surface engineering of metals, principles, equipment, technology. CRC Press, Boca Raton, Florida
4. Buta V (2008) Performance of plasma sprayed nano structured and conventional coatings. J ACS 44(2):56–62
5. Fervel V, Normand B, Coddet C (1999) Tribological behaviour of plasma sprayed $Al_2O_3$ – based cermet coating. Wear 230(1):70–77
6. Singh H, Puri D, Prakash S (2007) Advance material science 16:27–50
7. Huang XF, Sun HR, Guan ZQHu (2006) Oxid Met 65:207–222
8. Du H et al (2005) Influence of process variables on the qualities of D-Gun sprayed WC–Co coatings. Mater Sci Eng A 408:202–210

9. Stringer J (1998) Coatings in the electricity supply industry, opportunities for the future surface and coatings in electricity supply industries in past and present. 108/109:1–9
10. Zhang JS, Hu ZQ, Murata Y, Morinaga M, Yukawa N (1993) Met Trans A 24A:2443–2450
11. Rapp RA, Zhang YS (1994) Corros Sci 47:67–72
12. Rapp RA (2002) Corros Sci 44:209–211
13. Mishra SB, Prakash S, Chandra K (2006) Wear 260:422–432
14. Sidky PS, Hocking MG (1999) Review of inorganic coatings and coating processes for reducing wear and corrosion. Br Corros J 34:171–183
15. Buta S et al (2005) Mechanical and metallurgical properties of plasma sprayed and laser remelted Ni–20Cr and Stellite-6 coatings. J Mater Process Technol 159:347–355
16. Esmaeili S, Engler-Pinto CC Jr, Ilschner B, Rezai-Aria F (1995) Scripta Metall Mater 32:1777–1781
17. Kamal S, Jayaganthan R, Prakash S (2010) Bull Mater Sci 33:299–306
18. Kamal S, Jayaganthan R, Prakash S (2011) Hot corrosion Studies of detonation- gun-sprayed NiCrAlY+0.4 wt%$CeO_2$ coatings on superalloys in molten salt environment. J Mater Eng Perform 20(6):1068–1077
19. Sidhu TS, Agarwal RD, Prakash S (2005) Surf Coat Technol 198:441–446
20. Sidhu TS et al (2006) Characterizations of HVOF sprayed NiCrBSi coatings on Ni and Fe based Superalloys and evaluation of cyclic oxidation behaviour of some Ni-based superalloys in molten salt environment. Thin Solid Films 515:95–105
21. Kai T et al (2009) Behaviors of HVAF-sprayed conventional and nano structured NiCrC coatings oxidation and hot corrosion. Trans Nonferrous Met Soc China 19:1151–1160
22. Tomazek R, Pawlowski L, Zdanowski J, Grimblot J, Laureyns J (2004) Microstructural transformation of $TiO_2$, $Al_2O_3$-13$TiO_2$ and $Al_2O_3$-40$TiO_2$ at plasma spraying and laser engraving. Surf Coat Technol 185(2–3):137–149
23. Varacalle DJ Jr, Herman H, Bancke GA, RiggsII WL (1992) Vacuum plasma sprayed alumina titania coatings. Surf Coat Technol 54/55(1):19–24
24. Znamriowski Z et al (2004) Low microscopic field electron emission from surface of plasma sprayed and laser engraved $TiO_2$, $Al_2O_3$+13$TiO_2$, and $Al_2O_3$+40$TiO_2$ coatings. Surf Coat Technol 187(1):6–37

# Numerical Simulation of Crack Propagation in Double Torsion Test

**Karuppasamy Pandian Marimuthu, Kwangmin Lee and Hyungyil Lee**

## 1 Introduction

In the field of material science research, advanced and reliable testing techniques are required for material property evaluation of new and existing materials to improve the performance and reliability of end products. Double torsion (DT) test is a powerful, versatile, and relatively simple technique for material property evaluation in brittle materials. The basic configuration of DT test comprises a rectangular plate with a crack or a pre-notch at one end, where symmetric four-point loadings is applied. Whereas the other end of plate is simply supported at two points. As a result, the two plate halves are subject to equal *torsional* deformations due to symmetric loadings. From the bending moment, the resulting tensile stresses cause the crack propagation at the (pre-) crack tip.

Outwater and Gerry [1] applied the DT test to study subcritical crack growth (SCG) in soda-lime glass plates. Afterwards, the double torsion test has attracted significant interest to further develop the methods to understand the fracture in brittle materials and to evaluate the fracture toughness of tested materials [2–8]. So far, the DT testing methodology is applied to characterize the fracture behavior of a wide range of materials such as highly brittle materials [9], ceramics (e.g. silicon nitride [10], silicon carbides [11], glasses [12, 13], piezoelectric materials [14] and fuel cell materials [15]. DT test is also suitable to evaluate the fracture characteristics of opaque and non-reflective materials in which the crack length measurements might be difficult to perform [8]. In addition, mode I stress intensity factor $K_I$ is independent of crack length, which is one of the salient features of DT test, and only function of applied load for a range of crack lengths in the tested specimen [4]. However, the DT testing technique is not yet standardized (test procedure and

K. P. Marimuthu · K. Lee · H. Lee (✉)
Department of Mechanical Engineering, Sogang University,
Seoul 04107, Republic of Korea
e-mail: hylee@sogang.ac.kr



K. Antony and J. P. Davim (eds.), *Advanced Manufacturing and Materials Science*,
Lecture Notes on Multidisciplinary Industrial Engineering,
https://doi.org/10.1007/978-3-319-76276-0_36

specimen geometry) and therefore the results from this method should be interpreted with appropriate assumptions and clarifications [8].

For better use of DT testing technique, several unresolved questions about the specimen dimensions and assumptions such as the effect of (i) deformation of un-cracked part in the specimen, (ii) crack length and specimen dimension, (iii) the exact crack-front shape on stress-intensity factor calculations and (iv) linear compliance-crack length relationship need to be answered. Several studies including finite element method [7, 16–18], X-ray computed tomography and digital volume correlation [19] have been performed, and empirical correction factors were proposed to tackle the above-mentioned questions; however, these studies are not fully sufficient to obtain general conclusions. By means of extended finite element method (XFEM), the common hypothesis about the stress intensity calculation and crack propagation can be studied in detail. To the authors best knowledge, there is no numerical studies available in the literature to simulate the actual crack propagation in DT test.

In this present work, for better understanding of the cracking in DT test, XFEM is extensively used to simulate the crack propagation in DT fracture test and an attempt is made to tackle some of the above mentioned unresolved questions such as crack length independency of stress intensity factor (SIF), and crack propagation. The shape and direction of propagation crack is observed by assigning the material properties of silicon carbide (SiC) to the XFE model. The effect of initial crack parameters and friction on the crack-front shape are also investigated. In addition, the effect of exact crack-front shape and initial crack length on stress-intensity factor calculations are explained in detail Finally, this study provides suggestions to standardize the DT testing technique for fracture toughness evaluation.

## 2 Analytical Approaches

Based on the linear elastic fracture mechanics (LEFM), the stress intensity factor (SIF) $K_I$ [$= (E'G)^{1/2}$] for plane strain [$E' = E/(1-\nu^2)$] and plane stress ($E' = E$) conditions in double torsion test can be expressed as function of the applied load $P$ as follows [3]:

$$K_I = PW_m\left(\frac{3}{Wt^4(1-\nu)\psi}\right)^{1/2} \quad \text{for plane strain} \tag{1a}$$

$$K_I = PW_m\left(\frac{3(1+\nu)}{Wt^4\psi}\right)^{1/2} \quad \text{for plane stress} \tag{1b}$$

where $E$ is the Young's modulus and $\nu$ is the Poisson's ratio of the specimen; $W$, $t$ and $W_m$ are width, thickness, and moment arm, respectively. Finite beam thickness correction factor $\psi$ is experimentally validated in glass ceramic specimens [20]. As per Eqs. (1a, 1b), SIF is independent of crack length $a$; however, this independency is limited to a range of $0.25 < a/L < 0.75$ ratio [4]; $L$ is the length of the specimen. By substituting a critical load at failure ($P_c$) in Eqs. (1a, 1b), the fractures toughness of the tested material can be evaluated from DT fracture tests [9].

Equations (1a, 1b) is analytically expressed based on the major assumptions involved in the DT testing. They limit the fracture toughness values from DT test under certain circumstances. Equations (1a, 1b) assumes constant $W_m$; however, as the rectangular bar deflects, $W_m$ decreases as a result of changes in the contact points between the specimen and loading balls. Leevers [21] corrected this decrease in $W_m$ as $W_m|_{\text{LDC}}$ by proving a large deflection correction (LDC) factor. For load relaxation test, Chevalier et al. [5] proposed a correction to correct the effects from the deflection of un-cracked portion of the specimen. With the three-dimensional FE analyses including contact between the crack faces, Ciccotti et al. [18, 22] provided correction factors to incorporate the experimental variables *i.e.* crack dimensions, notch length and specimen geometry. Based on the analytical, empirical and FE derivations, these corrections aim to force the geometry to be crack length independent [9]. Becker et al. [9] used FE analyses and digital image correlation (DIC) to examine the aforementioned assumptions, and concluded that the SIF is dependent on crack length for considered specimen configurations. For the geometries studied, the calculated SIF deviates up to 10% when Chevaliers (overestimates) and Ciccotti (underestimates) corrections are used; these corrections enable a reproducible fracture toughness [9]. In this study, SIF at the crack-tip is calculated using XFEM and compared with analytical solutions. Stationary crack XFE analyses are carried out by introducing initial damages of various length and shape in the pre-notched specimens.

## 3 Extended Finite Element Modelling

Three-dimensional finite element (FE) model of double torsion specimen with length $L$, width $W$ and thickness $t$ is created by considering pre-notch with length $a_n$ in Abaqus/standard [23]. The FE model is then meshed with 81410 continuum 8-node brick elements (C3D8) and 149336 nodes as shown in Fig. 1. Center of the specimen, where the crack propagation is expected to occur, is refined with finer elements, and enriched with discontinuous functions for crack propagation analysis in extended finite element method. Since, Young's modulus of supporting and loading balls is usually much higher than that of the material to be tested, spherical balls with diameter $D = 5$ mm are modeled with rigid analytical surfaces. Fixed boundary conditions (BCs) are imposed to the supporting balls whereas the loading balls are only allowed to move in the $z$-direction. Symmetry about $x$-plane (XSYMM) is applied to the nodes at the back surface, while a single node each (right, left) side of the specimen is restricted to move in the $y$-direction.

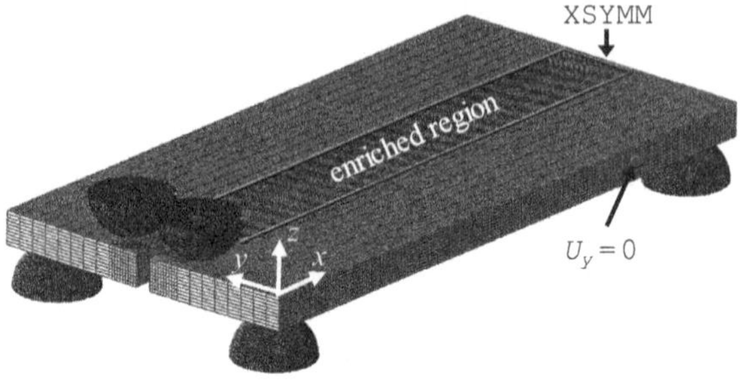


**Fig. 1** XFE model of double-torsion test specimen

Material with linear elastic behavior is assigned to the FE model of the specimen; the material properties of silicon carbide are considered in this study. Damage is assumed to obey bilinear continuum traction-separation law, which is governed by a damage-initiating stress threshold $\sigma_c$ and the fracture energy $\Gamma$. For the crack propagation analyses in XFEM, the enrichment option is activated. The contact between the balls and specimen is assumed to have surface-to-surface contact and the friction coefficient is taken as $f = 0.1$. All simulations are performed with a prescribed maximum displacement $h_{\max}$ of loading balls. To observe the effect of crack front shape on crack propagation, XFE analyses are performed by introducing initial damages with the shape of straight/curved crack front at the pre-notch front; the initial damages are included through the `INITIAL CONDITIONS` option in Abaqus/standard.

# 4 Results and Discussion

The crack initiation and propagation in double-torsion test is numerically simulated in XFEM by considering material properties of SiC ($E = 410$ GPa, $\nu = 0.164$, $\sigma_c = 800$ MPa, $\Gamma = 62.5$ N/m) [24]. Before the damage initiation, $P$ increases linearly with $h$; right after the damage initiation, the linearity of $P - h$ curve disappears at critical load $P_c$ as shown Fig. 2a. The pre-notch length $a_n$ has significant effect on $P_c$. As the crack start to grow, due to increase in compliance, the load reaches almost plateau without any sudden load drop. In experiments, depends on the loading rate, sudden load drop can be expected due to full specimen cracking resulting from crack propagation in high velocity. The effect of friction on the critical load and the crack propagation is analyzed by assuming frictionless contact between the contact surfaces of specimen and supporting balls. Comparison of XFE load-displacement curves between friction ($f = 0.1$) and frictionless ($f = 0$) case shows that the effect of friction on $P - h$ curve is negligible as shown in Fig. 2b. Similarly, the effect of friction on crack shape and length is also negligible as the stress distribution near the crack initiation point is not affected by friction deformation around the contact region.

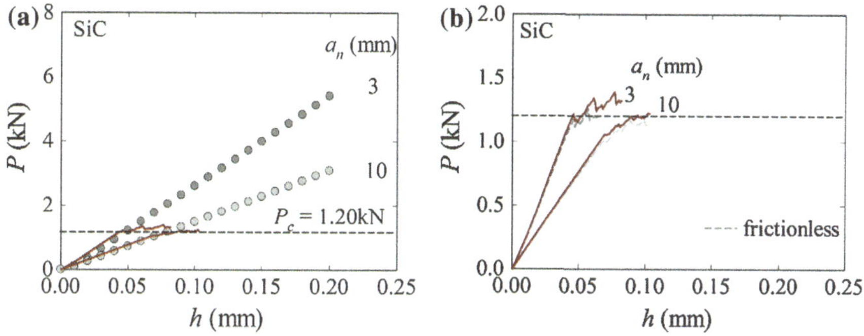


**Fig. 2** Comparison of XFE load-displacement curves from SiC specimens with (**a**) different $a_n$ (=3, 10 mm) and (**b**) with and without friction between the contact surfaces

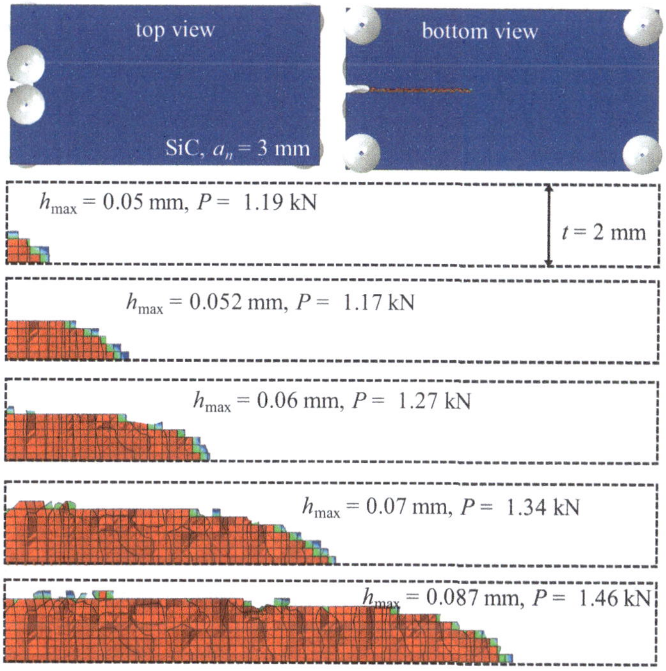


**Fig. 3** Plane view of propagated crack in double-torsion test specimen at various $h_{max}$. Formation of curved crack propagation is evident from XFEM analyses

Since the symmetry of loading confirms the mode I crack opening condition, the crack propagates along the center line of the specimen (Fig. 3). In XFE analysis, this is ensured by carefully applying boundary conditions to FE model. Owing to the different stress state between bottom (loaded in tension) and top (loaded in compression) surfaces, damage initiates at the bottom surface of the specimen. As

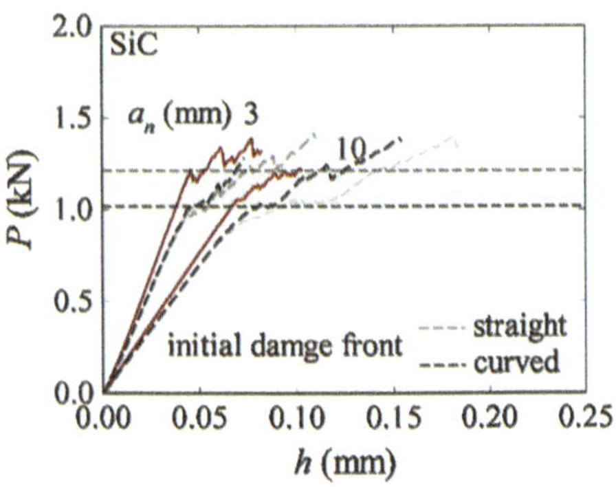


**Fig. 4** Comparison of XFE load-displacement curves from silicon carbide specimens with and without initial damages

the $P$ increases, the crack advances both $x$ and $z$-directions; it forms a curved crack front in DT specimens as shown in Fig. 3. However, due to larger compressive stress at the top surface, the crack could not reach up to the top surface.

Similarly, XFE analyses are performed by introducing initial damages with a shape of straight/curved crack front at the pre-notch to analyze the effect of crack front on crack propagation and critical load. Length of the initial damage $a_i$, in the bottom surface, is arbitrarily considered as 3 mm in all the cases. Based on the load-displacement curves illustrated in Fig. 4, the effect of initial damage on the load-displacement slope before damage initiation is easily explained as initial damage increases the compliance in the specimen by increasing the length of the torsional bar $a$. As a result, the slope of $P - h$ curve decreases. Also, the initial damage has comparatively sharp crack front than the pre-notch with circular radius; therefore, presence of initial damage significantly decreases the critical load for damage initiation and crack propagation as indicated in Fig. 4. This should be considered, if DT experiments is conducted with pre-test for making initial crack. In DT experiments, pre-tests are conducted with lower loading rate to reduce the load required for damage initiation; so that stable crack propagation is possible [13]. The (straight/curved) crack front shape of initial damage has negligible effect on load-displacement curves before the damage initiation; but it significantly affects the load-displacement curve during the crack propagation. The crack area in straight crack is larger than that of curved crack; this makes the difference in the load-displacement curves. The formation of curved crack front is inevitable in DT fracture tests whatever the initial condition i.e. initial damage with straight or curved front as shown Fig. 5. Whereas the shape of initial damage significantly affects the crack propagation DT fracture test. At specified $P_{max}$, initial damage with straight crack front enhance the crack propagation when compared with curved crack front.

For SIF calculation using stationary crack XFE analysis, initial damages of various length and shape are introduced in the pre-notched specimens. To obtain path independent contours, enrichment radius (ER) from the crack front is considered as 0.3 mm; thus sufficient number of elements are considered for SIF calculation [23]. Numerically calculated SIF at the crack-tip of initial damage

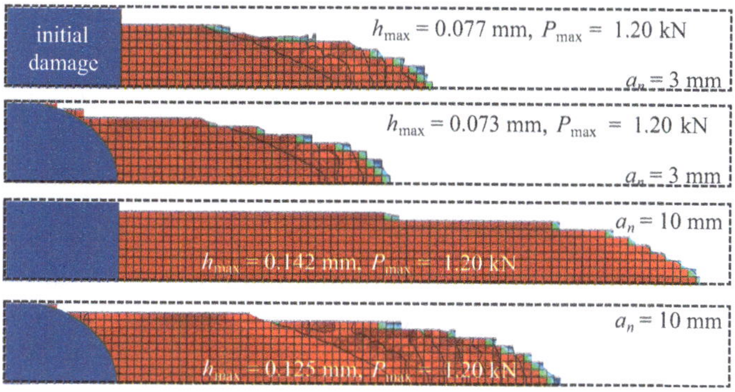


**Fig. 5** Effect of initial damages on crack propagation in SiC at specified $P_{max}$

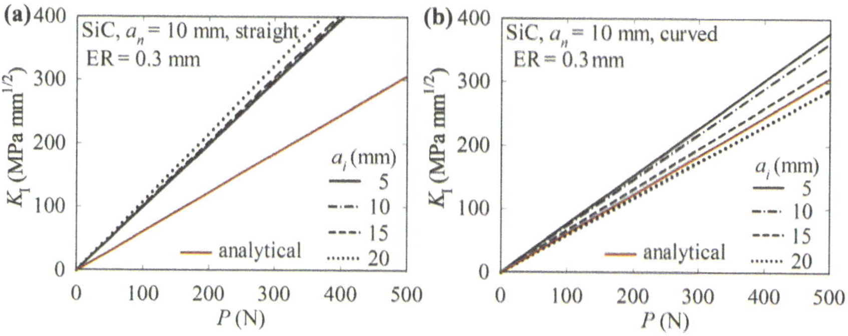


**Fig. 6** Variation of SIF with load in DT specimens at **a** straight and **b** curved crack front

(bottom surface) is compared with analytical solutions (Fig. 6); material properties of SiC is assigned to the XFE model. SIF at the crack-tip at the bottom surface (longest crack) linearly increases with increasing load. Since, the analytical solution [Eqs. (1a, 1b)] is derived based on the straight crack front assumption, obtained SIF from the XFE analyses are largely deviates from the analytical solutions as shown Fig. 6a. SIF is also calculated by varying initial damage length and (curved) shape in pre-notched specimen ($a_n$ = 10 mm), and then compared as shown in Fig. 6b. SIF at the tip of straight crack front is comparatively crack length independent, whereas SIF at the tip of curved crack front becomes dependent on crack length. The results confirm that the straight crack front assumption in the analytical solution is no longer valid. Profile of curved crack front have no significant effect on SIF at the crack tip. Since, previous XFE analyses showed that the formation of curved crack front in DT specimen is inevitable (Fig. 3); therefore, *analytical solution for SIF calculation should be revised by considering curved crack front* to obtain reliable fracture toughness from the DT experiments. The same factor should be considered while standardizing DT testing for fracture toughness evaluation.

## 5 Summary

The crack propagation in DT fracture test was simulated using XFEM for better understanding of fracture characteristics of tested materials. The shape and direction of propagation crack is observed by assigning the material properties of silicon carbide (SiC) to the XFE model. The effect of initial crack parameters and friction on the crack-front shape and crack propagation were also investigated. The formation of curved crack front is inevitable in DT fracture tests whatever the initial condition. The crack length independency of stress intensity factor was examined by comparing SIF from analytical solutions with those obtained using stationary crack XFE analyses. In addition, the effect of exact crack-front shape and initial crack length on stress-intensity factor calculations were explained. The results confirmed that the straight crack front assumption in the analytical solution is no longer valid. Finally, suggestions were provided to standardize the DT testing technique for fracture toughness evaluation.

**Acknowledgements** This research was supported by the Basic Science Research Programs through the Nation Research Foundation of Korea (NRF-2017R1A2B3009706).

## References

1. Outwater JO, Gerry DJ (1969) On the fracture energy, rehealing velocity and refracture energy of cast epoxy resin. J Adhes 1:290–298. https://doi.org/10.1080/00218466908072191
2. Evans AG (1972) A method for evaluating the time-dependent failure characteristics of brittle materials—and its application to polycrystalline alumina. J Mater Sci 7:1137–1146. https://doi.org/10.1007/BF00550196
3. Williams DP, Evans AG (1973) A simple method for studying slow crack growth. J Test Eval 1:264–270
4. Tait RB, Fry PR, Garrett GG (1987) Review and evaluation of the double-torsion technique for fracture toughness and fatigue testing of brittle materials. Exp Mech 27:14–22. https://doi.org/10.1007/BF02318858
5. Chevalier J, Saadaoui M, Olagnon C, Fantozzi G (1996) Double-torsion testing a 3Y-TZP ceramic. Ceram Int 22:171–177
6. Salem JA, Radovic M, Lara-Curzio E, Nelson G (2006) Fracture toughness of thin plates by the double-torsion test method. In: Tandon R, Wereszczak A, Lara-Curzio, E. (eds) Mechanical properties and performance of engineering ceramics II: ceramic engineering and science proceedings, vol 27(2), Wiley, Hoboken, NJ
7. Attia MS, Mahmoud AS, Megahed MM, Radwan AA (2010) Numerical investigations of the stepped double torsion fracture toughness specimen. Eng Fract Mech 77:3359–3367. https://doi.org/10.1016/j.engfracmech.2010.08.012
8. Dwivedi G, Viswanathan V, Sampath S, Shyam A, Lara-Curzio E (2014) Fracture toughness of plasma-sprayed thermal barrier ceramics: influence of processing, microstructure, and thermal aging. J Am Ceram Soc 97:2736–2744
9. Becker TH, Marrow TJ, Tait RB (2011) An evaluation of the double torsion technique. Exp Mech 51:1511–1526

10. Govila RK (1980) Indentation-precracking and double-torsion methods for measuring fracture mechanics parameters in hot-pressed $Si_3N_4$. J Am Ceram Soc 63:319–326. https://doi.org/10.1111/j.1151-2916.1980.tb10730.x
11. Mchenry KD, Yonushonis T, Tressler RE (1976) Low-temperature suberitical crack growth in SiC and $Si_3N_4$. J Am Ceram Soc 59:262–263. https://doi.org/10.1111/j.1151-2916.1976.tb10950.x
12. Madjoubi MA, Hamidouche M, Bouaouadja N et al (2007) Experimental evaluation of the double torsion analysis on soda-lime glass. J Mater Sci 42:7872–7881. https://doi.org/10.1007/s10853-007-1660-4
13. Ferri LE, Hoff KC, Baker SP (2016) Double-torsion measurements of the brittle-to-ductile transition temperature in calcium aluminosilicate glasses. Int J Appl Glas Sci 544:532–544. https://doi.org/10.1111/ijag.12188
14. Shin DC, Kim TG, Kim DW (2015) Electromechanical linear contribution at fracture of crack on pre-notched piezoelectric ceramics under double torsion. Exp Mech 55:1729–1744. https://doi.org/10.1007/s11340-015-0081-6
15. Pecanac G, Wei J, Malzbender J (2016) Fracture toughness of solid oxide fuel cell anode substrates determined by a double-torsion technique. J Power Sources 327:629–637. https://doi.org/10.1016/j.jpowsour.2016.07.103
16. Trantina GG (1977) Stress analysis of double-torsion specimen. J Am Ceram Soc 60:338–341
17. Tseng AA, Berry JT (1979) A three-dimensional finite element analysis of the double torsion test. J Press Vess Technol Trans ASME 101:328–335
18. Ciccotti M (2000) Realistic finite-element model for the double-torsion loading configuration. J Am Ceram Soc 83:2737–2744
19. Leplay P, Rethore J, Meille S, Baietto MC, Adrien J, Chevalier J, Maire E (2013) Three-dimensional analysis of an in situ double-torsion test by X-ray computed tomography and digital volume correlation. Exp Mech 53:1265–1275
20. Fuller Jr ER (1979) An evaluation of double-torsion testing—analysis. In: Freiman SW(ed) Fracture mechanics applied to brittle materials, ASTM STP 678. ASTM international, pp 3–18
21. Leevers PS (1986) Large deflection analysis of the double torsion test. J Mater Sci Lett 5: 191–192
22. Ciccotti M, Negri N, Gonzato G, Mulargia F (2001) Practical application of an improve methodology for the double torsion load relaxation method. Int J Rock Mech Min Sci 38:569–576. https://doi.org/10.1016/S1365-1609(01)00019-3
23. Abaqus (2014) User's manual-version 6.14. Dassault Systems Simulia Corp., Providence, RI, USA
24. Marimuthu KP, Rickhey F, Lee JH, Lee H (2017) Spherical indentation for brittle fracture toughness evaluation by considering kinked-cone-crack. J Eur Ceram Soc 37:381–391. https://doi.org/10.1016/j.jeurceramsoc.2016.08.014

# Structural and Dielectric Properties of Polycrystalline Calcium Copper Titanate (CCTO)

**J. Mohammed, Isah I. Garba, K. U. Yerima, A. Muhammad, U. F. Zulfatu, Mubarak S. Abubakar, Jyoti Sharma, T. Tchouank Tekou Carol, Sachin Kumar and A. K. Srivastava**

## 1 Introduction

Calcium copper titanate (CCTO) with its high room temperature dielectric constant and high thermal stability has attracted the attention of scientists over the past few years [1–4]. Dielectric constant of CCTO is independent in wide ranges of

J. Mohammed · J. Sharma · T. Tchouank Tekou Carol · A. K. Srivastava (✉)
Department of Physics, School of Physical Sciences and Chemical Engineering, Lovely Professional University, Phagwara 144411, Punjab, India
e-mail: srivastava_phy@yahoo.co.in

J. Mohammed · K. U. Yerima
Department of Physics, Faculty of Science, Federal University Dutse, P.M.B. 7156, Dutse, Jigawa, Nigeria

I. I. Garba
Department of Physics with Electronics, Faculty of Science, Federal University Birnin Kebbi, P.M.B. 1157, Dutse, Kebbi, Nigeria

A. Muhammad
Department of Physics, Faculty of Science, Sule Lamido University, P.M.B. 048, Kafin Hausa, Jigawa, Nigeria

A. Muhammad
Physics Department, Faculty of Arts and Sciences, Eastern Mediterranean University, G. Magusa, Cyprus

U. F. Zulfatu
Department of Satellite Application and Development, Nigerian Communications Satellite Ltd, Obasanjo Space Centre, Umaru Musa Yar'Adua Express Way, Lugbe, Abuja, Nigeria

M. S. Abubakar
Department of Physics, NIMS University, Jaipur 303121, Rajasthan, India

S. Kumar
Department of Chemistry, Guru Nanak Dev University, Amritsar 143005, India

 
K. Antony and J. P. Davim (eds.), *Advanced Manufacturing and Materials Science*, Lecture Notes on Multidisciplinary Industrial Engineering,
https://doi.org/10.1007/978-3-319-76276-0_37

temperature (100–600 K) and frequency ($10^2$–$10^6$ Hz) while other material with high dielectric constant such as barium titanate depends on temperature and frequency [5]. High dielectric constant and low dielectric loss at room temperature are needed for the miniaturization of electronic devices [6]. With such excellent properties, CCTO is a viable candidate for miniaturised and improved electronic devices. For better technological application, the dielectric properties of CCTO ceramic have to be tailored so as to achieve high dielectric constant and low dielectric loss. Li et al. achieved high dielectric constant (in the frequency range 1–20 kHz) by doping calcium (Ca) site with magnesium (Mg) i.e. $Ca_{1-x}Mg_xCu_3Ti_4O_{12}$ (x = 0.05) [7]. Wang et al. study the dependence of calcination temperature on the structural and dielectric properties of undoped CCTO ($CaCu_3Ti_4O_{12}$) and obtain high dielectric constant and low dielectric loss in the temperature range 20–200 °C [8]. In order to tailor the dielectric properties of CCTO, we have substituted cerium (Ce) at the calcium (Ca) site and nickel (Ni) at the titanium (Ti) site.

Solid state reaction method has been the traditional method for a long time for the synthesis of different nanomaterials, but this method suffers from disadvantages such as high sintering temperature and long reaction time. Sol-gel method on the other hand, offers low sintering temperature and less reaction time as compared to solid state reaction method [9, 10]. In this research, we have employed sol-gel method and synthesized $Ca_{1-x}Ce_xCu_3Ti_{4-y}Ni_yO_{12}$ (x = 0.0/y = 0.0, x = 0.25/y = 0.3, x = 0.45/y = 0.5, and x = 0.65/y = 0.7). The prepared samples were characterised using X-ray diffraction (XRD), High Resolution scanning electron microscope (HRSEM), and impedance analyser. High dielectric constant is observed in the sample $Ca_{1-x}Ce_xCu_3Ti_{4-y}Ni_yO_{12}$ (x = 0.25/y = 0.3) whereas low dielectric loss is observed in the sample $Ca_{1-x}Ce_xCu_3Ti_{4-y}Ni_yO_{12}$ (x = 0.45/y = 0.5).

## 2 Experimental and Characterization Details

Analytic reagent (AR) grade chemicals with high purity (99–99.5%) are used as raw material for sol-gel synthesis of polycrystalline calcium copper titanate with chemical composition $Ca_{1-x}Ce_xCu_3Ti_{4-y}Ni_yO_{12}$ (x = 0.0/y = 0.0, x = 0.25/y = 0.3, x = 0.45/y = 0.5, and x = 0.65/y = 0.7). Stoichiometric amount of the raw materials such as calcium nitrate, cerium nitrate, copper nitrate, nickel nitrate and titanium dioxide were dissolved in 100 ml of double distilled to form an aqueous solution. Thereafter, citric acid was added to the aqueous solution with cations to citric acid molar ratio of 1:1. The prepared solution was heated on a hot plate (magnetic stirrer) at 80–100 °C in order to evaporate water. A light blue gel was formed which upon further heating at a temperature of 280–300 °C gives a black precursor material. The precursor material was grounded with mortar and pestle so as to obtain fine CCTO powder. Pre-sintering of the fine CCTO powder was carried out at a temperature of 800 °C for 6 h so as to remove organic matter and impurities. Furthermore, the samples were sintered at 900 °C for 6 h. X-ray diffraction pattern (obtained in the range 20°–80° using Cu–K radiation operating at 40 kV and 35 mA with step size 0.02°) was used to study the structural properties of the prepared sample (Bruker AXS D8 advance diffractometer). High resolution scanning electron microscope (HRSEM) (BRUKER) operating at a voltage of 20 kV was used to study the surface features and elemental analysis. Impedance analyser (Wayne Kerr 6500B) was used to study the dielectric properties in the frequency range 1 kHz–5 MHz.

## 3 Result and Discussion

### *3.1 XRD Analysis*

Figure 1 presents X-ray diffraction (XRD) patterns of $Ca_{1-x}Ce_xCu_3Ti_{4-y}Ni_yO_{12}$ (x = 0.0/y = 0.0, x = 0.25/y = 0.3, x = 0.45/y = 0.5, and x = 0.65/y = 0.7) prepared by sol-gel method and calcinated at 900 °C. It has been confirmed from X-ray diffraction that the prepared sample exhibit crystalline structure of calcium copper titanate ($CaCu_3Ti_4O_{12}$). Traces of secondary phases of calcium titanate ($CaTiO_3$) and copper oxide (CuO) are observed in the prepared samples. A single perovskite structure of the material is formed in the absence of Ce–Ni substitution. The lattice parameters are calculated and presented in Table 1.

From Fig. 1, it can be seen that the intensity of the peaks of the secondary phases increases with Ce–Ni substitution. The XRD patterns of the prepared samples are indexed using JCPDS card number 75-2188. The lattice constant (*a*) is calculated

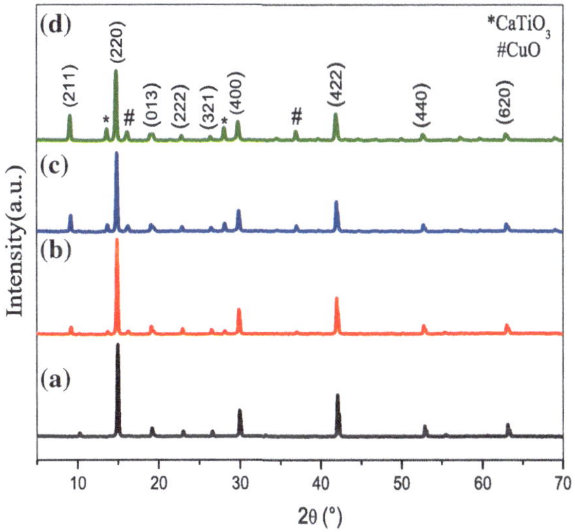


**Fig. 1** XRD spectra of $Ca_{1-x}Ce_xCu_3Ti_{4-y}Ni_yO_{12}$ (**a**) x = 0.0/y = 0.0, (**b**) x = 0.25/y = 0.3, (**c**) x = 0.45/y = 0.5, (**d**) x = 0.65/y = 0.7

**Table 1** Values of lattice constant ($a$), volume of unit cell ($V_{cell}$) and crystallite size ($D$) of $Ca_{1-x}Ce_xCu_3Ti_{4-y}Ni_yO_{12}$ (x = 0.0, 0.25, 0.45, 0.65 and y = 0.0, 0.3, 0.5, 0.7)

| Composition | $a$ [Å] | $V_{cell}$ [Å$^3$] | $D$ [nm] |
|---|---|---|---|
| x = 0.0/y = 0.0 | 7.390 | 403.583 | 50.732 |
| x = 0.25/y = 0.3 | 7.392 | 403.911 | 52.654 |
| x = 0.45/y = 0.5 | 7.393 | 404.075 | 54.013 |
| x = 0.65/y = 0.7 | 7.394 | 404.239 | 55.818 |

using the formula $a = d\sqrt{h^2 + k^2 + l^2}$ where $d$ is the inter planer spacing and hkl are the miller indices. It can be observed from Table 1 that the lattice constant increases with Ce–Ni substitution. The volume of the unit cell is calculated using $V_{cell} = a^3$ whereas the value of the crystallite size is calculated using Scherer formula $D = \frac{0.89\lambda}{\beta \cos\theta}$, where $\lambda$ = 1.54056 Å (X-ray wavelength), $\theta$ is the Braggs angle and $\beta$ is the full width at half maximum in radian. Similarly, both the volume of unit cell and crystallite size show increase as concentration of Ce–Ni is increased (Table 1). This could be attributed to the fact the ionic radius of $Ce^{3+}$ (1.034 Å) and $Ni^{2+}$ (0.690 Å) are larger than those of $Ca^{2+}$ (0.990 Å) and $Ti^{4+}$ (0.605 Å) respectively [11]. The calculated lattice parameters presented in Table 1 are in agreement with literature [10, 12].

## 3.2 HRSEM and EDX Analysis

Figure 2 presents the HRSEM micrograph of $Ca_{1-x}Ce_xCu_3Ti_{4-y}Ni_yO_{12}$ where x = 0.25/y = 0.3 at (a) at 2 $\mu$m (b) at 1 $\mu$m (c) 200 nm and (d) 100 nm. Large, regular and polyhedral grains of sizes between 902.4 nm and 1.562 $\mu$m were observed.

It is clear from the HRSEM micrographs that the prepared sample exhibit bi-modal distribution of grains, this shows that the prepared sample are agglomerated which upon calcination forms larger grains [13]. Energy dispersive X-ray (EDX) spectra and atomic percentage of $Ca_{1-x}Ce_xCu_3Ti_{4-y}Ni_yO_{12}$ where x = 0.25/y = 0.3 is presented in Fig. 3, the spectra show clearly the presence of Ca, Ce, Cu, Ti, Ni and O which is as per the stoichiometry of the prepared sample.

## 3.3 Dielectric Constant

Room temperature dielectric constant ($\varepsilon'$) as a function of frequency (log $f$) for the prepared samples $Ca_{1-x}Ce_xCu_3Ti_{4-y}Ni_yO_{12}$ (x = 0.0/y = 0.0, x = 0.25/y = 0.3, x = 0.45/y = 0.5, and x = 0.65/y = 0.7) are presented in Fig. 4.

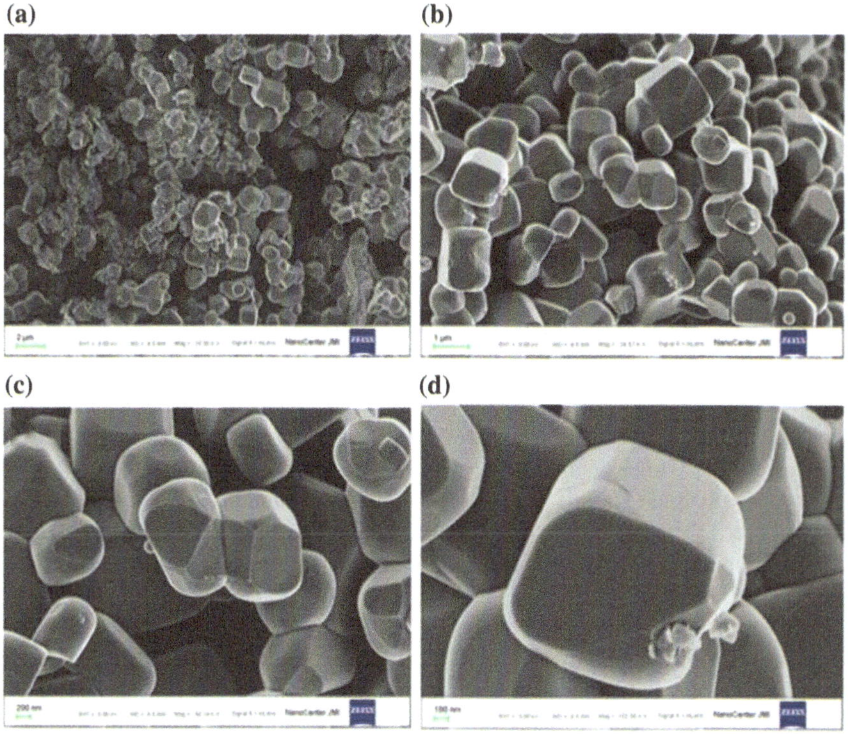


**Fig. 2** HRSEM micrograph of $Ca_{1-x}Ce_xCu_3Ti_{4-y}Ni_yO_{12}$ (x = 0.25/y = 0.3) (**a**) at 2 m (**b**) At 1 $\mu$m (**c**) 200 nm (**d**) 100 nm

| Elements | Atomic percentage [%] |
|---|---|
| Ca | 4.39 |
| Ce | 2.50 |
| Cu | 17.88 |
| Ti | 19.51 |
| Ni | 1.90 |
| O | 53.82 |
| Total | 100.00 |

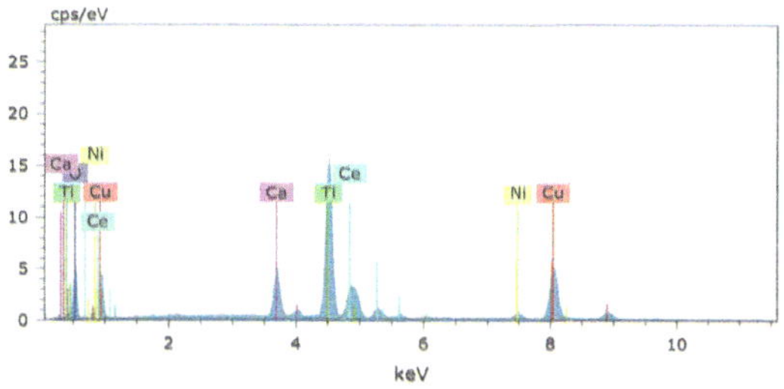


**Fig. 3** EDX spectra and atomic percentage of $Ca_{1-x}Ce_xCu_3Ti_{4-y}Ni_yO_{12}$ (x = 0.25/y = 0.3)

The dielectric constant of the prepared sample first increases with Ce–Ni substitution and then decreases. The dielectric constant of the prepared samples at frequency 100 Hz are 4590, 16000, 2130 and 2750 for the corresponding Ce–Ni substitution x = 0.0/y = 0.0, x = 0.25/y = 0.3, x = 0.45/y = 0.5, and x = 0.65/y = 0.7. We observed high dielectric constant (of value 16000 at 100 Hz) in the sample $Ca_{1-x}Ce_xCu_3Ti_{4-y}Ni_yO_{12}$ (x = 0.25/y = 0.3) which could be attributed to the

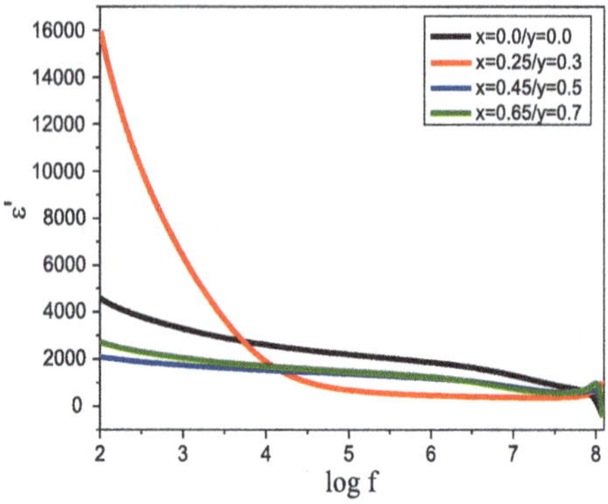


**Fig. 4** Dielectric constant of $Ca_{1-x}Ce_xCu_3Ti_{4-y}Ni_yO_{12}$ (x = 0.0/y = 0.0, x = 0.25/y = 0.3, x = 0.45/y = 0.5, and x = 0.65/y = 0.7) at room temperature

presence of thin re-oxidized grain boundary regions on the surfaces of grains with large sizes [14]. This is in agreement with HRSEM micrograph as they show grains having large sizes. Also, this high dielectric constant could be explained in terms of barrier layer formation which is due to the differences in conductivity of the grain and grain boundaries [9]. The decreasing behaviour of the dielectric constant in the samples with higher Ce–Ni concentration (x = 0.45/y = 0.5 and x = 0.65/y = 0.7) could be related to the presence of secondary phase ($CaTiO_3$ and CuO) at the grain boundaries of the prepared samples [7]. At higher frequencies, the dielectric constant for the undoped sample shows frequency independent behaviour. This could be ascribed to different sizes of the grains [15]. A Debye-type relaxation peak is observed in the sample $Ca_{1-x}Ce_xCu_3Ti_{4-y}Ni_yO_{12}$ (x = 0.65/y = 0.7) at higher frequency. This can be explained on the basis of Maxwell-Wagner relaxation between the grain and grain boundary [11].

## *3.4 Dielectric Loss*

The room temperature dielectric loss (tan $\delta$) versus frequency (log $f$) for the synthesized sample is presented in Fig. 5. The dielectric losses of the samples are 0.330, 1.140, 0.189, 0.227 at 100 Hz for the samples $Ca_{1-x}Ce_xCu_3Ti_{4-y}Ni_yO_{12}$ (x = 0.0/y = 0.0, x = 0.25/y = 0.3, x = 0.45/y = 0.5, and x = 0.65/y = 0.7) respectively. At much higher frequency (at 63.9 MHz), the dielectric loss increases rapidly. This rapid increase in dielectric loss might be due to resonance effect resulting from stray conductance of the contacts and leads [9]. The sample with the highest dielectric constant exhibits a relaxation peak at a frequency of 11.7 kHz. The sample $Ca_{1-x}Ce_xCu_3Ti_{4-y}Ni_yO_{12}$ (x = 0.65/y = 0.7) exhibits the lowest dielectric tangent loss (0.189).

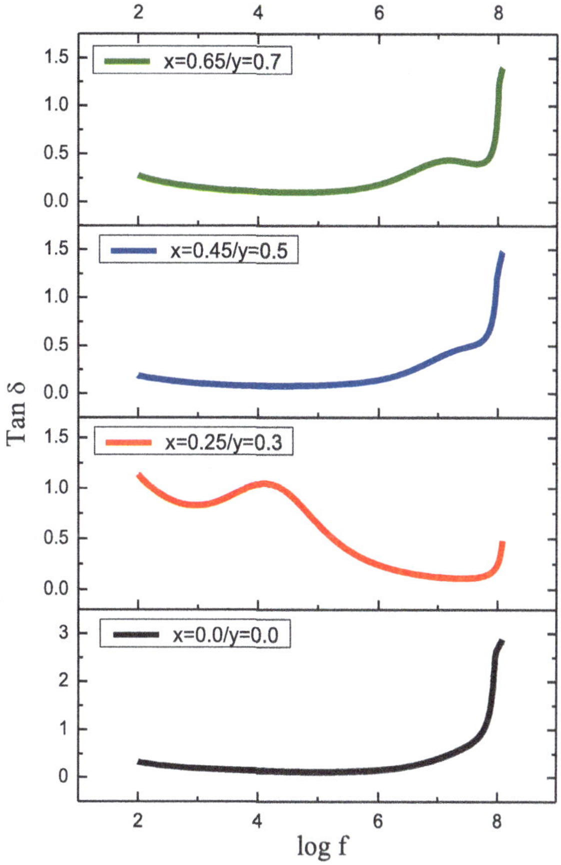


**Fig. 5** Dielectric loss of $Ca_{1-x}Ce_xCu_3Ti_{4-y}Ni_yO_{12}$ (x = 0.0/y = 0.0, x = 0.25/y = 0.3, x = 0.45/y = 0.5, and x = 0.65/y = 0.7) at room temperature

## 3.5 *Impedance Analysis*

We carry out impedance studies in order to understand the nature of grain boundary resistance of the prepared samples. The nature of the grain boundary resistance is determined by the nature of curve resulting from plotting $Z''$ against $Z'$. The complex impedance plot (Cole-Cole plot) is presented in Fig. 6.

The samples $Ca_{1-x}Ce_xCu_3Ti_{4-y}Ni_yO_{12}$ (x = 0.45/y = 0.5 and x = 0.65/y = 0.7) show high grain boundary resistance. These are also the samples with low dielectric tangent loss and their XRD peak show increasing intensity of secondary phases ($CaTiO_3$ and CuO). Li et al. reported the sample with high grain boundary resistance usually have low dielectric loss as a result of existence of secondary phases in the prepared samples [7]. The semi-circle arc at the higher frequency part of the complex impedance plot represents contribution from the grain while that in the

**Fig. 6** Cole-Cole plot of $Ca_{1-x}Ce_xCu_3Ti_{4-y}Ni_yO_{12}$ (x = 0.0/y = 0.0, x = 0.25/y = 0.3, x = 0.45/y = 0.5, and x = 0.65/y = 0.7)

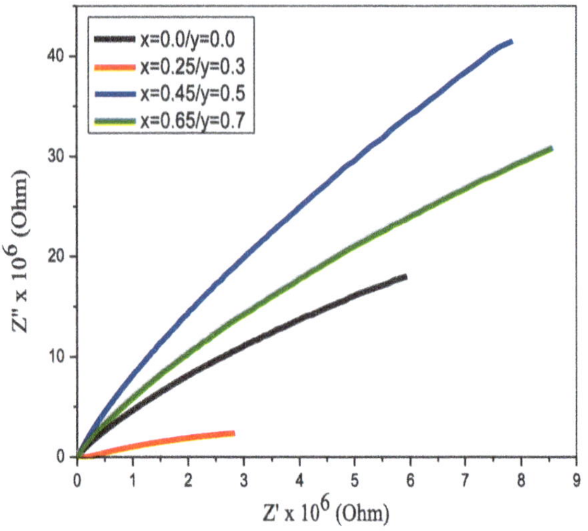


lower frequency part represents contributions from grain boundary [9]. Hence, the dielectric properties of the prepared samples can be ascribed to grain boundary resistance.

## 3.6 *Ac Conductivity*

Figure 7 presents the room temperature variation of AC conductivity with frequency (log *f*) of the prepared samples. The recorded AC conductivity of the prepared samples at a frequency of 100 Hz are $1.66 \times 10^{-8}$, $1.99 \times 10^{-7}$, $4.38 \times 10^{-9}$, and $8.33 \times 10^{-9} (\Omega m)^{-1}$ for $Ca_{1-x}Ce_xCu_3Ti_{4-y}Ni_yO_{12}$ (x = 0.0/y = 0.0, x = 0.25/y = 0.3, x = 0.45/y = 0.5, and x = 0.65/y = 0.7) respectively. The variation of the conductivity of the prepared sample with frequency can be understood using the equation representing total conductivity $\sigma_{total} = \sigma_{ac} + \sigma_{dc}$, where is the frequency dependent part of the conductivity and is the frequency independent part of the conductivity [2]. From Fig. 7, we can see that the conductivity first decreases and then later increases with increase in frequency. Also, it can be seen that the conductivity shows frequency independent behaviour at low frequency. This corresponds to DC conductivity ($\sigma_{dc}$). At higher frequencies, the conductivity increases in all the prepared samples showing that the prepared samples exhibit frequency dependent behaviour. This corresponds to AC conductivity ($\sigma_{ac}$) [16]. The transition from DC conductivity ($\sigma_{dc}$) at lower frequencies to AC conductivity ($\sigma_{ac}$) at higher frequencies started to occur after the DC conductivity ($\sigma_{dc}$) reaches a certain frequency called the cross over frequency.

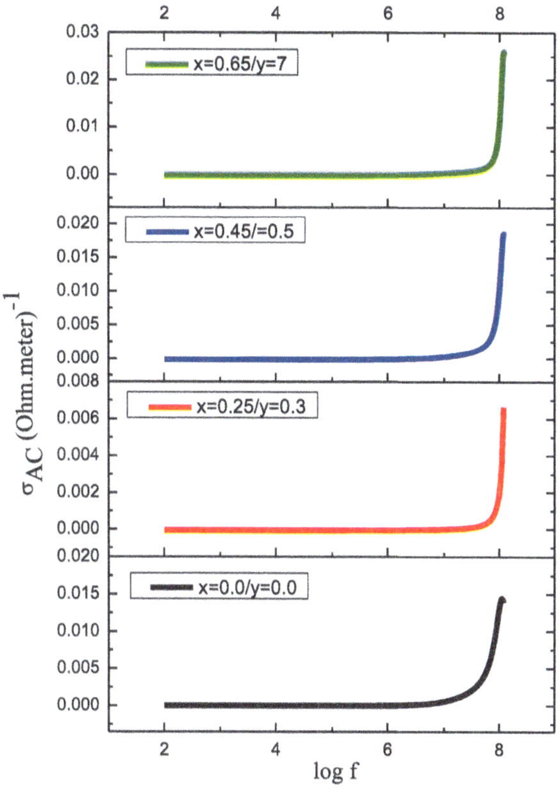


**Fig. 7** AC conductivity of $Ca_{1-x}Ce_xCu_3Ti_{4-y}Ni_yO_{12}$ (x = 0.0/y = 0.0, x = 0.25/y = 0.3, x = 0.45/y = 0.5, and x = 0.65/y = 0.7) at room temperature

# 4 Conclusion

We have successfully synthesized and characterised Ce–Ni substituted calcium copper titanate using sol-gel method. The XRD patterns of the prepared samples show trace of $CaTiO_3$ and CuO as secondary phase. The microstructure shows large grains of sizes ranging from 902.4 nm to 1.562 $\mu$m. High dielectric constant and low dielectric losses was observed in the prepared sample. The sample with Ce–Ni substitution of x = 0.25/y = 0.3 exhibits the highest dielectric constant. The prepared material has valuable application in electronics devices.

# References


1. Barman N, Tripathi S, Ravishankar N, Varma KBR (2016) Centrosymmetric tetragonal tellurium doped calcium copper titanate and its dielectric tunability. Solid State Commun 241:7–13
2. Khare A, Sundar S, Mandal KD, Mukhopadhyay NK (2016) Effect of sintering duration on the dielectric properties of $0.9BaTiO_3$—$0.1CaCu_3Ti_4O_{12}$ nanocomposite synthesized by solid state route. Microelectron Eng 164:1–6

3. Sharma S, Singh MM, Rai US, Mandal KD (2015) Materials science in semiconductor processing rationalization of dielectric properties of nano-sized iron doped yttrium copper titanate using impedance and modulus studies. Mater Sci Semicond Process 31:720–727
4. Singh L, Sin BC, Kim IW, Mandal KD, Chung H, Lee Y (2016) A Novel one-step flame synthesis method for tungsten-doped CCTO. J Am Ceram Soc 99(1):27–34
5. Chunhong MU, Huaiwu Z, Yingli LIU, Yuanqiang S, Peng LIU (2010) Rare earth doped $CaCu_3Ti_4O_{12}$ electronic ceramics for high frequency applications. J Rare Earths 28(1):43–47
6. Liang P, Yang Z, Chao X (2016) Enhanced intrinsic permittivity and bulk response in $Y_{2/3}$ $Cu_3Ti_{4+x}O_{12}$ ceramics. Ceram Int 12:1–6
7. Li W, Qiu S, Chen N, Du G (2010) Enhanced dielectric response in Mg-doped $CaCu_3Ti_4O_{12}$ ceramics. J Mater Sci Technol 26(8):682–686
8. Wang XW (2016) Calcining temperature dependence on structure and dielectric properties of $CaCu_3Ti_4O_{12}$ ceramics. J Mater Sci: Mater Electron 27(11):12134–12140
9. Mandal KD, Rai AK, Kumar D, Parkash O (2009) Dielectric properties of the $Ca_{1-x}La_xCu_3Ti_{4-x}Co_xO_{12}$ system (x = 0.10, 0.20 and 0.30) synthesized by semi-wet route. J. Alloys Compd. 478:771–776
10. Kumar A, Mandal KD, Kumar D, Parkash O (2010) Characterization of nickel doped CCTO: $CaCu_{2.9}Ni_{0.1}Ti_4O_{12}$ and $CaCu_3Ti_{3.9}Ni_{0.1}O_{12}$ synthesized by semi-wet route. J Alloys Compd 491:507–512
11. Dubey AK, Singh P, Singh S, Kumar D, Parkash O (2011) Charge compensation, electrical and dielectric behavior of lanthanum doped. J Alloys Compd 509(9):3899–3906
12. Jesurani S, Kanagesan S, Hashim M, Ismail I (2013) Dielectric properties of Zr doped $CaCu_3Ti_4O_{12}$ synthesized by sol—gel route. J Alloys Compd 551:456–462
13. Adams BTB, Sinclair DC, West AR (2002) Giant barrier layer capacitance effects in $CaCu_3Ti_4O_{12}$ ceramics. Ceramics Adv Mater 14:1321–1323
14. Jesurani S, Kanagesan S, Hashim M, Ismail I, Ibrahim IR (2014) Microstructural and dielectric properties of Zr doped microwave sintered $CaCu_3Ti_4O_{12}$ synthesized by sol-gel route. Adv Mater 2014:1–6
15. Liu L, Fan H, Fang P, Chen X (2008) Sol-gel derived $CaCu_3Ti_4O_{12}$ ceramics: synthesis, characterization and electrical properties. Mater Res Bull 43:1800–1807
16. Singh L, Kim IW, Sin BC, Woo SK, Hyun SH, Mandal KD, Lee Y (2015) Combustion synthesis of nano-crystalline $Bi_{2/3}Cu_3Ti_{3.90}Fe_{0.10}O_{12}$ using inexpensive $TiO_2$ raw material and its dielectric characterization. Powder Technol 280:256–265

# Investigation on Mechanical and Wear Properties of Aluminium Based Metal Matrix Composite Reinforced with $B_4C$, Gr and Fly Ash

**M. H. Faisal and S. Prabagaran**

## 1 Introduction

Nowadays industrial areas such as defense, aerospace, consumer and automotive industries seek advantages of transpiring metal matrix composites. Due to attractive material properties such as simpler fabricability, less density and better engineering characteristics, aluminium alloys are selected as matrix material. Better wear resistance, good specific stiffness, and lower coefficient of thermal expansion are some profits of aluminium metal matrix composites (AMCs) compared to unreinforced aluminium alloys. Aluminium matrix composites infused with ceramic particles have better wear resistance, creep resistance and stiffness compared to conventional aluminium alloys which makes their usage in the marine, aerospace and automobile industry [1]. AMCs are the excellent substances in the industrial world. Because of its superb mechanical characteristics, it is mainly employed in areas such as marine, automobiles and aerospace etc. [2]. In comparison with fiber reinforced composites, particle reinforced composites have a lot of advantages. Due to their enhanced wear resistance and light weight they can be easily fabricated with less cost and can be used for the automobile parts which are in transmission like connecting rods, cylinder liners, brake drum, pistons and cylinder block [3]. Comparing to Al–SiC composites, boron carbide reinforced aluminium matrix composites have better properties like less density, greater hardness and enhanced tensile strength. $CaC_2$ also has better mechanical properties with low density. Here $CaC_2$ and $B_4C$ are reinforced with 6063 Al alloy. Resulting composites have better properties such as toughness, shear strength and tensile strength. A Higher

M. H. Faisal (✉) · S. Prabagaran
Department of Mechanical Engineering, Karpagam Academy of Higher Education, Coimbatore, Tamilnadu, India
e-mail: faisalmh78@gmail.com

S. Prabagaran
e-mail: s.prabagaran.cbe@gmail.com

 
K. Antony and J. P. Davim (eds.), *Advanced Manufacturing and Materials Science*, Lecture Notes on Multidisciplinary Industrial Engineering,
https://doi.org/10.1007/978-3-319-76276-0_38

percentage of $CaC_2$ causes an increase in toughness. Reduction of weight without any damage to the mechanical properties is achieved by the addition of $CaC_2$ particles of composites [4]. AA6061–$B_4C$ and AA6061–$B_4C$–Gr composites were fabricated by employing the stir-casting method. The SEM analysis reveals the uniform distribution of boron-carbide and graphite particles in the aluminium matrix. The AA6061–$B_4C$ composite had a higher hardness compared to AA6061. The wear resistance of the AA6061–$B_4C$–Gr hybrid composite and Al–$B_4C$ composite increases steadily with the sliding load and velocity [5]. Fly ash is the waste product of coal which remains after the combustion of coal in the thermal power plants. It forms into temperature of 920–1200 °C [6]. Over 90 million tones of fly ash are produced each year in thermal power plants, most of which is land filled [7]. So it easily available with little or no cost as compared to other reinforcements and also hazardous if remains in the open environment.

## 2 Materials and Methodology

1. **Aluminium 7075**: Al7075 capable to withstand high loads and temperature and it have better wear resistance, good fatigue strength average machinability, average corrosion resistance. Its main alloy element is Zinc. Due to their high strength-to-density ratio Al7075 are often used in transport applications, including marine, automotive and aviation, it is also used in rock climbing equipment, bicycle components, and hang glider airframes.
2. **Fly ash**: Fly ash is the waste product of coal which remains after the combustion of coal in the thermal power plants. Fly ash is one of the most inexpensive discontinuous, low-cost waste by-products. $SiO_2$, $Al_2O_3$, $Fe_2O_3$ and occasionally CaO are the main chemical components present in fly ashes.
3. **Boron carbide**: Boron carbide is a very hard ceramic substance employed in bulletproof vests, tank armors and in other industrial areas. As per Mohr's ranking scale, it maintains 3rd position after CBN and diamond.
4. **Graphite**: Graphite and Diamond are two different allotropic forms of Carbon. Depending upon the carbon alignment, characteristics and structure vary. It exhibits both non metal and metal properties. It is also used in industrial applications such as thrust bearings, vanes, piston rings and journal bearings.
5. **Stir casting**: The stir casting apparatus is mainly used for the fabrication of particulate metal matrix composites. Uniform distribution of reinforcement particles can be achieved by this method which ensures a homogeneous composite. At first, equipment temperature is fixed at 500 °C and then it is slowly increased to 900 °C. The Required quantity of alloy ingots is cut down into rectangular pieces and deposited in the graphite crucible for melting. LM13 alloy is melted at 695 °C. Then the preheated boron carbide and graphite particles are added to the molten matrix and the uniform stirring is performed for

5 min to achieve the uniform reinforcement particle distribution in the composite. After that the molten metal is poured into the prepared sand moulds and kept for enough time to solidify. Composite specimens are taken out from the moulds after solidification. By changing particles same steps are repeated for all composites [8].

6. **Tension test**: Tension test is the conventional test used to determine the mechanical characteristics of a substance. Tensile strength (UTS), elastic modulus (E), yield strength ($\sigma y$), reduction in area (RA %) and percent elongation (L %) are the dominant variables that explain the stress-strain curve procured during the tension test. Using the obtained parameters Poisson's ratio ($\nu$), Resilience and Toughness can be calculated. A specimen of required dimensions is prepared and held in the machine jaws. Based on the standards, specimens are prepared from the work material.
7. **Rockwell Hardness Test**: Based on ASTM E18:2014 standards, specimens are produced. At first, a preliminary minor load of 10 kgf is applied on the specimen surface. As soon as the equilibrium occurs, the indicator is adjusted to a datum position. Indicating device is used to react to variations in indenter's penetration depth by following the motion of the indenter. An extra major load resulting penetration improvement is applied while the preliminary load is being exerted. During equilibrium, the major load is detached by conserving minor load. When the major load is withdrawn, partial spring back occurs. This decreases the penetration depth. So for determining Rockwell hardness, the increase in penetration depth after the expulsion of the major load is used.
8. **Pin-on-disk testing**: In these tests, we have the capability to regulate speed, normal load and other variables such as lubricants, gas type, pressure, temperature etc. With the help of transducers normal and frictional forces are evaluated. The transducer assesses the deflection of pin holder fixture and transforms it into force. A profilometer is used to identify volume/mass loss and to compute the wear rate and friction coefficient. Thus the system performance can be analyzed.

## 3 Results and Discussion

### 1. Micro structural analysis

Microstructures need higher magnification than macrostructures. The conventional light microscope is the most frequent microscope utilized. Generally, optical microscopes can be utilized to examine the specimens in reflection and transmission modes. We are confined to examine the specimen surface using optical microscope due to the opaque nature of many substances (Figs. 1, 2, 3).

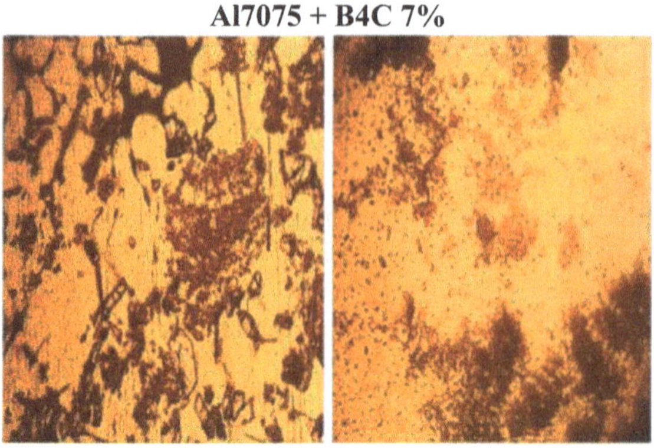


**Fig. 1** **a** Mag: 100×, **b** Mag: 500×

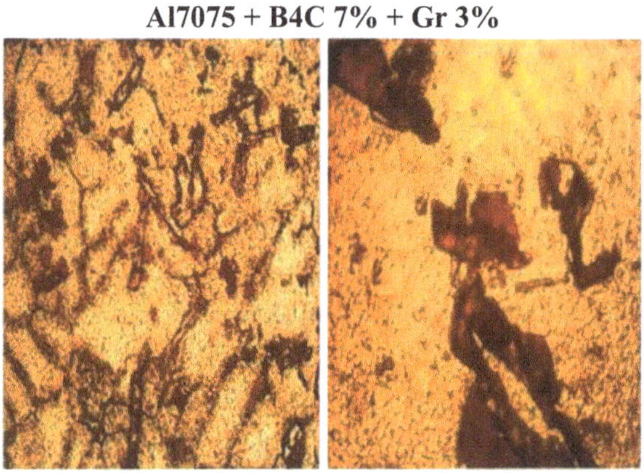


**Fig. 2** **a** Mag: 100×, **b** Mag: 500×

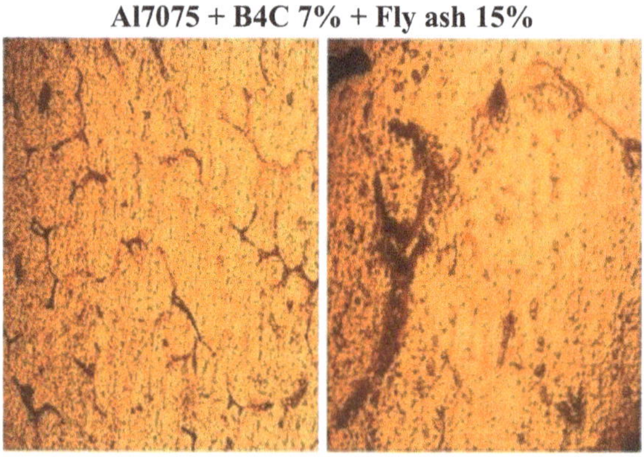


**Fig. 3** **a** Mag: 100×, **b** Mag: 500×

2. **Tensile test**

Figures 4 and 5 exhibits the tensile test outcomes of the Al7075, Al7075/ $B_4C$, Al7075/ $B_4C$/ Gr 3% and Al7075/ $B_4C$/ Fly ash composites. It is perceived that by adding the boron carbide and fly ash particles in the Al7075, both tensile strength and yield stress are intensified. This is due to the superior surface (interfacial) bonding formed between the reinforcement and the matrix which results the boron carbide particles to support the tensile force entrusting with the matrix [9]. 69% increase in tensile strength was perceived due to the addition of 7% $B_4C$ particles to the Al7075 matrix and the addition of 7% $B_4C$ and 15% Fly ash particles to the Al7075, 116% hike in tensile strength compared to the unreinforced alloy. In the same way, contingent upon yield stress, 7% of $B_4C$ and 15% Fly ash particles causes an increment in yield stress by 54 and 98% respectively.

3. **Rockwell hardness testing**

Table 1 represents the hardness test outcomes of the Al7075 alloy, Al7075/$B_4C$, Al7075/$B_4C$/Gr 3% and Al7075/$B_4C$/Fly ash composites. It is perceived that by incorporating the $B_4C$ and Fly ash particles in the composites, hardness is improved accompanied by declined atomic dislocation. The presence of hard boron carbide and Fly ash particles resists plastic deformation in the specimens and the grain boundary strength improves to a supreme level. Enhancement in reinforcement surface area and a decline in grain size is the discovered outcomes produced as a result of infusement of the boron carbide particles to the composites. Because of these outcomes harnesses of the composite gets improved [10]. 26.6% improvement in hardness was noticed due to the addition of 7% $B_4C$ particles. Similarly, 15% of Fly ash reinforcement cause 17% intensification in hardness of hybrid composite than the Al7075+$B_4C$ 7% composite.

**Fig. 4** Tensile strength of tensile test specimens

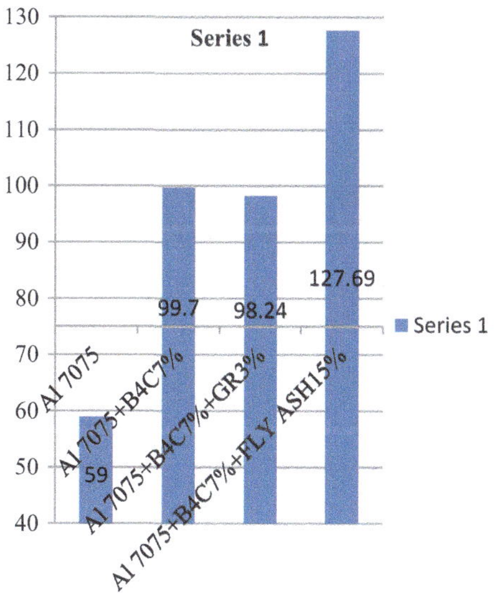

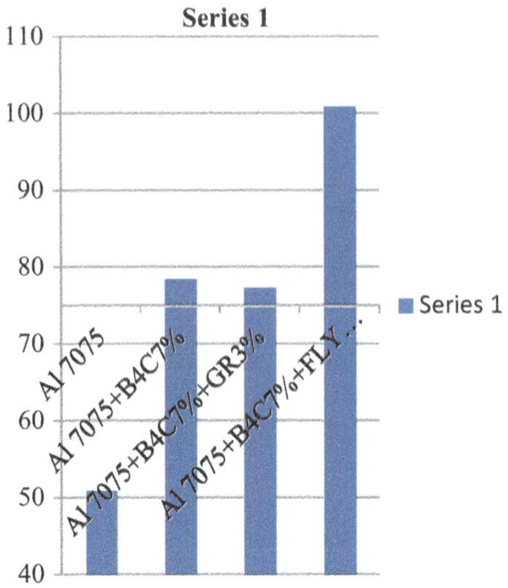


**Fig. 5** Yield stress of tensile test specimens

**Table 1** Rock well hardness test result

| SI. No. | Sample Id | Observed values, HRB | | | Average, HRB |
|---|---|---|---|---|---|
| | | 1 | 2 | 3 | |
| 1 | Al 7075 | 60 | 60 | 60 | 60 |
| 2 | Al 7075 + $B_4C$ 7% | 76 | 75 | 77 | 76 |
| 3 | Al 7075 + $B_4C$ 7% + Gr 3% | 81 | 80 | 79 | 80 |
| 4 | Al 7075 + $B_4C$ 7% + Fly ash 15% | 91 | 89 | 89 | 89 |

## 4. Wear testing

Table 2 shows the pin on disc test results of the Al7075/$B_4C$ 7%, Al7075/$B_4C$ 7%/Gr 3%, Al7075/$B_4C$7%/ Fly ash 15% and composites. By analyzing, it is perceived that by incorporating the boron carbide in the composite properties like wear, coefficient of friction (COF) and frictional force are declined accompanied by

**Table 2** Pin on disc test result Track Radius D = 0.05 M (50 mm)

| Sl. No. | Material | RPM | Load (N) | Time duration (Min) | Sliding velocity (m/s) | Wear in (μm) | COF | Frictional force (average) N |
|---|---|---|---|---|---|---|---|---|
| 1 | Sample1 | 500 | 40 | 10 | 2.617 | 260.79 | 0.2990 | 11.96 |
| **2** | Sample2 | 500 | 40 | 10 | 2.617 | 121.1 | 0.3069 | 12.27 |
| **3** | Sample3 | 500 | 40 | 10 | 2.617 | 91.7 | 0.2803 | 11.21 |

attainment of excellent tribological properties. 53.6% decline in wear was noticed for 3% Graphite particles infusion to Al7075/$B_4C$ composite and 15% of fly ash reinforcement results in 65% decline in wear than the Al7075/$B_4C$ composite.

## 4 Conclusion

The present work deals with the different tests performed on Al7075/$B_4C$, Al7075/$B_4C$/Gr and Al7075/$B_4C$/Fly ash composites in which diverse mechanical properties like yield stress, tensile strength, wear, hardness, COF and frictional force were analyzed using tensile, hardness and pin on disc tests. By adding of boron carbide and Fly ash in the Al7075 aluminium matrix, the hardness of the produced composites gets raised. In the case of tribological properties like wear, coefficient of friction and frictional force, a decline was observed with addition of boron carbide and Fly ash. Also, the presence of 3% Graphite helped to obtain better tribological as well as self-lubricating characteristics. The Negative influence of graphite on mechanical characteristics can be enhanced by including boron carbide into it and thus making a hybrid composite. The outcomes of this research can be properly employed for refinement of the automotive and structural components so that components with excellent mechanical characteristics can be developed.

## References

1. Surappa MK (2013) Aluminium matrix composites: challenges and opportunities, sadhana, vol 28, pp 319–334
2. Feng YC, Geng L, Zheng PQ, Zheng ZZ, Wang GS (2008) Fabrication and characteristic of Al-based hybrid composite reinforced with tungsten oxide particle and aluminum borate whisker by squeeze casting. Mater Des 29:2023–2026
3. Raviteja T, Radhika N, Raghu R (2013) Fabrication and Mechanical Properties of Stir Cast Al-Si$1_2$Cu/$B_4C$ Composites. Int J Res Eng Technol, pp 343–346
4. Madheswaran K, Sugumar S (2015) Mechanical characterization of Aluminium–Boron carbide composites with influence of Calcium carbide particles. Int J Emerg Technol Adv Eng 492–496
5. Prabagaran S, Chandramohan G, Shanmughasundaram P (2014) Influence of graphite on the hardness and wear behavior of AA6061–$B_4C$ Composite. Materiali in tehnologije/Mat Technol 48:661–667
6. Hrairi M, Ahmed M, Nimir Y (2009) Compaction of fly ash–aluminum alloy composites and evaluation off their mechanical and acoustic properties. Adv Powder Technol 20:548–553
7. Rohatgi PK, Weiss D, Gupta N (2006) Applications' of fly ash in synthesizing low-cost MMCs for automotive and other applications. J Mat Sci
8. Rama Rao S, Padmanabhan G (2012) Fabrication and mechanical properties of aluminium–boron carbide composites. Int J Mat Biomat Appl 2(3):15–18
9. Shankar MG, Jayashree PK, Shetty R, Kini A, Sharma SS (2013) Sharma individual and combined effect of reinforcements on stir cast aluminium metal matrix composites—a review. Int J Curr Eng Technol 3:922–934
10. Kalaiselvan SK, Murugan N, Parameswaran Siva (2011) Production and characterization of AA6061–$B_4C$ stir cast composite. Mater Des 32:4004–4009

# Aluminium Carbon Nanotube Composites—A Review on Latest Approaches

**A. H. Sajeeb Rahiman and D. S. Robinson Smart**

## 1 Introduction

Aluminium is rising as the most propitious metal in engineering application now a day. The high stiffness, low density, and energy/sound absorption properties makes it suitable for many applications [1, 2]. As Al is a very receptive metal it readily forms aluminium oxide in contact with oxygen. But unlike in corrosion of steel, the oxide formed chemically bond with Al surface and will be a cover for Al from further reactions. This phenomenon adds to the corrosion resistant property of aluminum [3]. But Al has got low mechanical strength and poor wear resistance. These are limitations for Al to be used as structural material [4]. Hence reinforcing Al to impart the suitable properties is very important. After Iigima [5] presented carbon nanotubes in 1991, it draws the consideration of many researchers. The cylindrical graphite structure of CNTs, with high young's modulus of 1 TPa and superior chemical stability makes them an ideal promising reinforcement for Al matrix [6]. They are the toughest of fibres with stiffness up to 1000 GPa, strength of 100 GPa and thermal conductivity up to 6000 $Wm^{-1}$ $K^{-1}$ [7].

Aluminium ball milling involves; (1) The powders getting formed to flakes (2) the flakes getting cold welded to lamellar structure (3) the cold welded particles getting into fine particles. In ordinary ball milling both CNT clusters and formation of $Al_4C_3$ in the interface are reasons which badly affect the ductility of CNT/Al composite [8]. When the time in milling is prolonged the dispersion will improve but it will damage the Al structure [8–10]. Uniform dispersion with minimum

A. H. Sajeeb Rahiman (✉)
Department of Mechanical Engineering, Karunya Institute of Technology and Sciences, Coimbatore 641114, India
e-mail: sajeebrahman@mgits.ac.in

D. S. Robinson Smart
Department of Mechanical Engineering, Karunya Institute of Technology and Sciences, Ernakulam 682308, India



K. Antony and J. P. Davim (eds.), *Advanced Manufacturing and Materials Science*, Lecture Notes on Multidisciplinary Industrial Engineering,
https://doi.org/10.1007/978-3-319-76276-0_39

damage to the CNT framework and establishment of a strong bonded interface amidst Al and CNT are the major challenges with CNT reinforcement [8]. $Al_4C_3$ has got rhombohedra structure whereas Al matrix is of cubic structure. The difference in the structures cause rise in interfacial strains between $Al_4C_3$ and Al, which influence the interfacial bonding strength [11]. The study details the latest approaches in the attempt of establishing high performance Al/CNT composites with latest approaches.

## 2 Literature Review

The major and the initial step in forming Al/CNT nanocomposite is to diffuse the reinforcement material with the Al matrix. The most popular method of fabrication of aluminium composite is the High Energy Ball Milling (HEBM) method. Bia and Shufeng et al. [12] studied the fractural performance of Al/CNT nanocomposites with powder metallurgy route. It is recorded that the CNTs retard the growth of crack by operating like a bridge. The high magnification of crack reveals CNTs with inner walls of decreased outer diameter indicating fracturing from outer diameter. The CNTs are also observed to have undergone multiple peeling with a few fractured stages. According to them the CNT fracture mode accounts for effectual load transmission with MWCNT and the matrix and among MWCNT walls in Al/CNT material during tensile failure. The Ultimate Tensile and Yield Strengths are reported to increase whereas elongation is decreased with CNT reinforcement. Carvalo and Miranda et al. [13] recorded that the temperature and frequency are factors which influence the damping behavior of the Al-CNT composites. From 1 to 50 Hz frequency range and from the room temperature to 200 °C temperature range the damping ability of the nanocomposite is observed to increase. With high temperature above 200 °C, the damping performance is reported to decline with increasing frequency. The dynamic young's modulus is found to decrease with temperature. Shin and Ko et al. [14] studied Al matrix composite with multi walled CNT at elevated temperature range obsereved that the yield stress is enhanced two times and the compressive yield stress is found approximately three times superior to the unreinforced Al 2024. Shin and Bae et al. [15] studied on fatigue performance of 2024Al/MWCNT composites. Tensile strength of the material is reported to increase with CNT content. It is the bridging behavior of MWCNT that enhances the fatigue strength for the material. As stated by them 4 vol% of MWCNT the composite material is found to overcome $2.5 \times 10^6$ cycles at an applied stress of 600 MPa.

Increased milling time for good dispersion condition by High Energy Ball Milling (HEBM) results in reduction in the length of CNT. The impact of CNT length is an important factor in strengthening by CNT reinforcement. Chen and Shen et al. [16] recorded that the solution coating methods induces high aspect ratio for CNT whereas HEBM resulted in considerable reduction in length of CNT which is pronounced in long milling time. It is reported that CNT strengthening in aluminium matrix induces considerable enhancement in strength regardless of the aspect ratio. The studies revealed a transition in strengthening mechanism with aspect ratios. With aspect ratio over 40, the dominant phenomenon of strengthening is load transmission mechanisms where as it is Orowan mechanism when the aspect ratio is below 10. The reinforcing influence of CNT in Al matrix is critically determined by the interface bonding with them. A strong transition interface is a determining factor for strong bonding. Chen and Shen et al. [17] studied Al-CNT composites in different interfacial reaction degrees at sintering temperatures of 700–900 K. From 700 to 875 K the quantity of $Al_4C_3$ is observed to be increasing in the interface amidst partially reacted CNTs and Al matrix. Accordingly they reported pronounced improvement in interface strength and load transmission efficiency when the sintering temperature is increased. With elevated temperature of 900 K, the strengthening response is reported to decrease with formation of mono crystal $Al_4C_3$ phases with rod shapes. The variations in structures of $Al_4C_3$ and Al causes rise in interfacial strains which influence the interface bonding strength. Instead of generating a stable phase of $Al_4C_3$, a non equilibrium state achieved in the interfacial area amidst Al and CNT may improve properties of nanocomposite. The method of consolidation is an important factor in achieving the required interfacial phase. Spark plasma Sintering (SPS) which allows rapid cooling rates and precise holding time at relatively low sintering temperatures is an encouraging method to achieve the desired interface. Baisong and Min et al. [18] reported an amorphous layer because of inter diffusion of C and Al in the interface. The faster cooling rates with short processing time of SPS makes it suitable for achieving the interface without $Al_3C_4$ phase. The yield strength of the composite is reported to enhance considerably which accounts for the non equilibrium interface formed. They recorded that the size of the initial Al powder is observed to have great influence on mechanical behaviour of the material. Housaer and Beclinet et al. [19] studied in detail on the impact of sintering temperature and consolidation techniques on the composite interface. For HP at sintering temperature of 580 °C a continuous amorphous $Al_2O_3$ is observed. The $Al_2O_3$ layer existing side by side with CNT serves as a barrier for the straight contact of Al and CNT at the interface. Traces of $Al_4C_3$ are also noted which accounts to the crushing down of some $Al_2O_3$ layer causing diffusion of Al and carbon. At 600 °C, growth of $Al_4C_3$ is reported which shows the diffusion of CNT to Al even below the melting point of Al. With 645 °C no isolated CNT is noted in the interface, instead several rods of $Al_4C_3$ and fine needle shaped $Al_2O_3$ is observed. $Al_4C_3$ rods are present whenever $Al_2O_3$ layers are broken from 610 to 645 °C and grow perpendicular to the matrix grain boundaries into the Al grains. It is recorded that CNT degradation by the generation of $Al_4C_3$ can be slowed down by the precise holding time in SPS technique and hence a promising solution to control the CNT–Al interfaces during sintering. The diffusion of CNTs is improved by introducing carboxyl or hydroxyl grades on the CNT surface. Xian and YuGuang et al. [20] studied 2014 Al matrix composite with untreated carboxyl functionalized carbon nano tubes by combination of sintering and hot extrusion. It is recorded that the addition of functional class of –COOH by carboxyl treatment improves the diffusion of CNT in the matrix and adds to the interfacial bonding with CNT and matrix. The Ultimate Strength of the material has an increment from 530 to 630 MPa with 0.5 wt% of treated CNT. The total elongation is reported to slightly decrease compared to the matrix. Xiang and Xie et al. [21] made use of molecular dynamics simulations (MD) for studying the reinforcement and deformation mechanics at atomic level of aligned dispersion of carbon nano tubes in aluminium matrix composite. They reported considerable enhancement in the elasticity and strength of CNT/Al composite with increased CNT diameter. Further the improvement in mechanical behavior is higher by armchair carbon nano tube than those reinforced by zigzag CNT. It is recorded that CNT bridges the fractural Al sections restricting the propagation of cracks.

The researchers are interested in optimizing the strength and ductility of the Aluminium-CNT nanocomposites which offers better diffusion of CNT with less damage in its structure. Run and Zhianqui et al. [22] reported 1.5 wt% CNT/Al composite by Shift Speed Ball Milling (SSBM). With 9 h LSBM ultra fine grain Al powders are formed. At 2 h of HSBM cold working of Al flakes started and by 9 h the particle size found to increase. The strength of SSBM composite is noticed to be the same and the ductility of the composite increased twice and triple times with LSBM and HSBM respectively. The SSBM material is reported to have high yield strength and ultimate tensile strength with total elongation of 12.4%. Kunning and Xudong et al. [23] recorded that the carbon nano tube reinforcement enhances the energy dissipation capacity by retarding dislocation movement and grain refinement. The damping characteristics are observed to be promising with 3.0 wt% of CNT within temperatures of 25–390 °C. As recorded by them the reinforcement content and porosity determines the loss factor of the nanocomposite. The damping mechanism owes to the great intrinsic damping of CNTs, improved interfacial area, number and bond force amidst CNTs and Al matrix. Xundong and Tianchun et al. [24] reported the hardness and tensile values for the material increased with the wt % of CNT reinforcement up to 4.5 wt%. Within the temperature order of 100–300 ° C the increment in CNT content resulted in decrease of CTEs of the composite. This is characterized to the high diffusion of CNT and effective bonding amidst Al and CNT which results in effective pinning inside the matrix retarding Al grain growth during heating. As recorded by them the CNT reinforcement thus improves the thermal stability of Al matrix. The ductility is also affected by the strain hardening influence of HEBM in case of Al. Ehab and Abbas et al. [25] reported dual matrix fabrication of Al/CNT composite which improves the ductility maintaining sufficient strength for the material. CNTs were dispersed in Al powders for synthesizing the Single matrix (SM) composites whereas pre-processed SM Al-CNT powders were embedded into a secondary matrix of soft aluminum for preparing the dual matrix (DM) composites. Uniform dispersion of reinforcement and firm bond with SM and DM phases are reported with 60 min of mechanical milling. Tensile strength is observed to be highest for 50% mixing ratio. Studies are attempted to improve the ductility by raising the sintering temperature as studied by Chen and Kondoh et al. [26]. They studied a three step fabrication of CNT/Al composite which includes (1) Planetary ball milling, (2) mechanical solution coating and (3) consolidation by SPS. The fragments of low angle GBs is observed to be increasing significantly with rise in sintering temperature from 800 to 900 K which is related to the increased bonding amidst matrix grains. The improved bonding conditions contribute to an enhancement of 18% in yield strength and 14% in Ultimate Tensile Strength and 82% increase in tensile elongation. Mokdad and Chen et al. [27] reported that the refining of grains is greatly affected by CNT reinforcement which contributed to Hall–Petch strengthening or GB strengthening as a result of interactions of GBs and dislocations. They reported a substantial increase in flow stress with decreasing temperature and increasing strain rate. The activation energy in plastic deformation is reported to increase from 274.5 to 322.3 kJ/mol. The compressive yield strength which indicates the high temperature deformation resistance is enhanced with increasing vol% of CNTs.

New approaches have been continuously introduced to unbundle carbon nanotubes via mechanical forces or chemical reactions with minimum destruction of CNTs structure. Rikhtegar and Shabestariet al. [28] investigated semi-wet (SW) and slurry based (SB) and the yield, compressive and tensile strength of Al is found to be significantly increased by both methods. According to them the tensile test specimen exhibits pull out phenomenon for SW products and fracture and bridging of CNTs in SB samples with fracture analysis. Mansoor and Shahid [29] studied induction melting to attain better diffusion of CNTs in molten Al. reported substantial refinement of crystallite size with increment in CNT amount. It is recorded that the mismatched coefficient of thermal expansion of the matrix and the strengthening material featured a rise in the lattice strain with increment in CNT content. The rise in lattice strains causes accumulation in dislocation density which accords to the strengthening of the composite. The synergetic effect of crystallite size refinement, increased lattice strain and reinforcement of nano tubes results in increase of yield strength by 77%, tensile strength by 52%, elongation by 44% and a 45% in hardness. CNT presence activates more slip systems which accounts to the synchronous increase in strength and ductility of the material. Another challenge in conventional liquid processing methods of fabrication is to run over the deformalities during solidification. Zhenglin and Ming et al. [30] recorded that the nano reinforcement caused pinning effect which retards the grain growth in the dynamic crystallization process resulting in pronounced grain refinement. The CNT pull out mechanism at submicron scale around $Al_2O_3$ attributes to 70% hike in yield strength of the material. Samuel and Robinson et al. [31] reported enhancement in corrosion resistance and hardness for AA5083/CNT with increment in wt% of CNT with stir casting process.

## 3 Conclusion

New approaches in achieving uniform dispersion of CNT in Al matrix are included in the study. The study affirms that CNT as reinforcement for Al matrix will improve the thermal and mechanical properties, fatigue resistance, corrosion resistance and damping capabilities of the material. The study contributes the researchers an opportunity to optimize the parameters in refining the grain size and establishing interfacial bonding to attain high performance composites with competent strength maintaining adequate ductility. Different property analyses have been stated in detail which makes Al-CNT composite a promising material to find utilization in many engineering fields.

## References


1. Bakshi SR, Agarwal A (2011) An analysis of the factors affecting strengthening in carbon nanotube reinforced aluminum composites. Carbon N. Y. 49:533–544
2. Torralba JM, Costa CE, Velasco F (2003) P/M aluminum matrix composites: an overview. J Mater Process Technol 133:203–206
3. Ing. Jongejan M, Mackintosh M, Msc. and Ing. de Kluijver D Bayards (2015), cost and weight reduction by choosing aluminium for offshore structures Ing., SPE-177449-MS
4. Lu K (2010) The future of metals. Science 328:319
5. Iijima S (1991) Nature 354:56–58
6. Shadakshari R, Mahesha K, Niranjan HB (2012) Carbon nanotube reinforced aluminium matrix composites—a review. Int J Innovative Res Sci, Eng Technol 2:206–213
7. Bakshi SR, Lahiri D, Agarwal A (2010) Carbon nanotube reinforced metal matrix composites —a review. Int Mat Rev 55:41–64
8. Suryanarayana C (2001) Mechanical alloying and milling. Prog Mater Sci 46:1–184
9. Esawi A, Morsi K (2007) Dispersion of carbon nanotubes (CNTs) in aluminum powder. Compos PartA: Appl Sci Manuf 38:646–650
10. Esawi AM, Morsi K, Sayed A, Gawad AA, Borah P (2009) Fabrication and properties of dispersed carbon nanotube–aluminum composites. Mater Sci Eng, A 508:167–173
11. Kwon H, Park DH, Silvain JF, Kawasaki A et al (2010) Investigation of carbon nanotube reinforced aluminum matrix composite materials. Compos Sci Technol 70:546–550
12. Chen B, Li S, Imai H, Jia L, Umeda J, Takahashi M, Kondoh K (2015) Load transfer strengthening in carbon nanotubes reinforced metal matrix composites via in-situ tensile tests. Compos Sci Technol 113:1–8
13. Carvalho O, Miranda G, Buciumeanu M, Gasik M, Silva FS, Madeira S (2016) PII: high temperature damping behavior and dynamic Young's modulus of AlSi- CNT-SiCp hybrid composite. Compos Struct 141:155–162
14. Shin SE, Ko YJ, Bae DH (2016) Mechanical and thermal properties of nanocarbon-reinforced aluminum matrix composites at elevated temperatures. Compos B 106:66–73
15. Shin SE, Bae DH (2017) Fatigue behavior of Al2024 alloy-matrix nanocomposites reinforced with multi-walled carbon nanotubes. Compos B Eng 134:61–68
16. Chen B, Shen J, Ye X, Jia L, Li S, Umeda J, Takahashi M, Kondoh K (2017) PII: length effect of carbon nanotubes on the strengthening mechanisms in metal matrix composites. Acta Mater 140:317–325
17. Chen B, Shen J, Ye X, Imai H, Umeda J, Takahashi M, Kondoh K (2017) Solid-state interfacial reaction and load transfer efficiency in carbon nanotubes (CNTs)-reinforced aluminum matrix composites. Carbon 114:198–208
18. Guo B, Song M, Yi J, Ni S, Shen T, Du Y (2017) Improving the mechanical properties of carbon nanotubes reinforced pure aluminum matrix composites by achieving non-equilibrium interface. Mat Des 120:56–65
19. Housaer F, Beclin F, Touzin M, Tingaud D, Legris A, Addad A (2015) Interfacial characterization in carbon nanotube reinforced aluminum matrix composites. Mater Charact 110:94–101

20. Zhu X, Zhao Y-G, Wu M, Wang H-Y, Jiang Q-C (2016) Fabrication of 2014 aluminum matrix composites reinforced with untreated and carboxyl-functionalized carbon nanotubes. J Alloys Compd 674:145–152
21. Junfeng Xiang, Lijing Xie, Meguid Shaker A, Siqin Pang, Jie Yi, Yua Zhang, Ruoa Liang (2017) An atomic-level understanding of the strengthening mechanism of aluminum matrix composites reinforced by aligned carbon nanotubes. Comput Mater Sci 128:359–372
22. Xu R, Tan Z, Xiong D, Fan G, Guo Q, Zhang J, Su Y, Li Z, Zhang D (2017) Balanced strength and ductility in CNT/Al composites achieved by flake powder metallurgy via shift-speed ball milling. Compos: Part A Appl Sci Manuf 97:57–66
23. Yang Kunming, Yang Xudong, He Chunnian, Liu Enzuo, Shi Chunsheng, Ma Liying, Li Qunying, Li Jiajun, Zhao Naiqin (2017) Damping characteristics of Al matrix composite foams reinforced by in-situ grown carbon nanotubes. Mater Lett 209:68–70
24. Yang Xudong, Zou Tianchun, Shi Chunsheng, Liu Enzuo, He Chunnian, Zhao Naiqin (2016) Effect of carbon nanotube (CNT) content on the properties of in-situ synthesis CNT reinforced Al composites. Mater Sci Eng, A 660:11–18
25. Salama EI, Abbas A, Esawi AM (2017) Preparation and properties of dual-matrix carbon nanotube-reinforced aluminum composites. Compos Part A: Appl Sci Manuf 99:84–93
26. Chen B, Kondoh K, Imai H, Umeda J, Takahash M (2016) Simultaneously enhancing strength and ductility of carbon nanotube/ aluminum composites by improving bonding conditions. Scripta Mater 113:158–162
27. Mokdad F, Chen DL, Liu ZY, Ni DR, Xiao BL, Ma ZY (2017) Hot deformation and activation energy of a CNT—reinforced aluminum matrix nanocomposite. Mater Sci Eng, A 695:322–331
28. Rikhtegar F, Shabestari SG, Saghafian H (2017) Microstructural evaluation and mechanical properties of Al-CNT nanocomposites produced by different processing methods. J Alloy Compd 723:633–641
29. Mansoor M, Shahid M (2016) Carbon nanotube-reinforced aluminum composite produced by induction melting. J Appl Res Technol 14:215–224
30. Dua Z, Tana MJ, Guob JF, Bi G, Wei J (2016) Fabrication of a new Al-Al2O3-CNTs composite using friction stir processing (FSP), Mat Sci Eng A 667:125–131
31. Samuel Ratna Kumara PS, Robinson Smart DS, John Alexis S (2017) Corrosion behaviour of aluminium metal matrix reinforced with multi-wall carbon nanotube. J Asian Ceram Soc 258:1–5

# Effect of Mechanical and Metallurgical Analysis of Magnesium and Aluminium Alloys Using Diffusion Bonding

**R. J. Golden Renjith Nimal, M. Sivakumar, S. Arungalai Vendan and G. Esakkimuthu**

## 1 Introduction

Diffusion bonding is a process of solid state joining in which two metals with neat surfaces are brought together in an ascending temperature and stress for a predefined time [1]. In the current study, diffusion bonding of magnesium AZ 80 alloy and aluminium 7075 alloy was investigated. Aluminium is considered as the most important metal available in the earth's crust [2]. This becomes a strong competitor for steel in various Engineering applications. Due to the excellent mechanical properties and corrosion resistance, aluminium 7075 alloys are extensively used [3]. Magnesium alloys usually compete with aluminium alloys in structural applications [4]. For determining the various mechanical and metallurgical properties of these Al and Mg alloys its thermodynamical properties should be flexible at lower temperatures [5]. The problem when joining both Magnesium and Aluminium, the formation of good hardness and intermetallic compounds (IMCs) brittleness [6]. Joseph Fernandus et al. concentrated on the central trouble when joining magnesium (Mg) and aluminum (Al) lies in the presence of development of oxide movies and weak intermetallic in the security area. However dissemination holding can be

R. J. Golden Renjith Nimal (✉)
Mechanical Department, Bharath University, Chennai 600073, India
e-mail: goldenrenjith@gmail.com

M. Sivakumar
Department of Mechanical Engineering, Sree Sowdambika College of Engineering, Aruppukottai, Tamil Nadu, India

S. Arungalai Vendan
Industrial Automation and Instrumentation Division, School of Electrical Engineering, VIT University, Vellore, India

G. Esakkimuthu
Department of Mechanical Engineering, National Engineering College, Kovilpatti, Tamil Nadu, India



K. Antony and J. P. Davim (eds.), *Advanced Manufacturing and Materials Science*, Lecture Notes on Multidisciplinary Industrial Engineering,
https://doi.org/10.1007/978-3-319-76276-0_40

utilized to join these composites without much trouble. In this examination, an endeavor was made to create Temperature–Time and Pressure–Time charts for dispersion holding of AZ80 magnesium (Mg) and AA6061 aluminum (Al) unique materials [7]. The holding nature of the joints was checked by microstructure examination and lap shear tractable testing. In light of the outcomes Temperature–Time and Pressure–Time outlines were developed [8]. These charts will go about as reference maps for selecting fitting dispersion holding process parameters to join AZ80 magnesium combination and AA6061 aluminum composite without trial tests [9]. The heating rate of 25 °C/min is the maximum elevating temperature with 1473 K-273 = 1200 °C. The highest pressure of 10 tons may be applied through hydraulic pressure and the vacuum degree −760 mm of Hg [10]. The vacuum system is composed of diffusion pump [11]. By studying the microstructure and shear strength of the joints, the effects on Mg/Al diffusion bonded joint was discussed.

## 2 Experimentation

Aluminium AA7075 alloy plate of 5 mm thickness and Magnesium AZ80 alloy plate of 10 mm thickness were cut into the required dimensions (45 × 45 mm) by power hacksaw and grinding. The highly polished and chemically treated specimens of Mg/Al was arranged in a die made up of 316 L stainless steel and the entire diffusion bonding die set up shown in Fig. 1 was kept in a vacuum chamber [12]. The constant load or constant diffusion bonding pressure was applied using a servo controlled hydraulic press. The maximum capacity of the press is 10 Ton. The chemical composition of the base metals are shown in Table 1.

Water cooling system is essential to maintain the temperature of the vacuum seals bellows at 473 K. The cooling system has two tanks connected to the heating chamber. The tepid water (at 325 K) popping out of the heating chamber is pumped up to the top cooling tank and recirculated to lower tank to keep the inlet

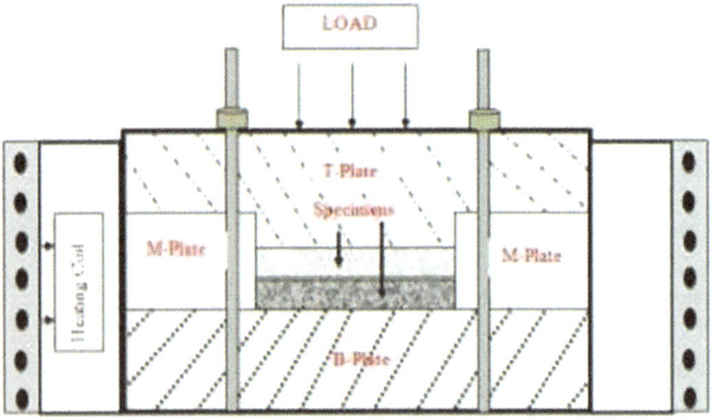


**Fig. 1** Configuration of the diffusion bonding die set up

**Table 1** Chemical composition of the base metal

| | Wt. (%) | |
|---|---|---|
| Elements | Magnesium AZ 80 alloy | Aluminium 7075 alloy |
| Mg | 90.591 | 2.1 |
| Al | 8.36 | 90.02 |
| Zn | 0.75 | 5.1 |
| Cu | 0.002 | 1.2 |
| Fe | 0.0037 | 0.5 |
| Si | 0.033 | 0.4 |
| Mn | 0.26 | 0.3 |
| Ni | 0.00056 | – |
| Cr | – | 0.18 |
| Ti | – | 0.2 |

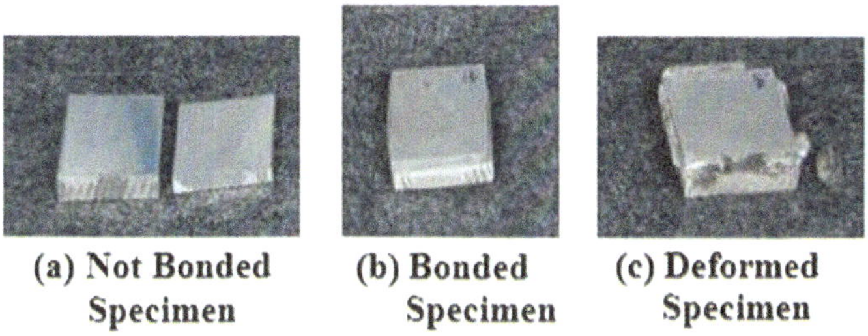


**Fig. 2** Diffusion bonded specimens

**Table 2** Trials conducted with the effect of temperature on Mg /Al alloys

| Effect of temperature | | | | |
|---|---|---|---|---|
| Sl. No. | Bonding temperature (°C) | Bonding pressure (MPa) | Bonding time (min) | Remarks |
| 1 | 350 | 10 | 15 | Not bonded |
| 2 | 375 | 10 | 15 | Bonded |
| 3 | 400 | 10 | 15 | Bonded |
| 4 | 425 | 10 | 15 | Bonded |
| 5 | 450 | 10 | 15 | Deformed |

temperature at 300 K. The cooled water is circulated to the heating furnace through a 0.25 hp pump. The process parameters of temperature, pressure and time were varied systematically. To identify the bonding quality, the effect of temperature on diffusion bonded specimens of magnesium AZ80 alloy and aluminium 7075 alloy was carried out for different temperatures ranging from 350 to 450 °C while keeping the bonding pressure at 10 MPa and bonding time at 15 min (Fig. 2; Table 2).

# 3 Results and Discussion

## *3.1 Mechanical Testing*

The joints fabricated under this condition to evaluate the bond quality, mechanical testing like lap shear tensile specimens were prepared. The tensile test results are given in Table 3, which indicates that the weld strength is slightly above the ultimate tensile strength in the base material of the specimen (Fig. 3).

The lap shear strength (LSS) with respect for the Mg/Al alloy interfaces for various temperatures. For the temperature below 375 °C, bonding between the specimens do not occur due to insufficient temperature. For temperature of 375 °C, the LSS was found to be around 15 MPa. As the temperature is increased to 400 °C further the lap shear strength increases to the value of 27 MPa and for the temperature of 425 °C, the LSS was found to decrease to 21 MPa. It can be clearly seen from the above graph that both the specimens stands apart at lower temperature and at higher temperature the deformation occurs. From these above trials, it was concluded that the optimum diffusion temperature is between 375 and 425 °C. The error in the experimental values assigned was around 5% and this is also indicated in the below figure with the error bar. The well separated samples at lower temperature inside the figure clearly depicts the un-bonded nature.

**Table 3** Results of lap shear test specimens

| Effect of temperature | | | | |
|---|---|---|---|---|
| Sl. No. | Bonding temperature (°C) | Bonding pressure (MPa) | Bonding time (min) | Lap shear strength (MPa) |
| 1 | 350 | 10 | 15 | – |
| 2 | 375 | 10 | 15 | 15 |
| 3 | 400 | 10 | 15 | 26 |
| 4 | 425 | 10 | 15 | 21 |
| 5 | 450 | 10 | 15 | – |

**Fig. 3** Photographs of lap shear test specimen **a** Before testing and **b** After testing

## 3.2 Metallurgical Testing

The interface layer could be visualized from microscopic analysis. Figure 4 represents the optical microscopic image at various sample temperatures of 375, 400 and 425 °C.

The increase in the hardness may be due to the formation of $Mg_2Al_3$ and $Mg_{17}Al_{12}$ intermetallic (IMC) compounds during the high degree of holding time and this IMC formation during solidification can lead to an increased mechanical properties than the parent alloys of aluminium and magnesium. The Field Emission Scanning Electron Microscope (FESEM) is used to investigate on the surface morphology of the diffusion bonded specimen (Fig. 5).

The optimized well bonded samples which were heated to 400 °C was used in the present investigation. The bonding pressure was maintained at 10 MPa with holding time of 15 min. The well bonded samples were cut into required sizes and the SEM image is observed to investigate the diffusion mechanism.

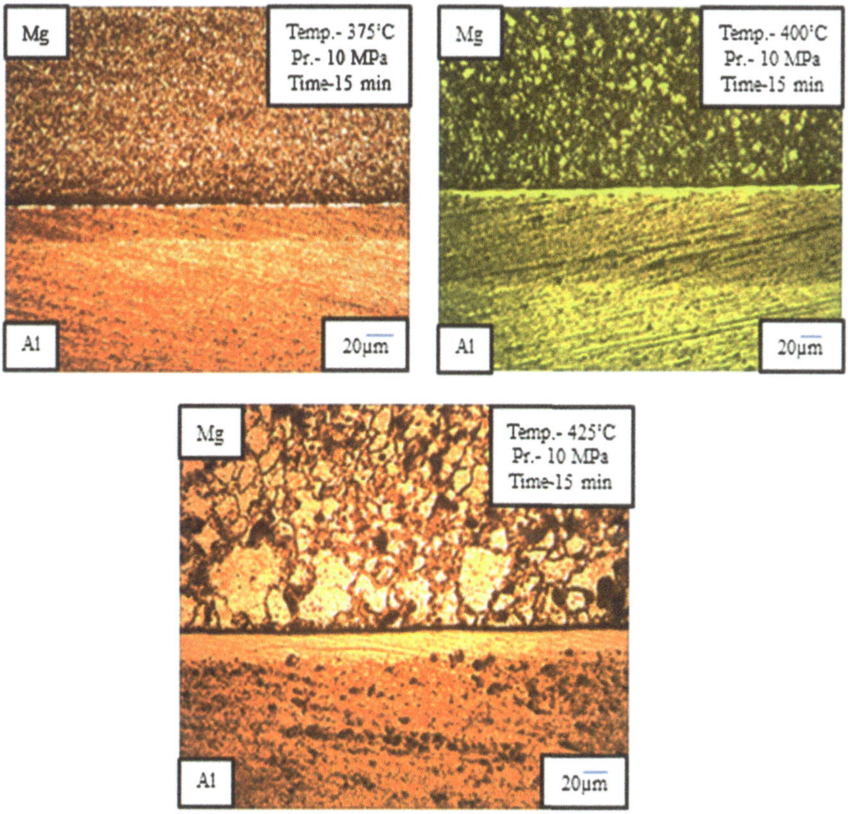


**Fig. 4** Optical microscope image of Mg/Al alloy at various sample temperatures

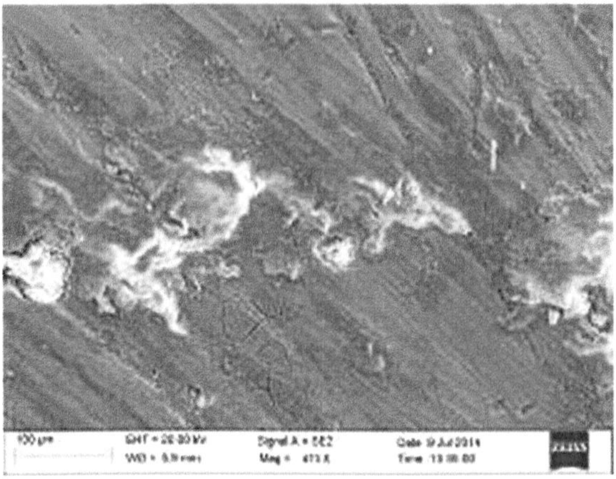

Fig. 5 FESEM fractograph of bonded specimen of Mg/Al at 100 X

# 4 Conclusion

Diffusion bonding of Magnesium AZ80 alloy and Aluminium 7075 alloy, exhibit maximum shear strength for the specimen bonded at 400 °C, 10 MPa and 15 min. The lap shear strength of the bonded specimens is determined to extend with increasing temperature till a most worth is reached beyond that it decreases. The thickness of the inter-metallic layer is narrower at a lower holding temperature and their widths increase with the rise within the change of integrity temperature attributable to increase within the inter-diffusion of chemical species. The thickness of IMC's is proportional to the temperature and holding time. Once the holding time is longer than the latent amount, the super saturated solid solution seems to be unstable. Inter metallic phases nucleated and grew, while AlMg phase was missing.

**Acknowledgements** We would like to thank Dr. V. Balasubramanian, Professor, Centre for Materials Joining and Research (CEMAJOR), Department of Manufacturing Engineering, Annamalai University for his support in performing diffusion bonding experiments and testing the bonded joints.

# References

1. Joseph Fernandus M, Senthilkumar T, Balasubramonian V, Rajakumar S (2012) Optimizing diffusion bonding parameters in AA6061-T6 aluminium and AZ80 magnesium alloy dissimilar joints. J Mater Eng Perform 21:2303–2315
2. Peng L, Yajiang L, Haoran G, Juan W (2006) Investigation of interfacial structure of Mg/Al vacuum diffusion-bonded joint. Vacuum 80:395–399
3. Mahendran G, Babu S, Balasubramanian V (2010) Analyzing the effect of diffusion bonding process parameters on bond characteristics of Mg-Al dissimilar joints. J Mat Eng Perf 19:657–665

4. Liu W, Long L, Ma Y, Wu L (2015) Microstructure evolution and mechanical properties of Mg/Al diffusion bonded joints. J Alloy Compd 643:34–39
5. Huang Y, Humphreys FJ, Ridley N, Wang ZC (1988) Diffusion bonding of hot rolled 7075 aluminium alloy. Mater Sci Technol 14:405–410
6. Zhang J, Shen Q, Luo G, Li M, Zhang L (2012) Microstructure and bonding strength of diffusion welding of Mo/Cu joints with Ni interlayer. Mat Des 39:81–86
7. Ziegelheim J, Hiraki S, Ohsawa H (2007) Diffusion bondability of similar/dissimilar light metal sheets. J Mater Process Technol 186:87–93
8. Mahendran G, Balasubramanian V, Senthilvelan T (2009) Developing diffusion bonding windows for joining AZ31B magnesium-AA2024 aluminium alloys. Mater Des 30: 1240–1244
9. Nishi H (2006) Notch toughness evaluation of diffusion bonded joints of alumina dispersion-strengthened copper to stainless steel. Fusion Eng Des 81:269–274
10. Orhan N, Khan TI, Eroglu M (2001) Diffusion bonding of a micro duplex stainless steel to Ti-6Al-4V. Scripta Mater 45:441–446
11. Padron T, Khan TI, Kabir MJ (2004) Modelling the transient liquid phase bonding behaviour of a duplex stainless steel using copper interlayers. Mater Sci Eng A 385:220–228
12. Golden Renjith Nimal RJ, Siva Kumar M, Arungalai Vendan S, Essaki Muthu G, Chinnadurai T (2015) Studies on mechanical and metallographical aspects of diffusion bonding of AA 7075 aluminium alloy and AZ80 magnesium alloy. Ann "Dunarea de Jos" University of Galati, Fascicle XII, Weld Equip Technol 26:19–26

# Operation Sequence Dependent Part Family Formation for Reconfigurable Manufacturing System (RMS)

**Rutuja Krishnat Shivads and Martand Tamanacharya Telsang**

## 1 Introduction

Today's manufactures face progressively challenges earlier than because of highly impulsive market changes and hence there are large fluctuations in product demand. To be to be competitive, it is mandatory for companies to design manufacturing systems to produce high quality products at low cost which responding to market changes in an economical way. RMS should be designed in such that it can be rapidly and cost effectively reconfigured to the exact capacity to meet a new market demand. Dedicated manufacturing system (DMS) are beneficial for high volume and low variety production while, Flexible manufacturing system (FMS) economical for large diversity of parts and low production volumes. FMSs are unable to provide robustness of DMS and can have excessive functionality. FMSs build using of inadequate system software and advancement of user specified software is very expensive. Due to development in computers optics, a control, processing i.e. advancement in technologies today's most capable system becomes incapable after short life. Koren et al. suggested RMS as solution to tackle changing market demands. RMS should be designed in such that it can be rapidly and cost effectively reconfigured to the exact capacity to meet a new market demand. Researchers at the NSF Engineering research centre for RMS (ERC/RMS) start addressing system scalability since 1990s, and issued patent that deals with strategies to change production capacity in RMS [8]. RMS can be defined as: "Reconfigurable Manufacturing Systems are designed at the outset for rapid change in structure, also in hardware and software components, in order to quickly adjust

R. K. Shivads (✉)
A/P-Umbraj (Surabhi Chauk), Karad, Satara, Maharashtra 415109, India
e-mail: rutujashivads7@gmail.com

M. T. Telsang
Department of Mechanical Engineering, Rajarambapu Institute of Technology, Rajaramnagar, Islampur, Maharashtra 415414, India



K. Antony and J. P. Davim (eds.), *Advanced Manufacturing and Materials Science*, Lecture Notes on Multidisciplinary Industrial Engineering,
https://doi.org/10.1007/978-3-319-76276-0_41

production capacity and functionality within a part family in response to sudden changes in market or regulatory requirements".

In this paper part family formation methodology for RMS suggested by Goyal et al. [2]; is applied for parts manufactured in gear manufacturing company. In this study part family is formed for 7 parts having some variations in their geometry and manufacturing processes. Study analyses benefits of using RMS over existing manufacturing system.

## 2 Literature Review

Koren et al. recommended RMS as solution for changing manufacturing systems since 1990s, and handful research have made for implementation of reconfigurable manufacturing system in practice. Design of RM is one of the very important stage in development of RMS. Researchers at the NSF Engineering research centre issued RMT design patent in the year 1999 [8]. Koren et al. commence features and principles of RMS. They evaluate RMS with conventional manufacturing system and address method to calculate no. of RMS configurations [4]. They identify need of method to determine right time to build RMS. Wang et al. introduces scalability planning scheme for reconfiguring practical manufacturing system and propose GA for optimization. Mathematical approach to minimize the total no. of machines required to achieve required production presented and case study that validate proposed approach presented [5]. History of reconfigurable manufacturing system is presented by comparing RMS with traditional manufacturing systems. Challenges in using RMS are also presented and address need of design concepts and methodologies expansion for RMS [7]. Putnik et al. brings in scalability as state of art in manufacturing [6]. They propose approach for basic design, operational aspects and management of scalability in manufacturing with explaining scalability is key character of Reconfigurable manufacturing system.

Reconfigurable machines designed considering part family of products; hence grouping products and parts in families plays important role in successive implementation of RMS. Galan et al. Focus on system level issues to reduce time needed to manufacture products in right quantity and with the right quality [1]. AHP methodology used to form unique matrix that consider the similarities in five products attributes. ALC approach proposed to form product family and Dendogram used to show product families. Goyal et al. developed operation sequence dependent BMIM (bypassing moves and idle machines) similarity coefficient for part family creation [2]. Kashkoush and Elmaraghy introduce product family formation method for reconfigurable assembly system. In similarity coefficients product assembly sequences, demand and commonality are considered. Techniques of phylogenetics and biology along with average linkage hierarchical clustering are used to group the products. Novel consensus tree based method is applied to find the best aggregation and proposed this system will increase system efficiency and productivity. Identifies further work need to be done for estimating quantitative benefits after application of proposed methodology [3].

# 3 Part Family Formation for RMS

Reconfigurable machine (RM) is designed considering a part family of products. RM machines are designed after formation of part family considering manufacturing operations need to perform on jobs. Hence, part family formation for RMS plays vital role in application of RMS. To check implementation feasibility of RMS in real life example the part family formation methodology is applied to parts in company. In this study part family has been formed by selecting 7 parts of company having some variations in their geometry and manufacturing processes. Operation sequence dependent BMIM (bypassing moves and idle machines) similarity coefficient for part family creation proposed by Goyal et al. [2] is applied for part family formation. This study analyzes benefits of using RMS over existing manufacturing system. The part family formation starts with identifying general operation sequence for parts. In second stage BMIM similarity coefficients are calculated and in last stage Average linkage clustering (ALC) applied to group parts in family. To understand the operation sequences of parts drawings of considered parts are firstly studied. The parts selected for study covers almost all the types of parts that are manufactured in company. Hence criteria applied for selection of parts is at least one part should consider for study from each type of parts.

## *3.1 Operation Sequence*

The operation sequence for the selected parts such as Spur gear, Worm shaft, helical gear etc. are considered. The operation sequences for parts observed in company are shown in Table 1. The nomenclature used to denote the operation name described in Table 2.

**Table 1** Operation sequence for parts

| Part no. | Part name | General operation sequence |
|---|---|---|
| 1 | Helical gear | [D,T,F,t,O,B,C,H,S] |
| 2 | Spur gear | [D,T,F,t,O,B,C,c,D,R,H,S] |
| 3 | Worm (a) | [D,T,F,t,O,B,C,M,b] |
| 4 | Worm (b) | [D,T,F,m,M,D,K] |
| 5 | Spur & worm wheel | [D,L,O,F,t,OFg,B,c,C,d,R,H,M,S] |
| 6 | Worm shaft | [d,O,F,k,G,M,D,P] |
| 7 | Eccentric pin | [f,O,F,E,C] |

**Table 2** Nomenclature

| No. | Nomenclature | Operation | No. | Nomenclature | Operation |
|---|---|---|---|---|---|
| 1 | D | Drilling | 12 | M | Thread milling |
| 2 | T | Turning | 13 | B | Broaching |
| 3 | F | Facing | 14 | K | Key slotting |
| 4 | t | Toughing | 15 | m | CNC machining finish |
| 5 | O | O.D turning | 16 | L | Blanking |
| 6 | B | Boring | 17 | g | Grooving |
| 7 | C | Chamfering | 18 | k | Key way |
| 8 | H | Hobbing | 19 | G | Grinding |
| 9 | S | Shaving | 20 | P | Tapping |
| 10 | C | Circlip cutting | 21 | E | Eccentric grinding |
| 11 | R | Threading | 22 | f | Forging |

## 3.2 Calculation of BMIM

The BMIM similarity coefficient computed using Eq. 1 [2]. Table 3 shows calculated parameters values of similarity coefficients

$$S_{ab} = \left\{ \left[ \frac{BMP_a}{2 \times |TM_a|} + \frac{BMP_b}{2 \times |TM_b|} \right] + \left[ \frac{IM_a}{2 \times |SCS_{ab}|} + \frac{IM_b}{2 \times |SCS_{ab}|} \right] \right\} \quad (1)$$

where,

*a, b* Operation sequence a and b.
*LCSab* Longest common subsequence between operation sequences a and b.
*SCSab* Shortest common supersequence between operation sequences a and b.
*NOBLa* Number of operations of operation sequence a to be appended before the to form *SCSab*.
*NOALa* Number of operations of operation sequence a to be appended after the to form *SCSab*.
*NOILa* Number of operations of operation sequence a to be appended in between the to form *SCSab*.

Table 4 shows BMIM similarity coefficient values for parts 1–7. The Average linkage clustering (ALC) methodology is used for grouping the products. Tables 5, 6, 7, 8 and 9 respectively shows submatrix-1 part family, Submatrix-2 part family, Submatrix-3 part family, Submatrix-4 and Final matrix part family obtained after Appling ALC methodology.

Final matrix shows that Part families formed are such that Part family I-(1, 2, 3 and 5) and Part family II-(4, 7 and 6) i.e. part family I constitutes parts such as Helical gear, Spur gear, Worm gear (a) and Spur & Worm gear while part family II constitutes Worm gear (b), Worm shaft and Eccentric pin.

**Table 3** BMIM similarity coefficient calculation

| ab | LCSab | LCSab sequence | $SCS_{ab}$ | ξa | $\Phi_a$ | $\xi_b$ | $\Phi_b$ | $BMP_a$ | $BMP_b$ | $TM_a$ | $TM_b$ | $IM_a$ | $IM_b$ | $S_{ab}$ |
|---|---|---|---|---|---|---|---|---|---|---|---|---|---|---|
| 12 | 9 | [DTFtOBCHS] | 12 | 0 | 0 | 0 | 0 | 3 | 0 | 13 | 13 | 3 | 0 | 0.77 |
| 13 | 7 | [DTFtOBC] | 11 | 0 | 2 | 0 | 2 | 2 | 2 | 12 | 12 | 2 | 2 | 0.65 |
| 14 | 3 | [DTF] | 13 | 0 | 4 | 0 | 0 | 4 | 0 | 14 | 8 | 2 | 6 | 0.55 |
| 15 | 8 | [DFtOBCHS] | 17 | 0 | 0 | 0 | 0 | 8 | 1 | 18 | 18 | 8 | 1 | 0.49 |
| 16 | 1 | [O] | 16 | 1 | 0 | 0 | 4 | 1 | 4 | 11 | 13 | 7 | 8 | 0.33 |
| 17 | 2 | [OC] | 12 | 1 | 0 | 0 | 0 | 3 | 1 | 13 | 7 | 3 | 7 | 0.4 |
| 23 | 7 | [DTFtOBC] | 14 | 0 | 2 | 0 | 0 | 2 | 0 | 15 | 10 | 2 | 5 | 0.68 |
| 24 | 4 | [DTFD] | 15 | 0 | 1 | 0 | 0 | 3 | 5 | 16 | 13 | 3 | 8 | 0.35 |
| 25 | 10 | [DFtOBcDRHS] | 18 | 0 | 0 | 0 | 0 | 6 | 2 | 19 | 19 | 6 | 2 | 0.57 |
| 26 | 2 | [OD] | 18 | 1 | 1 | 0 | 0 | 6 | 3 | 19 | 12 | 6 | 10 | 0.27 |
| 27 | 2 | [OC] | 15 | 1 | 0 | 0 | 0 | 3 | 1 | 16 | 7 | 3 | 8 | 0.47 |
| 34 | 4 | [DTFM] | 12 | 0 | 0 | 0 | 1 | 1 | 5 | 11 | 13 | 3 | 5 | 0.43 |
| 35 | 7 | [DFtOBCM] | 18 | 0 | 1 | 0 | 1 | 9 | 2 | 19 | 19 | 9 | 2 | 0.40 |
| 36 | 2 | [OM] | 15 | 1 | 0 | 0 | 1 | 4 | 3 | 14 | 18 | 6 | 7 | 0.34 |
| 37 | 2 | [OC] | 12 | 1 | 0 | 0 | 0 | 3 | 1 | 13 | 7 | 3 | 7 | 0.4 |
| 45 | 3 | [DFD] | 20 | 0 | 0 | 0 | 1 | 9 | 4 | 17 | 21 | 13 | 4 | 0.22 |
| 46 | 3 | [FMD] | 12 | 2 | 1 | 2 | 1 | 5 | 4 | 13 | 13 | 5 | 4 | 0.28 |
| 47 | 1 | [F] | 11 | 2 | 2 | 2 | 0 | 4 | 2 | 12 | 8 | 4 | 3 | 0.39 |
| 56 | 3 | [OFM] | 21 | 1 | 0 | 0 | 1 | 3 | 11 | 20 | 20 | 5 | 13 | 0.22 |
| 57 | 3 | [OFC] | 18 | 1 | 0 | 0 | 0 | 2 | 6 | 19 | 12 | 2 | 13 | 0.28 |
| 67 | 2 | [OF] | 11 | 1 | 1 | 2 | 0 | 3 | 1 | 12 | 7 | 3 | 6 | 0.39 |

**Table 4** BMIM similarity coefficient values for parts

| Part | 1 | 2 | 3 | 4 | 5 | 6 | 7 |
|---|---|---|---|---|---|---|---|
| 1 | – | 0.77 | 0.65 | 0.55 | 0.49 | 0.33 | 0.4 |
| 2 | | | 0.68 | 0.35 | 0.57 | 0.27 | 0.47 |
| 3 | | | | 0.43 | 0.40 | 0.34 | 0.4 |
| 4 | | | | | 0.22 | 0.28 | 0.39 |
| 5 | | | | | | 0.22 | 0.28 |
| 6 | | | | | | | 0.39 |

**Table 5** Submatrix-1 part family

| Part | (1, 2) | 3 | 4 | 5 | 6 | 7 |
|---|---|---|---|---|---|---|
| (1, 2) | | 0.67 | 0.45 | 0.53 | 0.3 | 0.44 |
| 3 | | | 0.43 | 0.40 | 0.34 | 0.4 |
| 4 | | | | 0.22 | 0.28 | 0.39 |
| 5 | | | | | 0.22 | 0.28 |
| 6 | | | | | | 0.39 |

**Table 6** Submatrix-2 part family

| Part | (1, 2, 3) | 4 | 5 | 6 | 7 |
|---|---|---|---|---|---|
| (1, 2, 3) | | 0.29 | 0.31 | 0.21 | 0.28 |
| 4 | | | 0.22 | 0.28 | 0.39 |
| 5 | | | | 0.22 | 0.28 |
| 6 | | | | | 0.39 |

**Table 7** Submatrix-3 part family

| Part | (1, 2, 3) | (4, 7) | 5 | 6 |
|---|---|---|---|---|
| (1, 2, 3) | | 0.095 | 0.31 | 0.21 |
| (4, 7) | | | 0.25 | 0.34 |
| 5 | | | | 0.22 |

**Table 8** Submatrix-4 part family

| Part | (1, 2, 3) | (4, 7, 6) | 5 |
|---|---|---|---|
| (1, 2, 3) | | 0.033 | 0.31 |
| (4, 7, 6) | | | 0.16 |

**Table 9** Final matrix part family

| Part | (1, 2, 3, 5) | (4, 7, 6) |
|---|---|---|
| (1, 2, 3, 5) | | 0.016 |

Figure 1 shows Dendogram for calculated data from which selection of part families can be made. The selection of part families will be decision of organization and it will depend upon the objectives of organization. Dendogram formed for part family shows that parts 1 and 2 i.e. helical gear and spur gear respectively are 77%

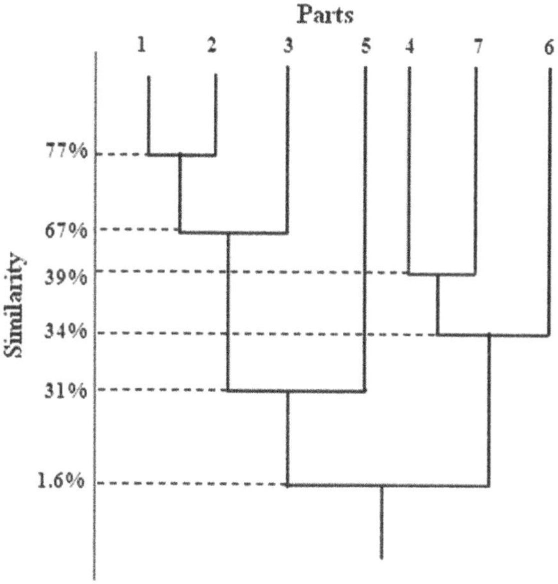


**Fig. 1** Dendogram part family

similar considering their operation sequence while grouping both of them with worm gear (a) gives 67% similarity.

Group of spur, helical, worm gear (a) and spur & worm gives 31% similarity. Similarly from dendogram it is observed group of worm gear (b), worm shaft and eccentric pin gives 34% similarity considering their operation sequences. These families if processed using RMS then benefits in terms of reduction in setup times on hobbing machines, CNCs will be reduced. Production rate will be increased cost effectively. The cost related with designing of fixtures, tools etc. for processing various parts will be reduced. Recognized benefits of using RMS over existing manufacturing system are presented in following section.

## 4 Conclusion

RMS is cost efficient solution for problems arising due to changeable manufacturing system. Application of RMS is in incubation period i.e. period required to convey knowledge and technology to industry. Production planning and control techniques for RMS plays important role in implementation of RMS. To check implementation feasibility of RMS in real life example the part family formation methodology is applied to parts in company. Dendogram formed for part family shows that helical gear and spur gear are 77% similar considering their operation sequence while grouping both of them with worm gear (a) gives 67% similarity. Group of spur, helical, worm gear (a) and spur & worm gives 31% similarity.

Similarly from dendogram it is observed group of worm gear (b), worm shaft and eccentric pin gives 34% similarity considering their operation sequences. This industrial study helps to analyze benefits of using RMS over existing manufacturing system. These families if processed using RMS then benefits in terms of reduction in setup times on hobbing machines, CNCs will be reduced. Production rate will be increased cost effectively. The cost related with designing of fixtures, tools etc. for processing various parts will be reduced. The industrial study identified there is need to apply part family formation methodology to all parts in company and further work need to carry out for designing reconfigurable machines for part families grouped.

## References


1. Galan R, Racero J, Eguia I, Garcia J (2007) A systematic approach for product families formation in reconfigurable manufacturing systems. Robot Comput-Integr Manuf 23:489–502
2. Goyal K, Jain P, Jain M (2013) A comprehensive approach to operation sequence similarity based part family formation in the reconfigurable manufacturing system. Int J Prod Res 51:1762–1776
3. Kadhkoush M, Elmaraghy H (2014) Product family formation for reconfigurable assembly systems. Procedia CIRP 17:302–307
4. Koren Y, Shpitalni M (2010) Design of reconfigurable manufacturing systems. J Manuf Syst 29:130–141
5. Koren Y, Wang W (2012) Scalability planning for reconfigurable manufacturing system. J Manuf Syst 31:83–91
6. Putnik G, Sluga A, Eimaraghy H, Teti R, Korean Y, Yolio T, Hon B (2013) Scalability in manufacturing systems design and operation: state-of-art the future developments roadmap. CIRP Ann 62:751–774
7. Singh A, Kumar P, Singh S (2013) Vision, principles and impact of reconfigurable manufacturing system. Int J Eng Adv Technol 3:238–240
8. Tillbury D, Kota S (1999) Integrated machine and control design for reconfigurable machine tools. In: Proceedings of IEEE/ASME, 1999, pp 629–634

# Modeling and Analysis of Tool Wear Based on Cutting Force and Chip-Tool Interface Temperatures in Turning

L. B. Abhang and M. Hameedullah

## 1 Introduction

Machining processes are very common in manufacturing technology. These operations are applied in manufacturing of almost every mechanical element. Because of their frequent use, these processes have to be efficient and economical. To lower manufacturing costs, many parameters need to be considered. Tool wear is one of the most important considerations in metal cutting operations as it affects surface quality, productivity, cost etc. In the modern era of intense competition, reduction of machining cost is the key to sustain. For this the, performance of cutting tool plays a vital role.

A tool life is directly related to the wear behavior of cutting tool. To predict tool life, it is necessary to identify the actual wear progress under practical conditions [1, 2]. Taking out the tool to the metrology lab and measuring the actual wear of tool is a difficult task and affects the cutting efficiency. If the tool wear rate prediction models are available in terms of chip-tool interface temperature or cutting forces, the machining efficiency can be improved. By changing the tool where the wear rate reaches a control values.

In turning operations, it is convenient to consider the forces as a three component system. These are main cutting force component, feed force component and radial force component. The components of the cutting forces and chip-tool interface temperatures in metal cutting operations provide a wealth of information on the metal removal process. Changes in these forces and cutting temperatures indicate changes in machining parameters and tool geometry, such as cutting speed, feed rate, depth of cut and condition of tool. Thus the accuracy of machining operations could be improves through the cutting force and chip-tool interface temperatures.

L. B. Abhang (✉) · M. Hameedullah
Mechanical Engineering Department, Aligarh Muslim University,
Aligarh 202002, India
e-mail: abhanglb@yahoo.co.in



K. Antony and J. P. Davim (eds.), *Advanced Manufacturing and Materials Science*,
Lecture Notes on Multidisciplinary Industrial Engineering,
https://doi.org/10.1007/978-3-319-76276-0_42

A rubbing action takes place between tool flank and workpiece which removes material at the flank side. As the cutting velocity and feed rate increase, the temperature in the chip formation region also rises [3].

Chaudhary and Ramesh [4] have used an optical displacement sensor for online tool wear monitoring. A feed back control system to provide compensation for the tool wear and keep the dimensions of the workpiece within the tolerance zone has also been suggested. Luo et al. [5] have studied relationship between tool flank wear and operational conditions (cutting) speed, feed rate and depth of cut) in metal cutting process using carbide inserts. The wear of the cutting inserts is calculated and recorded using zygo new view 5000 microscope. The experimental results indicate that cutting speed has a more dramatic effect on tool life than feed rate [5]. Ozel and Karpat [6] developed predictive model of tool wear and surface roughness in hard turning by CBN tool using neural network and regression method. Trained neural network models were used in predicting tool flank wear and surface roughness for other cutting conditions. They concluded that decrease in feed rate resulted in better surface roughness but slightly faster tool wear.

Ramesh et al. [7] used RSM for minimizing the surface roughness in machining titanium alloy. The surface roughness model has been developed in terms of cutting parameters such as cutting speed, feed rate and depth of cut. Machining test has been carried out using CVD coated carbide cutting tools under different machining conditions using Taguchi's orthogonal array. The experimental data have been investigated using analysis of variance (ANOVA). The results indicated that the feed rate is the main influencing factor on surface roughness. Surface roughness is increased with increasing feed rate, but decreased with increasing cutting speed and depth of cut. Felix et al. [8] developed mathematical model for tool wear in turning operations using multiple regression method. A non-traditional optimization algorithm, genetic algorithm has been employed to find the optimal cutting parameters such as cutting speed, feed rate and depth of cut.

Modeling and optimization are necessary for the control of the hard turning process to achieve improved product quality, high productivity and low cost. Empirical models for tool life, surface roughness and cutting force are developed for turning of AISI302 by Al-Ahmari [9]. Anthony et al. [10] determined the influence of different cutting fluids on wear and surface roughness in turning of stainless steel. Noordin et al. [11] studied the application of response surface methodology in describing the performance of coated carbide tools when turning AISI 1045 steel. The factors investigated were cutting speed, feed rate and side cutting edge angle. The response variables were surface roughness and tangential cutting force. ANOVA revealed that feed rate is the most significant factor influencing the response variables investigated. Huang et al. [12] observed that wear pattern depend on the tools used, workpiece material composition, and cutting conditions. They also concluded that generally, adhesion, abrasion and diffusion are considered to be main tool wear mechanisms in CBN hard turning. However, the individual effect of each mechanism depends on the combinations of the work materials, tools, machining conditions, tool geometry etc. It was found that the temperature rise during the machining could significantly reduce the strength of the tool and hence the wear resistance of the tool [13]. Palanisamy et al. [14] developed a regression and ANN mathematical model to predict tool flank wear in terms of machining parameters such as cutting speed, feed rate and depth of cut. Predicted values of tool flank wear of the mathematical model compared with the experimental values. The Grey-Taguchi method was adopted to optimize the milling parameters of aluminum alloy with multiple performance characteristics and found that the flank wear decreased from 0.177 to 0.067 mm [15]. Abhang and Hameedullah [16] used the grey relational analysis technique and determined the optimum turning process parameters. Various turning parameters such as cutting speed, feed rate, depth of cut, tool nose radius and concentration of solid-liquid lubricants were considered and optimized by the grey relational grades obtained from the grey relational analysis for multi-performance characteristics (chip-tool interface temperatures, main cutting force and tool wear).

In the present study article an attempt has been made to perform experimental testing on turning operations in order to establish a relationship between chip-tool interface temperature with tool wear rate and three components of cutting forces with tool wear rate during turning of En-31 steel alloy using tungsten carbide inserts. The predicted results are compared with the experimental results to find the nearness of the predictions to the experimental values. Further to take into considerations these wears also use of machining ratio has been proposed as a new criterion for tool life assessment. Machining ratio is the ratio of volume of material removed to the volume of tool wear.

## 2 Experimental Procedure

The experiments were carried out on a HMT, heavy duty lathe machine without any coolant. The workpiece EN-31 steel was turned for a certain amount of time, while measuring average chip-tool interface temperature and cutting forces by on line and then machining was stopped and tool wear was measured by offline. The cutting tool holder used for turning operation is WIDAX tool holder SCLCR 1212 Fog13 and diamond shape carbide (CNMA 120408). The work piece material used has a dimension of 400 mm in length and 60 mm in diameter. This material is suitable for a wide variety of automotive type applications. Experiments were carried out under different cutting speeds in the ranges of 47–134–226 m/min. Feed rates employed were 0.06–0.10–0.15 mm/rev and depth of cut and tool nose radius was 0.4 and 0.8 mm respectively. The chip-tool interface temperature is measured using tool-work thermocouple designed, fabricated and calibrated in the lab by author. Tool wear rate is measured on a sensitive single pan balance (maximum 300 g and minimum scale is 0.01 mg). Tool inserts were properly cleaned before weighing so that dust or any other adhered particles are removed. Online measurements of cutting forces are carried out using lathe tool dynamometer.

# 3 Experimental Results and Mathematical Modeling

We first present the experimental results and then discuss the use of polynomial equations to express the relationship between the cutting forces with tool wear rate and cutting temperature with tool wear rate.

## 3.1 Relationship Between Cutting Forces and Tool Wear Rate

From the experimental results, Figs. (1, 2 and 3) show variations of three cutting force components with tool wear rate for different cutting speed and feed rate with constant depth of cut and tool nose radius. It can be seen from theses figures that all cutting forces increase with the increase of total tool wear (flank and crater). Increasing tool wear results in an increasing area of contact between the tool-tip and the workpiece. The greater the value of tool wear, the higher the friction of the tool on the workpiece and high heat generation will occur; this ultimately causes the higher value of cutting forces. Table 1 shows experimental results for the first set of experiment, where the inserts begins to fail, all the three cutting forces increase rapidly. Figures (1, 2 and 3) show experimental results at stages where the tool insert begins to fail, for the cutting condition (cutting speed, feed rate, depth of cut and tool nose radius), thrust force and feed force increases slowly i.e. 8 and 17% and radial force increases (maximum) i.e. 70% during machining of EN-31 steel at present conditions. Figures (1, 2 and 3) show the relationship between tool wear and the cutting forces based on the experimental results. From the experimental results, we found that the radial cutting force was approximately (9–10%) larger than the feed force when the cutting tool begins to fail. This relationship is agrees with the results obtained by Noordin et al. [11]. Thrust force is the main force component responsible for material removal while feed force advances the cutting tool in the direction of feed. Radial force is the rubbing force. Polynomial mathematical models can be used to express the relationship between tool wear rate and the cutting forces based on the experimental results. Figure 4 shows the modeling results for the first set of models. It can be seen that the polynomial curves are very close to the experimental results.

## 3.2 Relationship Between Chip-Tool Interface Temperature and Tool Wear Rate

Experiments were continued until the tool wear became too high for the tool-work thermocouple technique to measure temperature accurately. Maximum tool wear was recorded. The results of the tool wear experiments are consistent with the

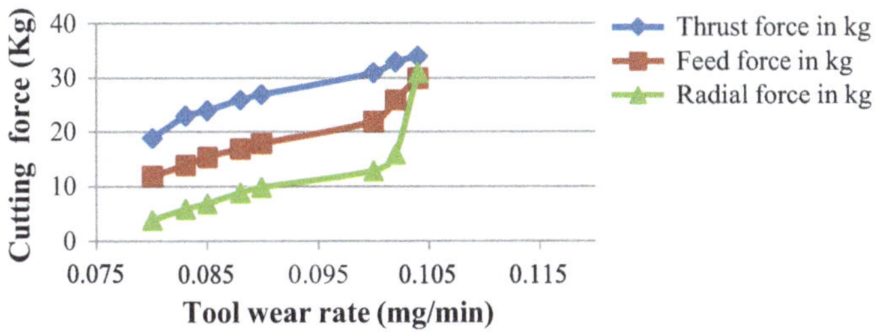


**Fig. 1** Cutting force versus tool wear rate (cutting speed 47 m/min, feed rate 0.06 mm/rev, depth of cut 0.4 mm and tool nose radius 0.8 mm)

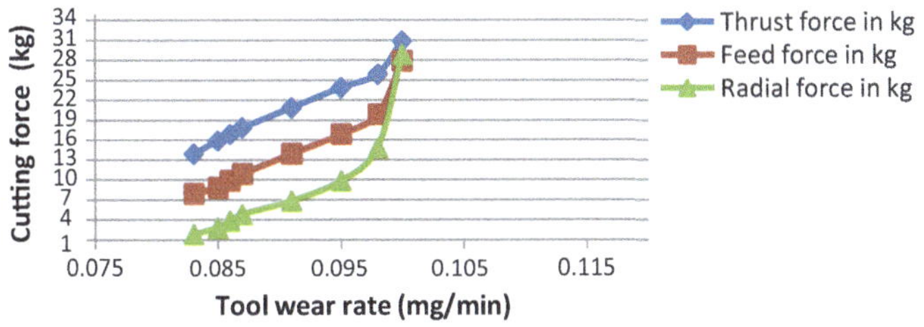


**Fig. 2** Cutting force versus tool wear rate (cutting speed 134 m/min, feed rate 0.06 mm/rev, depth of cut 0.4 mm and tool nose radius 0.8 mm)

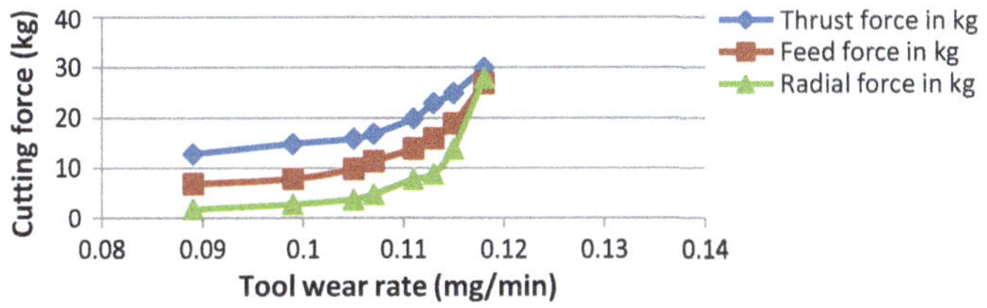


**Fig. 3** Cutting force versus tool wear rate (cutting speed 226 m/min, feed rate 0.06 mm/rev, depth of cut 0.4 mm and tool nose radius 0.8 mm)

results obtained by Tugrul et al. and Anthony et al. [6, 10]. Figures (5 and 6) clearly show that the wear increases with the increase cutting speed and feed rate. It can be seen from these figures that all chip-tool interface temperature increases with the increase in tool wear. When tool inserts begins to fail, all the chip-tool interface temperatures increase sharply. The experimental result shows in Fig. 5 is important when considering the temperature effects on tool wear, which is temperature dependent, i.e. when increasing tool, wear (flank and crater) then increases the chip-tool interface temperature during machining. Polynomial mathematical models

**Table 1** First set (cutting speed = 47 m/min, feed rate = 0.06 mm/rev, depth of cut = 0.4 mm, nose radius = 0.8 mm)

| Exp. no. | Cutting time (min) | Tool wear rate (mg/min) | Thrust force (Kg) | Feed force (kg) | Radial force (kg) |
|---|---|---|---|---|---|
| 1 | 10 | 0.08 | 19 | 12 | 4 |
| 2 | 20 | 0.083 | 23 | 14 | 6 |
| 3 | 30 | 0.085 | 24 | 16 | 7 |
| 4 | 40 | 0.088 | 26 | 17 | 9 |
| 5 | 50 | 0.089 | 27 | 19 | 10 |
| 6 | 60 | 0.10 | 29 | 20 | 11 |
| 7 | 70 | 0.102 | 33 | 26 | 16 |
| 8 | 80 | 0.104 | 34 | 30 | 31 |

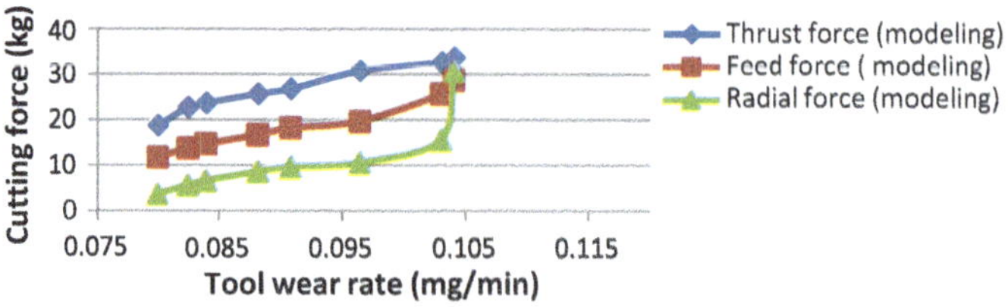


**Fig. 4** Cutting force versus tool wear rate (cutting speed 47 m/min, feed rate 0.06 mm/rev, depth of cut 0.4 mm and tool nose radius 0.8 mm)

also used to express the relationship between tool wear rate and the cutting temperature based on the experimental results. Figure 6 shows the modeling results for the first set of models. It can be seen that the polynomial curves are very close to the experimental results. Similar results were obtained for other sets of model. Therefore a good relationship between the chip-tool interface temperature and tool wear rate can be established from the experimental results. Since we obtained eight sets of data for each experiment, all in limited range, polynomials of the fourth degree were used. The polynomial coefficients were found by MATLAB.

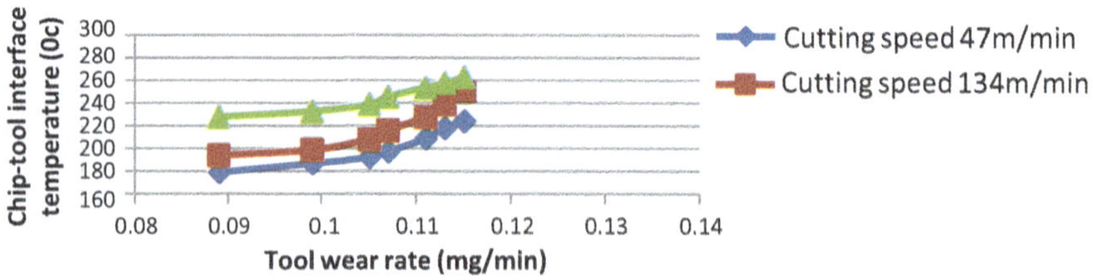


**Fig. 5** Chip-tool interface temperature versus tool wear rate (cutting speed 47 m/min, feed rate 0.06 mm/rev, depth of cut 0.4 mm and tool nose radius 0.8 mm)

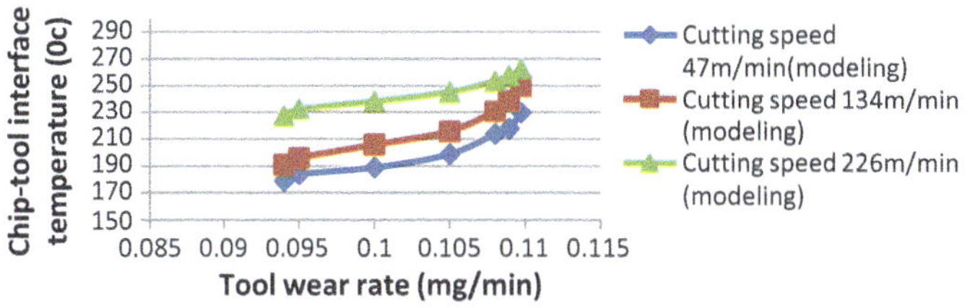


**Fig. 6** Chip-tool interface temperature versus tool wear rate (cutting speed 47 m/min, feed rate 0.10 mm/rev, depth of cut 0.4 mm and tool nose radius 0.8 mm)

### 3.2.1 Polynomial Models

Thrust force = (X):

1. Cutting speed = 47 m/min, feed rate 0.06 mm/rev.
   Tool wear rate = $-2.021 \times 10^{-6} X^4 + 0.0001931 X^3 - 0.006697 X^2 + 0.1012 X - 0.4863$
   Feed force (X):
   Tool wear rate = $1.47 \times 10^{-6} X^4 - 0.0001307 X^3 + 0.004155 X^2 - 0.05423 X + 0.3282$
   Radial force (X):
   Tool wear rate = $1.534 \times 10^{-6} X^4 - 9.474 \times 10^{-5} X^3 + 0.001828 X^2 - 0.0113 X + 0.1021$

2. Cutting speed = 134 m/min, feed rate 0.06 mm/rev. Thrust force (X):
   Tool wear rate = $-1.21 \times 10^{-6} X^4 + 0.0001007 X^3 - 0.003044 X^2 + 0.003193 X + 0.09383$
   Feed force (X):
   Tool wear rate = $-2.047 \times 10^{-8} X^4 - 5.159 \times 10^{-6} X^3 + 0.0002824 X^2 - 0.003193 X + 0.09383$
   Radial force (X):
   Tool wear rate = $3.664 \times 10^{-7} X^4 - 1.932 \times 10^{-5} X^3 + 0.0002609 X^2 + 0.0002727 X + 0.0818v$

3. Cutting speed = 226 m/min, feed rate = 0.06 mm/rev. Thrust force (X):
   Tool wear rate = $-1.275 \times 10^{-6} X^4 + 0.000124 X^3 - 0.00449 X^2 + 0.0726 X - 0.3318$
   Feed force (X):
   Tool wear rate = $-2.003 \times 10^{-6} X^4 + 0.0001394X^3 - 0.003525 X^2 + 0.03941 X - 0.05588$
   Radial force (X):
   $-2.218 \times 10^{-6} X^4 + 0.0001249 X^3 - 0.002332 X^2 + 0.01806 X + 0.06185$
   Chip-tool interface temperature:
   Cutting speed = 47 m/min, feed rate = 0.10 mm/rev.

Tool wear rate = $4.03 \times 10^{-10}$ $X^4 - 3.19210^{-7}$ $X^3 + 8.696 \times 10^{-5} -$ 0.008731 X + 0.2693
Cutting speed = 47 m/min, feed rate = 0.06 mm/rev
Tool wear rate = $4.19 \times 10^{-9}$ $X^4 - 3.88 \times 10^{-6}$ $X^3 + 0.001337$ $X^2 - 0.2024$ X + 11.45.

### 3.2.2 Maximum Machining Ratio

Machining ratio is given by the following equation.

$$MR = vdft/V = vdft\,\rho/w \tag{1}$$

where, v = cutting speed (m/min), f = feed rate (mm/rev), d = depth of cut (mm), t = machining time (minute), V = volume of tool wear ($cm^3$), w = weight of tool wear (mg) and $\rho$ = density of tool material (15.8 $g/cm^3$) etc. For different machining conditions the values of machining ratio are shown in Fig. 7. It indicates that machining ratio first increases rapidly and reaches up to maximum value, then this value remains constant for some period after which it falls sharply. Machining ratio is inversely proportional to the tool wear rate. The tool wear rate shows a decreasing trend with respect to machining time and then it reaches to a certain value after which it starts increasing rapidly. The maximum machining ratio can be used as a suitable criterion for tool life. The tool life in case of maximum machining ratio criterion will be defined as the time when the machining ratio reaches its maximum value and when it begins to fall sharply. It is a dimensionless quantity and is based on combined wear of flank, crater, chip notching, primary and secondary grooving.

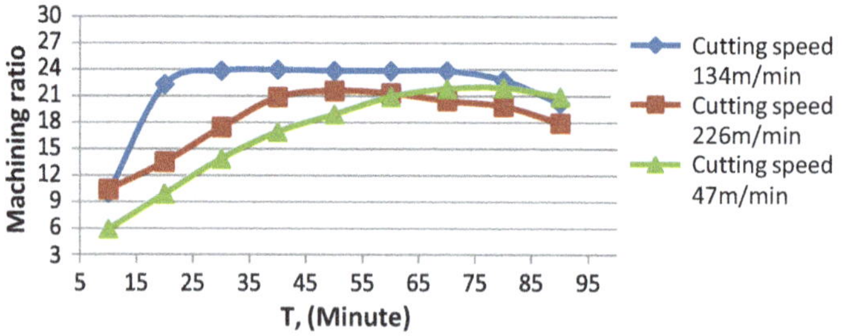


**Fig. 7** Machining ratio versus cutting time (feed rate 0.06 mm/rev, depth of cut 0.4 mm and nose radius 0.8 mm)

## 4 Conclusions

(1) All cutting forces increase with increase of tool wear. The rate of cutting force increase (thrust force increases by approximately 8%, feed force increases by approximately 17% and the radial force increases by approximately 70%) of feed force and radial force is higher than the thrust force. The radial force was also found to be slightly larger than the axial force when tool insert begins to fail.

(2) The study also presented the analysis of the relationship between tool wear and temperature by applying the tool work thermocouple technique. The temperature measured by the tool-work thermocouple method shows that temperature has a strong effect on tool wear.

(3) There is a maximum value of machining ratio and can be taken as a criterion of tool failure.

## References


1. Kalpakjian S (2005) Manufacturing engineering and technology, Fourth edition, llanos Institute of Technology, pp 551–557
2. Rao P (2000) Manufacturing technology, metal cutting and machine tools, Tata McGraw-Hill Publishing Company Limited, New Delhi, pp 37–44
3. Noordin MY, Venkatesh VC, Sharif S (2007) Dry turning of tempered martensitic stainless steel tool steel using coated cermets and coated carbide tools. J Mater Process Technol 185:83–90
4. Chaudhary SK, Ramesh S (1995) On line tool wear sensing and compensation in turning. J Mater Process Technol 49:247–254
5. Luo X, Cheng K, Holt R, Liu X (2005) Modeling flank wear of carbide tool inserts in metal cutting. Wear 259:1235–1240
6. Ozel T, Karpat Y (2005) Predictive modeling of surface roughness and tool wear in hard turning using regression and neural networks. Int J Mach Tools Manuf 45:467–479
7. Ramesh S, Karunamoorthy L, Palnikumar K (2008) Surface roughness analysis in machining of titanium alloy. Mater Manuf Process 23:174–181
8. Felix Prasad C, Jayabal S, Natarajan U (2007) Optimization of tool wear in turning using Genetic algorithm. Indian J Eng Mater Sci 14:403–407
9. Al-Ahmari AMA (2007) Predictive machinability models for a selected hard material in turning operations. J Mater Process Technol 190:305–311
10. Anthony Xavior M, Adithan M (2009) Determining the influence of cutting fluids on tool wear and surface roughness during turning of AISI304 austenitic stainless steel. J Mater Process Technol 209:900–909
11. Noordin MY, Venketesh VC, Sharif S, Elating A (2004) Application of R.S.M. in describing the performance of coated carbide tools when turning AISI 1045 steel. J Manuf Process Technol 145:46–58
12. Huang Y, Chou YK, Liang SY (2007) CBN tool wear in hard turning: a survey on research progress. Int J Adv Manuf Technol 35:443–453
13. Braghini A, Coelho RT (2001) An investigation of the wear mechanisms of PCBN tools when end milling hardened steel at low/medium cutting speeds. Int J Adv Manuf Technol 17:244–257

14. Palanisamy P, Rajendran I, Shanmugasundaram S (2008) Prediction of tool wear using regression and ANN models in end-milling operation. Int J Adv Manuf Technol 37:29–41. https://doi.org/10.1007/s00170-007-0948-5
15. Tsao CC (2009) Grey-Taguchi method to optimize the milling parameters of aluminum alloy. Int J Adv Manuf Technol 40:41–48. https://doi.org/10.1007/s0017-007-1314-3
16. Abhang LB, Hameedullah M (2011) Empirical modeling of turning parameters using grey relational analysis. J Appl Mech Mater 110–116(2011):2596–2603

# On the Mechanical and Microstructural Properties of Novel A356/Kevlar Surface Composite Fabricated via Friction Stir Processing

S. Pradeep Devaneyan, D. Peter Pushpanathan and Adhesh Parambath

## 1 Introduction

Surface metal matrix composites (SMMC) are applied in various field of engineering where properties in addition to that of the substrate materials are desired. A356-T6 cast aluminum (Al) alloy is a light weight structural material, which is preferred widely for manufacturing mechanical components especially in automobile industries where components of complex geometries are used [1]. The usage of this alloy is limited due to a decline in the mechanical properties induced by casting defects such as porosity. Hence, it is important to eliminate the casting defects in order to improve its properties. Surface composites could also be fabricated with ceramic carbide reinforcements like silicon carbide, carbides of transition metals and oxides of aluminium using the conventional surface modification techniques such as high energy laser melt treatment, high energy electron beam irradiation and thermal spray process which are basically fusion processes. These techniques yield the fusion related problems, such as interfacial reaction, pin holes, shrinkage cavities or voids and other casting related defects, and pave the way to need of an efficient technique which must be based on solid state. FSP is a solid state technique where a rotating tool consisting of a shoulder and a pin enters the substrate and moves with steady rotational and traverse speeds in the desired direction. The tool serves two primary functions of heating and deformation of work-piece material [2]. The heating is produced mainly by the friction of the rotating shoulder with the work-piece, while the rotating probe or pin deforms or stirs the locally heated material. The heated

S. Pradeep Devaneyan (✉)
School of Mechanical and Building Sciences, Christ College
of Engineering and Technology, Puducherry, India
e-mail: pr.signs@gmail.com

D. Peter Pushpanathan · A. Parambath
Department of Mechanical Engineering, Christ College
of Engineering and Technology, Puducherry, India



K. Antony and J. P. Davim (eds.), *Advanced Manufacturing and Materials Science*,
Lecture Notes on Multidisciplinary Industrial Engineering,
https://doi.org/10.1007/978-3-319-76276-0_43

material softens and flows around the rotating pin, and then fills the void at the rear of the tool. The surface composites with various reinforcement particles are fabricated successfully using FSP [3]. Defence components made of A356 alloy require wear resistance and wear is one of common failure modes for moving metallic mechanical components. Improved tribological properties are important to overcome the wear failure of above defence components. FSP also results in grain refinement and homogenization of a cast microstructure of A356. Metal matrix composites are found to exhibit superplastic behaviour under special conditions. Microstructural features that limit the application of as cast A356 Al–Si alloy for high performance applications are dendrite size, porosity, heterogeneous microstructures, intermetallics and coarse eutectic silicon particles. FSP has been used to improve the ductility of AZ31 magnesium alloys by performing multiples passes and rapid cooling [4]. The effect of tool rotation speed and tool traverse speed on the surface characteristics of cast hypereutectic Al-17% Si alloy has been studied. The results have shown significant grain refinement, homogeneous distribution of Si and reduction in porosity [5]. The influencc of grain refiner (Al–5Ti–B), Mn and Sn additions on the sliding wear behavior of A356 Aluminum alloys was investigated. The results indicated a significant reduction in crack formation and improved wear resistance [6]. The microhardness and wear resistance of A356 alloy were improved by incorporating boron carbide and $MoS_2$ powders using FSP. SiC particles were distributed uniformly in AA2014 matrix and the surface hardness was improved. Surface metal matrix composites fabricated using fusion techniques lead to problems such as interfacial reactions, pin holes, shrinkage cavities, etc. due to the high temperature. These problems could be avoided if a solid state processing route such as FSP is used for fabricating SMMCs. The strength, ductility and machinability of cast Al–Si alloys used in automobile industry were improved significantly using FSP due to the dissolution of iron rich intermetallics [7]. FSP carried out on commercial AA2024 Al alloy at different tool rotation rates produced significant reduction in size of the constituent particles. Severe plastic deformation (SPD), peak temperature and constituent particle characteristics due to FSP were found to influence their evolution. From the literature survey it is understood that the limited application of cast Al–Si alloys is mainly due to poor mechanical and tribological properties. The main motive of this study is to modify the surface properties of as cast A356 alloy by reinforcing with Kevlar using FSP, and investigate its effect on the microstructural and mechanical properties as well.

## 2 Experimental Procedure

In this work, a eutectic 7 Si, 0.3 Mg ingot was procured from the local distributor. The ingot was cast into a plate of size 150 mm × 72 mm × 10 mm. The chemical composition of the A356 plate was found to be 0.25% Copper, 0.2–0.4% Mg, 6.5–7.5% Si, 0.25% Ti, 0.35 Mn, 0.35% Zn, 0.6% Fe and the rest Al. The cast plate was then cut into 8 plates of equal dimensions in order to accommodate it in the fixture used in the modified vertical milling machine. The surfaces of these plates were machined and smoothened to make it ideal to FSP operation. The plates thus prepared are shown in Fig. 1.

The FSP tool was fabricated using EN31 tool material with a threaded cylindrical tool pin. EN31 tool material was used for fabrication. The tool was hardened and tempered subsequently. The tool shoulder diameter was 18 mm and the probe diameter was 6 mm. The tool shoulder was given a concavity of 2° and the tool pin length was set as 3 mm [8]. The tool used for FSP is shown in Fig. 2.

The Kevlar aramid pulp manufactured by DuPont®, was used as a reinforcement. Several trial runs were carried out in order to set the upper and lower limit of the process parameters. The tool rotation speed was set as 1000 and 1400 rpm based on the trial runs. However the tool traverse speed was kept constant at 21 mm/min to analyse the effect of tool rotation speed on the properties of the friction stir processed surface metal matrix composite. Rectangular grooves of size 2 mm × 2 mm were machined on the plates and Kevlar pulp was carefully filled inside the grooves. The plates loaded with Kevlar were held in fixture and a pin less tool was used to compress the fibres further into the groove. Then the pin less tool was replaced with the threaded cylindrical tool and FSP was performed [9]. Ten experimental runs were carried out with various tool rotation speed and tool traverse speed combinations as specified in Table 1. Trial runs were conducted and the lowest and highest limits of tool rotation speed and tool traverse speed required to produce defect free surface composites were determined. The tool rotation speed started from 1000 rpm and incremented in steps of 100 to reach the final value of 1400 rpm. The tool traverse speed was varied between 19 and 21 mm/min.

Tests to assess the tensile strength were carried out on the specimens machined out perpendicular to the FSP direction as per ASTM E8M04. The specimens were extracted from the FSPed area for microstructural examination and microhardness testing. The extracted specimens were polished and etched with Keller's reagent.

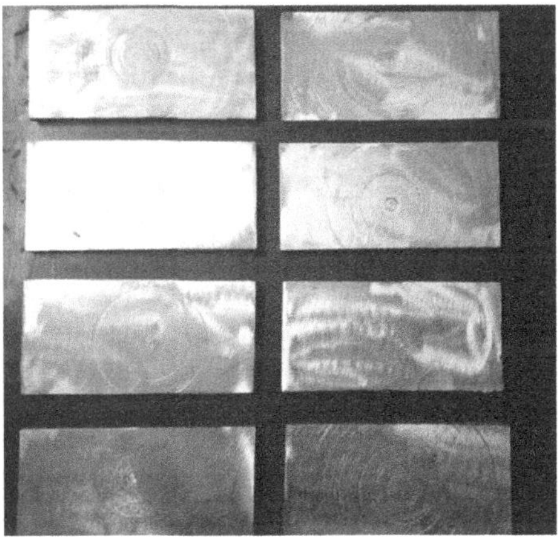

**Fig. 1** Plates prepared for FSP

**Fig. 2** Threaded cylindrical tool

**Table 1** Process parameters used for friction stir processing

| Sl. No. | Tool | Tool rotation speed (mm/min) | Tool traverse speed (rpm) |
|---|---|---|---|
| 1 | Base metal | – | – |
| 2 | Threaded cylindrical | 1400 | 21 |
| 3 | Threaded cylindrical | 1400 | 19 |
| 4 | Threaded cylindrical | 1300 | 21 |
| 5 | Threaded cylindrical | 1300 | 19 |
| 6 | Threaded cylindrical | 1200 | 21 |
| 7 | Threaded cylindrical | 1200 | 19 |
| 8 | Threaded cylindrical | 1100 | 21 |
| 9 | Threaded cylindrical | 1100 | 19 |
| 10 | Threaded cylindrical | 1000 | 21 |
| 11 | Threaded cylindrical | 1000 | 19 |

Microstructural examination was carried out using optical microscope (Make: Meiji, Japan) with image analysis software. The microhardness of the samples were then tested using Vickers microhardness tester (Make: Shimadzu: Japan, Model: HMV-2T) employing Vickers indentation at 0.3 kgf load for 15 s [10].

# 3 Results and Discussion

## *3.1 Microstructure*

The microstructure of the as cast A356 aluminium alloy is characterized by the presence of a primary α aluminium solid solution and eutectic mixture of silicon and aluminium. The dendritic arms of the primary phase are clearly seen in the

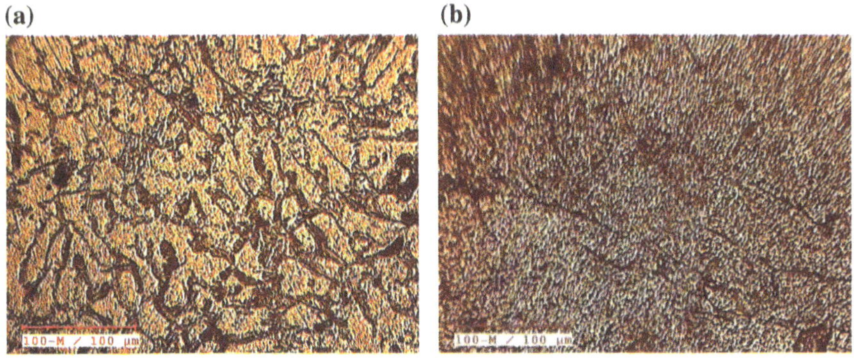


**Fig. 3** **a** Microstructure of the base metal. **b** Microstructure of the FSPed SMMC

microstructure of the base metal. Fibrous morphology, instead of globular is noticed. Also, microshrinkage, interdendritic voids and porosities are also observed in the micrograph of the base metal presented in Fig. 3a.

From Fig. 3b, it could be inferred that the size of the chopped fibers are significantly reduced and significant grain refinement also occurred in the stir zone. The stir zone depth was observed to be around 3 mm which is the depth of the tool pin. The interface between the aramid fibers and the matrix plays a major role in deciding the properties of the FSPed surface metal matrix composite.

## 3.2 *Tensile Strength*

The photographs of the specimen before and after tensile testing are presented in Fig. 4a, b.

From the results presented in Fig. 5a, b, it could be inferred that the base metal specimen was subjected to tensile testing and the ultimate tensile strength was found to be 199 MPa. The FSPed SMMC specimen processed with a tool rotation speed of 1400 rpm exhibited the tensile strength of 191 MPa. The specimen processed at 1000 rpm exhibited a tensile strength of 182 MPa. The failure of the

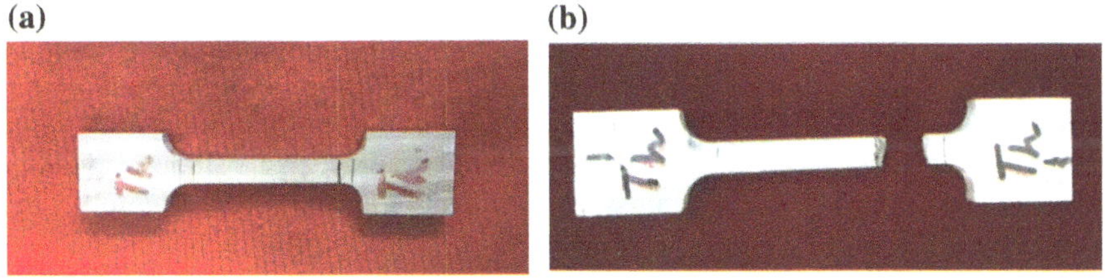


**Fig. 4** **a** Tensile specimen before testing. **b** Tensile specimen after testing

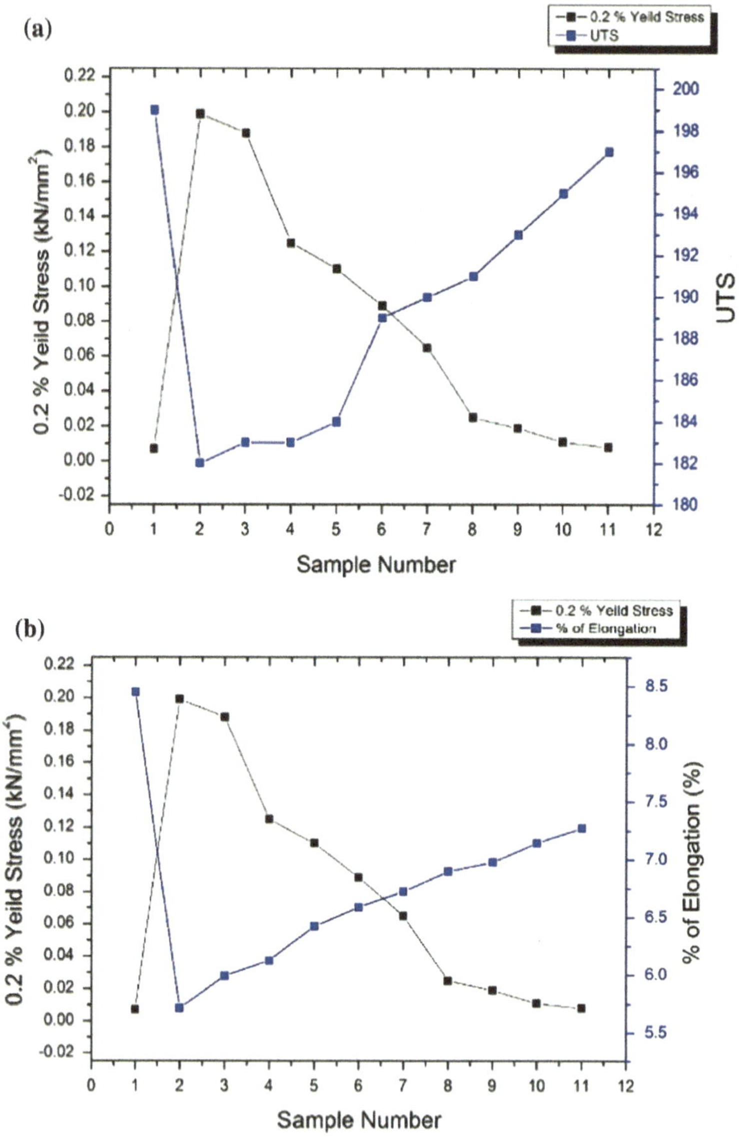


**Fig. 5** **a** Yield strength versus elongation. **b** Yield strength versus tensile strength

specimen near the fillet indicates that the FSPed region possesses strength at par with that of the base metal [11]. The slight reduction in the tensile strength could be due to the breaking up of the grains due to the inhomogeneous local deformation caused by the addition of aramid Fibers.

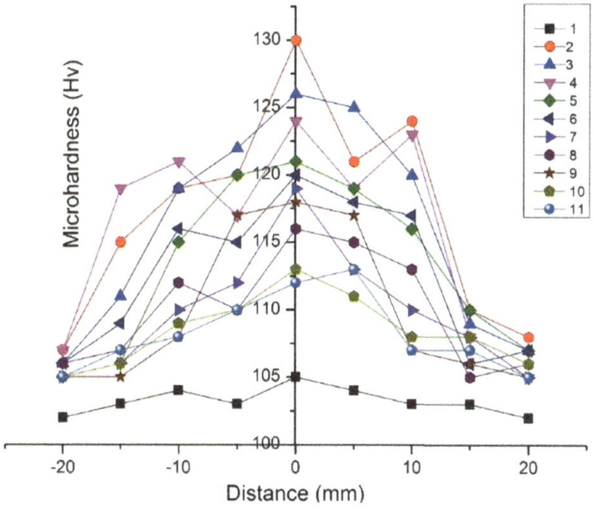


**Fig. 6** Microhardness at different locations

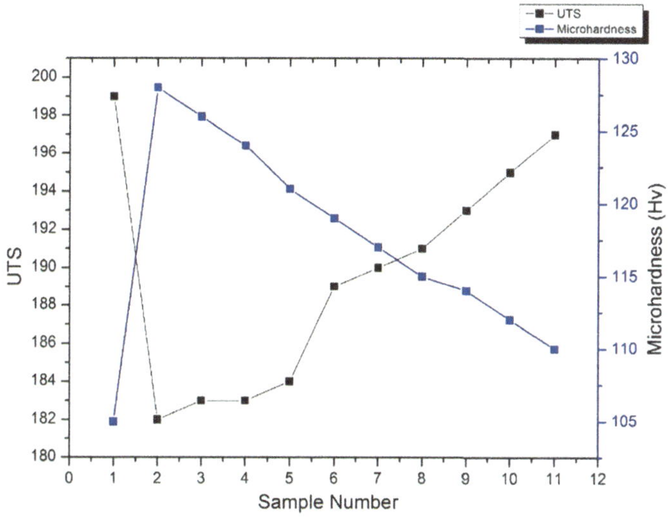


**Fig. 7** Microhardness versus tensile strength

## 3.3 *Microhardness*

The microhardness of the FSPed surface composite was tested at four locations on either side of the nugget zone and in the stir zone itself. From the results plotted in Figs. 6 and 7, it is understood that the hardness in the stir zone has improved considerably for the threaded cylindrical pin processed at 1400 rpm as compared to the base metal. The hardness of the sample processed at 1000 rpm was 110 which was higher than the base metal hardness. However it was less than the hardness of the sample processed at 1400 rpm. The first reason for the increase in microhardness is due to the considerable reduction in grain size due to intense stirring effected by the threaded cylindrical tool pin. This caused the grain boundary to rupture forming high angle grain boundaries which hindered the movement of dislocations and thereby enhancing the hardness [12]. The second reason is the grain boundary pinning caused by the chopped Kevlar pulp and dispersion hardening which caused improvement in the hardness of the composite layer [13]. The interaction of dislocations and the chopped fibers could also have impeded the movement of dislocations and enhanced the hardness.

## 4 Conclusions

A356/Kevlar Surface metal matrix composites were successfully fabricated without any defects. Significant grain refinement and homogeneous distribution of Kevlar aramid chopped fibers were achieved. The tensile strength of the SMMC was found to be a little lower than the base metal. This could be attributed to the inhomogeneous local deformation of the aramid fibers. However, the failure which did not happen in the processed area indicate that the processed area is as strong as the base metal. The microhardness of the stir zone area showed significant improvement as compared to the base metal. This proves that the aramid fibers and grain refinement effected by FSP has played a major role in enhancing the hardness. It is implied that the enhancement of hardness would result in significant wear resistance as well.

## References


1. Santella ML, Engstrom T, Storjohann D, Pan TY (2005) Effects of friction stir processing on mechanical properties of the cast aluminum alloys A319 and A356. Scripta Mater 53(2):201–206
2. Sharma SR, Ma ZY, Mishra RS (2004) Effect of friction stir processing on fatigue behavior of A356 alloy. Scripta Mater 51(3):237–241
3. Ma ZY, Sharma SR, Mishra RS, Mahoney MW (2003) Effect of friction stir processing on the microstructure of cast A356 aluminum. Mater Sci Forum 426–432:2891–2896
4. Mishra RS, Ma ZY (2005) Friction stir welding and processing. Mater Sci Eng: R: Rep 50(1–2):1–78
5. Ma ZY, Mishra RS, Mahoney MW (2004) Superplasticity in cast A356 induced via friction stir processing. Scripta Mater 50:931–935
6. Ma ZY (2008) Friction stir processing technology: a review. Metall Mater Trans A 39(3):642–658
7. AlaviNia A, Omidvar H, Nourbakhsh SH (2014) Effects of an overlapping multi-pass friction stir process and rapidcooling on the mechanical properties and microstructure of AZ31 magnesium alloy. Mater Des 58:298–304

8. Saini* N, Dwivedi DK, Jain PK, Singh H (2015) Surface modification of cast Al-17% Si Alloys using friction stir processing. Procedia Eng 100:1522–1531
9. Qui K, Wang R-C, Peng C-Q, Wang N-G, Cai Z-Y, Zhang C (2015) Effect of individual and combined additions of Al–5Ti–B, Mn and Sn on sliding wear behavior of A356 alloy. Trans. Nonferrous Met. Soc. China 25:3886–3892
10. Srinivasu R, Sambasiva Rao A, Madhusudhan Reddy G, Rao S (2015) Friction stir surfacing of cast A356 aluminium silicon alloy with boron carbide and molybdenum disulphide powders. Def Technol 11(2):140–146
11. Sudhakar I, Madhusudhan Reddy G, Srinivasa Rao K (2016) Ballistic behavior of boron carbide reinforced AA7075 aluminium alloy using friction stir processing. An experimental study and analytical approach. Def Technol 12(1):25–31
12. Gurua PR, Khan F, Panigrahia SK, Janaki Ram GD (2015) Enhancing strength, ductility and machinability of a Al–Si cast alloy by friction stir processing. J Manuf process 18:67–74
13. Pasebani S, Charit I, Mishra RS (2015) Effect of tool rotation rate on constituent particles in a friction stir processed 2024 Al alloy. Mater Lett 160:64–67

# Production of Biodiesel from High FFA Non-edible Nahar Oil and Optimization of Yield

**Santosh Kumar Dash and Pradip Lingfa**

## 1 Introduction

Nahar (*mesua ferrea lin.*) is an underutilized tree abundantly available in northeast part of India. Abundance of Nahar in northeast part of India and higher oil content (65–70% of seed kernel) can be seen in earlier literature [1–3]. Bora et al. used Nahar seed oil for production of Nahar oil methyl ester by single stage transesterification and utilized it in diesel engine [2]. Two years later Kushwah et al. reported, due to high acid value (34 mg KOH/g) and moisture content both single stage and double stage (acid-base) transesterification failed and suggested up to 15% blend with high speed diesel can be used to run a diesel engine [3]. However earlier literatures reported for FFA value more than 3%, transesterification reaction is not possible [4]. For high FFA feedstock pretreatment is necessary [5, 6]. High FFA in seed oil, by chemically reacting with alkaline base catalyst in transesterification process forms soap, which inhibit chemical reaction [7–9]. In light of the above ambiguous reports in earlier published literature pertinent to biodiesel production from Nahar oil, present investigation was undertaken to produce biodiesel from Nahar seed oil. Heterogeneous catalyst has several benefits over homogeneous catalyst in terms of cost, elimination of water-wash step; which account for 10–15% loss of alkyl ester, environmental benign and reusability characteristics unlike homogeneous catalyst which partially get consumed to produce gel or soap [4, 10, 11]. Cost is the prime reason of biofuel not being popular in market. Non-edible feedstock with heterogeneous catalyst combinedly could reduce the overall production cost of biodiesel. A heterogeneous catalyst CaO is used in present study for transesterification. Chavan et al. studied biodiesel synthesis from Jatropha curcas oil using CaO as solid base heterogeneous catalyst derived from

S. K. Dash (✉) · P. Lingfa
Department of Mechanical Engineering, North Eastern Regional Institute of Science and Technology, Itanagar 791109, Arunachal Pradesh, India
e-mail: 90nash@gmail.com



K. Antony and J. P. Davim (eds.), *Advanced Manufacturing and Materials Science*, Lecture Notes on Multidisciplinary Industrial Engineering,
https://doi.org/10.1007/978-3-319-76276-0_44

waste eggshell. The evaluated properties were of superior quality and falls within limits given in ASTM standards [12]. Veljkovic et al. obtained 98% yield of biodiesel with using 1 wt% CaO catalyst at 60 °C for 2 h with a molar ratio of 1:6 [13]. Demirbas used CaO for methanolysis of sunflower oil in supercritical process [14].

The objective of the present study is to synthesis biodiesel from Nahar seed oil using CaO as heterogeneous catalyst, optimizing the process parameters (molar ratio of oil to alcohol, reaction time, catalyst amount, reaction temperature, speed of reaction) for both esterification and transesterification for maximum yield of biodiesel and evaluating fuel properties for quality checkup as per ASTM specifications.

## 2 Materials and Methods

Nahar seed was collected from NERIST college campus, Arunachal Pradesh, India. The collected seed kernels were dried in the sun light for 3 days at college campus to reduce moisture content. The dried seed kernels were expelled using a mechanical expeller in Assam, India. After filtration the raw oil obtained was 67% of total weight of Nahar seed kernel. The filtered raw oil is used for the present study. Analytical grade methanol, Calcium oxide (CaO) powder, Anhydrous sodium sulfate ($Na_2SO_4$), Sulphuric acid, Isopropyl alcohol, Phenolphthalein indicator were purchased from Merck Ltd. Mumbai, Maharashtra.

## 3 Characterization of Nahar Seed Oil

The raw Nahar oil has been characterized for improving biodiesel production process and properties of biodiesel. Various physico-chemical properties and fatty acids content were evaluated. Fatty acid compositions of Nahar oil are listed in Table 1. The physico-chemical properties are listed in Table 2. The FFA content of Nahar oil was obtained 16%, with a corresponding acid value of 31.97 mg KOH/g. From the gas chromatograph (GC), it was observed oleic acid (46%) is the major fatty acid constituent followed by Linoleic acid 22.099%. Palmitic (16.690%) and stearic (8.8%) acids shared the saturated portion of fatty acid profile and are responsible for high cloud and pour point.

**Table 1** Fatty acid composition (%) of Nahar oil

| Sr. No | Fatty acid name | Formula | Composition (%) |
|---|---|---|---|
| 1 | Myristic acid | C14:0 | 2.3 |
| 2 | Palmitic acid | C16:0 | 16.690 |
| 3 | Stearic acid | C18:0 | 8.8 |
| 4 | Behenic acid | C22:0 | 0.445 |
| 5 | Tricosanoic acid | C23:0 | 0.199 |
| 6 | Oleic acid | C18:1n9c | 45.859 |
| 7 | Cis-11 Ecosenoic acid | C20:1 | 0.992 |
| 8 | Linoleic acid | C18:2n6c | 22.099 |
| 9 | Alpha-Linolenic acid | C20:3n6 | 0.013 |
| 10 | Cis-8, 11, 14-Ecosatrienoic acid | C18:3n3 | 0.197 |
| 11 | Arachidonic acid | C20:4n6 | 0.331 |
| 12 | Cis-13, 16 Docosadienoic acid | C22:2 | 0.103 |

**Table 2** Properties of Nahar oil

| Sr. No | Parameter | Nahar oil |
|---|---|---|
| 1 | Density (gm/cc) | 0.925 |
| 2 | Viscosity ($mm^2/s$) | 49.7 |
| 3 | Calorific value (MJ/Kg) | 37.5 |
| 4 | Cetane no. | 53 |
| 5 | Flashpoint (°C) | 163 |
| 6 | Fire point (°C) | 176 |
| 7 | Cloud point (°C) | 12 |
| 8 | Pour point (°C) | 6.9 |
| 10 | Iodine value (g/100 g of oil) | 82 |
| 11 | Acid value (mg KOH/g) | 31.97 |
| 12 | Saponification value (mg KOH/g) | 191 |

# 4 Results and Discussion

## *4.1 Biodiesel Production*

Several attempts were failed to get optimum biodiesel from Nahar oil by alkali catalyzed transesterification process. This may be due to the presence of high free fatty acid which hinders the progress of the reaction by forming soap and water. Hence a two step acid-base transesterification process was developed to reduce the high free fatty acid of Nahar oil in the 1st stage esterification.

## 4.2 First Stage Acid Esterification

In this stage of reaction, the FFA present in the oil was converted into fatty acid methyl ester. Several sets of experiments were conducted to know the optimum value of methanol quantity, sulphuric acid ($H_2SO_4$), reaction time, reaction temperature, mixing speed of reaction mixture. From the Fig. 1, can be observed that at 60 °C reaction temperature with 1:6 molar ratio of oil to methanol by employing 0.7% of $H_2SO_4$ the acid value reduced to 23 mg KOH/g in 15 min from the start of reaction. The conversion slowed down for next 15 min with reduction of 7 mg KOH/g acid value. With further addition of 0.5% (v/v) $H_2SO_4$ the conversion gets accelerated and reduces to 2.83 mg KOH/g in 75 min. The conversion was faster for the first 60 min compared to the rest of the reaction time. With further increasing the amount of sulphuric acid the conversion was negatively affected by forming a sudden change of color of reaction mixture. This may be attributed to the fact that, burning of preheated triglyceride at high concentration of sulphuric acid. Further addition of 0.5% (v/v) sulphuric acid after 30 min from the start of reaction results better conversion in lesser time than direct addition of 1.2% (v/v) sulphuric acid in the oil-methanol mixture. In 1 h 45 min the acid value of the oil reduced to 2.83 mg KOH/g from 31.97 mg KOH/g with keeping other parameter at optimum condition at 60 °C 500 rpm and (0.7 + 0.5) % (v/v) sulphuric acid ($H_2SO_4$). Slightly formed moisture in esterification reaction was removed by passing the mixture over anhydrous sodium sulfate ($Na_2SO_4$). Then the esterified oil was immediately taken for calcium oxide (CaO) catalyzed transesterification reaction.

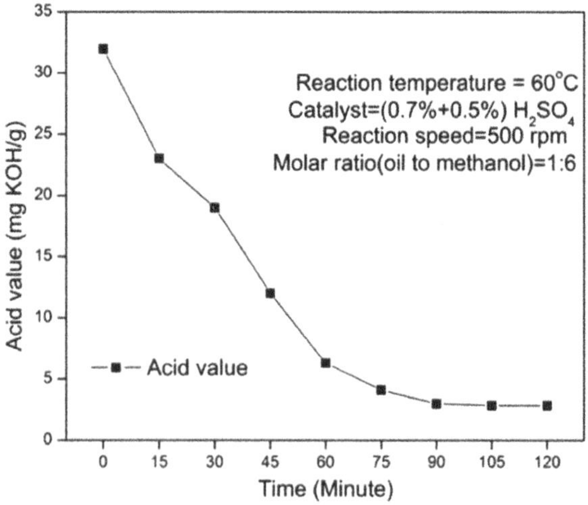


**Fig. 1** Effect of time on reduction of acid value during first stage esterification

## 4.3 Second Stage Alkali Catalyzed Transesterification

Combined effect of molar ratio of oil to alcohol (1:4 to 1:12) and catalyst amount (0.5–3.00 wt%) were studied. The other reaction parameters were kept constant at reaction temperature of 60 °C, reaction time of 2 h with speed of reaction 500 rpm. It can be seen from Fig. 2 for catalyst amount 0.5 to 1.5 wt% with increase in molar ratio the biodiesel yield increases. For catalyst amount 2 to 3 wt% maximum yield was observed at molar ratio of 1:8. For 2 wt% of loading of CaO catalyst maximum yield of 74% observed for both molar ratio of 1:8 and 1:10. Maximum biodiesel yield (82%) was obtained at 2.5 wt% loading of catalyst with molar ratio of 1:8. Hence for the transesterification process, the optimum value of molar ratio is 1:8 and loading of catalyst amount is 2.5 wt% of CaO.

Other reaction parameters were optimized to further enhance biodiesel yield. Reaction temperature was varied from 45 to 70 °C to study the effect of temperature on yield. From the Fig. 3, it can be seen with increase in temperature from 45 to 60 °C, the yield increases rapidly and then only 2% yield increases for increasing temperature from 60 to 65 °C. Further increasing temperature (at 70 °C) showed negative effect on yield of biodiesel. This may be due to the evaporation of methanol from round bottom flask to condenser, contributing to a net lesser methanol in the reaction mixture. Maximum yield of 84% observed at 65 °C reaction temperature. The optimum reaction time was obtained as 2.5 h, which accounts for maximum 87% biodiesel yield. With increase in mixing speed, the yield increased and a maximum yield of 88% observed for 600 rpm. Further increase in stirring speed would not give any higher yield (Figs. 4, 5).

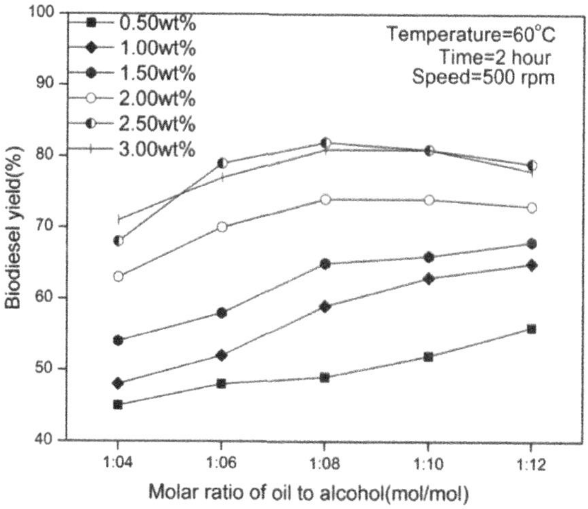


**Fig. 2** Effect of molar ratio and catalyst amount on yield of biodiesel in second stage transesterification

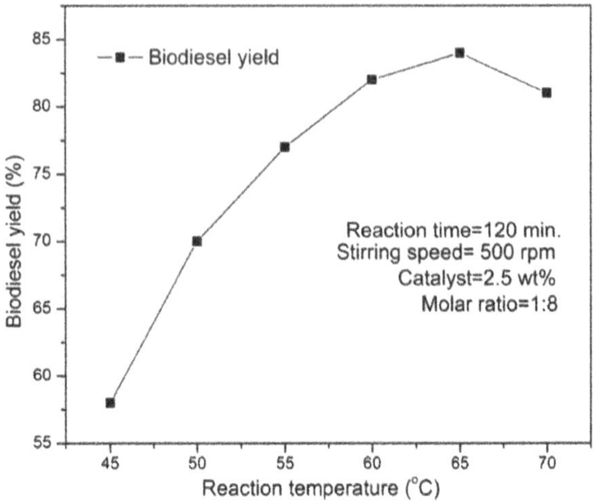


**Fig. 3** Effect of reaction temperature on yield of Nahar biodiesel during second stage transesterification

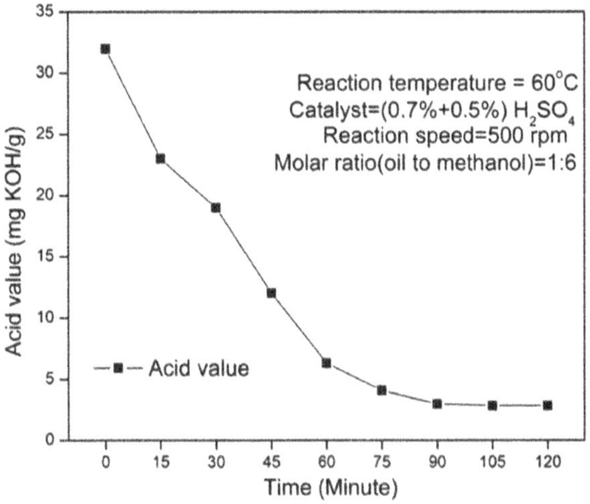


**Fig. 4** Effect of reaction time on biodiesel yield during alkali catalyzed transesterification

# 5 Evaluation of Fuel Properties

This work can be compared to the work of Boro et al. [15]. They prepared biodiesel by using 2% Li and 5% loading of CaO derived from eggshell in transesterification process. However a total reaction period of 10 h for esterification, catalyst activation and transesterification may not be considered for production of biodiesel at an industrial scale due to higher power consumption. Owing to the high cost of $Li_2CO_3$, adding 2% Li would further hit the problem.

Several important fuel properties of Nahar biodiesel are listed in Table 3. Both the density and viscosity reduced to safer limit after double stage transesterification. The calorific value was obtained 38.012 MJ/kg for Nahar methyl ester, which is

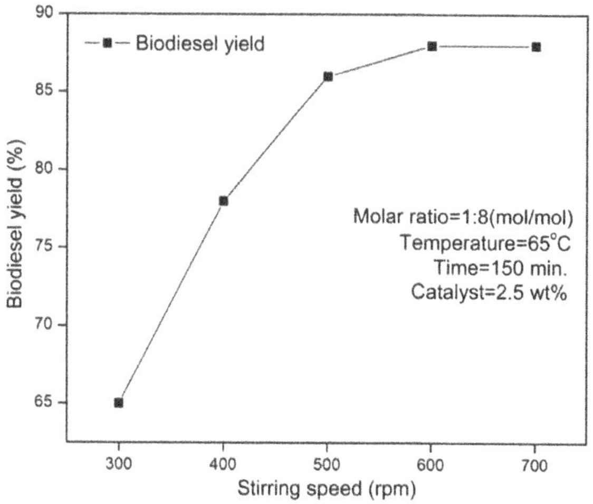


**Fig. 5** Effect of stirring speed on biodiesel yield during second stage transesterification process

**Table 3** Fuel properties of Nahar biodiesel

| Fuel property | Test method | Limits | Biodiesel quality | Diesel |
|---|---|---|---|---|
| Acid value (mg KOH/g) | ASTM D664 | 0.5 max | 0.43 | – |
| Density @ 30 °C (gm/cc) | ASTM D1448 | 0.800–0.900 | 0.882 | 0.830 |
| Viscosity @ 40 °C (mm$^2$/s) | ASTM D445 | 1.9–6.00 | 5.9 | 2.7 |
| Calorific value (MJ/kg) | ASTM D6751 | 34–45 | 38.012 | 42 |
| Cetane number | ASTM D613 | 47 min | 51 | 48 |
| Flash point (°C) | ASTM D93 | 130 min | 145 | 64 |
| Fire point (°C) | ASTM D93 | – | 155 | 71 |
| Cloud point (°C) | ASTM D2500 | – | 7 | −4 |
| Pour point (°C) | ASTM D2500 | – | 4 | −9 |
| Sulphated ash (mass %) | ASTM D874 | 0.02 max | 0.005 | – |
| Free glycerin (mass %) | ASTM D6584 | 0.00–0.02 | 0.01 | – |
| Total glycerin (mass %) | ASTM D6584 | 0.00–0.24 | 0.017 | – |
| Methanol content (vol. %) | EN14100 | 0.05 | 0.05 | – |
| Carbon content (wt%) | Elemental analysis | | 77.19 | 86.4 |
| Oxygen content (wt%) | Elemental analysis | | 12.02 | 0 |
| Hydrogen content (wt%) | Elemental analysis | | 10.37 | 13.1 |
| Nitrogen content (wt%) | Elemental analysis | | 0.03 | 0.14 |

comparatively lower than high speed diesel but is in the range of ASTM prescribed limits. High cetane number (51) would help in burning characteristics of fuel and reduce delay period when injected into the combustion chamber of diesel engine. Storing of Nahar biodiesel alone or biodiesel diesel blended form would be safer than diesel fuel alone due to high flash and fire point characteristics of Nahar biodiesel. Poor cloud point (7 °C) and pour point (4 °C) characteristic of Nahar biodiesel may be due to the presence of palmitic and stearic acid. However using biodiesel in blended form (biodiesel diesel blend) in a country like India, where high average temperature (more than 20 °C) prevails throughout the year would not cause any trouble. Other parameters like, presence of free glycerin, sulphated ash, total glycerin and methanol content were in the limit. From elemental analysis the carbon (C), hydrogen (H), nitrogen (N) and oxygen (O) compositions were obtained as 77.19, 10.37, 12.02 and 0.03 wt% respectively. The presence of oxygen in the biodiesel would result better combustion of fuel in the compression ignition engine compared to diesel fuel.

## 6 Conclusion

Abundantly available non-edible high oil content (nearly 70%) underutilized seed has been investigated to produce biodiesel. Owing to high FFA content, double stage (acid-base) transesterification process was developed after several attempts failed with simple alkali catalyzed transesterification process to yield biodiesel. CaO used as a heterogeneous catalyst easily separated from the biodiesel after end of the reaction. The usual water washing step was avoided in this study, due to the use of heterogeneous catalyst which saves time, energy and cost both in direct and indirect way. The first stage acid esterification reduces FFA to less than 1.5% in 1 h. 45 min with a molar ratio of oil to alcohol 1:6, at 60 °C reaction temperature, loading of (0.7 + 0.5) % (v/v) acid catalyst ($H_2SO_4$) with mixing speed of 500 rpm. The optimum conditions for alkali catalyzed transesterification reaction are; 65 °C reaction temperature, 1:8 molar ratio of oil to alcohol, 600 rpm reaction speed, 2.5 wt% of catalyst loading and time of 2.5 h. Maximum 88% biodiesel yield was obtained from transesterification reaction with using optimum conditions of reaction parameters. Fuel properties of Nahar biodiesel are observed to be comparable to that of diesel fuel and in the limits prescribed by ASTM specification. Hence it is quite clear that; biodiesel prepared from Nahar oil can be utilized in diesel engine as an alternative fuel. Use of biodiesel derived from Nahar oil in conjunction with other (Gov. of India) identified potential feedstock, (Jatropha and Karanja) would generate rural employment, reduce dependency on fossil based fuel and curb the carbon emissions as per the commitment of India towards the 'National biofuel policy (Gov. of India)' and inspiring 'Paris climate agreement'.

**Acknowledgements** The authors are thankful to Indian Biodiesel Corporation (IBDC), Baramati, Maharashtra, for help in testing of prepared samples.

## References


1. Azam MM, Waris A, Nahar NM (2005) Prospects and potential of fatty acid methyl esters of some non-traditional seed oils for use as biodiesel in India. Biomass Bioenerg 29:293–302
2. Bora DK, Nath R (2007) Use of nahar oil methyl ester (NOME) in CI engines. J Sci Ind Res 66:256–258
3. Kushwah YS, Mahanta P, Mishra SC (2008) Some Studies on Fuel Characteristics of MesuaFerrea. Heat Transf Eng 29(4):405–409. https://doi.org/10.1080/01457630701825788
4. Aransiola EF, Ojumu TV, Oyekola OO, Madzimbamuto TF, Ikhu-Omoregbe DIO (2014) A review of current technology for biodiesel production: state of the art. Biomass Bioenerg 61:276–297
5. Dash SK, Lingfa P (2017) A review on production of biodiesel using catalyzed transesterification. AIP conference proceedings 1859, p020100, https://doi.org/10.1063/1.4990253 (http://dx.doi.org/10.1063/1.4990253)
6. Dash SK, Dash S, Lingfa P (2017) Comparative assessment of performance and emission analysis of a diesel engine fueled with biodiesel prepared from different sources. J Ind Pollut Control 33(2):1114–1119
7. Moser BR (2009) Biodiesel production, properties, and feedstocks. In Vitro Cell Dev Biol Plant 45:229–266. https://doi.org/10.1007/s11627-009-9204-z
8. Gerpen JV (2005) Biodiesel processing and production. Fuel Process Technol 86:1097–1107
9. Naik M, Meher LC, Naik SN, Das LM (2008) Production of biodiesel from high free fatty acid karanja (*Pongamia pinnata*) oil. Biomass Bioenerg 32:354–357
10. Shahid EM, Jamal Y (2011) Production of biodiesel: a technical review. Renew Sustain Energy Rev 15:4732–4745
11. Lee HV, Juan JC, Taufiq-yap YH, Kong PS, Rahman NA (2015) Advancement in heterogeneous base catalyzed technology: an efficient production of biodiesel fuels. J Renew Sustain Energy 7(032701):1–46. http://dx.doi.org/10.1063/1.4919082
12. Chavan SB, Kumbhar RR, Madhu D, Singh B, Sharma YC (2015) Synthesis of biodiesel from Jatropha Curcas oil using waste eggshell and study of its properties. RSC Adv 5:63596–63604
13. Veljkovic VB, Stamenkovic OS, Todorovic ZB, Lazic ML, Skala DU (2009) Kinetics of sunflower oil methanolysis catalyzed by calcium oxide. Fuel 88:1554–1562
14. Demirbas A (2007) Biodiesel from sun flower oil in supercritical methanol with calcium oxide. Energy Convers Manag 48:937–941
15. Boro J, Konwar LJ, Deka D (2014) Transesterification of non edible feedstock with lithum incorporated eggshell derived CaO for biodiesel production. Fuel Process Technol 122:72–78

# Microstructural Evaluation, Tribological Behavior and Mechanical Properties of Cu–Ni Alloy After Addition of Chromium Under Spinodal Conditions

Sagar C. Jirapure and Atul B. Borade

## 1 Introduction

Copper and copper alloys are widely used for engineering applications. Due to wide range of use Copper needs some heat treatment for strengthening its alloys. To develop the mechanical properties of Cu based alloys various researches were carried out. A new developed Spinodal decomposition will enhances the properties of Cu-based alloys. Cu–Ni–Cr alloys are subject to embrittlement when aged at temperatures in the range of 300–800 °C. Vintaykin et al. [1] showed relationship between experimental results and the Cahn-Hilliard theory of spinodal decomposition. The miscibility gap in the Cu–Ni–Cr system was first reported by Meijering et al., by means of metallographic and X-ray diffraction methods at 930 °C an isothermal section was determined Fig. 1 [2].

The phase field method [3, 4] is a powerful tool to analyze the phase decomposition in binary and ternary alloy systems. This will help to study microstructure evolution and growth kinetics of the decomposed phases in the early stages of aging. Cahn [5, 6] developed a linear model, which is applicable in three dimensions. It refers to a method in which a supersaturated solid solution decomposes into solute-rich and solute-poor regions when it is aged at a suitable temperature [7]. It forms the ordered microstructure of alloy which delays the dislocation motion which is the source of hardening. Aging time will plays an important role for deciding in remarkable strength of the alloy. Zhang et al. [8] have calculated the wear performance of Cu-15Ni-8Sn spinodal alloy with hardness whereas S. Ilangovan et al investigated wear behavior of Cu–Ni–Sn spinodal alloy by varying the Sn composition [9]. Deyong et al. have informed about the effect of Sn content on the hardness of Cu–Ni–Sn alloys maintaining Ni fixed at 10 wt% [10].

S. C. Jirapure (✉) · A. B. Borade
Mechanical Engineering Department, Jawaharlal Darda Institute of Engineering and Technology, Yavatmal, MS, India
e-mail: sagarjirapure@rediffmail.com



K. Antony and J. P. Davim (eds.), *Advanced Manufacturing and Materials Science*, Lecture Notes on Multidisciplinary Industrial Engineering,
https://doi.org/10.1007/978-3-319-76276-0_45

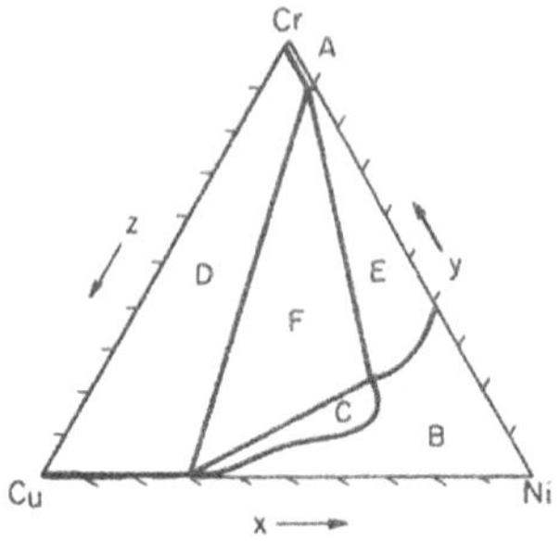


**Fig. 1** Cu-Ni-Cr ternary phase diagram at 930 °C. Atomic concentrations- **A** homogeneous b.c.c.; **B** homogeneous f.c.c.; **C** 2 f.c.c. phases; **D** b.c.c. and Cu-Rich f.c.c.; **E** b.c.c. and Ni-rich f.c.c.; **F** b.c.c. and 2 f.c.c. phases [2]

The intention of this research is to investigate the phase decomposition in Cu–Ni–Cr alloy aged at 6000C. So, in the present work, the effects of Cr on hardness, wear rate and the coefficient of friction of Cu–Ni–Cr is discussed. Microstructure and strength of the alloy has been changed via spinodal decomposition [11].

## 2 Experimental Procedure

### (1) The Sample

The required alloy sample elements were procured in the commercial market in pure form. The sample is a commercial Cu-based alloy of following composition by weight %:

$$\begin{aligned} &\text{Cu} - \text{balance C} - 0.014\%,\\ &\text{Ni} - 28.9\%\,\text{Ti} - 0.051\%,\\ &\text{Cr} - 2.84\%\,\text{Si} - 0.091\%,\\ &\text{Mn} - 0.55\%\,\text{Fe} - 0.32\%,\\ &\text{Zr} - 0.22\%. \end{aligned}$$

### (2) Heat Treatment

The alloy was placed in an electric arc furnace with an inert atmosphere; Aragon was supplied during the heat treatment. The temperature within the hot zone of the furnace was controlled to an accuracy of ±5%. Homogenization was carried out at a temperature above the miscibility gap for the Cu–Ni–Cr and cooled in the furnace itself.

Solution heat treatment involves heating the alloy to a temperature in the single phase solid solution region and rapidly quenched in water.

Aging heat treatment was carried out after the specimens were subjected to solution heat treatment. The aged specimens withdrawn from the furnace were quenched in water.

The microscopic examinations were conducted using Carl-Zeiss metallurgical microscope.

(3) **Hardness Testing**

To study the hardness of prepared alloys the hardness of the entire alloy specimens were measured using micro-hardness tester of HMV G21. Numerous readings were taken at different position for each specimen and a typical value was calculated.

(4) **Tribological Testing**

The friction and wear tests were performed on entire alloy specimens. Pin-on-Disc type wear tester was used for tribological testing. Hemispherical pins of 5 mm radius of curvature were mated against 130 mm diameter and 10 mm thick discs of the same material. The pin was held stationary and the disc was rotated. All the tests were carried out at a constant load of 20 N for 800 s each at constant sliding velocity of 0.8 m/s resulting in total sliding distance of 640 m in each test.

# 3 Results and Discussion

## *3.1 Microstructural Evolution*

Microstructural evolution, based on Cr concentration, is shown in Fig. 2a–c for the alloy aged at 600 °C for different times.

Long time ageing will results in Cu-rich and Cu-poor regions. In microscopic observations, bright regions correspond to the Cu-rich phase and dark regions correspond to the Cu-poor phase since the copper rich regions preferentially thin in the electrolyte used. The morphology of the Cr-rich phase is interconnected and rounded irregular without any preferential alignment in the alloy. Both phases contain all three alloying elements.

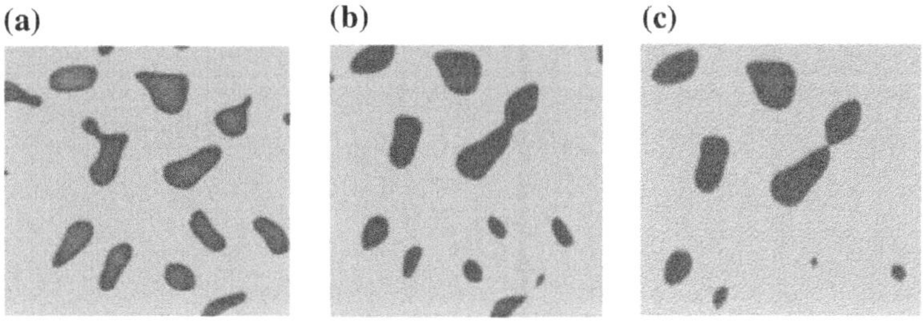


**Fig. 2** Simulated microstructural evolution

## 3.2 Coarsening Behavior

In the early stages of ageing, the wavelength was showing a constant value; however after long time ageing a particle coarsening behavior was obtained. However, clear periodic structures are observed above 500 °C Fig. 3.

## 3.3 Hardening Behavior

The plot of Vickers hardness as a function of aging time is shown in Fig. 4 for the present alloy aged at 600 °C.

From the observations it is cleared that at the early stages of aging there is a sudden increase in hardness and then hardness increases gradually for delayed aging. The increase in hardness related with the spinodal decomposition of the supersaturated solid solution. The decomposed phases are consistent with nanometre size. Also, the hardness is proportional to the wavelength of modulation

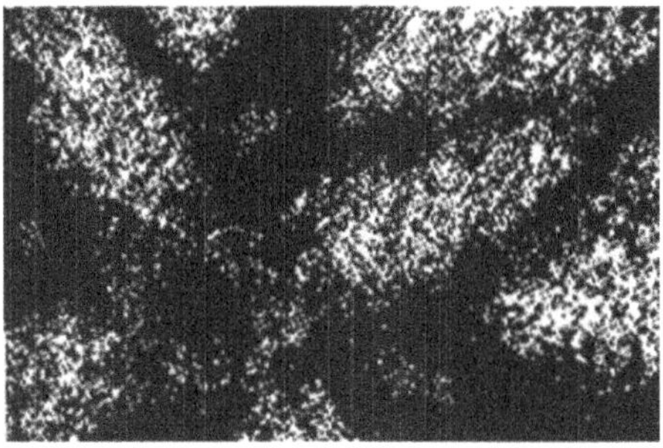

**Fig. 3** Clear periodic structure

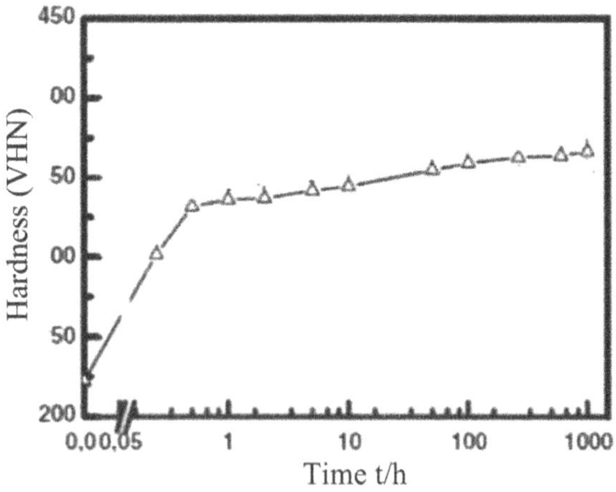


**Fig. 4** Plot of vickers hardness versus aging time

composition [12]. Size of the decomposed phases increases with aging time. These two features causing an increase in hardness. For delayed aging, the decomposed phases are still coherent and nanometre size and thus there is no decrease in hardness.

## 3.4 *Tribological Behavior with Cr Content*

Chromium contributing the spinodal decomposition. The hardness of the alloys was greatly influenced by the presence of Cr. The hardness in Cu–Ni–Cr is because of presence of Cr in Cu–Ni binary system which is effective in increasing the resistance to dislocation motion and increasing the work-hardening rate of the alloy.

The dendritic structure was fully removed by the homogenization and the solution heat treatments (Fig. 5).

The modulated structure and an ordering reaction take place during the aging process. The modulated and the ordered structures are very fine.

It is observed from the plot that hardness of Cu–Ni–Cr alloys increases with Cr content (Fig. 6).

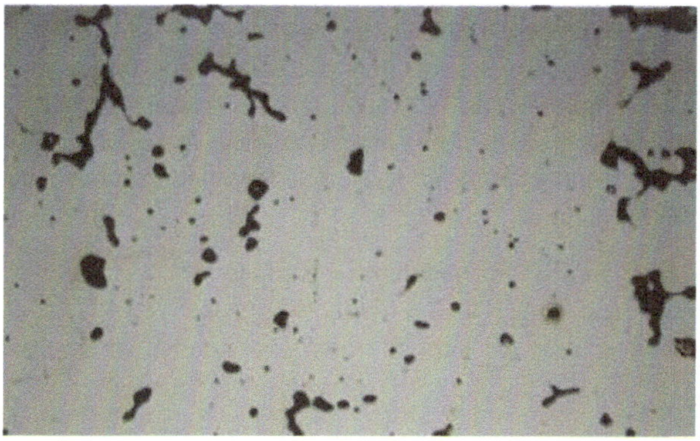

**Fig. 5** Homogenized & solution heat treated microstructure

**Fig. 6** Variation of hardness with Cr content

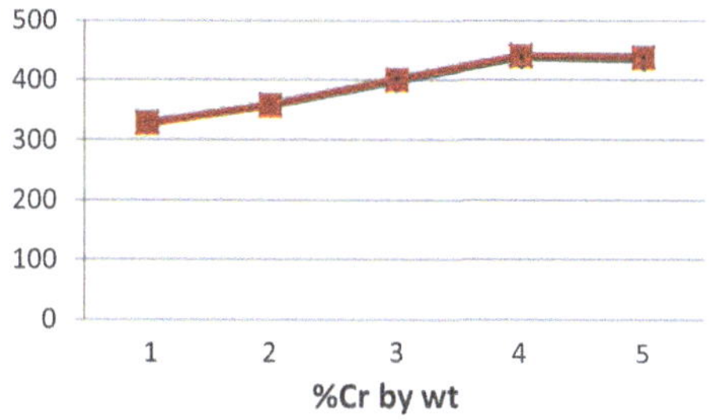

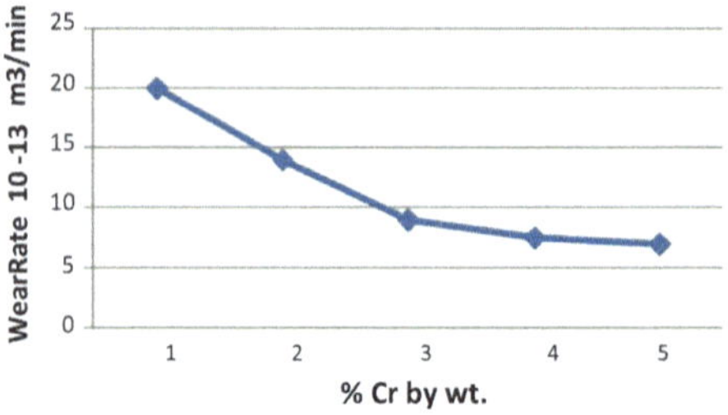


**Fig. 7** Variation of wear rate with Cr content

It is concluded from the result that the addition of Cr content reduces wear rate by increasing the hardness (Fig. 7).

## 4 Conclusions

Following conclusions are made on the basis of the present study.

The growth kinetics of phase decomposition was observed with very slow growth kinetics during the spinodal decomposition of the supersaturated solid solution.
Added aging caused an increase in raise kinetics due to diffusion-controlled coarsening of the decomposed phases.
The alloy showed morphological features such as aligned precipitates and absence of preferential precipitations at microstructural in-homogeneities.
The increase in hardness seems to be associated with the coherency and nanometre size of the spinodally-decomposed phases in the aged alloy. Cr contributes significantly to the hardening.

Coefficient of friction is independent of hardness.
Wear rate decreases linearly with hardness irrespective of Cr content.
Friction and wear may be kept to some lower value to improve mechanical processes.

Any definite relationship between the mechanical properties and microstructure can be seen to depend on the compositional fluctuations.
The spinodal decomposition method forms homogeneous microstructure, no influencing by section thickness and uniform all over the section, over the conventional hardening method.

In the present alloy, only FCC particles are obtained.

**Acknowledgements** The authors are immensely thanking Heads & associates of Research lab, Mechanical Engineering Department, Engineering Metallurgy laboratory, JD Institute of Engineering & Technology to carry out the present work.

## References

1. Vintaykin Ye Z, Dmitriyev VN (1970) Fiz Metallov Metalloved 29:1257–1267
2. Meijering JL, Rathenau GH (1955–1956) J lnst Metals 118
3. Baburaj EG, Kulkarni UD (1979) Initial stages of decomposition in Cu-9Ni-6Sn. J Appl Crystallogr 12(5):476–480
4. Honjo M, Saito Y (2000) ISIJ Int 40:914–919
5. Cahn JW (1961) Acta Metall 9:795–801
6. Cahn JW (1962) Acta Metall 10:179–183
7. Kuwajima T, Saito Y (2003) Intermetallics 11:1279–1285
8. Zhang S et al (2010) Dry sliding wear of Cu-15Ni-8Sn alloy. Tribol Int 43(1–2):64–68
9. Ilangovan S, Sellamuthu R (2016) Effects of tin on hardness, wear rate and coefficient of friction of cast Cu-Ni-Sn alloys. J Eng Sci Technol 8(1):34–43
10. Deyong L et al (1990) Microstructural and mechanical properties of rapidly solidified Cu-Ni-Sn alloys'. Mater Sci Eng 124(2):223–231
11. Findik F, Rehman F (1993) A spinodal Cu-Ni-Cr alloy. Trans Eng Sci, 2, WIT Press, 219–228
12. Kostorz G (1996) Phase transformations in materials. Wiley-VCH, Germany, pp 309–407
13. Archard JF (1953) Contact and rubbing of flat surfaces. J Appl Phys 24(8):981–988
14. Kato M, Schwartz LH (1979) The temperature dependence of yield stress and work hardening in spinodally decomposed Cu-10Ni-6Sn alloy. Mater Sci Eng 41(1):137–142
15. Chou A, Datta, A. G. (1978) Microstructural behaviour and mechanical hardening in a Cu-Ni-Cr alloy. J Mater Sci 13:541–552 CrossRef
16. Jantzen CMF, Herman H (1978) Phase diagrams. Materials science and technology. Academic Press New York, vol 6, pp 127–184
17. Ilangovan S, Sellamuthu R (2016) Measurement of the variation of mechanical properties with aging temperatures for sand cast Cu-5ni-5sn alloy. J Eng Sci Technol 11(11):31609–31619 © School of Engineering, Taylor's University
18. Findik Fehim (2013) Modulated alloys. Periodicals Eng Nat Sci 1(1):47–55
19. Rao PP, Agrawal BK (1991) Comparative-study of spinodal decomposition in symmetrical and asymmetric Cu-Ni-Cr alloys. J Mater Sci 26:1485–1496
20. Xicheng Wei et al (2009) Evolution of friction-induced microstructure of SUS 304 meta-stable austenitic stainless steel and its influence on wear behaviour. Wear 267:1386–1392
21. Hemant et al (2010) The sliding friction and wear behaviour of 15Cr–15Ni–2Mo titanium-modified austenitic stainless steel (alloy D9) was studied in high purity liquid sodium at high temperature viz. 473 and 823 K respectively. Wear 270:1–4
22. Palásti-Kovács et al (2004) Microtopography changes in wear process. Acta Polytechnica Hungarica 1:108–119
23. Yuan CQ et al (2004) The surface roughness evolutions of wear particles and wear components under lubricated rolling wear condition. Wear 259:512–518
24. Yuan CQ et al (2008) Surface roughness evolutions in sliding wear process. Wear 265:341–348
25. Zum Gahr K (1987) Microstructure and wear of materials. Elsevier, Tribol, pp 174–176
26. Archard JF (1980) Wear theory and mechanisms. In: Peterson MB, Winer WO (eds) Wear control handbook. ASME, New York, NY, pp 35–80
27. Jirapure SC, Borade AB (2014) A new approach of strengthening- spinodal decomposition. Int J Appl Sci Eng Res 3(5):1021–1029
28. Jirapure SC, Borade AB (2017) Measurement of mechanical properties and microstructure of Cu-Ni-Cr spinodal alloy. Int J Adv Eng Manag 2(3):78–81

# Development of Fire Resistant Polymeric Nano Composites Using Plasma Modified Calcium Silicate

**G. Ajeesh, P. Sanjana, M. Raji Sivani, S. Govardhan, A. Sudhin and Philip Goerge**

## 1 Introduction

Materials have been the basis for human development right from the start of civilization. Anthropologists have said that the historical epochs were developed by materials such as the Stone and Iron ages. Composites have gained tremendous importance because of their properties like high strength to weight ratio, High thermal and mechanical properties and its resistance to corrosion. Composite was first developed by man using mud and straw. Mud was used as matrix and straw was used as reinforcement. Composites are further classified as metal matrix, ceramic matrix and polymer matrix composites. Polymeric matrix composites are used for primarily applications pertaining to aerospace, defense, automobile and structural. Polymer matrix composites are further classified as thermosetting and thermoplastic composites. To develop a tougher, stronger and thermally stable composite, the present work is focused on using thermoplastic composites. Very little research has been performed on thermoplastic composites compared to thermosetting composites. Thermoplastics are expensive and difficult of develop considering their high thermal properties. In this investigation polypropylene (PP) was considered because of its ease of manufacturing and its superior thermal and mechanical properties [1]. Polypropylene is a semi-crystalline engineering thermoplastic and is known for its high strength, high thermal and mechanical properties. PP is used widely in application where high creep resistance is important. Non reinforced PP has low fire resistant. To improve its strength and thermal properties, different fillers (such as fibres, clay, silica, graphene and CNT) are

G. Ajeesh (✉) · P. Sanjana · M. R. Sivani · S. Govardhan · A. Sudhin
Department of Aerospace Engineering, Amrita School of Engineering,
Amrita Vishwa Vidyapeetham, Coimbatore 641112, India
e-mail: g_ajeesh@cb.amrita.edu

P. Goerge
Vijnan Institute of Science and Technology, Elanji, Ernakulam, India

 
K. Antony and J. P. Davim (eds.), *Advanced Manufacturing and Materials Science*,
Lecture Notes on Multidisciplinary Industrial Engineering,
https://doi.org/10.1007/978-3-319-76276-0_46

reinforced. The reinforcement of fillers has been proved to be an effective way to improving the physical and thermal properties. Nano Calcium silicate has been used in this investigation as it possesses high thermal stability and hence, enhances the fire resistant property of composites [2].

Calcium silicate is used for passive fire protection and fireproofing in the form of tiles or bricks. This material has been extensively used in Europe. Calcium-silicate retains their properties under moisture and damp environment [3]. In some cases calcium silicate is being treated so as to enhance its property towards chemical environments and moisture.

# 2 Experimental

## 2.1 Materials

Non reinforced polypropylene of grade H110MA was procured from Repol, India. Nano-calcium silicate was procured from Sigma Aldrich, India.

## 2.2 Plasma Modification

Plasma modification is a type of modification technique that modifies the surface of the components using plasma. Very high voltage is supplied to the gas to ionize the gas and use it as plasma. The whole process is carried out in vacuum conditions. Oxygen plasma removes organic contaminants by reacting chemically with highly reactive oxygen radicals. Oxygen plasma surface treatment promotes surface oxidation and hydroxylation that increases the wetting property of the surface.

## 2.3 Preparation of PP/(Unmodified/Modified) Nano Calcium Silicate Composites

Samples of unfilled polymer was developed using an injection molding machine. The composites were done in two steps. In the first step, the filler was reinforced into matrix using a twin screw extruder at different ratios. The extruder screw rpm was maintained at 71 and the die temperature was kept constant at 232 °C. The outcome was in the form of strands which was then cut to smaller pieces using a shear cutter and then fed into the injection molding machine. Weight ratio of 0.5, 1, 1.5, 2 and 5 were being used for this study. The compounded samples were then injection molded to achieve the required shape. The mold temperature was set as 250 °C.

# 3 Characterization

## 3.1 FTIR Spectroscopy

The effect of modification in the nano filler was observed using a Nicolet iS10 machine manufactured by Thermo Scientific, USA.

## 3.2 Tensile Test

Tensile test was done using a Universal Testing Machine. The machine used was Zwick/Roell Z010. The gauge length of the specimen was 50 mm and the cross-head speed was 5 mm/min according to ASTM D-638.

## 3.3 Scanning Electron Microscopy

SEM is performed to study the effect of modification on the dispersion of the nano fillers in the composites.

## 3.4 Limiting Oxygen Index

LOI test, is a very widely used method for determining the relative flammability of polymeric materials. ASTM D-2863 standard was followed. Oxygen content was increased in steps of 1%. The test was done for 0.5, 1, 2 and 5% mass ratios for composites containing the unmodified and modified filler.

## 3.5 Horizontal Flammability

Flammability is the ability of a substance to burn or ignite, causing fire or combustion. The horizontal flammability test method is intended for use in determining the resistance of materials to flame and the test standard followed was ISO 3795. The flame height was 38 mm and it was applied for 15 s.

# 4 Results and Discussion

## 4.1 FTIR Spectroscopy

It is observed that on comparing the FTIR spectrum of the unmodified and plasma modified nano calcium silicate as shown in Figs. 1 and 2, there is an increase of 18% of the transmittance value of the hydroxyl group (O–H) and aromatic group (C=C). This is attributed to the effect of plasma modification on the surface of calcium silicate. Research in the past has revealed that on modification of nano

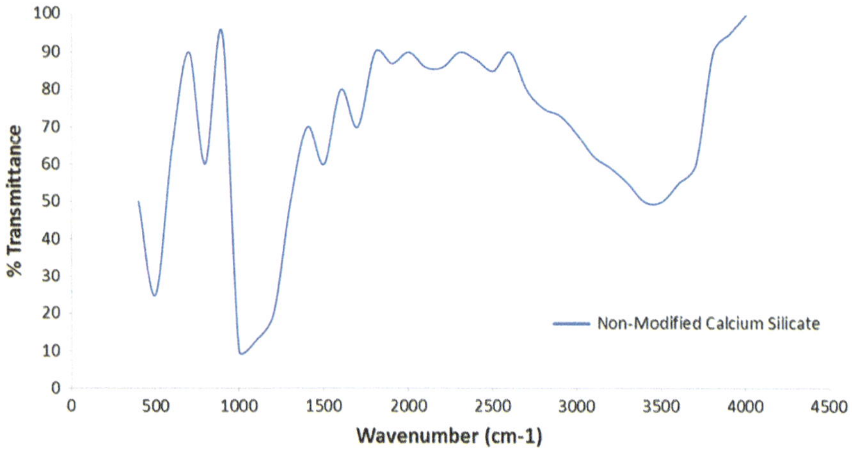


**Fig. 1** FTIR spectra of unmodified calcium silicate

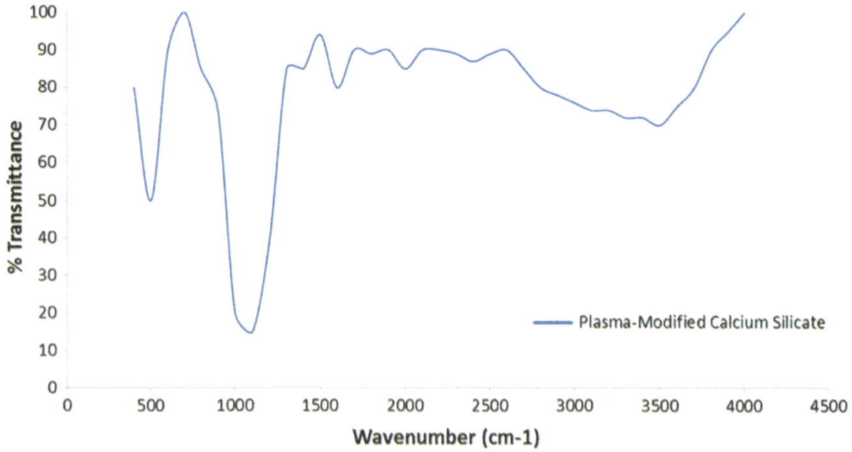


**Fig. 2** FTIR spectra of plasma modified calcium silicate

fillers, a tremendous enhancement in the dispersion and interfacial adhesion between matrix and reinforcement is observed [4, 5]. It was also observed that on modification, there is a covalent linking of matrix with the reinforcement.

## 4.2 Tensile Tests

It is observed from Fig. 3, that there is a marginal increase in the tensile strength of composites on reinforcing higher ratio of calcium silicate. It is also observed that there is a marginal increase in the tensile strength of plasma modified filler based compared to non-modified composites. This may be attributed to the high surface area of the nano filler have [6].

## 4.3 SEM

SEM was carried out to study the dispersion of nano filler in the composites. Figure 4 shows the micrographs of non-reinforced composites and Fig. 5 shows the figure of plasma modified calcium silicate composites.

Based on these findings it is evident that plasma modified nano filler shows better dispersion.

## 4.4 LOI

It can be observed from Fig. 6 that the LOI value increases with increasing mass ratio of the composites containing the unmodified filler. A significant change is also

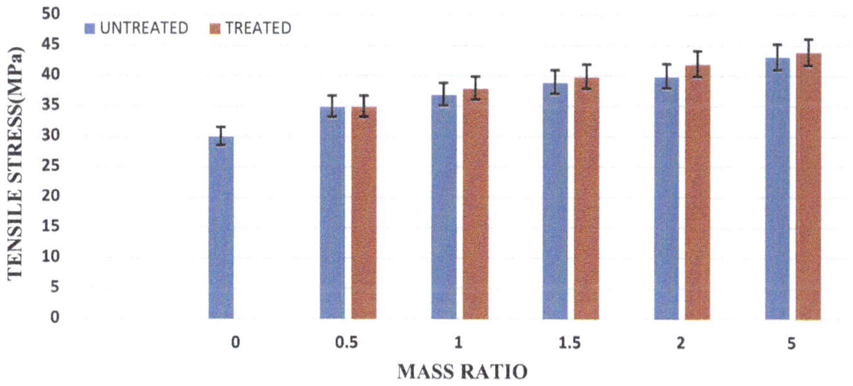


**Fig. 3** Tensile strength of composites

**Fig. 4** SEM micrographs of non-reinforced polymer

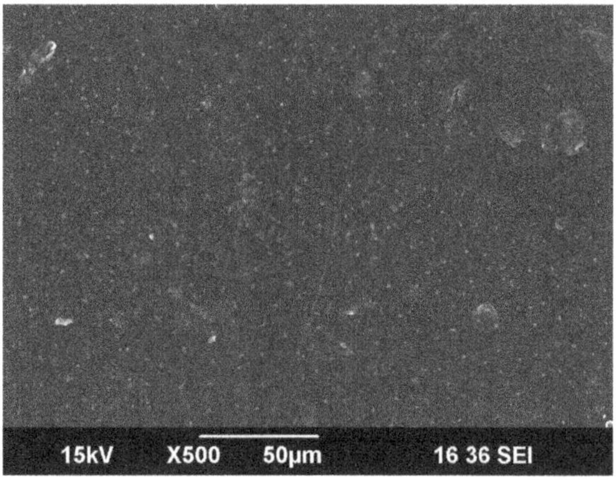


**Fig. 5** SEM micrographs of plasma modified calcium silicate composites

noted for composites with plasma modified nano filler. This is attributed to the uniform dispersion.

This gives a clear indication that plasma modification plays a major role in enhancing the fire resistance of composites.

## *4.5 Horizontal Flammability*

It can be observed from Fig. 7 that the burn rate value decreases with increasing mass ratio of the composites containing fillers. Significant enhancement is observed in case of plasma modified nano particles. The burn rate has significantly reduced for plasma modified composites compared to non-modified composites.

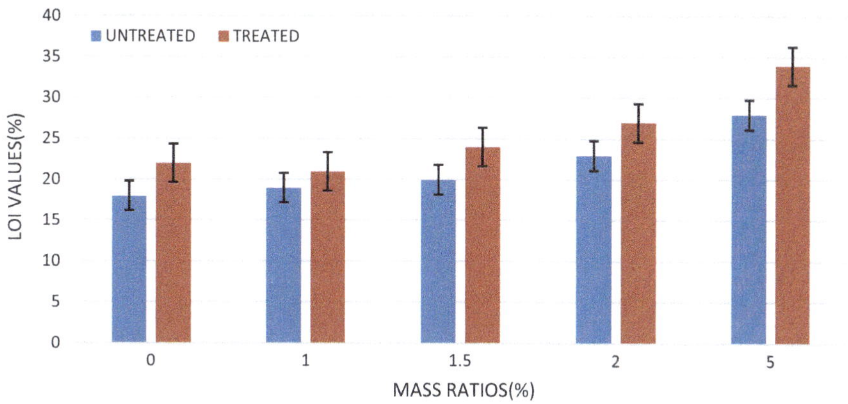


**Fig. 6** LOI values for polymeric composite samples

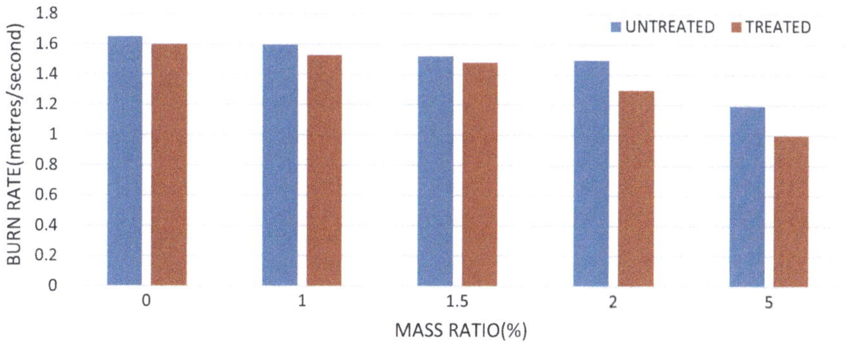


**Fig. 7** Burn rate values for polymeric composite samples

## 5 Conclusion

The study revealed the following conclusions:

- The mechanical properties of the composite containing fillers increased with increasing mass ratio.
- The oxygen plasma modification done on the filler increased its surface energy. Hence there was proper bonding between the filler and the polymer. Therefore the composites containing modified calcium silicate gives marginally better results with regard to tensile and dispersion properties.

- LOI and flammability test clearly indicate that there is a significant improvement in the fire resistant property of composites when reinforced with plasma modified calcium silicate.
- Based on these investigation, it can be concluded that plasma modified calcium silicate is most suitable for fire resistant application.

## References

1. Gabr M, Okumura W, Ueda H, Kuriyama W, Uzawa K, Kimpara I (2015) Mechanical and thermal properties of carbon fiber/polypropylene composite filled with nano-clay. Compos B Eng 69:94–100
2. George P, Bhowmik S, Abraham M, Sriram PK, Pitchan MK, Ajeesh G (2016) High-performance fire-resistant polymeric nanocomposite for aerospace applications. Proc Inst Mech Eng Part L J Mater Des Appl 1464420716660874
3. Ariyanayagam AD, Mahendran M (2017) Fire tests of non-load bearing light gauge steel frame walls lined with calcium silicate boards and gypsum plasterboards. Thin Walled Struct 115:86–99
4. Ajeesh G, Bhowmik S, Sivakumar V, Varshney L, Kumar V, Abraham M (2017) Influence of surface activated carbon nano fiber on thermo-mechanical properties of high performance polymeric nano composites. J Compos Mater 51(8):1057–1072
5. Ajeesh G, Bhowmik S, Venugopal S, Varshney L, Baluch A, Park Y, Gilsang S, Kim CG (2016) Influence of chemically and plasma-functionalized carbon nanotubes on high-performance polymeric nanocomposites. High Perform Polym 28(5):570–580
6. Ajeesh G, Bhowmik S, Sivakumar V, Varshney L (2017) Influence of surface activated carbon nano fibres on mechanical properties of poly ether ketone (PEK). In IOP conference series: materials science and engineering, vol 204, no 1, p 012010. IOP Publishing

# Welding Metallurgy of Corrosion Resistant 21st Century Ni-Based Superalloy 686

**B. Arulmurugan, M. Sathishkumar, K. Gokulkumar, K. Mageshkumar, P. Subramani, M. Venkateshkannan, Manikandan and Arivazhagan**

## 1 Introduction

Alloy 686 is a single-phase, solid solution strengthened, austenitic, nickel-chromium-molybdenum-tungsten (Ni-Cr-Mo-W) based superalloy. The presence of chromium (Cr) and molybdenum (Mo) provides the good resistance to oxidizing and reducing environment. The alloy 686 is superior corrosion resistance alloy commercially viable now [1]. It has been emerged in the year 1992, from conventionally employed alloy C-22 and C-276 as a contextual. The alloy 686 is derived from the alloys C-22 and C-276. It endows with exceptional resistance to oxidizing, reducing and mixed acid environments comprising those containing halides [2]. This alloy is extensively used in the chemical processing, marine and petrochemical, oil and gas, pulp and paper & air pollution control (FGD) industries. The presence of carbon (C) content is restricted to 0.01% which facilitates in diminishing grain boundary precipitation in the heat-effected zone (HAZ) of welded areas to uphold the corrosion resistance [3].

The imperative attribute about this alloy is that it posses' high PREN value than, other alloys in the Ni-Cr-Mo ternary system, which determines that indicates the material's resistance to localized corrosion. It is capable of being suitable for the applications correspond to corrosive atmospheres of greater than 100,000 ppm of chloride along with oxidizing species like ferric and cupric chloride in very low pH (=1) conditions as well as an extensive diversity of concentrations of sulfuric acid. Crevice corrosion resistance of a Ni-based alloy is being examined by the maximum measured depth of attack at increasing test temperatures. Alloy 686 shows

B. Arulmurugan · M. Sathishkumar · K. Gokulkumar · K. Mageshkumar · P. Subramani
M. Venkateshkannan · Manikandan (✉) · Arivazhagan
School of Mechanical Engineering, VIT, Vellore 632 014, India
e-mail: mano.manikandan@gmail.com



K. Antony and J. P. Davim (eds.), *Advanced Manufacturing and Materials Science*, Lecture Notes on Multidisciplinary Industrial Engineering,
https://doi.org/10.1007/978-3-319-76276-0_47

very low depth of attack and minimum sites of attack when compared with its reference materials (C-22 and C-276) in a solution saturated sodium chloride and sulfur dioxide at 80 °C test results [4].

Alloy 686 was generally supplied in the solution annealed condition. Solution annealing is an imperative heat treatment practice which is used to melt any precipitates available in the material and alter the material at the solution annealing temperature into a soft single phase structure. Welding of this alloy plays a major role in many industrial applications as mentioned above. In general, arc welding methods like gas tungsten arc welding (GTAW) and gas metal arc welding (GMAW) are commonly used in the industrial practice. These welding techniques are most economical as compared to other welding techniques. In general, selection of filler wire is a most important weldment design. Two types of filler wire have been employed (i) base metal filler wire and (ii) over alloyed filler wire. Base metal filler wire is commonly used for the fabrication process, and overmatching filler wire is suitable for highly aggressive environments [5]. The chemical composition of alloy 686 and filler wire ERNiCrMo-10 are listed in Table 1 [1, 6].

The performance of Ni-Cr-Mo based superalloy adversely affected under some circumstances. The experience of different workers is documented in the literature. Cieslak et al. [7] investigated the welding metallurgy of Ni-Cr-Mo based super alloys C-4, C-22, and C-276. The authors accounted among these three alloys C-4 had minimum hot cracking susceptibility than other two alloys because of minimum segregation. The formation of secondary intermetallic, brittle and topologically close-packed (TCP) phases were being observed in alloy C-22 & C-276 at the end of solidification. The morphology of secondary phases being noticed in C-22 belonged to σ, P & μ phases. Perricone and DuPont [8] investigated the effect of alloying element in welding metallurgy of alloy C-22. The Mo levels escalating from 12 to 24% diminish the solute-depleted dendrite core, which causes hot cracking susceptibility, but the authors acknowledged this lay down the path to the formation of P and μ phases in the weld microstructure. Gallagher and Lippold [9] investigated the effect of changes in compositional range and formation of secondary phases using Schil simulation analysis, and they concluded that there is a positive exponential relationship between combined Mo+W level and secondary phase formation.

**Table 1** Chemical composition of Inconel 686 base metal & filler wire ERNiCrMo-10

| Base/filler metal | Chemical Composition (% Wt.) | | | | | | | | |
|---|---|---|---|---|---|---|---|---|---|
| | Ni | Mo | Cr | W | Mn | Fe | S | Co | Others |
| Inconel 686 | Bal | 15.83 | 19.88 | 3.215 | 0.312 | 0.858 | 0. 17 | – | 0.008 (P), 0.076 (Si), 0.005(Ti) 0.01(C) |
| ERNiCrMo-10 | Bal | 13.5 | 21.5 | 3 | 0.02 | 3.1 | 0.002 | 1.8 | 0.005 (P), 0.015 (S), 0.04 (Si) 0.01(C) |

Cieslak et al. [7] reported that at the terminal stage of solidification of alloy C-276 weldments were augmented with TCP phases and the constituents were relating to P & μ phases. Akhtar et al. [10] reported that there is a diminution in impact energy of C-276 due to the formation of Mo-rich μ phase precipitates after aging at 850 °C. Raghavan et al. [11] investigated theformation of three phases when the alloy is aged in the temperature range of 650–900 °C. They acknowledged that two phases TCP in nature (μ & P) and the third phase is associated with molybdenum-rich $M_6C$ carbide. So both the alloys C-22 and C-276 were having the problem of formation of TCP phase at the terminal stage solidification which leads to hot cracking in the welded region. Since the alloy 686 chemical composition is intimately matched (slight modification in Cr & Mo content) with C-22 & C-276 a similar kind of effect can be anticipated. Silva et al. [12] analyzed the microstructure characteristics of dissimilar weld overlays of Ni-based alloy 686 with Stainless Steel. The authors exposed that, in alloy 686, weldments were enriched with three TCP phases like σ, P & μ phases after carrying out SEM, TEM, EDAX and XRD analysis. A Mo-rich precipitate of the P-phase type is a major one. In general, the TCP phases in Ni-based superalloys produce a detrimental result on mechanical properties and in corrosion characteristics [13].

## 2 Problem Statement and Critical Review of the Work Done

It is imperative to stifle the microsegregation of alloying elements in the fusion zone of alloy 686 during welding. Otherwise, it will deteriorate corrosion resistance, tensile ductility, impact toughness and rupture strength, through hot cracking and that leads to premature failure during decisive application. The clue to diminish is that it is the too faster rate of cooling during solidification. Several techniques have been employed to control the microsegregation. Many researchers recommended that Pulsed current gas Tungsten arc welding (PCGTAW) is a potential and economical approach to bring down the microsegregation in the arc welding technique [14–16].

EDS analysis results of major elemental values employed in this study are shown in Table 2 (in wt%). The investigation has been commenced in alloy 686 with autogenous mode, i.e., without filler wire. In that particular, GTAW, shows significant segregation of Mo in the welded zone. The interdendritic region is having a higher amount of Mo content, impoverished with Ni compared to the dendritic core which is clearly observed from Table 2. In case of autogenous PCGTAW, this kind of segregation was not observed.

Subsequently, the welding was done with matching filler wire ERNiCrMo-10. Even at this juncture also the same results have been accomplished and are shown in Table 2. GTAW with filler wire ERNiCrMo-10 is perceived that the interdendritic region is enriched with Mo and impoverished in Ni and W compared to

**Table 2** Weld center EDS analysis results of weldments

| Types of welding | Zone | in weight % | | | |
|---|---|---|---|---|---|
| | | Ni | Mo | Cr | W |
| GTAW-Autogenous | Weld Centre Dendritic Core | 63.29 | 15.05 | 18.89 | 2.76 |
| | Weld Center Interdendritic Zone | 34.31 | 39.9 | 19.47 | 6.01 |
| PCGTAW-Autogenous | Weld Centre Dendritic Core | 65.69 | 12.49 | 18.52 | 3.24 |
| | Weld Center Interdendritic Zone | 56.28 | 20.13 | 19.65 | 3.91 |
| GTAW-ERNiCrMo-10 | Weld Centre Dendritic Core | 65.38 | 11.44 | 20.21 | 2.97 |
| | Weld Center Interdendritic Zone | 35.38 | 39.02 | 17.17 | 7.65 |
| PCGTAW-ERNiCrMo-10 | Weld Centre Dendritic Core | 65.56 | 12.72 | 18.07 | 3.63 |
| | Weld Center Interdendritic Zone | 42.42 | 32.32 | 19.88 | 5.31 |

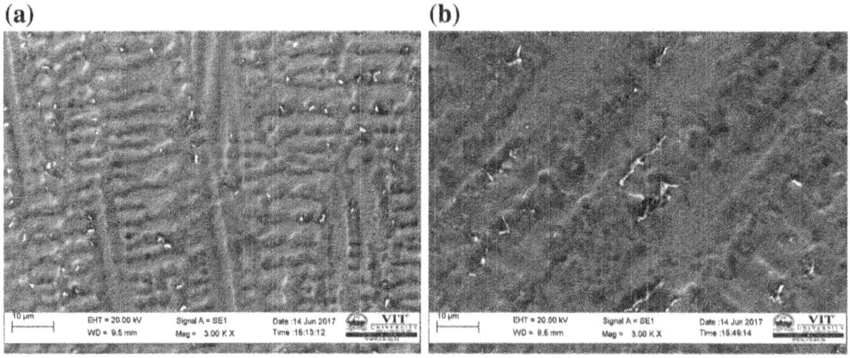


**Fig. 1** SEM image of weld center Microstructure of **a** GTAW Autogenous **b** PCGTAW Autogenous

dendritic core. In contrast with GTAW, in PCGTAW the alloying elemental values and their disparity in the interdendritic region and dendritic core are very less. From EDS analysis, it is apparent that the occurrence of secondary phases σ, P & μ were deleterious to a large extent by switch over from GTAW to PCGTAW. At that same time, weld joint produced with GTAW, both with autogenous and filler wire, the weld interface were also identified with Mo enriched secondary TCP phases. But in PCGTAW interfaces did not show any elemental segregation. Figure 1a, b being evidence that higher magnification scanning electron microscope weld zone photographs of alloy 686 fictitious by autogenous welding using GTAW and PCGTAW. Figure 2a, b represents SEM images of GTAW and PCGTAW produced by filler wire ERNiCrMo-10.

The fusion zone microstructure of GTAW comprises of cellular structure and columnar dendrites in the weld interface regions. In contrast with GTAW, fusion zone microstructure of PCGTAW was decorated with fine equiaxed dendrites, and the interface consists of columnar dendrites. The refinement in the microstructure of

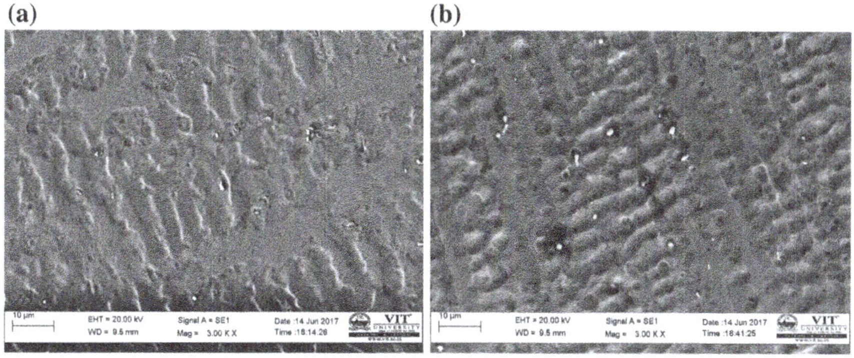


**Fig. 2** SEM image of weld center Microstructure of **a** GTAW-ERNiCrMo-10 **b** PCGTAW-ERNiCrMo-10

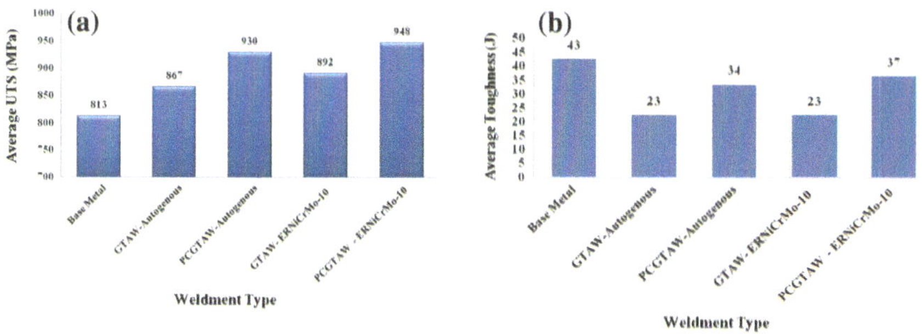


**Fig. 3** a&b Tensile & impact test results of weldments produced with autogenous & with filler wire ERNiCrMo-10

PCGTAW is due to combined effect of lower heat input, enhanced fluid flow, higher instantaneous cooling rate during solidification and reduced thermal gradients. These parameters lower the diffusion of alloying elements and not giving any time for intensification and precipitation of secondary phases. At that same time, PCGTAW is also considerably contributing towards the mechanical properties of weld joints. The fine and equiaxed dendrites are largely accountable for the properties improvement. Figure 3a, b indicates the values of weldments tensile strength and impact toughness values respectively.

The exceptional corrosion resistance of alloy 686 in the marine application has been highlighted in the Introduction. Table 3 shows the corrosion rate of alloy 686 weldments fabricated by GTA and PCGTA welding. The potentiodynamic polarization test as per standard ASTM D1141-2013 [17] and ASTM G59-97 [18] shows the weldment produced with PCGTAW autogenous shows highest corrosion resistance because of small and finer grains observed in the microstructure.

**Table 3** Corrosion rates of weldments tested in synthetic sea water environment

| Weldment type | Corrosion rate [mils/year] |
|---|---|
| Base Metal | 0.01 |
| GTAW-Autogeneous | 0.5 |
| PCGTAW-Autogeneous | 0.06 |
| GTAW-ERNiCrMo-10 | 0.16 |
| PCGTAW-ERNiCrMo-10 | 0.11 |

# 3 Conclusions

1. The level of microsegregation and the occurrence of secondary constituent's σ, P & μ are tremendously diminished in the welded zones of PCGTAW when compared to GTAW. So the hot cracking susceptibility of Alloy 686 plunges to a large extent because of current pulsation.
2. Mechanical properties of PCGTAW also improved significantly. The combination of high tensile strength and enhanced impact toughness are the outcome of PCGTAW. The refined grain structure with reduced severity of microsegregation is believed to be responsible for superior mechanical properties of PCGTAW weld joint.
3. Welded joints fictitious with PCGTAW recommend more corrosion resistance than GTAW, because of finer equiaxed grains. This has been effectively proved by the potentiodynamic technique which synthesis the sea water environment.

# References


1. Special Metals Corporation (2004) Inconel alloy 686 technical data sheet [Internet]. http://216.71.103.52/documents/Inconel%20alloy%20686,%20a%20new%20alloy.pdf
2. Experience in effective application of metallic materials for construction of FGD systems [Internet].http://www.pccforgedproducts.com/web/user_content/files/wyman/Metallic%20Materials%20for%20Construction%20of%20FGD%20Systems.pdf
3. Shoemaker LE, Crum JR Nickel-chromium-molybdenum superalloys: the solution to corrosion problems in wet limestone FGD air pollution control systems [Internet]. http://www.pccforgedproducts.com/web/user_content/files/wyman/Ni%20Cr%20Mo%20Alloys%20-%20Solution%20to%20Corrosion%20problems%20in%20Wet%20Limestone %20FGD%20Air%20Pollution%20Control%20Systems.pdf
4. Special Metals Corporation, Inconel alloy 686 technical data sheet [Internet]. http://www.specialmetals.com/assets/smc/documents/alloys/inconel/inconel-alloy-686.pdf
5. Internet http://www.pccforgedproducts.com/web/user_content/files/wyman/Special%20Alloys%20and%20Overmatched%20Welding%20Products%20Solve%20FGD%20Corrosion%20Problems.pdf Special Alloys and Overmatching Welding Products Solve FGD Corrosion Problems
6. Internet http://www.pinnaclealloys.com/wp/wp-content/uploads/2015/11/Pinnacle-Alloys-ERNiCrMo-10-622.pdf, filler wire ERNiCrMo-10technical data sheet
7. Cieslak MJ, Headley TJ, Romig AD (1986) The welding metallurgy of Hastelloy alloys C-4, C-22 and C-276. Metall Trans A 17A:2035–2047

8. Perricone MJ, DuPont JN (2006) Effect of composition on the solidification behavior of several Ni–Cr–Mo and Fe–Ni–Cr–Mo alloys. Metall Mater Trans A 37A:1267–1280
9. Gallagher ML, Lippold J (2011) Weld cracking susceptibility of alloy C-22 Weld-Metal. In: Lippold J, Bollinghaus T, Carl (eds) Hot cracking phenomena in welds III, 1st edn. Springer Science & Business Media, pp 367–392
10. Akhter JI, Shaikh MA, Ahmad M, Iqbal M, Shoaib KA, Ahmad WJ (2001) Effect of aging on the hardness and impact properties of Hastelloy C-276. J Mater Sci Lett 20:333–335
11. Raghavan M, Berkowitz BJ, Scanlon JC (1982) Electron microscopic analysis of heterogeneous precipitates in hastelloy C-276. Metall Trans A 13:979–984
12. Silva CC, Conrado RM, Afonso, Ramirez AJ, Motta MF, Miranda HC, Farias JP (2016) Assessment of microstructure of alloy Inconel 686 dissimilar weld claddings. J Alloys Compd 684:628–642
13. Dupont JN, Lippold JC, Kiser SD (2009) Welding metallurgy and weldability of nickel-base alloys, 1st edn. USA, Wiley, p 72
14. Janakiram GD, Venugopal Reddy A, Prasad Rao K, Madhusudhan Reddy G (2005) Control of Laves phase in Inconel 718 GTA welds with current pulsing. Sci Technol Weld Joining 9 (5):390–398
15. Manikandan M, Arivazhagan N, Nageswara Rao M, Reddy GM (2005) Improvement of microstructure and mechanical behavior of gas tungsten arc weldments of alloy C-276 by current pulsing. Acta Metall Sinica 28(2):208–215
16. Manikandan M, Arivazhagan N, Nageswara Rao M, Reddy GM (2014) Microstructure & mechanical properties of alloy C-276 weldments fabricated by continuous and pulsed current gas tungsten arc welding techniques. J Manuf Processes 16:563–572
17. Website (2013) https://www.astm.org/Standards/D1141.htm, ASTM D1141–98
18. Website (2014) https://www.astm.org/Standards/G59.htm ASTM G59-97

The manufacturer's authorised representative in the EU is Springer Nature Customer Service Centre GmbH, Europaplatz 3, 69115 Heidelberg, Germany. If you have any concerns regarding our products, please contact ProductSafety@springernature.com

Printed and bound by CPI Group (UK) Ltd, Croydon, CR0 4YY
30/09/2026
02225230-0016